COLOR CONFINEMENT AND HADRONS IN QUANTUM CHROMODYNAMICS

International Conference on

COLOR CONFINEMENT AND HADRONS IN QUANTUM CHROMODYNAMICS

The Institute of Physical and Chemical Research (RIKEN), Japan

21 – 24 July 2003

Editors

H Suganuma

N Ishii

M Oka

Tokyo Institute of Technology, Japan

H Enyo

RIKEN/RBRC, Japan

T Hatsuda

Tokyo University, Japan

T Kunihiro

Kyoto University, Japan

K Yazaki

TWCU/RIKEN, Japan

World Scientific

NEW JERSEY • LONDON • SINGAPORE • BEIJING • SHANGHAI • HONG KONG • TAIPEI • CHENNAI

Published by

World Scientific Publishing Co. Pte. Ltd.

5 Toh Tuck Link, Singapore 596224

USA office: Suite 202, 1060 Main Street, River Edge, NJ 07661

UK office: 57 Shelton Street, Covent Garden, London WC2H 9HE

British Library Cataloguing-in-Publication Data
A catalogue record for this book is available from the British Library.

COLOR CONFINEMENT AND HADRONS IN QUANTUM CHROMODYNAMICS
Proceedings of the International Conference on Confinement 2003

ISBN 981-238-781-1

Printed in Singapore by World Scientific Printers (S) Pte Ltd

ORGANIZING COMMITTEE

H. Suganuma	(IRCS, Tokyo Institute of Technology, Chair)
H. Enyo	(RIKEN/RIKEN-BNL Research Center)
T. Hatsuda	(The University of Tokyo)
N. Ishii	(RIKEN/Tokyo Institute of Technology)
T. Kunihiro	(YITP, Kyoto University)
M. Oka	(Tokyo Institute of Technology)
K. Yazaki	(Tokyo Woman's Christian Univ./RIKEN)

HOST INSTITUTE

Interactive Research Center of Science (IRCS),
Tokyo Institute of Technology

The Institute of Physical and Chemical Research (RIKEN)

PREFACE

The International Conference on "Color Confinement and Hadrons in Quantum Chromodynamics" (Confinement 2003) was held from July 21 to 24 in 2003 at the Institute of Physical and Chemical Research (RIKEN) with the host of Interactive Research Center of Science (IRCS) at Tokyo Institute of Technology. This is the successor of the International Symposia on "Color Confinement and Hadrons" (Confinement '95) and "Quantum Chromodynamics and Color Confinement" (Confinement 2000) both held in Osaka.

Quantum chromodynamics (QCD) has been established as the fundamental theory of the strong interaction three decades ago. Nevertheless, there remain lots of interesting problems related to nonperturbative dynamics in the infrared region of QCD such as color confinement, dynamical chiral-symmetry breaking and nontrivial topologies.

To understand the nonperturbative physics of QCD, numerous theoretical studies have been done with various effective models and mathematical approaches. In particular, the lattice QCD Monte Carlo simulation has been established as a powerful method for the analysis of nonperturbative QCD. From the experimental viewpoint, QCD physics encounters a new phase owing to the interesting experimental data at the RHIC project in Brookhaven National Laboratory.

The purpose of the present Conference is to summarize the current status of nonperturbative QCD in mathematical physics, lattice QCD physics and quark-hadron physics, and to promote the mutual understanding among these fields, as a necessary step for the global understanding of nonperturbative QCD.

The subjects covered in this Conference are as follows:

- Quark Confinement Mechanism
- Dynamical Chiral-Symmetry Breaking
- Topologies in QCD (Instantons, Monopoles and Vortices)
- Confinement in SUSY QCD
- Nonperturbative Analysis ($1/N_c$, ladder QCD, AdS/CFT)
- QCD Phase Transition at Finite Temperature and Density
- Quark Gluon Plasma

This Conference consists of plenary sessions, parallel sessions and a poster session. The number of participants was 131 in which 64 were from overseas from 18 different countries. For the active communication of participants, the poster session was held combined with a reception party. We had lively and active discussions and useful communications among the participants from different fields of physics.

These proceedings consist of 28 papers for the invited talks at the plenary sessions, 20 papers for the selected talks at the parallel sessions and 33 one-page abstracts for selected posters. For the plenary talks, we have sorted the papers into 6 categories in terms of their theme of physics. In each category, the papers are arranged in the order of the presentation at the Conference. For the parallel and the poster presentations, the papers are arranged in alphabetical order of the presenter's family name.

We believe that these proceedings will be useful for the global understanding of the current studies on the nonperturbative aspects of QCD from wide viewpoints of mathematical physics, lattice QCD physics and quark-hadron physics.

The Conference organizing committee would like to express its cordial thanks to all the participants, the speakers, and the chairpersons for making the Conference very successful. This Conference was partially supported by Inoue Foundation for Science in Japan.

We are extremely grateful to Ms. Noriko Kiyama at RIKEN for her devoted works for all aspects on the Conference. We acknowledge the members of TokyoTech theory group and RIKEN for their help during the Conference. We particularly thank Drs. T.T. Takahashi, T. Doi, H. Iida, K. Tanida and Y. Watanabe for their devoted helps.

We thank Prof. K.-I. Aoki for his wonderful photographs at Confinement 2003. The cover design is based on the recent lattice QCD results for the three-quark flux-tube formation by Drs. H. Ichie, V. Bornyakov, T. Streuer and G. Schierholz. We acknowledge them for their help in designing the proceedings cover.

Finally, H.S. thanks Mrs. Aigo Suganuma for many helps on the preparation of the Conference.

The Organizing Committee of Confinement 2003

Hideo Suganuma (IRCS, Tokyo Institute of Technology, Chair)
Hideto Enyo (RIKEN/RIKEN-BNL Research Center)
Tetsuo Hatsuda (The University of Tokyo)
Noriyoshi Ishii (RIKEN/Tokyo Institute of Technology)
Teiji Kunihiro (YITP, Kyoto University)
Makoto Oka (Tokyo Institute of Technology)
Koichi Yazaki (Tokyo Woman's Christian University/RIKEN)

CONTENTS

Parallel Session Presentations **381**

Supersymmetric QCD

SUPERSYMMETRIC GLUODYNAMICS AND ITS NONSUPERSYMMETRIC COUSINS: PLANAR EQUIVALENCE

M. SHIFMAN*

William I. Fine Theoretical Physics Institute, University of Minnesota, Minneapolis, MN 55455, USA

A. ARMONI, G. VENEZIANO

Theory Division, CERN, CH-1211 Geneva 23, Switzerland

We review the newly emerged notion of planar equivalence between $\mathcal{N} = 1$ gluodynamics and the non-supersymmetric "orientifold field theory" and discuss its consequences. We describe the genesis of the idea and its early implementations which turned out to be valid only at the perturbative level. A breakthrough development, the focus of this talk, occured after the idea was applied to orientifold field theories.

1. Introduction

Gauge field theories at strong coupling are of great importance in particle physics. Needless to say, exact results in such theories have a special weight. Supersymmetry proved to be an extremely powerful tool allowing one to solve problems (intractable otherwise) in supersymmetric gauge theories at strong coupling (for a review see e.g. Ref. 1). A significant number of results were obtained in this direction in the last twenty years, starting from the gluino condensate and exact β functions, to, perhaps, the most famous example — the Seiberg-Witten solution of $\mathcal{N} = 2$ supersymmetric theory, which (upon small perturbation breaking $\mathcal{N} = 2$ to $\mathcal{N} = 1$) explicitly exhibits the dual Meißner mechanism of color confinement conjectured by 't Hooft and Mandelstam in the 1970's. This review is devoted to a recent addition to the supersymmetry-based kit, which goes under the name of

*Talk by M. Shifman.

planar equivalence. Some *non*-supersymmetric gauge theories — quite close relatives of QCD — where shown[2,3] to be equivalent to supersymmetric gluodynamics in a bosonic sub-sector, provided the number of colors $N \to \infty$. Needless to say, if true, this statement has far-reaching consequences. An infinite number of supersymmetry-based predictions, such as spectral degeneracies, low-energy theorems, and so on, hold (to leading order in $1/N$) in strongly coupled nonsupersymmetric gauge theories!

The outline of the talk is as follows. We will start from summarizing basic features of the parent theory — supersymmetric gluodynamics also known as supersymmetric Yang-Mills (SYM) theory. We review the genesis of the idea of planar equivalence and its physical foundation, both at the perturbative and nonperturbative levels. After a brief introduction on perturbative planar equivalence, we proceed to the *orientifold field theory* whose bosonic sector was argued[2] to be fully equivalent to SYM theory at $N \to \infty$.

Among other aspects, this talk will demonstrate cross-fertilization between field theory and string theory. The idea of planar equivalence born in the depths of string theory, was reformulated and adapted to field theory, where its got its own life and was developed to a point where practical applications are looming.

2. Genesis of the idea

Kachru and Silverstein studied[4] various orbifolds of R^6 within the AdS/CFT correspondence. Starting from $\mathcal{N} = 4$, they obtained distinct four-dimensional (daughter) gauge field theories with matter, with varying degree of supersymmetry, $\mathcal{N} = 2, 1, 0$, all with vanishing β functions. In the latter case, $\mathcal{N} = 0$, Kachru and Silverstein predicted the first coefficient of the Gell-Mann–Low function β_0 to vanish in the large-N (planar) limit in the daughter theories, while for $\mathcal{N} = 2, 1$ they predicted $\beta_0 = 0$ even for finite N. Shortly after, this analysis was expanded in Ref. 5 where generic orbifold projections of $\mathcal{N} = 4$ were considered, and projections preserving the conformal invariance of resulting — less supersymmetric — gauge theories were identified (at two loops).

A decisive step was made[6] by Bershadsky *et al.* These authors abandoned the AdS/CFT limit of large 't Hooft coupling. They considered the string perturbative expansion in the presence of D branes embedded in orbifolded space-time in the limit $\alpha' \to 0$. In this limit the string perturbative expansion coincides with the 't Hooft expansion in $1/N$. The genus-zero string graphs can be identified with the leading (planar) term in the 't Hooft expansion, the genus-one is the next-to-leading correction, and so on. In Ref. 6 one can find a proof that in the large-N limit the β functions of all orbifold theories considered previously[4,5] vanish to any finite order in the gauge coupling constant. Moreover, a remarkable theorem was derived in Ref. 6. The authors showed that, not only the β functions, but a variety of amplitudes which can be considered in the orbifold theories coincide in the large-N limit (upon an appropriate rescaling of the gauge coupling) with the corresponding amplitudes of the parent $\mathcal{N} = 4$ theory, order by order in the gauge coupling.

In a week or so, Bershadsky and Johansen abandoned the string theory set-up altogether. They proved[7] the above theorem in the framework of field theory *per se*.

In a parallel development, Kakushadze suggested[8] considering orientifold rather than orbifold daughters of $\mathcal{N} = 4$ theory. His construction was close to that of Ref. 6, except that instead of the string expansion in the presence of D branes he added orientifold planes. The main emphasis was on the search of $\mathcal{N} = 2, 1$ daughters with the gauge groups other than SU(N) and matter fields other than bifundamentals.

The first attempt to apply to idea of orbifoldization of non-conformal field theories was carried out by Schmaltz who suggested[9] a version of Seiberg duality between a pair of non-supersymmetric large N orbifold field theories. Later on, the Z_2 version of Schmaltz suggestion was realized in a brane configuration of type 0 string theory.[10]

After a few years of a relative oblivion, the interest to the issue of planar equivalence was revived by Strassler.[11] In the inspiring paper entitled "On methods for extracting exact non-perturbative results in non-supersymmetric gauge theories," he shifted the emphasis away from the search of the conformal daughters, towards engineering QCD-like daughters. Similarly to Schmaltz, Strassler noted that one does not need to limit oneself to the $\mathcal{N} = 4$ parent; $\mathcal{N} = 1$ gluodynamics which is much closer to actual QCD gives rise to orbifold planar-equivalent daughters too. Although this equivalence was proven only at the perturbative level, Strassler formulated a "nonperturbative orbifold conjecture" (NPO). According to this conjecture, for "those vacua which appear in both theories (i.e. parent and orbifold daughter) the Green's functions which exist in both theories are identical in the limit $N \to \infty$ after an appropriate rescaling of the coupling constant."

Unfortunately, it turned out[12] that the NPO conjecture could not be valid. The orbifold daughter theories "remember" that they have less vacua than the parent one which results in a mismatch[12] in low-energy theorems. In the string theory language the killing factor is the presence of tachyons in the twisted sector. The appeal of the idea was so strong, however, that the searches continued. They culminated in the discovery of the orientifold field theory[2] whose planar equivalence to SYM theory, both perturbative and nonperturbative, rests on a solid footing. The orientifold daughter has no twisted sector. The SU(N) orientifold daughter and its parent, SU(N) $\mathcal{N}=1$ gluodynamics, have the same numbers of vacua.

3. The parent theories

Let us describe the class of the parent theories. It will include generic supersymmetric gauge theories of the following form:

$$\mathcal{L} = \left\{ \frac{1}{4g^2} \int \mathrm{d}^2\theta \, \mathrm{Tr}\, W^2 + \mathrm{H.c.} \right\} + \mathcal{L}_{\mathrm{matter}} \,, \tag{1}$$

where the spinorial superfield W_α in the Wess-Zumino gauge is

$$W_\alpha = \frac{1}{8} \bar{D}^2 \left(e^{-V} D_\alpha e^V \right) = i \left(\lambda_\alpha + i\theta_\alpha D - \theta^\beta \, G_{\alpha\beta} - i\theta^2 \mathcal{D}_{\alpha\dot\alpha} \bar\lambda^{\dot\alpha} \right) \,, \tag{2}$$

the Grassmann integration is normalized according to the formula

$$\int \theta^2 \, \mathrm{d}^2\theta = 2 \,,$$

and $\mathcal{L}_{\mathrm{matter}}$ is the matter part of the Lagrangian,

$$\mathcal{L}_{\mathrm{matter}} = \sum_{\mathrm{matter}} \left\{ \frac{1}{4} \int \mathrm{d}^2\theta \mathrm{d}^2\bar\theta \, \bar{Q} e^V Q + \left(\frac{1}{2} \int \mathrm{d}^2\theta \, \mathcal{W}(Q) + \mathrm{H.c.} \right) \right\} \,, \tag{3}$$

where Q is a generic notation for the matter chiral superfields, $\mathcal{W}(Q)$ is a superpotential.

The gauge group is assumed to be SU(N). The matter fields Q can belong to various representations R of the gauge group. The definitions of the Casimir operators to be used below are

$$\mathrm{Tr}(T^a T^b)_R = T(R)\delta^{ab} \,, \qquad (T^a T^a)_R = C_R \, I \,, \tag{4}$$

where R stands for the given representation, and I is the unit matrix in this representation. It is quite obvious that

$$C_R = T(R) \, \frac{\dim(G)}{\dim(R)} \,, \tag{5}$$

where dim(G) is the dimension of the group (= the dimension of the adjoint representation). Note that $T(R)$ is also known as (one half of) the Dynkin index, or the dual Coxeter number. For the adjoint representation $T(R)$ is denoted by $T(G)$. Moreover, $T(\mathrm{SU}(N)) = N$, while for other relevant representations the Casimir operators for SU(N) are collected in Table 1.

Table 1. The Casimir coefficients for relevant representations of SU(N).

Representations→ / Casimirs↓	fundamental	adjoint	2-index antisymmetric
$T(R)$	$\frac{1}{2}$	N	$\frac{N-2}{2}$
$C(R)$	$\frac{N^2-1}{2N}$	N	$\frac{(N-2)(N+1)}{N}$

3.1. *The simplest SUSY theory*

We start from $\mathcal{N} = 1$ super-Yang-Mills (SYM) theory, also known as supersymmetric gluodynamics — the theory of gluons and gluinos. The Lagrangian of the theory is obtained from Eq. (1) by omitting the matter part. In the component form

$$\mathcal{L} = -\frac{1}{4g^2} G^a_{\mu\nu} G^a_{\mu\nu} + \frac{i}{g^2} \lambda^{a\alpha} \mathcal{D}_{\alpha\dot\beta} \bar\lambda^{a\dot\beta} , \tag{6}$$

where λ is the Weyl (Majorana) fermion in the adjoint representation, and spinorial notations are used in the fermion sector. One can add, if one wishes, a θ term,

$$\mathcal{L}_\theta = \frac{\theta}{32\pi^2} G^a_{\mu\nu} \widetilde{G}^a_{\mu\nu} . \tag{7}$$

For the time being we will set $\theta = 0$. As was mentioned, the gauge group is assumed to be SU(N). Later on we will tend $N \to \infty$, following the 't Hooft law,[a]

$$\lambda \equiv \frac{g^2 N}{8\pi^2} \equiv \frac{N\alpha}{2\pi} = \text{const.} \tag{8}$$

[a]Here and below we follow the standard convention $g^2 = 4\pi\alpha$.

The distinction between QCD and SUSY gluodynamics lies in the fermion sector — the QCD quarks are the Dirac fermions in the fundamental representation of the gauge group.

The theory (6) is supersymmetric. The conserved spin-3/2 current J^μ_β has the form (in the spinorial notation)

$$J_{\beta\alpha\dot\alpha} \equiv (\sigma_\mu)_{\alpha\dot\alpha} J^\mu_\beta = \frac{2i}{g^2}\, G^a_{\alpha\beta}\bar\lambda^a_{\dot\alpha}\,. \tag{9}$$

Moreover, the theory is believed to be confining, with a mass gap. Some new semi-theoretical-semi-empirical arguments substantiating this point will be provided later.

The spectrum of supersymmetric gluodynamics consists of composite (color-singlet) hadrons which enter in degenerate supermultiplets. Note that there is no baryon charge in the theory at hand. The only U(1) symmetry of the classical Lagrangian (6),

$$\lambda \to e^{i\alpha}\lambda\,, \quad \bar\lambda \to e^{-i\alpha}\bar\lambda\,,$$

is broken at the quantum level by the chiral anomaly. A discrete Z_{2N} subgroup, $\lambda \to e^{\pi i k/N}\lambda$, is non-anomalous. It is known to be dynamically broken down to Z_2. The order parameter, the gluino condensate[b] $\langle\lambda\lambda\rangle$, takes N values,

$$\langle\lambda^a_\alpha \lambda^{a\,,\alpha}\rangle = -6N\Lambda^3 \exp\left(\frac{2\pi i k}{N}\right)\,, \quad k = 0, 1, ..., N-1\,, \tag{10}$$

labeling N distinct vacua of the theory (6), see Fig. 1. Here $\langle ... \rangle$ means averaging over the given vacuum state, and Λ is a dynamical scale, defined in the standard manner (i.e. in accordance with Ref. 18) in terms of the ultraviolet parameters,

$$\Lambda^3 = \frac{2}{3}\, M^3_{\rm UV}\, \frac{8\pi^2}{N g_0^2}\, \exp\left(-\frac{8\pi^2}{N g_0^2}\right) = \frac{2}{3}\, M^3_{\rm UV}\, \frac{1}{\lambda_0}\, \exp\left(-\frac{1}{\lambda_0}\right)\,, \tag{11}$$

[b]The gluino condensate in supersymmetric gluodynamics was first conjectured, on the basis of the value of his index, by E. Witten.[13] It was confirmed in an effective Lagrangian approach by G. Veneziano and S. Yankielowicz,[14] and exactly calculated (by using holomorphy and analytic continuations in mass parameters) by M. A. Shifman and A. I. Vainshtein.[15] The exact value of the coefficient $6N$ in Eq. (10) can be extracted from several sources. All numerical factors are carefully collected for SU(2) in the review paper.[16] A weak-coupling calculation for SU(N) with arbitrary N was carried out in Ref. 17. Note, however, that an unconventional definition of the scale parameter Λ is used in Ref. 17. One can pass to the conventional definition of Λ either by normalizing the result to the SU(2) case[16] or by analyzing the context of Ref. 17. Both methods give one and the same result.

where $M_{\rm UV}$ is the ultraviolet (UV) regulator mass, while g_0^2 and λ_0 are the bare coupling constants.

Note that since Λ is expressible in terms of the 't Hooft coupling, it is explicitly N independent. Equation (11) is exact[19] in supersymmetric gluodynamics. If $\theta \neq 0$, the exponent in Eq. (10) is replaced by

$$\exp\left(\frac{2\pi i k}{N} + \frac{i\,\theta}{N}\right).$$

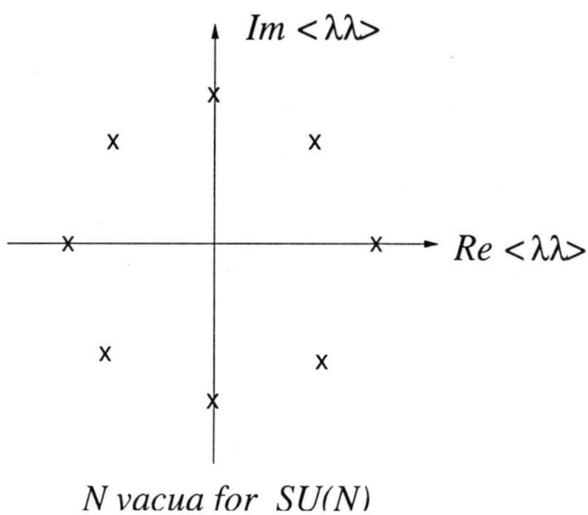

Figure 1. The gluino condensate $\langle\lambda\lambda\rangle$ is the order parameter labeling distinct vacua in supersymmetric gluodynamics. For SU(N) gauge group one has N discrete degenerate vacua.

All hadronic states are arranged in supermultiplets. The simplest is the so-called chiral supermultiplet which includes two (massive) spin zero mesons (with the opposite parities), and a Majorana fermion (with the Majorana mass). The interpolating operators producing the corresponding hadrons from the vacuum are G^2, $G\tilde{G}$ and $G\lambda$. The vector supermultiplet consists of a spin-1 massive vector particle, a 0^+ scalar and a Dirac spinor. All particles from one supermultiplet have degenerate masses. Two-point functions are degenerate too (modulo obvious kinematical spin factors). For instance,

$$\langle G^2(x)\,,\,G^2(0)\rangle = \langle G\tilde{G}(x)\,,\,G\tilde{G}(0)\rangle = \langle G\lambda(x)\,,\,G\lambda(0)\rangle\,. \qquad (12)$$

Unlike conventional QCD, both the meson and fermion masses in SUSY gluodynamics are expected to scale as N^0.

4. Orbifoldization and perturbative equivalence

The technique of orbifoldization was developed in the context of string/brane theory and it will be discussed in the string part of this review. Starting from supersymmetric gluodynamics with the gauge group SU(kN), where k is an integer, one can perform a Z_k orbifoldization.

The only thing we need to know at the moment, is that orbifolding the parent theory (6) one obtains a self-consistent daughter field theory by judiciously discarding a number of fields in the Lagrangian (6). This is illustrated in Fig. 2 corresponding to $k = 2$. The square presents the color contents of the fields of the daughter theory. The gauge fields which are *retained* in the daughter theory belong to two blocks on the main diagonal. Thus, the daughter theory has gauge symmetry SU(N)×SU(N). (There is no distinction between SU(N) and U(N) at large N). This specific theory has a realization in type 0 string theory.[10] To emphasize the fact that the gauge group is a direct product it is convenient to use "tilded" indices for one SU(N), e.g. $(A_\mu)^{\tilde{i}}_{\tilde{j}}$, and "untilded" for another, e.g. $(G_{\mu\nu})^i_j$, where $i, j, = 1, 2, ..., N$ while $\tilde{i}, \tilde{j} = N + 1, ..., 2N$.

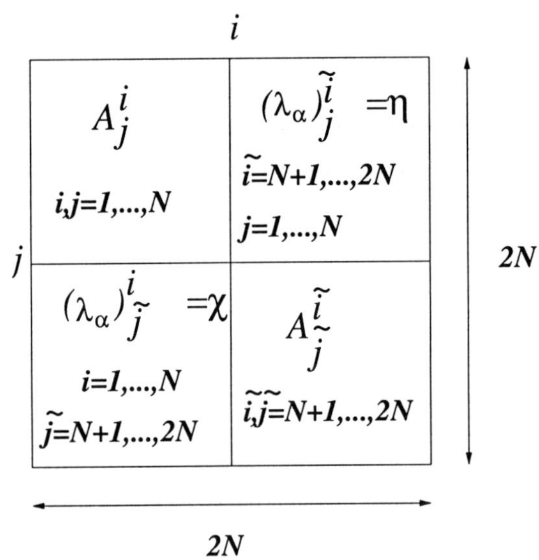

Figure 2. Color decomposition of fields in the Z_2-orbifold daughter.

The fermion fields to be retained in the daughter theory belong to two off-diagonal blocks in Fig. 2. Thus, each fermion carries one tilded and one untilded color index. In other words, they are fundamentals with respect to the first SU(N) and antifundamentals with respect to the second SU(N). Such fields are called ***bifundamental.*** Thus, in the daughter theory we deal with two Weyl bifundamentals, $(\lambda_\alpha)^i_{\tilde{j}}$ (we will call it χ) and $(\lambda_\alpha)^{\tilde{i}}_j$ (we will call it η). It is quite clear that one can combine them to form one four-component (Dirac) bifundamental field, $\eta,\ \bar{\chi} \to \Psi^{\tilde{i}}_j$. The Lagrangian of the daughter theory is

$$\mathcal{L} = -\frac{1}{4g^2} G^i_j G^j_i - \frac{1}{4g^2} G^{\tilde{i}}_{\tilde{j}} G^{\tilde{j}}_{\tilde{i}} + \frac{\theta}{32\pi^2}\left(G^i_j \widetilde{G}^j_i + G^{\tilde{i}}_{\tilde{j}} \widetilde{G}^{\tilde{j}}_{\tilde{i}}\right)$$

$$+ \overline{\Psi^{\tilde{i}}_j}\, i\, \not{D}\, \Psi^{\tilde{i}}_j\,, \tag{13}$$

where

$$D_\mu = \partial_\mu - i\, A^a_\mu T^a - i\, \tilde{A}^a_\mu \tilde{T}^a\,, \tag{14}$$

and we have explicitly indicated the θ term. Note that the perturbative planar equivalence requires the gauge couplings of both SU(N)'s to be the same. Below they will be denoted by g^2_D. We will assume then that the vacuum angles are the same too.

The the perturbative planar equivalence which takes place between supersymmetric gluodynamics and its orbifold daughter (13) requires a rescaling of the gauge couplings in passing from the parent to the daughter theory, namely

$$g^2_D = 2g^2_P\,. \tag{15}$$

In the general case of Z_k orbifoldization the relation is

$$g^2_D = kg^2_P\,. \tag{16}$$

The orbifold daughter is *not* supersymmetric. The color-singlet supercurrent (9) cannot be formed now since the color indices of the gauge fields and the fermion fields cannot be contracted. In this way the orbifold daughter seems to be closer to genuine QCD than SUSY gluodynamics. In fact, for $N = 3$ this is nothing but three-color/three-flavor "QCD" with a gauged flavor group. (Note, that flavor SU(2) is gauged anyway in the electroweak theory).

The absence of supersymmetry is not the only feature of (13) which distinguishes this theory from SUSY gluodynamics. The daughter theory possesses a global U(1). The conserved fermion number current is $J_\mu = \overline{\Psi}\gamma_\mu\Psi$. Applying this vector current to vacuum, one produces a vector meson which has no counter-partners in the parent theory. They say that the operator J_μ has no projection onto the parent theory. One can construct other operators with similar properties. There are many hadronic states in the daughter theory which have no analogs in the parent one. For instance, for odd N one expects composite states with the quark content $\Psi\Psi...\Psi\,\varepsilon\,\tilde{\varepsilon}$ and masses scaling as N. For even N the daughter theory has no color-singlet fermions at all.

Inverse is also true. For instance, the operator of the supercurrent produces a spin-3/2 hadron from the vacuum state of SUSY gluodynamics, which has no counterpart in the daughter theory. Therefore, in confronting these theories it makes sense to compare only those sectors which have projections onto one another.

Under the Z_k orbifoldization the gauge symmetry of the orbifold daughter is $\mathrm{SU}(N)^k$, and all fermion fields in the daughter theory are bifundamentals. Note also that the daughter theory is chiral for $k > 2$.

4.1. *Perturbation theory*

As was mentioned in Sect. 2, in 1998 it was discovered that the Green's functions for operators which exist in both theories have identical planar graphs[c] provided the rescaling (15) is implemented. In other words, perturbatively these Green's functions in the parent (supersymmetric) and daughter (non-supersymmetric) theories coincide order by order in the large N limit. Motivation for the work[7] came from the string side. A detailed discussion of orbifoldization in various gauge theories can be found in Ref. 9.

Rather than presenting a general proof (it can be found in the original publications, see e.g. Ref. 7) we will illustrate this statement (see Fig. 3) by a simple example of the correlation function of two axial currents,

$$(2/g^2)(\bar{\lambda}_{\dot\alpha})^i_j(\lambda_\alpha)^j_i\,.$$

For the computational purposes it is convenient to keep the fermion fields in the daughter theory in the same form $(\lambda_\alpha)^i_j$, just constraining the color indices. In the parent theory i, j run from 1 to $2N$. In the daughter theory either $i \in [1, N]$, $\quad j \in [N+1, 2N]$ or *vice versa*, $i \in [N+1, 2N]$, $\quad j \in [1, N]$. In the first case we deal with $\lambda^i_{\bar j}$ in the second $\lambda^{\bar i}_j$.

The easiest way to check the match is analyzing relevant graphs in the 't Hooft representation — each gluon or fermion line carrying two indices

[c]It is assumed that the daughter theory inherits the vacuum of the parent theory. Perturbation theory in the daughter theory is carried out around the Z_2 symmetric vacuum.

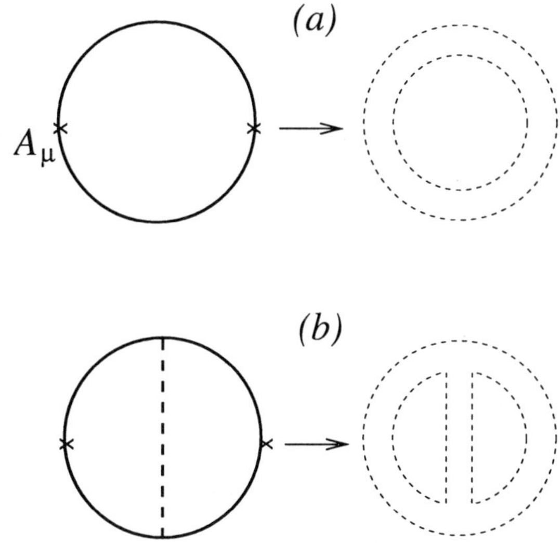

Figure 3. Counting N factors in the correlation function of the axial currents. The 't Hooft diagrams are depicted in dashed lines.

is represented by a double line, see Fig. 3. In this way we readily get that the weight of the graph 3a is $4N^2$ (parent), $2N^2$ (daughter), while that of 3b is $8N^3 g_P^2$ (parent), $2N^3 g_D^2$ (daughter). To get the latter estimate one notes that if the first index of λ — the one forming the outside loop — is untilded, the second index (corresponding to two inside loops in Fig. 3b) is tilded, and *vice versa.* The parent-to-daughter ratio is 2 for both graphs, 3a and 3b, provided $g_D^2 = 2\, g_P^2$. Proceeding along these lines it is quite easy to see that the parent-to-daughter ratio stays the same for any planar graph. This proves the perturbative planar equivalence in the case at hand. The proof can be readily extended to any correlation function which exists both in supersymmetric gluodynamics and its orbifold daughter.

4.2. *Consequences of the perturbative planar equivalence*

Thus, a relationship is established between two theories in the large-N limit — one is supersymmetric, the other is not. Although this relationship is admittedly perturbative, a natural question to ask is whether one can benefit from it, and if yes, how.

The *exact* β function of SU($2N$) SUSY gluodynamics established[19] almost twenty years ago is as follows:

$$\beta(\alpha_P) \equiv \frac{d\alpha_P}{d\ln\mu} = -\frac{1}{2\pi}\,\frac{6N\alpha_P^2}{1-(N\alpha_P)/\pi}\,. \tag{17}$$

Due to the planar equivalence it implies that the full β function in the (non-supersymmetric) daughter theory is also known. Namely, in the 't Hooft limit,

$$\beta(\alpha_D) = -\frac{1}{2\pi}\,\frac{3N\alpha_D^2}{1-(N\alpha_D)/(2\pi)}\,, \tag{18}$$

where the relation (15) between the coupling constants in the parent and orbifold theories is taken into account. The β function (18) is exactly the same as in SU(N) SUSY gluodynamics.

5. Nonperturbative orbifold conjecture

Needless to say, it is nonperturbative dynamics which is our prime concern in strongly coupled gauge theories. Nonperturbative effects are at heart of key phenomena such as color confinement, chiral symmetry breaking, and so on. Strassler conjectured[11] that the planar Feynman graph equivalence discussed in Sect. 4 extends to nonperturbative phenomena, namely Green's functions for the color-singlet operators which appear in both theories are identical at $N \to \infty$, including nonperturbative effects. This suggestion got the name *nonperturbative orbifold (NPO) conjecture.*

Unfortunately, the Strassler's suggestion, in its original form, is not valid,[12] as we will see shortly. It played an important role, however, in paving the way to further developments. Analyzing why orbifold daughters fail to retain planar equivalence at nonperturbative level one can guess how one can engineer daughters free from these drawbacks, thus having more solid chances to be nonperturbatively planar-equivalent to their supersymmetric parents (Sect. 6).

5.1. *The vacuum structure and low-energy theorems*

Passing to nonperturbative analysis let us first compare the number of vacua in the SU(kN) supersymmetric gluodynamics to that of its Z_k daughter.

The vacua in the parent theory are counted by Witten's index.[13] They are labeled by the gluino condensate, see Eq. (10), which for SU(kN) takes kN distinct values, so that we have kN degenerate vacua. For $k = 2$, the simplest case with which we will deal below, there are $2N$ supersymmetric vacua in the parent theory.

At the same time, the number of vacua in the orbifold theory is twice smaller. This is easy to see from the discrete chiral invariance of the daughter theory. Generally speaking, the theory (13) is *not* invariant under the chiral rotation

$$\chi \to \chi\, e^{i\alpha}\,, \qquad \eta \to \eta e^{i\alpha}\,, \tag{19}$$

since under this rotation the vacuum angle θ is shifted, $\theta \to \theta + 2\,N\,\alpha$. However, if

$$\alpha = k\pi/N\,, \qquad k \text{ integer}\,, \tag{20}$$

then $\delta\theta = 2\pi\, k$, and such discrete rotation *is* the invariance of the theory. The integer parameter k in Eq. (20) runs from 1 to N, since the sign of the fermion field is unobservable, and it is the phase of the bilinear operator $\chi\eta$ that counts the vacua. Alternatively, one can say that the condensate $\langle\chi\eta\rangle$ breaks (spontaneously) the discrete symmetry Z_{2N} down to Z_2. Thus, we conclude that we have N vacua in the daughter theory.

The phase of the bifermion operator *per se* is unobservable. One makes it observable by introducing a small mass term (which we will need anyway in order to regularize fermion determinants, see Sect. 6),

$$\mathcal{L}_m = \begin{cases} -\frac{m}{2g_P^2}\,\lambda^a\lambda^a + \text{H.c.} \\[2ex] -\frac{m}{g_D^2}\,\chi\eta + \text{H.c.} \end{cases} \tag{21}$$

The vacua are no more degenerate; the vacuum energy densities split,

$$\mathcal{E}_{\text{vac}} = \begin{cases} \frac{m}{2g_P^2}\,\langle \text{vac}_k|\lambda^a\lambda^a|\text{vac}_k\rangle_0\, e^{i\theta/(2N)} + \text{H.c.} \\[2ex] \frac{m}{g_D^2}\,\langle \text{vac}_k|\chi\eta|\text{vac}_k\rangle_0\, e^{i\theta/N} + \text{H.c.} \end{cases} + O(m^2)\,, \tag{22}$$

where $|\text{vac}_k\rangle$ denotes the k-th vacuum, while the subscript 0 marks the value of the condensate at $\theta = 0$. Finally, besides the vacuum energy densities, we will consider topological susceptibilities defined as

$$\mathcal{T} = -\left.\frac{\partial^2 \mathcal{E}_{\text{vac}}}{\partial\theta^2}\right|_{\theta=0} = i\int d^4x \left\langle \frac{1}{32\pi^2}\, G\widetilde{G}(x),\ \frac{1}{32\pi^2}\, G\widetilde{G}(0)\right\rangle_{\text{conn}}\,, \tag{23}$$

see Eqs. (7) and (13) for the definition of $G\widetilde{G}$ in the parent and orbifold theories, respectively.

Now, we can assemble all the above elements to prove that the NPO conjecture does not work. Indeed, assume it does. Then $\mathcal{E}^D_{\text{vac}} = \mathcal{E}^P_{\text{vac}}$. Here we choose a pair of "corresponding" vacua. Remember, the parent theory has twice more vacua. We take the one which can be projected onto the daughter theory. Equation (22) implies that in the given vacuum $\mathcal{T}_P = \frac{1}{4}\mathcal{T}_D$.

On the other hand, the topological susceptibilities in the given vacuum are measurable, for instance, on lattices. The statement that $\mathcal{T}_P = \frac{1}{4}\mathcal{T}_D$ is in disagreement with the definition (23) by a factor of two.[12] Thus, the non-perturbative sector of the orbifold theory remembers the mismatch between the numbers of vacua in the parent and daughter theories.

5.2. *What makes orbifold daughters unsuitable?*

The underlying reason for the failure of the NPO conjecture is most clearly seen in the string-theory language. It is the presence of *tachyons* in the twisted sector. Time does not permit us to discuss the string aspect; we will limit ourselves to some field-theoretical arguments illustrating killing consequences of the presence of the twisted sector.

First, let us explain what the twisted sector is in the field-theoretical language. To this end we will recast Fig. 2, summarizing the content of the Z_2 orbifold daughter, in a slightly different way, see Table 2.

Table 2. The Z_2 orbifold theory.

	$\mathrm{U}_e(N)$ ×	$\mathrm{U}_m(N)$
A_e^e	Adj	1
A_m^m	1	Adj
λ_m^e	$\square$	$\overline{\square}$
λ_e^m	$\overline{\square}$	$\square$

We replaced the untilded indices in Fig. 2 by generic sub/superscripts e (meaning electric) while the tilded indices by m (meaning magnetic). The Z_2 orbifold theory has a realization in type 0A theory.[10] It lives on a brane configuration of type 0A which consists of "electric" and "magnetic" D branes. This is the origin of the sub/superscripts e and m in Table 2. The Z_2 orbifold theory is obviously Z_2 symmetric under the exchange of all indices $e \to m$ and *vice versa*.

Let us divide all color-singlet operators of the Z_2 daughter in two classes: (i) those that are invariant (even) under the above Z_2 symmetry, and (ii) those that are non-invariant (odd). The operators from the second class are called twisted. For instance, $G_e^e G_e^e - G_m^m G_m^m$ is a twisted operator whereas $G_e^e G_e^e + G_m^m G_m^m$ is untwisted.

The perturbative relation between the orbifold theory and its supersymmetric parent concern only the untwisted sector.[6,7] The parent theory does not carry information about the twisted sector of the daughter theory. The NPO conjecture assumes that the vacuum of the daughter theory is Z_2 invariant. However, this need not be the case.

A possible sign of the Z_2 instability can be obtained just in perturbation theory. Indeed, let us assume that at some UV scale where perturbation theory is applicable, the gauge couplings of the two U(N) factors in the orbifold theory are slightly different. We will denote $8\pi^2/(Ng_e^2) = 1/\lambda_e$ and $8\pi^2/(Ng_m^2) = 1/\lambda_m$. It is not difficult to find the renormalization group flow of $\delta\lambda$ towards the infrared (IR) domain. As long as $\delta\lambda \ll \lambda_{e,m}$ we have

$$\frac{d(\delta\lambda)}{d\ln\mu} = -3\,\lambda^2\,(\delta\lambda) + \text{higher orders}. \tag{24}$$

Neglecting the weak logarithmic μ dependence of λ we get

$$\delta\lambda(\mu) = \delta\lambda(\mu_0)\left(\frac{\mu_0}{\mu}\right)^{3\lambda^2}. \tag{25}$$

If $\delta\lambda$ is small in UV, it exponentially grows towards the IR domain.

A similar analysis of the conformal orbifold daughter of $\mathcal{N} = 4$ SUSY theory, with similar conclusions, was carried out in Ref. 20.

The advantage of the orientifold theory over the orbifold theory is the absence of the twisted sector. Moreover, the gauge groups are the same in the parent and daughter theories, and so are the patterns of the spontaneous breaking of the global symmetry and the numbers of vacua. This is also seen from the string theory standpoint. The orbifold field theory originates from the type 0A string theory which is obtained by a Z_2 orbifold of type IIA. The result is a bosonic string theory with a tachyon in the twisted sector. In contrast, the orientifold field theory originates from a configuration with an orientifold which removes the twisted sector (the result is very similar to the bosonic part of the type I string).

Another way to search for instabilities in the twisted sector was suggested in Ref. 12. The logic is as follows. Since the perturbative planar equivalence between the SUSY parent and orbifold daughter does not depend on the geometry of space-time, one can compactify one or more dimensions, making the compactified dimension small, so that the theory in question becomes weakly coupled. Then one can compare the parent and the orbifold theories at weak coupling.

David Tong considered[21] the $R^3 \times S$ compactification. On $R^3 \times S$ the third spatial component of the gluon field becomes a scalar field with a flat (vanishing) potential at the classical level, $(A_3)^i_j \to \phi^i_j$. (Alternatively, one can speak of the Polyakov line in the x_3 direction.) In the parent theory the flatness is perturbatively maintained to all orders due to supersymmetry. Nonperturbatively, the flatness is lifted by $R^3 \times S$ instantons (monopoles) which generate a superpotential. As it turns out, in supersymmetric vacua the non-Abelian gauge symmetry is completely broken by $\langle \phi^i_j \rangle = v_i \delta^i_j$, down to $U(1)^{2N}$, so that the theory is in the Coulomb phase.[17]

In the daughter theory the situation changes drastically: a potential *is generated at one loop*, making the point of the broken gauge symmetry unstable. Shifting from this point, for a trial, one finds oneself in the Z_2-noninvariant (twisted) sector, for which no planar equivalence exists, and which proves to be energetically favored in the orbifold theory on $R^3 \times S$. In this true vacuum the energy density is negative rather than zero, and the full gauge symmetry is restored. The daughter theory is in the confining phase. Obviously, there can be no equivalence.

6. Orientifold field theory and $\mathcal{N} = 1$ gluodynamics

Having concluded, with deep regret, that the planar equivalence of the orbifold daughters does not extend to nonperturbative level, we move on to another class of theories, called *orientifold*, which lately gave rise to great expectations. In this section we will argue that in the $N \to \infty$ limit there is a sector in the orientifold theory exactly identical to $\mathcal{N} = 1$ SYM theory, and, therefore, exact results on the IR behavior of this theory can be obtained.

The parent theory is $\mathcal{N} = 1$ SUSY gluodynamics with the gauge group U(N). In the large N limit the U(1) factor is irrelevant. The daughter theory has the same gauge group,[d] and the same gauge coupling. The gluino field λ^i_j is replaced by two Weyl spinors $\lambda_{[ij]}$ and $\xi^{[ij]}$, with two antisymmetrized indices. We can combine the Weyl spinors into one Dirac spinor, either $\Psi_{[ij]}$ or $\Psi^{[ij]}$. Note that the number of the fermion degrees of freedom in $\Psi_{[ij]}$ is $N^2 - N$, as in the parent theory in the large-N limit.

There is another version of the orientifold daughter. Instead of the anti-symmetrization of the two-index spinors, we can perform symmetrization,

[d]To be more precise, the gauge groups are almost the same. The U(1) factor completely decouples in the parent theory, while it does not decouple in the daughter one. Moreover, in the former theory the Z_k center of the gauge group acts trivially while in the latter one it is only Z_2 that acts trivially. These distinctions are unimportant in the large-N limit.

so that $\lambda^i_j \to (\lambda_{\{ij\}}\,,\ \xi^{\{ij\}})$. The number of degrees of freedom in $\Psi_{\{ij\}}$ is N^2+N. The field content of the orientifold theory is shown in Table 3. We will mostly focus on the antisymmetric daughter since it is of more practical interest.

Table 3. The field content of the orientifold theories. λ and ξ are two Weyl fermions, while A_μ stands for the gauge bosons. In the left (right) parts of the table the fermions are in the two-index symmetric (antisymmetric) representation of the gauge group SU(N). Moreover, $U_V(1)$ is the conserved global symmetry while the $U_A(1)$ symmetry is lost at the quantum level due to the chiral anomaly. A discrete subgroup remains unbroken.

	SU(N)	$U_V(1)$	$U_A(1)$
$\lambda_{\{ij\}}$	$\square\square$	1	1
$\xi^{\{ij\}}$	$\overline{\square\square}$	-1	1
A_μ	Adj	0	0

	SU(N)	$U_V(1)$	$U_A(1)$
$\lambda_{[ij]}$	$\begin{smallmatrix}\square\\\square\end{smallmatrix}$	1	1
$\xi^{[ij]}$	$\overline{\begin{smallmatrix}\square\\\square\end{smallmatrix}}$	-1	1
A_μ	Adj	0	0

The hadronic (color-singlet) sectors of the parent and daughter theories are different. In the parent theory composite fermions with mass scaling as N^0 exist, and moreover, they are degenerate with their bosonic SUSY counterparts. In the daughter theory any interpolating color-singlet current with the fermion quantum numbers contains a number of constituents growing with N. Hence at $N=\infty$ the spectrum contains only bosons.

Classically the parent theory has just an R symmetry corresponding to chiral rotations of the gluino field. Instantons break this symmetry down to Z_{2N} . The daughter theory has, on top, a conserved anomaly free current

$$\bar\lambda_{\dot\alpha}\lambda_\alpha - \bar\xi_{\dot\alpha}\xi_\alpha\,. \tag{26}$$

In terms of the Dirac spinor this is the vector current $\bar\Psi\gamma_\mu\Psi$. If the corresponding charge is denoted by Q, in the color-singlet bosonic sector $Q=0$, with necessity. Then the only global symmetry which remains in both theories is Z_{2N} spontaneously broken down to Z_2 by the respective bifermion condensates. This explains the existence of N vacua[e] in both cases.

We will compare the bosonic sectors of the parent and daughter theories. Note that in the daughter theory the part of the bosonic sector probed by

[e]At finite N the parent theory has N vacua while the orientifold daughters have $N-2$ and $N+2$ in the antisymmetric and symmetric versions, respectively.

the operators of the type (26), which have no analogs in the parent theory, is inaccessible.

6.1. *Perturbative equivalence*

Let us start from perturbative consideration. The Feynman rules of the planar theory are shown in Fig. 4.

The difference between the orientifold theory and $\mathcal{N} = 1$ gluodynamics is that the arrows on the fermionic lines point in the same direction, since the fermion is in the antisymmetric representation, in contrast to the supersymmetric theory where the gaugino is in the adjoint representation and the arrows point in the opposite directions. This difference between the two theories does not affect planar graphs provided each gaugino line is replaced by the sum of $\lambda_{[..]}$ and $\lambda^{[..]}$.

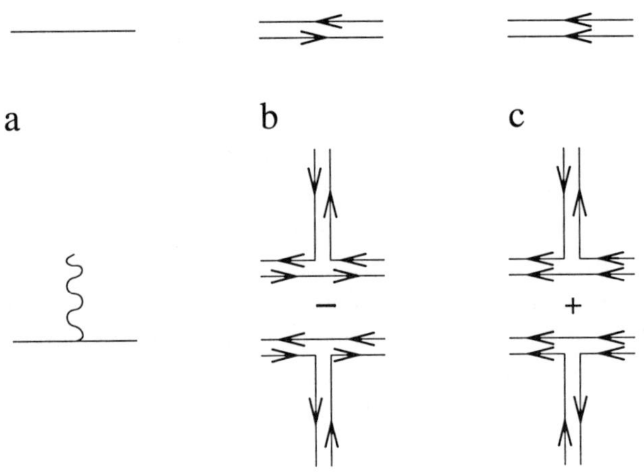

Figure 4. (a) The fermionic propagator and the vertex. (b) $\mathcal{N} = 1$ SYM theory. (c) Orientifold daughter.

There is a one-to-one correspondence between the planar graphs of the two theories. Diagrammatically this works as follows (see for example Fig. 5). Consider any planar diagram of the daughter theory: by definition of planarity it can be drawn on a sphere. The fermionic propagators form closed, non-intersecting loops that divide the sphere into regions. Each time we cross a fermionic line the orientation of color-indices loops (each one producing a factor N) changes from clock to anti-clock wise, and vice-versa, as easily seen in Fig. 5c. Thus the fermionic loops allow to attribute to each one of the above regions a binary label (say ± 1) according to whether the color loops go clock or anti-clock-wise in that region. Imagine now that one cuts out all the regions with a -1 label and glues them again on the sphere after having flipped them upside down. We will get a planar diagram of the SYM theory in which all color loops go, by convention, clock-wise. The number associated with both diagrams will be the same since the diagrams inside each region always contain an even number of powers of g so that the relative minus signs of Fig. 4 do not matter.

In fact, in the above argument, we cut corners, so that some people in the audience might be puzzled. For instance, in the parent theory gluinos are the Weyl (Majorana) fermions while fermions in the daughter theory are the Dirac fermions. Therefore, we hasten to add a few explanatory remarks which will hopefully satisfy the careful listeners.

First, let us replace the Weyl (Majorana) gluino of $\mathcal{N}=1$ gluodynamics by a Dirac spinor Ψ^i_j. Each fermion loop in the original theory is obtained from the Dirac loop by multiplying the latter by $1/2$. Let us keep this factor $1/2$ in mind.

On the daughter theory side, instead of considering the antisymmetric spinor $\Psi_{[ij]}$ *or* the symmetric one, $\Psi_{\{ij\}}$, we will consider a Dirac spinor in the *reducible* two-index representation Ψ_{ij}, without imposing any conditions on i, j. Thus, this reducible two-index representation is a sum of $\square\!\square + \begin{smallmatrix}\square\\\square\end{smallmatrix}$. It is rather obvious that at $N \to \infty$ any loop of $\Psi_{[ij]}$ yields the same result as the very same loop with $\Psi_{\{ij\}}$, which implies, in turn, that in order to get the fermion loop in, say, antisymmetric orientifold daughter, one can take the Dirac fermion loop in the above reducible representation, and multiply it by $1/2$.

Given the same factor $1/2$ on the side of the parent and daughter theories, what remains to be done is to prove that the Dirac fermion loops for Ψ^i_j and Ψ_{ij} are identical at $N \to \infty$. Thus, we will focus on the color factors.

Let the generator of SU(N) in the fundamental be T^a while that in the antifundamental $\bar{T}^a$,

$$T^a = T^a_{\square}, \qquad \bar{T}^a = T^a_{\bar{\square}}. \tag{27}$$

Then the generator in the adjoint is

$$\begin{aligned} T^a_{\text{Adj}} &\sim T^a_{\square\times\bar{\square}} = T^a_{\square}\otimes 1 + 1\otimes T^a_{\bar{\square}} \\ &\equiv T^a \otimes 1 + 1 \otimes \bar{T}^a\,, \end{aligned} \tag{28}$$

where we made use of large N neglecting the singlet. Moreover, in the daughter theory the generator of the reducible $\square\otimes\square$ representation can be written as

$$T^a_{\rm two-index} = T^a_{\square}\otimes 1 + 1\otimes T^a_{\square} \equiv T^a\otimes 1 + 1\otimes T^a$$
$$\text{or} \quad \bar{T}^a\otimes 1 + 1\otimes \bar{T}^a\,. \tag{29}$$

One more thing which we will need to know is the fact that

$$\bar{T} = -\tilde{T} = -T^*\,, \tag{30}$$

where the tilde denotes the transposed matrix.

Let us examine the color structure of a generic planar diagram for a gauge invariant quantity. For example, Fig. 5a exhibits a four-loop planar graph for the vacuum energy. The color decomposition (28) and (29) is equivalent to using the 't Hooft double-line notation, see Figs. 5b,c. In the parent theory each fermion-gluon vertex contains $T^a_{\rm Adj}$; in passing to the daughter theory we replace $T^a_{\rm Adj}\to T^a_{\rm two-index}$.

Upon substitution of Eqs. (28) and (29) the graph at hand splits into two (disconnected!) parts: the inner one (inside the dotted ellipse in Fig. 5c), and the outer part (outside the dotted ellipse in Fig. 5c). These two parts do not "talk to each other," because of planarity (large-N limit). The outer parts are the same in Figs. 5b,c. They are proportional the trace of the product of two T's in the both cases,

$${\rm Tr}(\bar{T}^a\bar{T}^a)\,.$$

This is the first factor. The second one comes from the inner part of Figs. 5b,c. In the parent theory the inner factor is built of 6 T's — one in each fermion-gluon vertex, and three T's in the three-gluon vertex ${\rm Tr}([A_\mu A_\nu]\,\partial_\mu A_\nu)$ where $A_\mu\equiv A^a_\mu T^a$. In the daughter theory the inner factor is obtained from that in the parent one by replacing all T's by $\bar{T}$'s. According to Eq. (30), $\bar{T}=-\tilde{T}$ (remember, the tilde marks the transposed matrix). This fact implies that the only difference between the inner blocks in Figs. 5b,c is the reversal of the direction of color flow on each of the 't Hooft lines. Since the inner part is color-singlet by itself, the above reversal has no impact on the color factor — they are identical in the parent and daughter theories.

It is, perhaps, instructive to illustrate how this works using more conventional notations. For the the inner part of the graph in Fig. 5b we have the color factor Tr $(T^aT^bT^c)\,f^{abc}$ while in the daughter one Tr $(\bar{T}^a\bar{T}^b\bar{T}^c)\,f^{abc}$.

Using the fact that

$$[T^a T^b] = i f^{abc} T^c \quad \text{and} \quad [\bar{T}^a \bar{T}^b] = i f^{abc} \bar{T}^c$$

we immediately come to the conclusion that the above two expressions coincide.

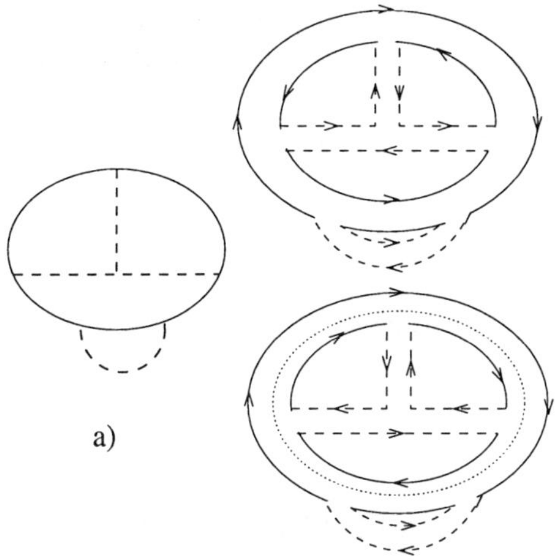

Figure 5. (a) A typical planar contribution to the vacuum energy. (b and c) The same in the 't Hooft notation for (b) $\mathcal{N} = 1$ SYM theory; (c) orientifold daughter.

Thus, all perturbative results that we are aware of in $\mathcal{N} = 1$ SYM theory apply in the orientifold model as well. For example, the β function of the orientifold field theory is

$$\beta = -\frac{1}{2\pi} \frac{3N\alpha^2}{1 - (N\alpha)/(2\pi)} \left\{ 1 + O\left(\frac{1}{N}\right) \right\}. \tag{31}$$

In the large N limit it coincides with the $\mathcal{N} = 1$ SYM theory result.[19] Note that the corrections are $1/N$ rather than $1/N^2$. For instance, the exact first coefficient of the β function is $-3N - 4/3$ versus $-3N$ in the parent theory.

6.2. *Nonperturbative equivalence proof*

Now we will argue that the perturbative argument can be elevated to nonperturbative level in the case at hand. A heuristic argument in favor of the nonperturbative equivalence is that the coincidence of all planar graphs of the two theories implies that the relevant Casimir operators of the two representations are equivalent in the large N limit. The partition functions of the two theories depend on the Casimir operators and therefore must coincide as well.

A more formal line of reasoning is as follows. It is essential that the fermion fields enter bilinearly in the action, and that for any *given* gauge field configuration in the parent theory there is exactly the same configuration in the daughter one. Our idea is to integrate out fermion fields for any *fixed* gluon field configuration, which yields respective determinants, and then compare them.

Consider the partition function of $\mathcal{N}=1$ SYM theory,

$$\mathcal{Z}_0 = \int \mathcal{D}A\,\mathcal{D}\lambda\, \exp\left(iS[A,\lambda,J]\right)\,, \tag{32}$$

where J is any source coupled to color-singlet gluon operators. (Appropriate color-singlet fermion bilinears can be considered too.)

For any given gluon field, upon integrating out the gaugino field, we obtain

$$\mathcal{Z}_0 = \int \mathcal{D}A\, \exp\left(iS[A,J]\right) \det\left(i\,\not\partial + \not A^a T^a_{\rm Adj}\right)\,, \tag{33}$$

where $T^a_{\rm Adj}$ is a generator in the adjoint representation.

If one integrates out the fermion fields in the non-supersymmetric orientifold theory, at fixed A, one arrives at a similar expression, but with the generators of the anti-symmetric (or symmetric) representation instead of the adjoint, $T^a_{\rm Adj} \to T^a_{\rm asymm}$ or $T^a_{\rm symm}$.

To compare the fermion determinants in the parent and daughter theories (assuming that the gauge field configuration $A^a_\mu(x),\ \ a=1,...,N^2$, is the same and fixed) we must cast both fermion operators in similar forms. To this end we will extend both theories. In the parent one we introduce the second adjoint Weyl fermion ξ^i_j and combine two Weyl fermions into one Dirac adjoint fermion Ψ^i_j. The determinant in this extended theory is the square of the original one.

In the daughter theory instead of $\Psi^{[ij]}$, we will work with the reducible representation, combining both symmetric and antisymmetric, Ψ^{ij} (no (anti)symmetrization over the color indices). Then the number of the fermion Dirac fields is the same as in the extended parent. Again, as in the previous case, the determinant in the extended daughter theory is the square of the original one. For infrared regularization we will introduce small mass terms, assuming that the vanishing mass limit is smooth.

The determinant in Eq. (33) is a gauge invariant object and, thus, it can be expanded in Wilson loops operators

$$\mathcal{W}_C[A_{\text{Adj}}] = \operatorname{Tr} P \exp\left(i\int_C A_\mu^a T_{\text{Adj}}^a \, dx^\mu\right) . \tag{34}$$

For other representations, the Wilson loops operator is defined in a similar manner. Thus one can write

$$D \equiv \det\left(i\,\partial\!\!\!/ + A\!\!\!/^a T^a_{\text{Adj}}\right) = \sum_C \alpha_C \, \mathcal{W}_C[A_{\text{Adj}}] \, . \tag{35}$$

Equation (28) implies then that

$$D = \sum_C \alpha_C \operatorname{Tr} P \exp\left(i\int_C A_\mu^a \left(T^a \otimes 1 + 1 \otimes \bar{T}^a\right) dx^\mu\right) . \tag{36}$$

Moreover, due to the fact that

$$[(T^a \otimes 1),\ (1 \otimes \bar{T}^a)] = 0 \, ,$$

the determinant (36) can be rewritten as

$$D = \sum_C \alpha_C \operatorname{Tr} P \exp\left(i\int_C A_\mu^a T^a \, dx^\mu\right) \operatorname{Tr} P \exp\left(i\int_C A_\mu^a \bar{T}^a \, dx^\mu\right) . \tag{37}$$

As a result, the partition function takes the form

$$\mathcal{Z}_0 = \sum_C \alpha_C \, \langle\, \mathcal{W}_C[A_\square] \, \mathcal{W}_C^*[A_\square] \,\rangle \, . \tag{38}$$

One of two most crucial points of the proof is the applicability of factorization in the large-N limit,

$$\mathcal{Z}_0 = \sum_C \alpha_C \, \langle\, \mathcal{W}_C(A_\square) \,\rangle \langle\, \mathcal{W}_C^*(A_\square) \,\rangle = \sum_C \alpha_C \, \langle\, \mathcal{W}_C(A_\square) \,\rangle^2 \, . \tag{39}$$

In the second equality in Eq. (39) we used the second most crucial point, the reality of the Wilson loop,

$$\langle\, \mathcal{W}_C \,\rangle = \langle\, \mathcal{W}_C^* \,\rangle \, . \tag{40}$$

The partition function (39) is *exactly* the same one gets in the orbifold theory upon exploiting Eq. (29) and factorization,

$$\mathcal{Z}_{\text{orbi}} = \sum_C \alpha_C \, \langle\, \mathcal{W}_C(A_\square) \,\rangle \langle\, \mathcal{W}_C(A_\square) \,\rangle \, . \tag{41}$$

Let us ask ourselves: "What is the critical difference between the orbifold and orientifold theories, that makes possible elevation of the perturbative planar equivalence to nonperturbative level in the latter case, while this does not work in the former?" The answer is quite transparent. In the orientifold theories one can choose the very same gauge field configuration as in the parent theory, one by one, and compare the fermion determinants in the given gauge filed background. This is due to the fact that the gauge groups are the same in the parent and the daughter. In the orbifold theories the gauge group is different from that of the parent ($SU(N)^k$ vs. $SU(kN)$).

7. Summary of the main results

The remarkable novel findings, which were the focus of the present talk, can be summarized as follows.

• Planar equivalence between supersymmetric and nonsupersymmetric gauge theories is a fruitful idea which can be implemented at the nonperturbative level.

•• The partition functions of the orientifold field theory and $\mathcal{N}=1$ SYM theory coincide at large N, both perturbatively and nonperturbatively. All color-singlet correlation functions which involve gluons as sources are the same in the parent and the daughter theory. This statement is also valid for the fermion-bilinear sources that can be projected onto each other in both theories.

••• The glueball spectrum in the daughter theory is identical to that of SUSY gluodynamics. In particular, there is a degeneracy between $0^{\pm}$ glueballs. Unlike $\mathcal{N}=1$ gluodynamics, there are no color-singlet fermions at $N=\infty$ in the orientifold daughter theory.

References

1. N. Seiberg, Proceedings 4th Int. Symposium on *Particles, Strings, and Cosmology (PASCOS 94)*, Syracuse, New York, May 1994, Ed. K. C. Wali (World Scientific, Singapore, 1995), p. 183 [hep-th/9408013]; *Int. J. Mod. Phys.* **A12** (1997) 5171 [hep-th/9506077].
2. A. Armoni, M. Shifman and G. Veneziano, *Nucl. Phys.* **B667**, 170 (2003) [hep-th/0302163].
3. A. Armoni, M. Shifman and G. Veneziano, *SUSY relics in one-flavor QCD from a new 1/N expansion,* hep-th/0307097; *QCD Quark Condensate from SUSY and the Orientifold Large-N Expansion*, hep-th/0309013.
4. S. Kachru and E. Silverstein, *Phys. Rev. Lett.* **80**, 4855 (1998) [hep-th/9802183].
5. A. E. Lawrence, N. Nekrasov and C. Vafa, *Nucl. Phys.* **B533**, 199 (1998) [hep-th/9803015].
6. M. Bershadsky, Z. Kakushadze and C. Vafa, *Nucl. Phys.* **B523**, 59 (1998) [hep-th/9803076].

7. M. Bershadsky and A. Johansen, *Nucl. Phys.* **B536**, 141 (1998) [hep-th/9803249].
8. Z. Kakushadze, *Nucl. Phys.* **B529**, 157 (1998) [hep-th/9803214]; *Phys. Rev.* **D58**, 106003 (1998) [hep-th/9804184].
9. M. Schmaltz, *Phys. Rev.* **D59**, 105018 (1999) [hep-th/9805218].
10. A. Armoni and B. Kol, *JHEP* **9907**, 011 (1999) [hep-th/9906081].
11. M. J. Strassler, *On methods for extracting exact non-perturbative results in non-supersymmetric gauge theories,* hep-th/0104032, unpublished.
12. A. Gorsky and M. Shifman, *Phys. Rev.* **D67**, 022003 (2003) [hep-th/0208073].
13. E. Witten, *Nucl. Phys.* **B202**, 253 (1982).
14. G. Veneziano and S. Yankielowicz, *Phys. Lett.* **B113**, 231 (1982),
15. M. A. Shifman and A. I. Vainshtein, *Nucl. Phys.* **B296**, 445 (1988).
16. M. A. Shifman and A. I. Vainshtein, *Instantons versus supersymmetry: Fifteen years later,* in M. Shifman, *ITEP Lectures on Particle Physics and Field Theory,* (World Scientific, Singapore, 1999) Vol. 2, p. 485-647 [hep-th/9902018].
17. N. M. Davies, T. J. Hollowood, V. V. Khoze and M. P. Mattis, *Nucl. Phys.* **B559**, 123 (1999) [hep-th/9905015].
18. I. Hinchliffe, *Phys. Rev.* **D66**, # 1-I, 010001-89 (2002).
19. V. A. Novikov, M. A. Shifman, A. I. Vainshtein and V. I. Zakharov, *Nucl. Phys.* **B229**, 381 (1983); *Phys. Lett.* **B166**, 329 (1986).
20. I. R. Klebanov, *Phys. Lett.* **B466**, 166 (1999) [hep-th/9906220].
21. D. Tong, *JHEP* **0303**, 022 (2003) [hep-th/0212235].

NON-ABELIAN SUPERCONDUCTOR AND QCD: LESSONS FROM SUPERSYMMETRIC THEORIES

K. KONISHI

Dipartimento di Fisica, "E. Fermi"
Università di Pisa,
Via Buonarroti, 2, Ed. C
56127 Pisa, Italy
E-mail: konishi@df.unipi.it

Much about the details of confinement mechanism and dynamical symmetry breaking in QCD might be learned from models with supersymmetry. In this talk, we discuss the idea that the ground state of QCD is a sort of (dual) nonabelian superconductor, suggested by a class of $N = 2$ supersymmetric theories.

1. Confinement in $SU(N)$ YM Theory

The test charges in $SU(N)$ YM theory take values in $(Z_N^{(M)}, Z_N^{(E)})$ where Z_N is the center of $SU(N)$ and $Z_N^{(M)}, Z_N^{(E)}$ refer to the magnetic and electric center charges. $(Z_N^{(M)}, Z_N^{(E)})$ classification of phases follows.[1,2] (See Figures) Namely,

(1) If field with $x = (a, b)$ condense, particles $X = (A, B)$ with

$$\langle x, X \rangle \equiv a\,B - b\,A \neq 0 \quad (mod\ N)$$

are confined. (e.g. $\langle \phi_{(0,1)} \rangle \neq 0 \to$ Higgs phase.)

(2) Quarks are confined if some field χ exist, such that

$$\langle \chi_{(1,0)} \rangle \neq 0.$$

(3) In the softly broken $N = 4$ (to $N = 1$) theory (often referred to as $N = 1^*$) all different types of massive vacua, related by $SL(2, Z)$, appear; the chiral condensates in each vacua are known.

(4) *Confinement index* [3] is equal to the smallest possible $r \in Z_N^{(E)}$ for which Wilson loop displays no area law. For instance, for $SU(N)$ YM, $r = N$ in the vacuum with complete confinement; $r = 1$ in the totally Higgs vacuum, etc.

(5) In softly broken $N = 2$ Gauge Theories, dynamics turns out to be particularly transparent.

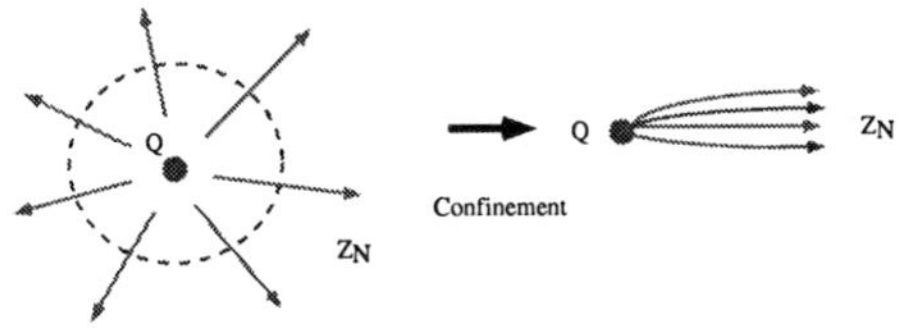

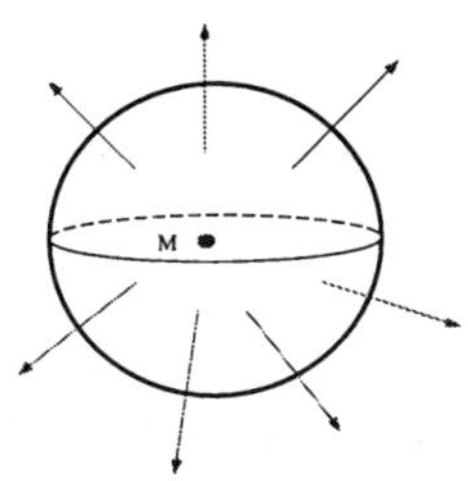

$\Pi_1(SU(N)/Z_N) \sim Z_N$

The questions we wish to address here are: What is χ in QCD ? How do they interact ? Is chiral symmetry breaking related to confinement ? θ vacua ?; $\frac{\epsilon'}{\epsilon}$; $\Delta I = \frac{1}{2}$?

A familiar idea is that the ground state of QCD is a dual superconductor.[2] There exist no elementary nor soliton monopoles in QCD; however, monopoles can be detected as topological singularities (lines in $4D$) of Abelian gauge fixing, $SU(3) \to U(1)^2$. Alternatively, one can assume that certain configurations close to the Wu-Yang monopole ($SU(2)$)

$$A_\mu^a = \tilde{\sigma}(x)(\partial_\mu n \times n)^a + \dots , \qquad n(r) = \frac{r}{r} \quad \Rightarrow \quad A_i^a = \epsilon_{aij}\frac{r^j}{r^3}$$

dominate.[4] Although there are some evidence in lattice QCD[5] for "Abelian dominance", there are several questions to be answered. Do abelian monopoles carry flavor ? (L_{eff} ?) What about the gauge dependence ? Most significantly, does dynamical $SU(N) \to U(1)^{N-1}$ breaking occur ? That would imply a richer spectrum of mesons ($T_1 \neq T_2$, etc) not seen in Nature and not expected in QCD. Both in Nature and presumably in QCD there is only one "meson" state

$$\text{Meson} \sim \sum_{i=1}^{N} |\, q_i\, \bar{q}_i \,\rangle$$

i.e., 1 state vs $\left[\frac{N}{2}\right]$ states. (See Figure.) Assuming $SU(N) \to U(1)^{N-1} \times$ *Weyl* symmetry is not enough to solve the problem.

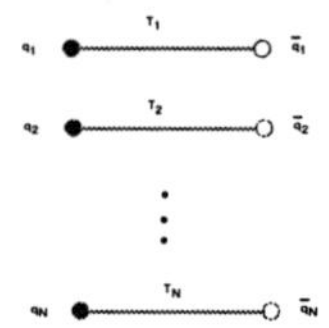

In an attempt to answer these questions we address ourselves to the world of supersymmetric gauge theories, where the nonabelian gauge dynamics can be analyzed to depth. Supersymmetric theories in fact continue to surprise us for rich insight they give us about the dynamics of nonabelian interactions. Although not a main subject of this talk, let me point out that much is going on at present in the exploration of various $N = 1$ or $N = 2$ susy gauge theories.[6]

Let us start by recalling some basic facts about supersymmetric gauge theories.[7]

2. SQCD

2.1. *Basics of Susy gauge theories*

The basic susy algebra contains

$$\{Q_\alpha, \bar{Q}_{\dot\alpha}\} = 2\sigma^{\mu}_{\alpha,\dot\alpha} P_\mu.$$

In order to construct supersymmetric theories it is convenient to introduce superfields

$$F(x,\theta,\bar\theta) = f(x) + \theta\psi(x) + \ldots$$

$$Q_\alpha = \frac{\partial}{\partial\theta^\alpha} - i\sigma^{\mu}_{\alpha,\dot\alpha}\bar\theta^{\dot\alpha}\partial_\mu, \quad \bar{Q}_{\dot\alpha} = \frac{\partial}{\partial\bar\theta^{\dot\alpha}} - i\theta^\alpha\sigma^{\mu}_{\alpha,\dot\alpha}\partial_\mu,$$

In general they are reducible with respect to supersymmetry transformations. We construct smaller irreducible multiplets. Chiral superfields are defined by the constraint $\bar{D}\Phi = 0$ ($D\Phi^\dagger = 0$) so that

$$\Phi(x,\theta,\bar\theta) = \phi(y) + \sqrt{2}\theta\,\psi(y) + \theta\theta\, F(y), \quad y = x + i\theta\sigma\bar\theta$$

$$D_\alpha = \frac{\partial}{\partial \theta^\alpha} + i\sigma^\mu_{\alpha,\dot\alpha}\bar\theta^{\dot\alpha}\partial_\mu, \qquad \bar D_{\dot\alpha} = -\frac{\partial}{\partial \bar\theta^{\dot\alpha}} - i\theta^\alpha \sigma^\mu_{\alpha,\dot\alpha}\partial_\mu,$$

Verctor superfields are defined to be real $V^\dagger = V$. They are conveniently expressed in terms of a chiral (fermionic) superfield

$$W_\alpha = -\frac{1}{4}\bar D^2 e^{-V} D_\alpha e^V = -i\lambda + \frac{\mu,}{2}(\sigma^\mu \bar\sigma^\nu)_\alpha^\beta F_{\mu\nu}\theta_\beta + \dots$$

Supersymmetric Lagrangian ($\int d\theta_1\, \theta_1 = 1$, etc) can then be written simply as

$$\L = \frac{1}{8\pi}\mathrm{Im}\,\tau_{cl}\left[\int d^4\theta\, \Phi^\dagger e^V \Phi + \int d^2\theta\, \frac{1}{2} W W\right] + \int d^2\theta\, W(\Phi) \qquad (1)$$

where $W(\Phi)$ is the superpotential and $\tau_{cl} = \frac{\theta}{2\pi} + \frac{4\pi i}{g^2}$. The scalar potential is the sum of the F-term and D-term:

$$V_{sc} = \sum_{mat}\left|\frac{\partial W}{\partial \phi}\right|^2 + \frac{1}{2}\sum_a \left|\sum_{mat}\phi^* t^a \phi\right|^2.$$

For SQCD, $\{\Phi\} \to Q \sim \underline{N},\ \tilde Q \sim \underline{N}^*$ of $SU(N)$

$$G_F = SU(n_f)\times SU(n_f)\times U_V(1)\times U_A(1)\times U_\lambda(1).$$

The theory has a characteristic continuous vacuum degeneracy - flat directions or classical moduli space (CMS). For instance, it looks like for

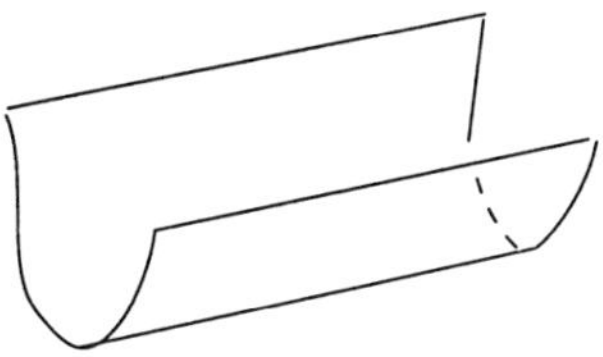

Φ

$n_f < n_c$,

$$Q = \tilde Q^\dagger = \begin{pmatrix} a_1 & 0 & \dots & 0 \\ 0 & \ddots & & \\ 0 & \dots & & a_{n_f} \\ 0 & 0 & \dots & 0 \\ \dots & & & \dots \end{pmatrix}. \qquad (2)$$

The problem is: is superpotential generated dynamically ? Is CMS modified ?

2.2. *Phases of SQCD; Seiberg's duality*

The analyses by use of various supersymmetry, Ward-Takahashi identities, nonrenormalization theorems,[8] symmetry considerations, dynamical (instanton) calculations[9,10] and Seiberg's duality[11] have established the following picture of the vacuum in the massless SQCD.

- The dynamically generated superpotential implies the vacuum runaway for $n_f < n_c$, while no superpotential is generated for $n_f > n_c$.
- For $n_f = n_c$ the moduli space (space of vacua) is quantum mechanically modified as

$$\det M - B\,\tilde{B} = \Lambda^{2n_f}.$$

- $\frac{3n_c}{2} < n_f < 3n_c$ (conformal window), the system is in an infrared fixed point (superconformal theory - SCFT): the low energy phsyics is described either as the original SQCD or as the dual $SU(\tilde{n}_c) = SU(n_f - n_c)$ theory with dual quarks.

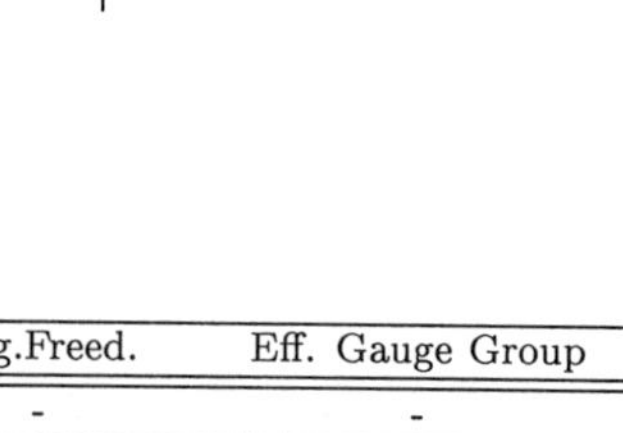

N_f	Deg.Freed.	Eff. Gauge Group	Phase	Symmetry
0 (SYM)	-	-	Confinement	-
$1 \leq N_f < N_c$	-	-	no vacua	-
N_c	$M, B, \tilde{B}$	-	Confinement	$U(N_f)$
$N_c + 1$	$M, B, \tilde{B}$	-	Confinement	Unbroken
$N_c + 1 < N_f < \frac{3N_c}{2}$	$q, \tilde{q}, M$	$SU(\tilde{N}_c)$	Free-magnetic	Unbroken
$\frac{3N_c}{2} < N_f < 3N_c$	$q, \tilde{q}, M$ or $Q, \tilde{Q}$	$SU(\tilde{N}_c)$ or $SU(N_c)$	SCFT	Unbroken
$N_f = 3N_c$	$Q, \tilde{Q}$	$SU(N_c)$	SCFT (finite)	Unbroken
$N_f > 3N_c$	$Q, \tilde{Q}$	$SU(N_c)$	Free Electric	Unbroken

3. Confining vacua in $N = 2$ Supersymmetric Gauge Theories

Although dynamical properties of $N = 1$ supersymmetric gauge thoeries are quite well known by now, as in the example of SQCD discussed above, a detailed understanding of the working of confinement and dynamical symmmetry breaking is still lacking. $N = 2$ theories appear to allow for a deeper level of understanding of the nonperturbative dynamics, by displaying the quantum behavior of magnetic monopoles and vortices very clearly. We start from afar: Dirac's monopoles in QED.

3.1. *Dirac's monopoles*

As is well known QED admits pointlike magnetic monopoles if Dirac's quantization condition

$$g\,e = \frac{n}{2}, \qquad n \in \ Z, \tag{3}$$

is satisfied. In the presence of a magnetic monopole, there cannot be a gauge vector potential which is everywhere regular. A possible singularity (Dirac string) along $(0,0,0) \to (0,0,-\infty)$ is invisible if (3) is satisfied. A proper formulation is to cover S^2 by two regions a $(0 \le \theta < \frac{\pi}{2} + \epsilon)$ and b $(\frac{\pi}{2} - \epsilon < \theta \le \pi)$ [16]

$$(A_\phi)^a = \frac{g}{r\sin\theta}(1-\cos\theta), \qquad (A_\phi)^b = -\frac{g}{r\sin\theta}(1+\cos\theta),$$

so that in each neighborhood the vector potential is regular. The two descriptions are related along the equator by a gauge transformation

$$A_i^a = A_i^b - U^\dagger \frac{i}{e}\partial_i U, \qquad U = e^{2ige\phi}.$$

The gauge transformation is well-defined if the condition (3) is met. More generally, for dyons (e_1, g_1), $(e_2 g_2)$, the quantization condition reads

$$e_1\,g_2 - e_2\,g_1 = \frac{n}{2}, \qquad n \in \ Z. \tag{4}$$

The topology involved is: $\Pi_1(U(1)) = \ Z$.

3.2. *Non-Abelian gauge theories*

In the case of a nonabelian gauge group, one might embed Dirac's monopole in a $U(1)$ subgroup. However, the homotopy group properties such as

$$SU(2) \sim S^3, \qquad \Pi_1(SU(2)) = \ 1, \tag{5}$$

$$SO(3) \sim \frac{S^2}{Z_2}, \qquad \Pi_1(SO(3)) = Z_2, \tag{6}$$

show that there are no topologically stable monopoles in $SU(2)$, $SU(N)$; there is only one type of monopole in $SO(3)$, and so on.[16]

In spontaneously broken gauge theories, instead, there are ('t Hooft-Polyakov) monopoles[17]

$$SU(2) \stackrel{\langle\phi\rangle\neq 0}{\longrightarrow} U(1)$$

$$D\phi \stackrel{r\to\infty}{\longrightarrow} 0, \quad \Rightarrow \quad \phi \sim U\cdot\langle\phi\rangle\cdot U^{-1};$$

$$A_i \sim U\cdot\partial_i U^\dagger \quad \Rightarrow \quad F_{ij} = \epsilon_{ijk}\frac{r_k}{r^3}m\frac{\tau_3}{2}$$

which are regular, finite energy soliton-like configurations, with topology $\Pi_2(SU(2)/U(1)) = \Pi_1(U(1)) = Z$. The static energy can be written as (Bogomolny)

$$H = \int d^3x \left[\frac{1}{4}(F^a_{ij})^2 + \frac{1}{2}(D_i\phi^a)^2 + \frac{\lambda}{2}(\phi^2 - v^2)^2\right].$$

$$= \int d^3x \left[\frac{1}{4}(F^a_{ij} - \epsilon_{ijk}D_k\phi^a)^2 + \frac{1}{2}F^a_{ij}\epsilon_{ijk}D_k\phi^a + \text{pot.}\right]$$

where $\frac{1}{2}F^a_{ij}\epsilon_{ijk}D_k\phi^a = \partial_i S_i$; $S_i = \frac{1}{2}\epsilon_{ijk}F^a_{ij}\phi^a$: the second term in the square bracket is a topological invariant. It follows that in a given sector

$$H \geq \int d^3x \nabla\cdot S = \frac{4\pi v}{g}m, \qquad m = 1, 2, \ldots$$

If $\lambda = 0$ the configuration of the minimum energy is given by the solution of the linear (Bogomolny) equations

$$F^a_{ij} - \epsilon_{ijk}D_k\phi^a = 0; \qquad B^a_i = D_i\phi^a$$

whose solutions are known in analytic form.

A more general situation is that of a spontaneously broken gauge theory

$$G \stackrel{\langle\phi\rangle\neq 0}{\longrightarrow} H$$

where H is nonabelian.[18,19] The asymptotic behavior is

$$D\phi \stackrel{r\to\infty}{\longrightarrow} 0, \quad \Rightarrow \quad \phi \sim U\cdot\langle\phi\rangle\cdot U^{-1} \sim \Pi_2(G/H) = \Pi_1(H);$$

$$A_i \sim U\cdot\partial_i U^\dagger \to \quad F_{ij} = \epsilon_{ijk}\frac{r_k}{r^3}\beta_\ell T_\ell.$$

$SU(N)/Z_N$	$\Leftrightarrow$	$SU(N)$
$SO(2N)$	$\Leftrightarrow$	$SO(2N)$
$SO(2N+1)$	$\Leftrightarrow$	$USp(2N)$

where $T_i \in$ Cartan S.A. of H. Topological quantization leads to $2\ \alpha \cdot \beta \subset Z$, where $\beta_i =$ weight vectors of $\tilde{H} =$ dual of H. Examples of the dual of several groups are given in the Table below.

The point of this discussion is that these objects – abelian and non-abelian monopoels – appear naturally in $N = 2$ gauge theories as dynamical degrees of freedom and play a crucial role in the infrared physics. Before coming to this central issue, however, let us first briefly review the celebrated Seiberg-Witten solution of $N = 2$ gauge theory.

3.3. *Seiberg-Witten, $N = 2$ Gauge Theory*

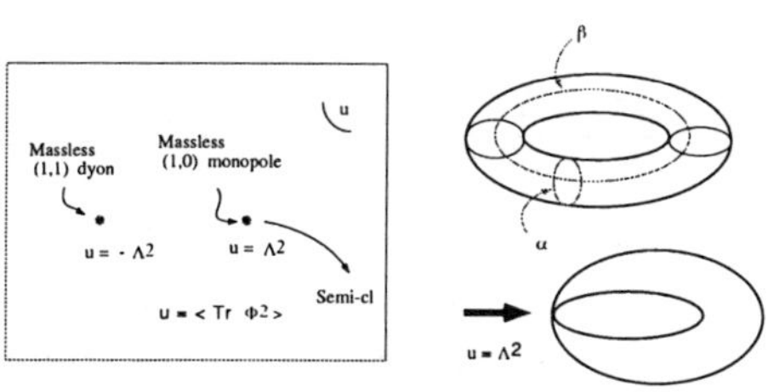

The Lagrangian of a $N = 2$ YM theory is Eq.(1) with $W(\Phi) = 0$, where Φ is a chiral superfield in the adjoint representation of the gauge group. For $SU(2)$ the vacuum degeneracy (moduli space) is parametrized as

$$\langle \Phi \rangle = \begin{pmatrix} a & 0 \\ 0 & -a \end{pmatrix}, \tag{7}$$

$a \neq 0$ breaks the gauge symmetry as $SU(2) \to U(1)$. In the infrared,

$$L_{eff} = \mathrm{Im} \left[\int d^4\theta\, \bar{A}\, \frac{\partial F_p(A)}{\partial A} + \int \frac{1}{2} \frac{\partial^2 F_p(A)}{\partial A^2} W_\alpha W^\alpha \right]$$

where W_α, A describe $N = 2$, $U(1)$ theory. $F_p(A)$ is called prepotential. Define the dual of A, $A_D \equiv \frac{\partial F_p(A)}{\partial A}$: then

$$\frac{d\,A_D}{d\,u} = \oint_\alpha \frac{dx}{y}, \qquad \frac{d\,A}{d\,u} = \oint_\beta \frac{dx}{y},$$

where the curve ($u \equiv \mathrm{Tr}\,\langle\Phi^2\rangle$ describes the quantum moduli space - QMS) is

$$y^2 = (x-u)(x+\Lambda^2)(x-\Lambda^2).$$

The exact mass formula (BPS) following from the $N=2$ susy algebra is

$$m_{n_m,n_e} = \sqrt{2}\,|\,n_m A_D + n_e A\,|.$$

The above four formulae consitute the Seiberg-Witten solution.[20,21,22]

The adjoint scalar mass ($\mu\,\Phi^2$ perturbation) leads to the low-energy effective superpotential near the singularity, $u \simeq \Lambda^2$:

$$W_{eff} = \sqrt{2}\,A_D\,M\tilde{M} + \mu\,U(A_D)$$

Minimization of the potential leads to the condensation of the monopole $\langle M\rangle \sim \sqrt{\mu\,\Lambda}$ (Confinement).

It is interesting to note that at the singularities $u = \pm\Lambda^2$, instanton sum diverges

$$\langle \mathrm{Tr}\Phi^2\rangle = \frac{a^2}{2} + \frac{\Lambda^4}{a^2} + \ldots = \ldots + 1 + 1 + 1 + \ldots$$

The discussion can be generalized to $N=2$ pure YM theories with other gauge groups. In general, dynamical abelianization occurs near the monopole singularities, for instance, $SU(N)$ gauge group gets dynamically broken as $SU(N) \to U(1)^{N-1}$ (*cfr* QCD).

It is important to realize that these light "monopoles" are indeed 't Hooft-Polyakov monopoles becoming light by quantum corrections. This can be proven by studying the charge fractionalization.[23] For instance, the electric charge of the monopole is known to behave as

$$\frac{2}{g}Q_e = n_e + \left[-\frac{4}{\pi}\mathrm{Arg}\,a + \frac{1}{2\pi}\sum_{f=1}^{N_f}\mathrm{Arg}\,(m_f^2 - 2a^2)\right]\,n_m + \ldots$$

in the semi-classical region. The Seiberg-Witten exact solution, when extrapolated back to the semi-classical domain, reproduces exactly this results. An anaologous check has been done for the quark number fractionalization. There is an interesting phenomenon of quantum quenching of quark numbers of massless, condensing monopole. Also, the non-abelian flavor quantum numbers of the monopoles as encoded in the Seiberg-Witten solution are consistent with the well-known Jackiw-Rebbi mechanism.

3.4. *More general $N = 2$ models*

The study of the more general class of $N = 2$ theories[24,25] has shown that there are variety of confining vacua (see Fig. 3.4):

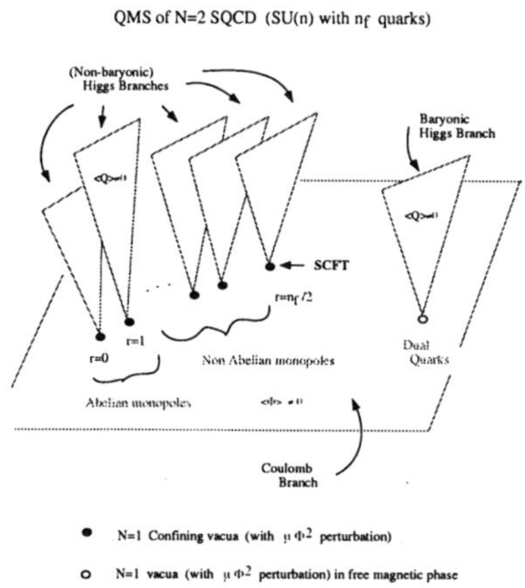

(1) There are vacua ($r = 0, 1$) in which the low-energy effective action is an abelian (dual) gauge theory. Upon the adjoint scalar mass perturbation $\mu\,\Phi^2$, the magnetic monopole condenses (confinement); the system displays dynamical abelianization, a feature not shared by the real world QCD;
(2) In a series of r-vacua, the effective action is $G_{eff} \sim SU(r) \times U(1)^{n_c-r-1}$; with n_f dual quarks in $\underline{r}$ of the low-energy $SU(r)$ group. The "dual quarks" can be identified with the standard non-abelian monopoles. These r-vacua exist for $r \leq \frac{n_f}{2}$;
(3) Superconformal theory (SCFT) occurs at $r = \frac{n_f}{2}$. Here the question is what (mutually nonlocal) degrees of freedom describe the SCFT (which confines upon the perturbation $\mu\,\Phi^2$).

Physics of $USp(2\,n_c)$ ($SO(n_c)$ similar) theory is also very interesting. All r vacua (at finite m) collapse into a single SCFT at $m \to 0$; in other words, all confining vacua (with $\mu\,\Phi^2$) are deformed SCFT, with mutually nonlocal dyons in the infrared physics. Also, the global $SO(2\,n_f) \to U(n_f)$ symmetry breaking pattern (in the case of $USp(2\,n_c)$ theory) is very reminiscent of what happens in QCD (*cfr* $\langle\bar{\psi}\,\psi\rangle^{(QCD)} \neq 0$).

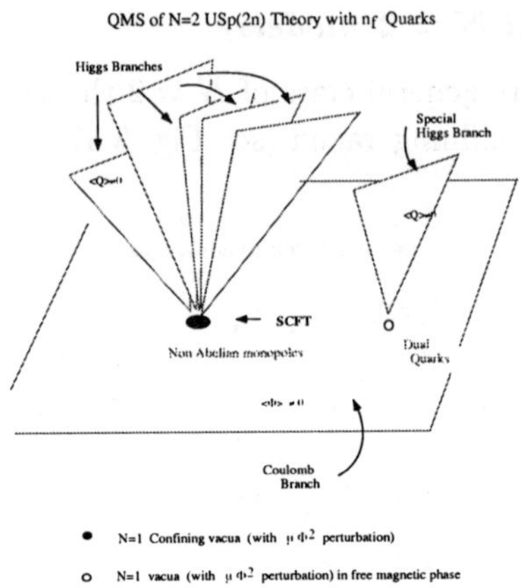

3.5. *More about Nonabelian Monopoles*

Consider the system with gauge symmetry breaking

$$SU(3)\xrightarrow{\langle\phi\rangle}SU(2)\times U(1), \qquad \langle\phi\rangle = \begin{pmatrix} v & 0 & 0 \\ 0 & v & 0 \\ 0 & 0 & -2v \end{pmatrix},$$

By making use of the 't Hooft-Polyakov solutions in $SU_U(2), SU_V(2) \subset SU(3)$ one finds two degenerate $SU(3)$ solutions. Analogously, for the sys-

monopoles	$\tilde{SU}(2)$	$\tilde{U}(1)$
$\tilde{q}$	$\underline{2}$	1

tem with symmetry breaking

$$SU(n)\xrightarrow{\langle\phi\rangle}SU(r)\times U^{n-r}(1), \qquad \langle\phi\rangle = \begin{pmatrix} v_1 \mathbf{1}_{r\times r} & 0 & \dots & 0 \\ 0 & v_2 & 0 & \dots \\ 0 & 0 & \ddots & \dots \\ 0 & 0 & \dots & v_{n-r+1} \end{pmatrix},$$

one finds the following set of the minimal monopoles:

We note[26] that they represent a degenerate r-plet of monopole solutions (q); that they have *the same charge structure as that of the "dual quarks" appearing in the* r *-vacua of* $N = 2$ *SQCD*. The latter have also the correct flavor quantum numbers as expected from the Jackiw-Rebbi mechanism.

monopoles	$\tilde{SU}(r)$	$\tilde{U}_0(1)$	$\tilde{U}_1(1)$	$\tilde{U}_2(1)$	...	$\tilde{U}_{n-r-1}(1)$
q	$\underline{r}$	1	0	0	...	0
e_1	$\underline{1}$	0	1	0	...	0
e_2	$\underline{1}$	0	0	1	0	0
$\vdots$	$\underline{1}$	0	...			0
e_{n-r-1}	$\underline{1}$	0	0	...	...	1

3.6. *Subtle are nonabelian monopoles*

There are however certain subtleties about the nonabelian monopoles. First of all, "colored dyons" are known not to exist.[27] In other words, no charge fractionalization is possible for non-Abelian quantum numbers. This being so, it does not preclude the nonabelian monopoles of our interest: magnetic particles having abelian and nonabelian charges, *both* magnetic, can perfectly well exist, and do appear in the r-vacua of the sofly broken $N = 2$ SQCD [26] ! The main point is that the nonabelian monopoles are multiplets of the dual $\tilde{H}$ group, not of H itself. Transformations among the members of a multiplet of monopoles are nonlocal transformations with respect to the original gauge field variables.

The no-go theorem nonetheless does mean that the true gauge symmetry of the theory is not $H \otimes \tilde{H}$, as sometimes suggested, but $\tilde{H}$, H or some other combination, according to which physical degrees of freedom are relevant in a particular situation.

Also, it is not justified to study the system $G \stackrel{\langle\phi\rangle\neq 0}{\longrightarrow} H$ as a limit of maximally broken cases as sometimes made in the literature. Nonabelian monopoles are never really semi-classical, even if

$$\langle\phi\rangle \gg \Lambda_H.$$

For H were broken it would produce simply an *approximately* degenerate set of monopoles as, for instance, in the pure $N = 2$, $SU(3)$ theory. Only if H remains unbroken do nonabelian monopoles in irreducible representation of $\tilde{H}$ appear. Which option is realized is a dynamical question which cannot be determined from a semiclassical consideration.

The important fact is that the second option *is* realized in the r vacua of $N = 2$ SQCD with $SU(r) \times U(1)^{n_c - r+1}$ gauge group, where $r < \frac{N_f}{2}$. This last constraint can be understood from a renormalization-group consideration: for $r < \frac{N_f}{2}$ there is a sign flip in the beta functions of the dual magnetic gauge group, with respect to that in the underlying theory:

$$b_0^{(dual)} \propto -2\,r + n_f > 0, \qquad b_0 \propto -2\,n_c + n_f < 0.$$

In fact, when such a sign flip is not possible, *e.g.*, pure $N = 2$ YM, dynamical abelianization occurs ! *The quantum behavior of nonabelian monopoles thus depends critically on the presence of massless fermions in the underlying theory.* $r = \frac{n_f}{2}$ is a boundary case: the corresponding vacua are SCFT (nontrivial IR fixed point). Nonabelian monopoles and dyons still show up as low-energy degrees of freedom, but their interactions are nonlocal and strong. The possible mechanism of confinement in these vacua has been recently studied.[29]

3.7. $\mathbf{Z}_N$ *Vortices*

Once the relevant degrees of freedom which act as the order parameter of confinement are identified, we are interested in the dominant field configurations which are capable of actually confining quarks. In the abelian dual superconductor picture of confinement in a $SU(N)$ YM theory, the quarks would be confined by abelian Abrikosov-Nielsen-Olesen vortices of $U(1)^{N-1}$. However, this leads to the difficulty mentioned at the end of the section 1. The quarks must be confined by some sort of nonabelian chromoelectric vortices.

The simplest type of vortices involving a nonabelian gauge group is the $\mathbf{Z}_N$ vortex, which occurs in a system with gauge symmetry breaking as

$$SU(N) \Rightarrow \mathbf{Z}_N.$$

An analogous vortex appear in a system with a general symmetry breaking pattern, $H \Rightarrow C$, a discrete center. Vortices represent nontrivial elements of $\Pi_1(H/C)$, e.g. $\Pi_1(SU(N)/\mathbf{Z}_N) = \mathbf{Z}_N$. The asymptotic behavior of the fields is

$$A_i \sim \frac{i}{g}\,U(\phi)\partial_i U^\dagger(\phi); \quad \phi_A \sim U\phi_A^{(0)}U^\dagger, \qquad U(\phi) = \exp i\sum_j^r \beta_j T_j \phi$$

where T_j are the generators of the Cartan subalgebra of H. The quantization condition is (α = root vectors of H)

$$U(2\pi) \in \mathbf{Z}_N, \qquad \alpha\cdot\beta \in \mathbf{Z}:$$

the vortices are characterized by the *weight vectors* of the group $\tilde{H}$, dual of H. It seems as though the vortex solutions were classified according to the irreducible representations of $\tilde{H} = SU(N)$. Actually, the fact that the topology involved is $\Pi_1(SU(N)/\mathbf{Z}_N) = \mathbf{Z}_N$ means that the stable vortices are characterized by $\mathbf{Z}_N$ charge (N-ality) only.[28]

These $\mathbf{Z}_N$ vortices are non BPS and this makes the analysis of these objects so far relatively little explored. However there are interesting quantities which characterize these systems such as the tension ratios for different N-ality sources: an intriguing proposal (sine formula) [30,31] is

$$T_k \propto \sin \frac{\pi k}{N}$$

which can be measured on the lattice.

3.8. *BPS vortices; Nonabelian Superconductors*

Systems with BPS vortices with a nonabelian flux - nonabelian superconductors - have been shown to exist only recently.[32,33] Consider a gauge theory in which the gauge group is broken at two different scales

$$G \stackrel{\langle\phi\rangle\neq 0}{\longrightarrow} H \stackrel{\langle\phi'\rangle\neq 0}{\longrightarrow} \emptyset, \qquad \langle\phi\rangle \gg \langle\phi'\rangle,$$

where the unbroken (non-Abelian) group H gets broken at a much lower scale, $\langle\phi'\rangle$. We are interested in the physics at scales between the two scales $\langle\phi\rangle$ and $\langle\phi'\rangle$. When $\Pi_1(H) \neq \emptyset$ the system develops vortices. If the theory contains an exact continuous symmetry G_F, respected both by the interactions and by the vacuum (not spontaneously broken), but broken by a vortex solution, then there will be a nontrivial degeneracy of vortex solutions (zero modes).

An example[33] is the $SU(3)$ $N = 2$ theory with $n_f = 4, 5$ quark flavors with large common (bare) mass m, with the $N = 2$ symmetry broken softly to $N = 1$ by the adjoint mass term, $\mu \operatorname{Tr} \Phi^2$. We consider a particular vacuum, the "$r = 2$" vacuum, of this system, which is characterized by the VEVs

$$\Phi = -\frac{1}{\sqrt{2}} \begin{pmatrix} m & 0 & 0 \\ 0 & m & 0 \\ 0 & 0 & -2m \end{pmatrix}, \quad < q^{kA} >=< \bar{\tilde{q}}^{kA} >= \sqrt{\frac{\xi}{2}} \begin{pmatrix} 1 & 0 \\ 0 & 1 \end{pmatrix}$$

where $\xi = \mu m$. We take the (bare) quark mass m much larger than μ so that $m \gg \sqrt{\xi}$. At the mass scales between m and $\sqrt{\xi}$, the system has an exact $SU(2) \times U(1)/Z_2$ gauge symmetry as well as an $SU(n_f)$ global symmetry. The action has the form, after the Ansatz $\Phi = \langle\Phi\rangle$; $q = \tilde{q}^\dagger$; and $q \to \frac{1}{2} q$:

$$S = \int d^4x \left[\frac{1}{4g_2^2} \left(F_{\mu\nu}^a\right)^2 + \frac{1}{4g_1^2} \left(F_{\mu\nu}^8\right)^2 + \left|\nabla_\mu q^A\right|^2 \right.$$

$$+\frac{g_2^2}{8}\left(\bar{q}_A\tau^a q^A\right)^2+\frac{g_1^2}{24}\left(\bar{q}_A q^A-2\xi\right)^2\Big].\tag{8}$$

The tension can be rewritten à la Bogomolny:

$$T=\int d^2x\Big(\sum_{a=1}^{3}\left[\frac{1}{2g_2}F_{ij}^{(a)}\pm\frac{g_2}{4}\left(\bar{q}_A\tau^a q^A\right)\epsilon_{ij}\right]^2+$$

$$+\left[\frac{1}{2g_1}F_{ij}^{(8)}\pm\frac{g_1}{4\sqrt{3}}\left(|q^A|^2-2\xi\right)\epsilon_{ij}\right]^2+\frac{1}{2}\left|\nabla_i\,q^A+i\,\varepsilon\,\epsilon_{ij}\nabla_j\,q^A\right|^2\pm\frac{\xi}{2\sqrt{3}}\tilde{F}^{(8)}\Big)$$

where the first three terms are positive definite and the fourth term is a topological invariant ($U(1)$ flux). The nonabelian Bogomolny equations

$$\frac{1}{2g_2}F_{ij}^{(a)}\pm\frac{g_2}{4}\left(\bar{q}_A\tau^a q^A\right)\epsilon_{ij}=0,\qquad (a=1,2,3);$$

$$\frac{1}{2g_1}F_{ij}^{(8)}\pm\frac{g_1}{4\sqrt{3}}\left(|q^A|^2-2\xi\right)\epsilon_{ij}=0,$$

$$\nabla_i\,q^A+i\varepsilon\epsilon_{ij}\nabla_j\,q^A=0,\qquad A=1,2.\tag{9}$$

follow from the last formula. The equations (9) have abelian (n,k) solutions of the type (where n,k are integers) studied in [34]

$$q^{kA}=\begin{pmatrix}e^{i\,n\,\varphi}\phi_1(r) & 0\\ 0 & e^{i\,k\,\varphi}\phi_2(r)\end{pmatrix},$$

$$A_i^3(x)=-\varepsilon\epsilon_{ij}\,\frac{x_j}{r^2}\,\left((n-k)-f_3(r)\right),$$

$$A_i^8(x)=-\sqrt{3}\,\varepsilon\epsilon_{ij}\,\frac{x_j}{r^2}\,\left((n+k)-f_8(r)\right)\tag{10}$$

where $\phi_1(r)$, $\phi_2(r)$, $f_3(r)$, $f_8(r)$ are profile functions with appropriate boundary consitions.

The crucial observation is that the system (8) has an exact $SU(2)_{C+F}$ symmetry, which is neither broken by the interactions nor by the squark VEVS. However, an individual vortex configuration breaks it as $SU(2)_{C+F}\to U(1)$ therefore the vortex acquires zero modes parametrizing

$$\frac{SU(2)}{U(1)}\sim CP^1\sim S^2.$$

For instance, minimum vortices of generic orientation (all degenerate) can be explicitly constructed as

$$q^{kA}=U\begin{pmatrix}e^{i\varphi}\phi_1(r) & 0\\ 0 & \phi_2(r)\end{pmatrix}U^{-1}=e^{\frac{i}{2}\varphi(1+n^a\tau^a)}\,U\begin{pmatrix}\phi_1(r) & 0\\ 0 & \phi_2(r)\end{pmatrix}U^{-1},$$

$$A_i(x) = U[-\frac{\tau^3}{2}\,\epsilon_{ij}\,\frac{x_j}{r^2}\,[1-f_3(r)]]U^{-1} = -\frac{1}{2}\,n^a\tau^a\epsilon_{ij}\,\frac{x_j}{r^2}\,[1-f_3(r)],$$

$$A_i^8(x) = -\sqrt{3}\,\epsilon_{ij}\,\frac{x_j}{r^2}\,[1-f_8(r)], \tag{11}$$

where U is an $SU(2)$ matrix, which smoothly interpolate between the abelian $(1,0)$ and $(0,1)$ vortices. Explicitly, if $n^a = (\sin\alpha\cos\beta, \sin\alpha\sin\beta, \cos\alpha)$, the rotation matrix is given by $U = \exp -i\beta\,\tau_3/2 \cdot \exp -i\alpha\,\tau_2/2$.

The (massive) nonabelian monopoles resulting from the gauge symmetry breaking $SU(3) \to SU(2) \times U(1)/Z_2$ by the adjoint Φ VEV, are confined by these nonabelian monopoles.

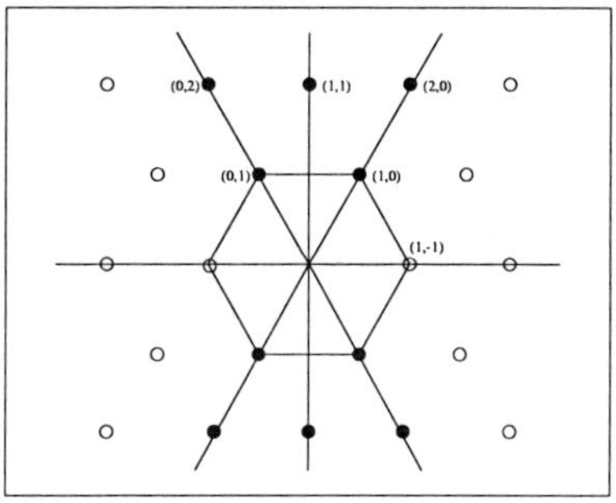

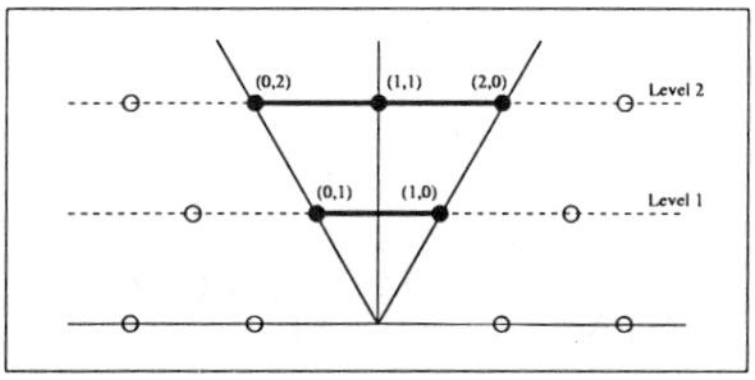

3.9. *Subtle are (also) nonabelian vortices :*

The reduction of the vortex spectrum - meson spectrum (see Figures above) - is due to the topology change

$$\Pi_1(\frac{U(1)\times U(1)}{\mathbf{Z}_2}) = \mathbf{Z}^2 \quad \to \quad \Pi_1(\frac{SU(2)\times U(1)}{\mathbf{Z}_2}) = \mathbf{Z};$$

which occurs in the limit of equal masses $m_i \to m$. The transition from the abelian ($m_i \neq m_j$) to the nonabelian ($m_i = m$) superconductivity is here reliably and quantum mechanically analyzed as the $SU(2) \times U(1)$ subgroup is non asymptotically free for $n_f = 4$ or $n_f = 5$. *Note that the quantum behavior of the nonabelian vortices also depends crucially on the massless fermions present in the underlying theory.* For instance, in the $N = 2$, $SU(3)$ theory with n_f less or equal to 3, there are no quantum vacua with unbroken $SU(2)$ gauge group.

Existence of degenerate, nonabelian vortices which continuously interpolate from $(1,0)$ to $(0,1)$ vortices, imply a corresponding $SU(2)$ doublet of monopoles, as the latter act as the sources (or sink) of the former (see Figure below) when the full $SU(3)$ interactions are taken into account.

The dynamics of vortex zero modes can be shown to be equivalent to the two-dimensional $O(3) = \mathbf{CP}^1$ sigma model ($\mathbf{n} \to \mathbf{n}(z,t)$):

$$S_\sigma^{(1+1)} = \beta \int dz\, dt\, \frac{1}{2}\, (\partial\, n^a)^2 + \text{fermions}.$$

It is dual[35,36] to a chiral theory with two vacua. The exact $SU(2)_{C+F}$ symmetry is not spontaneously broken. The dual ($N = 1$) $SU(2)$ theory is in confinement phase and has, correctly, two vacua (Witten index).

The whole picture generalizes naturally to the sytem with the symmetry breaking

$$SU(N) \to \frac{SU(N-1) \times U(1)}{\mathbf{Z}_{N-1}} \to \emptyset \tag{12}$$

with $2N > N_f \geq 2(N-1)$ flavors. The system at intermediate scales has vortices with $2(N-2)$-parameter family of zero modes representing

$$\frac{SU(N-1)}{SU(N-2) \times U(1)} \sim \mathbf{CP}^{N-2}.$$

The analysis of [33] was made at large m (large $\langle \phi \rangle$) where the system is semiclassical. Though more difficult to analyze, the situation at small m where the nonabelian monopoles condense and the quarks are confined by nonabelian chromoelectric vortices, is related smoothly to the nonabelian superconductor studied here, via holomorphic dependence of the physics on m and though the isomonodromy (in which quarks become monopoles and vice versa).

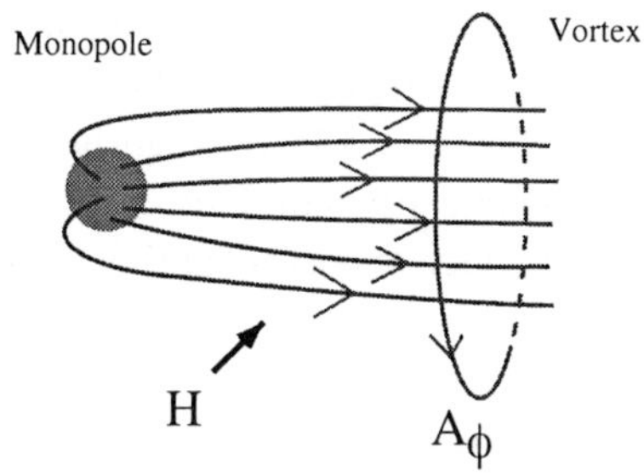

$\mu \neq 0$

3.10. *Lessons from N = 2 SQCD*

Summarizing, softly broken $N = 2$, $SU(n_c)$ gauge theories with n_f quarks with $m = 0$, show different types of confining vacua:

(1) $r = 0, 1$ vacua are described by weakly coupled abelian monopoles;
(2) $r < \frac{n_f}{2}$: The ground state is a nonabelian superconductor; nonabelian monopoles condense and confines the quarks;
(3) $r = \frac{n_f}{2}$ is a boundary case: the ground state is a deformed SCFT, with strongly coupled nonabelian monopoles and dyons.

In the $USp(2n_c)$ and $SO(n_c)$ gauge theories all confining vacua are of the third type.
Both at generic r - vacua and at the SCFT ($r = \frac{n_f}{2}$) vacua of $SU(n_c)$ SQCD, nonabelian monopoles condense as

$$\langle M_\alpha^i \rangle = \delta_\alpha^i \, v \neq 0, \qquad (\alpha = 1, 2, \ldots, r; \quad i = 1, 2, \ldots, n_f)$$

("Color-Flavor-Locked phase"). There are some indication that a similar result holds in the $r = \frac{n_f}{2}$, almost conformal vacua.[29]

4. Hints for QCD

What can one learn from these studies for QCD ? First of all, dynamical abelianization neither is observed in the real world nor is believed to occur in QCD. On the other hand, QCD with n_f flavors and its possible dual have the beta function coefficients ($\tilde{n}_c = 2, 3$, $n_f = 2, 3$) such that

$$b_0 = 11\, n_c - 2\, n_f \quad vs \quad \tilde{b}_0 = 11\, \tilde{n}_c - n_f,$$

where we assumed that in the standard QCD, the flavor carrying monopoles are scalars. Because of the large coefficient (eleven) in front of the color multiplicity, a sign flip (weakly-coupled nonabelian monopoles) is hardly possible, even though in nonsupersymmetric theories higher loop contributions are also important. These two facts, together, leave the option of a strongly-interacting nonabelian superconductor, as in the almost superconformal vacua in the $N = 2$ gauge theories, as the most likely picture of the ground state of QCD.

Taking a more detailed hint from supersymmetric models one might assume that nonabelian magnetic monopoles of QCD condense in a color-flavor-locked form

$$\langle M^i_{L,\alpha} \rangle = \delta^i_\alpha \, v_R \neq 0, \quad \langle M^\alpha_{R,i} \rangle = \delta^\alpha_i \, v_L \neq 0,$$

($\alpha = 1, 2, \ldots \tilde{n}_c; i = 1, 2, \ldots n_f$). As they are strongly coupled, a better physical picture might be

$$\langle M^i_{L,\alpha} \, M^\alpha_{R,j} \rangle = \text{const.} \, \delta^i_j \neq 0;$$

which yields for $\tilde{n}_c = 2$, $n_f = 2$ the correct symmetry breaking pattern

$$G_F = SU_L(2) \times SU_R(2) \Rightarrow SU_V(2),$$

observed in Nature.

References

1. G. 't Hooft, *Nucl. Phys.* **B138**, 1 (1978); *ibid* **B153**, 141 (1979), **B190**, 455 (1981).
2. G. 't Hooft, *Nucl. Phys.* **B190**, 455 (1981); S. Mandelstam, *Phys. Lett.* **53B**, 476 (1975).
3. F. Cachazo, N. Seiberg and E. Witten, *JHEP* **0302**, 042 (2003), hep-th/0301006.
4. T.T. Wu and C.N. Yang, in *Properties of Matter Under Unusual Conditions*, ed. H. Mark and S. Fernbach, (Interscience, New York, 1969); Y.M. Cho, *Phys. Rev.* **D21**, 1080 (1980); L.D. Faddeev and A.J. Niemi, *Phys. Rev. Lett.* **82**, 1624 (1999), hep-th 9807069.
5. L. Del Debbio, A. Di Giacomo and G. Paffuti, *Nucl. Phys.* **B** (Proc. Suppl.) **42**, 231 (1995); A. Di Giacomo, hep-lat/0206018.
6. R. Dijkgraaf and C. Vafa, hep-th/0208048; F. Cachazo, M. R. Douglas, N. Seiberg and E. Witten, *JHEP* **0212**, 071 (2002), hep-th/0211170.
7. J. Wess and J. Bagger, *Supersymmetry and Supergravity*, (Princeton, 1992).
8. V. Novikov, M. Shifman, A. Vainshtein and V. Zakharov, *Phys. Lett.* **B157**, 169 (1985); *Phys. Lett.* **B166**, 329 (1986); M. Shifman and A. Vainshtein, *Nucl. Phys.* **B277**, 456 (1986).
9. I. Affleck, M. Dine and N. Seiberg, *Nucl. Phys.* **B241**, 493 (1984); V. A. Novikov, M. A. Shifman, A. I. Vainshtein, and V. I. Zakharov, *Nucl. Phys.* **B260**, 157 (1985); D. Amati, G. C. Rossi and G. Veneziano, *Nucl. Phys.*

B249, 1 (1985); M. A. Shifman, A. I. Vainshtein, *ITEP Lectures in Particle Physics and Field Theory*, Vol. 2, pp.485, ed. M. Shifman, (World Scientific, Singapore, 1999), hep-th/9902018.
10. D. Amati, K. Konishi, Y. Meurice, G.C. Rossi and G. Veneziano, *Phys. Rept.* **162**, 170 (1988).
11. N. Seiberg, *Nucl. Phys.* **B435**, 129 (1995), hep-th/9411149; D. Kutasov, A. Schwimmer and N. Seiberg, *Nucl. Phys.* **B459**, 455 (1996), hep-th/9510222.
12. G. 't Hooft, *Phys. Rev.* **D14**, 3432 (1976), *Erratum-ibid.* **D18**: 2199, (1978).
13. N. M. Davies, T. J. Hollowood, V. V. Khoze and M. P. Mattis, *Nucl. Phys.* **B559**, 123 (1999), hep-th/9905015; N. M. Davies, T. J. Hollowood and V. V. Khoze, hep-th/0006011.
14. K. Konishi and A. Ricco, *Phys. Lett.* **B570**, 118 (2002), hep-th/0306128.
15. K. Konishi, *Phys. Lett.* **B135**, 439 (1984); K. Konishi and K. Shizuya, *Nuovo Cim.* **90A**, 111 (1985), and references therein.
16. T. T. Wu and C. N. Yang, *Phys. Rev.* **D12**, 3845 (1975).
17. G. 't Hooft, *Nucl. Phys.* **B79**, 276 (1974); A. M. Polyakov, *JETP Lett.* **20**, 194 (1974).
18. P. Goddard, J. Nuyts and D. Olive, *Nucl. Phys.* **B125**, 1 (1977).
19. E. Weinberg, *Nucl. Phys.* **B167**, 500 (1980); *Nucl. Phys.* **B203**, 455 (1982).
20. N. Seiberg and E. Witten, *Nucl. Phys.* **B426**, 19 (1994), hep-th/9407087.
21. N. Seiberg and E. Witten, *Nucl. Phys.* **B431**, 484 (1994), hep-th/9408099.
22. P. C. Argyres and A. F. Faraggi, *Phys. Rev. Lett.* **74**, 3931 (1995), hep-th/9411047; A. Klemm, W. Lerche, S. Theisen and S. Yankielowicz, *Phys. Lett.* **B344**, 169 (1995), hep-th/9411048; *Int. J. Mod. Phys.* **A11**, 1929 (1996), hep-th/9505150; A. Hanany and Y. Oz, *Nucl. Phys.* **B452**, 283 (1995), hep-th/9505075; P. C. Argyres, M. R. Plesser and A. D. Shapere, *Phys. Rev. Lett.* **75**, 1699 (1995), hep-th/9505100; P. C. Argyres and A. D. Shapere, *Nucl. Phys.* **B461**, 437 (1996), hep-th/9509175; A. Hanany, *Nucl. Phys.* **B466**, 85 (1996), hep-th/9509176.
23. K. Konishi and H. Terao, *Nucl. Phys.* **B511**, 264 (1998), hep-th/9707005.
24. P. Argyres, M. Plesser and N. Seiberg, *Nucl. Phys.* **B471**, 159 (1996), hep-th/9603042.
25. G. Carlino, K. Konishi and H. Murayama, *JHEP* **0002**, 004 (2000), hep-th/0001036; *Nucl. Phys.* **B590**, 37 (2000), hep-th/0005076; K. Konishi, Proc. of *Continuous Advances in QCD*, Minneapolis, Minnesota, May 2000, hep-th/0006086; G. Carlino, K. Konishi, S. Prem Kumar and H. Murayama, *Nucl. Phys.* **B608**, 51 (2001), hep-th/0104064.
26. S. Bolognesi and K. Konishi, *Nucl. Phys.* **B645**, 337 (2002), hep-th/0207161.
27. A. Abouelsaood, *Nucl. Phys.* **B226**, 309 (1983); P. Nelson and A. Manohar, *Phys. Rev. Lett.* **50**, 943 (1983); A. P. Balachandran *et al.*, *Phys. Rev. Lett.* **50**, 1553 (1983); A. P. Balachandran *et al.*, *Phys. Rev.* **D29** (1984) 2919, 2936; P. Nelson and S. Coleman, *Nucl. Phys.* **B227**, 1 (1984); Ki-M. Lee, E. J. Weinberg and P. Yi, *Phys. Rev.* **D54**, 1633 (1996), hep-th/9602167.
28. K. Konishi and L. Spanu, *Int. J. Mod. Phys.* **A18**, 249 (2003), hep-

th/0106075.
29. R. Auzzi, R. Grena and K. Konishi, *Nucl. Phys.* **B653**, 204 (2003), hep-th/0211282.
30. M. Douglas and S. Shenker, *Nucl. Phys.* **B447**, 271 (1995), hep-th/9503163.
31. A. Hanany, M. Strassler and A. Zaffaroni, *Nucl. Phys.* **B513**, 87 (1998), hep-th/9707244; B. Lucini and M. Teper, *Phys. Rev.* **D64**, 105019 (2001), hep-lat/0107007; L. Del Debbio, H. Panagopoulos, P. Rossi and E. Vicari, *Phys. Rev.* **D65**, 021501 (2002), hep-th/0106185; *JHEP* **0201**, 009 (2002), hep-th/0111090; C. P. Herzog and I. R. Klebanov, *Phys. Lett.* **B526**, 388 (2002), hep-th/0111078; R. Auzzi and K. Konishi, *New J. Phys.* **4**, 59 (2002), hep-th/0205172.
32. A. Hanany and D. Tong, *JHEP* **0307**, 037 (2003), hep-th/0306150.
33. R. Auzzi, S. Bolognesi, J. Evslin, K. Konishi and A. Yung, *Nucl. Phys.* **B** to appear, hep-th/0307287.
34. A. Marshakov and A. Yung, *Nucl. Phys.* **B647**, 3 (2002), hep-th/0202172.
35. K. Hori and C. Vafa, hep-th/0002222.
36. V. Novikov, M. Shifman, A. Vainshtein and V. Zakharov, *Phys. Rept.* **116**, 103 (1984).

Confinement Mechanism in QCD

CONFINEMENT OF COLOR: RECENT PROGRESS. *

ADRIANO DI GIACOMO
Dipartimento di Fisica, Università di Pisa,I.N.F.N. Sezione di Pisa,
Via Buonarroti 2, 56127 Pisa, Italy
E-mail: adriano.digiacomo@df.unipi.it

Recent progress done in Pisa on the subject is presented. It is shown that dual superconductivity of the vacuum or absence of it is an intrinsic property of QCD vacuum, independent of the choice of the abelian projection. The order of the deconfining phase transition in $N_f = 2$ QCD is studied as a key to understand the mechanism of confinement.

1. Dual superconductivity: dependence on the abelian projection

The mechanism of confinement by dual superconductivity of the vacuum [1]requires the identification in QCD of a U(1) fiber bundle, which has to be color gauge invariant and color singlet if color is not broken by monopole condensation in the vacuum. The procedure for that is known as "Abelian Projection" [2].

A disorder parameter is then introduced[3,4], $\langle\mu\rangle$ which is the vev of a magnetically charged operator μ . $\langle\mu\rangle \neq 0$ means Higgs breaking of the magnetic U(1) gauge symmetry, or dual superconducticity, and is expected to hold in the confined phase $T < T_c$. For $T > T_c$ (deconfined phase) instead the $U(1)$ is not broken and $\langle\mu\rangle = 0$.

It can be instructive to start from the SU(N) Higgs model, which is defined by the Lagrangean

$$L = -\frac{1}{4}Tr\{G_{\mu\nu}G_{\mu\nu}\} - Tr\{D_\mu\phi^\dagger D_\mu\phi\} - V(\phi) \tag{1}$$

The Higgs field ϕ is a N by N matrix and transforms in the adjoint representation. In the Higgs phase $\langle\phi\rangle = \phi^a \neq 0$. The index a labels different minima of $V(\phi)$.

*This work is partially supported by miur progetto teoria delle interazioni fondamentali.

The relevant quantity for monopole solutions is[5]

$$F_{\mu\nu} = Tr\left\{\phi G_{\mu\nu}\right\} - \frac{i}{g} Tr\left\{\phi\left[D_\mu \phi, D_\nu \phi\right]\right\} \tag{2}$$

A theorem can be proved on $F_{\mu\nu}$ [6]:

A necessary and sufficient condition (i) for the cancellation of bilinear terms in $A_\mu A_\nu$ between the two terms on the rhs of eq(2) and (ii) for the validity of the Bianchi identities $\partial_\mu F^*_{\mu\nu} = 0$ is that

$$\phi^a(x) = U(x)\phi^a_{diag} U^\dagger(x) \tag{3}$$

with

$$\phi^a_{diag} = diag\left(\overbrace{\frac{N-a}{N}, .., \frac{N-a}{N}}^{a}, \overbrace{-\frac{a}{N}, .., -\frac{a}{N}}^{N-a}\right) \tag{4}$$

and $U(x)$ an arbitrary gauge transformation. The little group of ϕ^a_{diag} is $SU(a) \otimes SU(N-a) \otimes U(1)$. It identifies a symmetric space[7]. If L_0 is the corresponding Lie algebra , L is the Lie algebra of $SU(N)$,and $L_1 = L - L_0$, $[L_0, L_0] \subset L_0$, $[L_0, L_1] \subset L_1$, $[L_1, L_1] \subset L_0$. It can be shown that all possible symmetric subspaces of $SU(N)$ have the form eq(3).

Viceversa, if the Higgs field ϕ belongs to the adjoint representation then the breaking identifies a symmetric space,i.e. ϕ^a has the form eq(3)[8].

For a Higgs field of the form eq(3) one has identically

$$F^a_{\mu\nu} = \partial_\mu Tr\left\{\phi^a A_\nu\right\} - \partial_\nu Tr\left\{\phi^a A_\mu\right\} - \frac{i}{g} Tr\left\{\phi^a\left[\partial_\mu \phi^a, \partial_\nu \phi^a\right]\right\} \tag{5}$$

$F^a_{\mu\nu}$ is gauge invariant by construction. In the unitary gauge $\partial_\mu \phi^a = 0$, the second term of eq(5) vanishes and

$$F^a_{\mu\nu} = \partial_\mu Tr(\Phi^a A_\nu) - \partial_\nu Tr(\Phi^a A_\mu) \tag{6}$$

assumes an abelian form. The transformation to the unitary gauge is called abelian projection.

Expanding the diagonal part of the fields $A^\mu_{diag} = \alpha^i A^i_\mu$, in terms of the roots

$$\alpha^i = diag(0, 0, 0 \ldots \overset{i}{1}, \overset{i+1}{-1}, 0 \ldots 0) \qquad tr\{\alpha^i \phi^a_{diag}\} = \delta^{ia} \tag{7}$$

gives

$$F^a_{\mu\nu} = \partial_\mu A^a_\nu - \partial_\nu A^a_\mu \tag{8}$$

A monopole solution exists in the $SU(2)$ subspace spanned by the diagonal elements i and $i+1$, of the form of the $SU(2)$ solution[5,9]. For this solution $E_i = F_{0i} = 0$ and

$$\vec{H} = \frac{2\pi}{g}\frac{\vec{r}}{r^3} + \text{Dirac string}$$

whence the name of monopole to the soliton.

A magnetic current can be defined

$$J^a_\nu = \partial_\mu F^{a*}_{\mu\nu} \tag{9}$$

Bianchi identities require that $J^a_\mu = 0$ but it can be different from zero in a compact formulation like lattice, in which the Dirac string is invisible. In any case

$$\partial_\mu J^a_\mu = 0 \tag{10}$$

$N-1$ $U(1)$ magnetic symmetries are thus defined, which are topological symmetries, which do not correspond directly to invariances of the lagrangean. In QCD they will be the magnetic symmetries which will eventually be Higgs broken in the confined phase, producing dual superconductivity.

The construction of the magnetically charged operators μ^a which will provide the disorder parameters for dual superconductivity is the following[4]. The denomination "disorder" comes from statistical mechanics and means that $\langle\mu^a\rangle$ is expected to be different from zero in the confined phase,which is strong coupling (disordered), and to be equal to zero in the ordered phase.

$$\mu^a(\vec{x},t) = e^{i\int d^3\vec{y}\,Tr\left(\phi^a \vec{E}(\vec{y},t)\right)\vec{b}_\perp(\vec{x}-\vec{y})} \tag{11}$$

with

$$\vec{\nabla}\vec{b}_\perp = 0 \quad , \vec{\nabla}\wedge\vec{b}_\perp = \frac{2\pi}{g}\frac{\vec{r}}{r^3} + \text{Dirac string} \tag{12}$$

μ^a is gauge invariant if ϕ^a belongs to the adjoint representation.

In the abelian projected gauge $\phi^a = \phi^a_{diag}$ $Tr(\vec{E}(y,t)\phi^a_{diag}) = \vec{E}^a$ and

$$\mu^a(\vec{x},t) = \exp\left\{i\int d^3\vec{y}\,\vec{E}^a_\perp(\vec{y},t)\vec{b}_\perp(\vec{x}-\vec{y})\right\} \tag{13}$$

where only the transverse part of $\vec{E}^a$ survives in the convolution with $\vec{b}_\perp$.

In whatever quantization scheme $\vec{E}^a_\perp$ is the conjugate momentum to $\vec{A}^a_\perp$, and hence

$$\mu^a(\vec{x},t)|\vec{A}^a_\perp(\vec{y},t)\rangle = |\vec{A}^a_\perp(\vec{y},t) + \vec{b}_\perp(\vec{x}-\vec{y})\rangle \tag{14}$$

μ^a creates a monopole in the U(1) generated by α^a in the abelian projected gauge.

The construction can be repeated unchanged in the Coulomb phase, in spite of the fact that there are no monopoles as solitons, by taking any $\phi^a(x) = U(x)\phi^a_{diag}U(x)\dagger$ in the adjoint representation. If $U(x)$ is defined as the gauge transformation which diagonalizes the Higgs field $\phi(x)$, then $\phi^a(x)$ is diagonal with $\phi(x)$, but any other choice provides an abelian projection : for example $\phi^a(x)$ can be diagonal in the maximal abelian gauge.

μ^a depends on the choice of U(x)

$$\mu^a(\vec{x},t) = e^{i\int d^3\vec{y}\, Tr\left(\phi^a_{diag}U^\dagger(\vec{y},t)\vec{E}(\vec{y},t)U(\vec{y},t)\right)\vec{b}_\perp(\vec{x}-\vec{y})} \tag{15}$$

If $U(x)$ does not depend on the gauge field configuration, when computing correlators of μ^a it can be reabsorbed by a change of variables which leaves the measure invariant[10],and

$$\mu^a(\vec{x},t) = e^{i\int d^3\vec{y}\, Tr\left(\phi^a_{diag}\vec{E}(\vec{y},t)\right)\vec{b}_\perp(\vec{x}-\vec{y})} \tag{16}$$

All memory of $U(x)$ has disappeared, and $\langle\mu^a\rangle \neq 0$ or $\langle\mu^a\rangle = 0$ are statements independent on the abelian projection.

If $U(x)$ depends on $A_\mu(x)$ in general the measure is not invariant and a non trivial jacobian can appear after gauge transformation by $U(x)$, so that the correlation functions of μ^a , and in particular its vev are projection dependent. However if the number density of monopoles is finite the operator $\mu^a(x)$ defined by eq(16) will create a monopole in all abelian projections, since the gauge transformation to any abelian projection will be continuous in a neighbourhood of x and will preserve topology. Hence, if the number density of monopoles is finite the statement $\langle\mu^a\rangle \neq 0$ and $\langle\mu^a\rangle = 0$ are abelian projection independent[10,11].

Dual superconductivity (or non-superconductivity) of the vacuum is an intrinsic property, independent on the particular choice of the abelian projection.

In QCD there are no Higgs fields, but, as discussed above, any operator in the adjoint representation $O(x)$ can provide an abelian projection, in the sense that the operator $U(x)$ of eq(2) can be chosen as the one which diagonalizes $O(x)$. Again, if the number density of monopoles is finite dual supercondutivity is an intrinsic property, independent of the choice of the abelian projection. The density of monopoles can be estimated by looking at the distribution of the difference of the eigenvalues of any operator in the adjoint representation on the sites of a lattice. The location of monopoles coincides indeed with such zeros[2]. We have studied that distribution on samples of lattice configurations , with different lattice spacings and for a number of operators. A typical distribution is shown in fig 1, which refers to the Polyakov line as operator, 10^3 configurations on a 16^4 lattice ,quenched SU(3) and $\beta = 6.4$ The number of sites on which there is a monopole is zero.

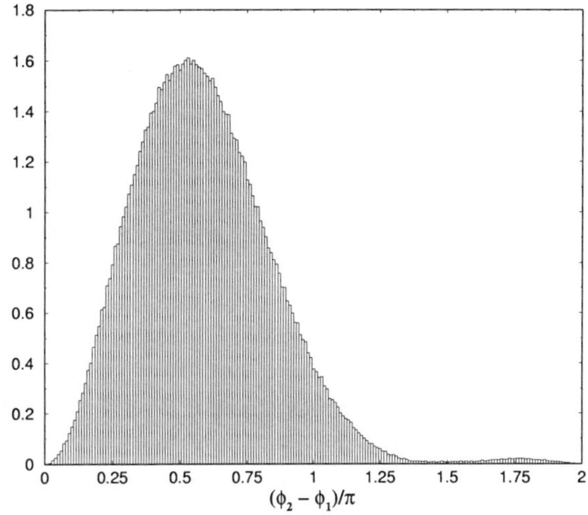

Figure 1. An example of probability distribution of the difference of the two highest eigenvalues of the phase Φ of the Polyakov line $e^{i\Phi}$, at the lattice sites. $SU(3)$ gauge group, $\beta = 6.4$, lattice 16^4, 10^3 configurations.

An independent test can be made by numerical comparison of the order parameter for different abelian projections, which confirms the independence of dual superconductivity on the choice of the abelian projection.[13]

The measurement of the disorder parameters $\langle\mu^a\rangle$ in the quenched case works as follows[4]. Instead of $\langle\mu^a\rangle$ one determines the quantity $\rho^a \equiv \frac{d}{d\beta} ln\langle\mu^a\rangle$. ρ^a is a susceptibility. At the deconfining transition where $\langle\mu^a\rangle$ has a sharp drop ρ^a has a negative peak. A phase transition can only take place in the infinite volume limit[12]. As the volume increases the drop of $\langle\mu^a\rangle$ becomes sharper and sharper and the negative peak of ρ higher and higher. Since [4]

$$\langle\mu^a\rangle = exp(\int_0^T \rho^a(\tau')d\tau') \tag{17}$$

and for $T < T_c$ ρ^a becomes volume independent within numerical errors

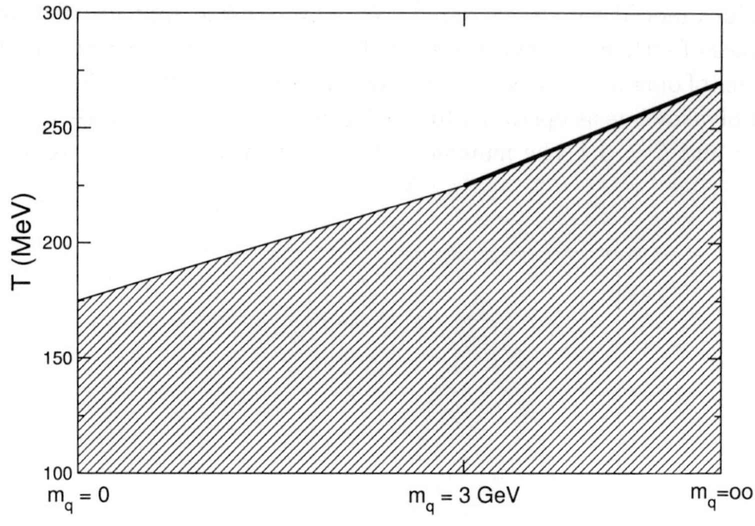

Figure 2. The phase diagram of two flavor QCD.

with increasing volume, one concludes that for $T < T_c$ $\langle \mu^a \rangle \neq 0$, implying dual superconductivity. For $T > T_c$ $\rho \sim -|c|N_s$ with N_s the spatial extension of the lattice, implying that $\langle \mu^a \rangle = 0$ strictly in the thermodynamical limit (normal vacuum). In the critical region the correlation length ξ goes large ,the ratio of the lattice spacing a to ξ, a/ξ can be put to zero, and then the disorder parameter only depends on the ratio N_s/ξ

$$\langle \mu^a \rangle = \tau^\delta \phi(N_s/\xi) \tag{18}$$

whence the scaling law follows

$$\rho^a / N_s^{1/\nu} = f(\tau N_s^{1/\nu}) \tag{19}$$

In particular the peak height scales as $N_s^{1/\nu}$) whence ν can be determined. The result is consistent[4] with the values obtained by use of the Polyakov line[14,15], or $\nu = .62$ for pure gauge $SU(2)$ (3d ising universality class) , and .33 for $SU(3)$, first order transition.

2. Two flavor QCD

In quenched theory one uses the Polyakov criterion to define confinement, which refers to the static potential between a quark and an antiquark.

The order parameter is $\langle L \rangle$ the Polyakov line : when $\langle L \rangle = 0$ the potential grows linearly with distance, when $\langle L \rangle \neq 0$ it goes to a constant.

Of course one should in principle show that confinement defined in this way implies the absence of any colored particle in asymptotic states, which is not easy to do, but the criterion is reasonable anyway. As shown above it fits with identifying confinement with dual superconductivity of the vacuum.

In the presence of dynamical quarks Z_3 symmetry is explicitely broken and $\langle L\rangle$ cannot be an order parameter. Moreover string breaking is expected to occur: the potential energy stops growing with distance, due to the instability for production of quark antiquark pairs,even if there is confinement. At $m_q = 0$ there is chiral symmetry, which is known to be spontaneously broken at zero temperature, the pseudoscalar mesons being the Goldstone particles. The symmetry is restored at some temperature T_c, where the order parameter $\langle\bar{\psi}\psi\rangle$ goes to zero. It is not known what is the relation between chiral symmetry breaking and confinement. In any case at $m_q \neq 0$ chiral symmetry is explicitely broken and $\langle\bar{\psi}\psi\rangle$ is not an order parameter either. The situation for $N_f = 2$ $m_u = m_d = m$ is depicted in fig 2.

A number of susceptibilities can be measures on the lattice as functions of the temperature T at given m (The susceptibility of $\langle\bar{\psi}\psi\rangle$, that of $\langle L\rangle$,the specific heat[17,18]). All of them show a peak at the same $T(m)$, which defines the curve in the phase diagram of fig 2. By convention the region below that curve is called confined, the region above it deconfined.

A renormalization group analysis can be made[16] of the chiral transition assuming that the Goldstone particles are the relevant degrees of freedom at the transition , with the following result. For $N_f = 3$ the chiral transition is first order and such is the transition at $m \neq 0$. For $N_f = 2$ if the anomaly of the U(1) axial current vanishes below T_c the transition is first order and such is the transition at $m \neq 0$; if instead the anomaly persists up to T_c the transition is second order in the universality class of O(4) and the line at $m \neq 0$ is a crossover. Lattice data are not yet conclusive on this issue, but for some reason the second possibility is usually assumed to be true.

A possible criterion for confinement could be dual superconductivity of the vacuum, which is already valid in the quenched case. Indeed the disorder parameter $\langle\mu^a\rangle$ can equally well be defined in the presence of dynamical quarks. Lattice simulations show[19] in fact that $\langle\mu^a\rangle$ is non zero below the critical line of fig 2, and is stricly zero above it in the thermodynamical limit. Of course in principle one should show that dual superconductivity implies absence of colored particles in asymptotic states, which is not trivial to do: but the situation is not different from that of the Polyakov criterion, as discussed in the previous section.

A finite size scaling analysis around T_c can be performed to get information on the order of the phase transition. The issue is very relevant to understand confinement. Indeed if the determination gives a result consistent with what is obtained by studying the specific heat a legitimation results for $\langle\mu^a\rangle$ as an order parameter and for dual superconductivity as a mechanism of confinement. Preliminary data[20] indicate that the chiral transition is first order and certainly not in the universality class of O(4). A careful analysis is being completed, which will give an unambiguous answer to the question. A careful analysis of the anomaly around T_c is also on the way to check consistency with ref.[16]. Some details on the analysis. A new scale is present in the problem with respect to the quenched case. Eq(18) now reads

$$\langle\mu^a\rangle = \phi(N_s/\xi, mN_s^{y_h}) \tag{20}$$

The problem can be reduced to a single scale by choosing masses and sizes such that $mN_s^{y_h} = constant$, assuming for y_h alternatively the value corresponding to O(4) universality class ($y_h = 2.49$) or the value for a first order transition $y_h = 3$. For the same values different susceptibilities and the specific heat can be determined, and the critical indices can be measured consistently. This program is being completed.

Thanks are due J.M. Carmona, L. Del Debbio, M. D'Elia, B. Lucini, G. Paffuti and C. Pica who collaborated to obtain the results presented here .

References

1. G.'tHooft in High Energy Physics EPS Conference Palermo 1975, A.Zichichi ed.; S.Mandelstam, *Phys.Rep.* **23C** , 245 (1976).
2. G.'tHooft , *Nucl.Phys.* **B190** ,455 (1981).
3. A. Di Giacomo, *Acta Physica Polonica* **B25**, 227 (1994).
4. L. Del Debbio, A. Di Giacomo, G. Paffuti, *Phys. Lett.* **B349** , 513 (1995); A. Di Giacomo, B. Lucini, L. Montesi, G. Paffuti, *Phys. Rev.* **D61** , 034503 (2000); A. Di Giacomo, B. Lucini, L. Montesi, G. Paffuti, *Phys. Rev.***D61**, 034504 (2000).
5. G.'tHooft, *Nucl.Phys.* **B79**, 276 (1974).
6. L. Del Debbio, A. Di Giacomo, B. Lucini, G. Paffuti hep-lat/0203023.
7. S.Weinberg *The quantum theory of fields*, Chapt.19, Cambridge Univ.Press (1990).
8. L. Michel *Rev.Mod.Phys.* **52**, 617 (1980).
9. A. M. Polyakov, *JETP Lett.* **20**, 894 (1974).
10. A. Di Giacomo, hep-lat/0206018

11. A. Di Giacomo, G. Paffuti, *The abelian projection revisited* (LATTICE 2003) hep-lat/0309019
12. C.N.Yang, T.D. Lee, *Phys.Rev.* **87**, 404 (1952).
13. J.M. Carmona, M. D'Elia, A. Di Giacomo, B. Lucini, G. Paffuti, *Phys.Rev.* **D64** , 114507 (2002).
14. J. Fingberg, U. M. Heller, F. Karsch, *Nucl. Phys.* **B392**, 493 (1993).
15. M. Fukugita, M. Okawa, A. Ukawa, *Phys.Rev. Lett.* **63**, 1768 (1989).
16. R. Pisarski and R. Wilczek, *Phys. Rev.* **D29**, 338 (1984).
17. F. Karsch and E. Laermann, *Phys. Rev.* **D50**, 6954 (1994)
18. S. Aoki et al. (JLQCD collaboration), *Phys. Rev.* **D57**, 3910 (1998).
19. J.M. Carmona, M. D'Elia, L. Del Debbio, A. Di Giacomo, B. Lucini, G. Paffuti, *Phys. Rev.* **D66**, 011503 (2002).
20. J.M. Carmona, M. D'Elia, L. Del Debbio, A. Di Giacomo, B. Lucini, G. Paffuti, C. Pica ,*Deconfining transition in two-flavor QCD.* (LATTICE2003) hep-lat/0309035.

CONSISTENT DEFINITIONS OF FLUX AND ELECTRIC AND MAGNETIC CURRENT IN ABELIAN PROJECTED $SU(2)$ LATTICE GAUGE THEORY

RICHARD W. HAYMAKER* AND TAKAYUKI MATSUKI†

Department of Physics and Astronomy, Louisiana State University, Baton Rouge, Louisiana 70803-4001, USA

E-mail: haymaker@rouge.phys.lsu.edu,matsuki@tokyo-kasei.ac.jp

Through the use of a lattice $U(1)$ Ward-Takahashi identity, one can find a precise definition of flux and electric four-current that does not rely on the continuum limit. The magnetic four-current defined for example by the DeGrand-Toussaint construction introduces order a^2 errors in the field distributions. We advocate using a single definition of flux in order to be consistent with both the electric and magnetic Maxwell's equations at any lattice spacing. In a $U(1)$ theory the monopoles are slightly smeared by this choice, i.e. are no longer associated with a single lattice cube. In Abelian projected $SU(2)$ the consistent definition suggests further modifications. For simulations in the scaling window, we do not foresee large changes in the standard analysis of the dual Abrikosov vortex in the maximal Abelian gauge because the order a^2 corrections have small fluctuations and tend to cancel out. However in other gauges, the consequences of our definitions could lead to large effects which may help in understanding the choice of gauge. We also examine the effect of truncating all monopoles except for the dominant cluster on the profile of the dual Abrikosov vortex.

1. Introduction

Dual superconductivity has long been suggested as a possible mechanism for quark confinement signaled by a spontaneously broken $U(1)$ gauge symmetry and manifested by a dual Abrikosov vortex between quark and anti quark. This picture was verified some time ago in Abelian projected SU(2) lattice gauge theory in the maximal Abelian gauge[1]. More recently further studies have elaborated this picture[2,3,4].

---

*Talk presented at Confinement 2003, RIKEN, July 21 - 24, 2003 by R. Haymaker

†Permanent address: Tokyo Kasei University, 1-18-1 Kaga, Itabashi, Tokyo 173-8602, Japan

As in all lattice calculations, there is freedom in choosing lattice operators, requiring only that they agree with the continuum definition to lowest order in the lattice spacing a. However we have the opportunity in these studies to be more precise by incorporating the lattice Ward-Takahashi identity derived from the residual $U(1)$ gauge symmetry[5]. This gives an Ehrenfest relation for the expectation value of the fields and currents giving the electric Maxwell equations exactly at finite lattice spacing. Interestingly, this defines a unique lattice expression for the field strength or flux and the electric and magnetic currents.

In the present work, we examine the impact of this on the study of the dual Abrikosov vortex. To be consistent, the magnetic Maxwell equation must use the same definition as the electric one. However the standard DeGrand-Toussaint[6] [DT] definition of the magnetic current is based on a different definition of flux, resulting in inconsistencies in the magnetic Maxwell equation. We argue here that one should alter the DT construction, using a single definition of flux throughout. A consequence is that the magnetic current no longer contains discrete monopoles but rather a more general magnetic charge distribution. In effect, the monopoles are smeared in our picture.

This consistency question is only relevant at finite lattice spacing and all these concerns go away in the continuum limit. Nevertheless it is desirable to have a consistent treatment of total flux in the vortex determined from the electric Maxwell equation, and the profile shape determined by the magnetic Maxwell eqution at fixed lattice spacing. Further we note that the finite lattice spacing effects are significant for the values of β that we often use for calculations.

We also report on the effect of truncating DT monopole loops, keeping only the one large connected cluster. This truncation is expected to have no effect on the confinement signal[9,10]. This should manifest itself here in that the tail of the profile of the magnetic current of the vortex are unaffected by the truncation. We find this to be the case. This procedure requires that the magnetic current consists of discrete monopoles. Truncation is not well defined in the above smeared monopole picture. Hence we take the conventional view in presenting these results that the continuum limit is required to obtain the consistency described above.

2. Three definitions of flux

Let us consider three definitions of field strength, all agreeing to lowest order in a. The first definition was used by DT to define monopoles:

$$\begin{aligned}\widehat{F}^{(1)}_{\mu\nu} &= \theta_\mu(\boldsymbol{m}) - \theta_\mu(\boldsymbol{m}+\nu) - \theta_\nu(\boldsymbol{m}) + \theta_\nu(\boldsymbol{m}+\mu) - 2\pi n_{\mu\nu}, \\ &\equiv \theta_{\mu\nu}(\boldsymbol{m}) - 2\pi n_{\mu\nu},\end{aligned}$$

where $\theta_\mu(\boldsymbol{m})$ refers to the $U(1)$ link angle in the domain $-\pi < \theta_\mu < +\pi$. The integers $n_{\mu\nu}$ are determined by requiring that $-\pi < \widehat{F}^{(1)}_{\mu\nu} < +\pi$. That is $\widehat{F}^{(1)}_{\mu\nu}$ is a periodic function of $\theta_{\mu\nu}$ with period 2π. Here quantities with $\widehat{\ }$ mean those which appear in the lattice numerical calculation without appending factors of e and a.

The second and third definitions gives the exact electric Maxwell equation for lattice averages

$$\Delta^-_\mu \langle \widehat{F}^{(i)}_{\mu\nu} \rangle_W = \langle \widehat{J}^{e(i)}_\mu \rangle_W, \qquad i = 2, 3, \tag{1}$$

where

$$\langle \cdots \rangle_W = \frac{\langle \cdots e^{i\theta_W} \rangle}{\langle e^{i\theta_W} \rangle},$$

for the cases of $U(1)$ gauge theory and $SU(2)$ gauge theory respectively. We give the $U(1)$ derivation in detail since it is straight forward and contains the essential points of the argument. There are significant complications in the $SU(2)$ case and hence we just sketch that derivation.

2.1. *Flux in the U(1) gauge theory*

Consider

$$Z_W(\epsilon_\mu(\boldsymbol{m})) = \int [d\theta] \sin\theta_W \exp(\beta S,)$$

$$S = \sum_{n,\ \mu>\nu} \cos\theta_{\mu\nu}(\boldsymbol{n}), \qquad \beta = \frac{1}{e^2}.$$

The subscript of $Z_W(\epsilon_\mu(\boldsymbol{m}))$ refers to the incorporation of the source into the partition function and the argument is a variable defined as the shift of one particular link, $\theta_\mu(\boldsymbol{m}) \to \theta_\mu(\boldsymbol{m}) + \epsilon_\mu(\boldsymbol{m})$. This translation can be transformed away since the measure is invariant under such an operation.

Therefore

$$0 = \delta Z_W = \int [d\theta] \sin\theta_W \exp(\beta S)\Big|_{\theta_\mu \to \theta_\mu + \epsilon_\mu} - \int [d\theta] \sin\theta_W \exp(\beta S)$$

$$= \epsilon_\mu \int [d\theta] \left(\delta_\mu(\boldsymbol{m}) \cos\theta_W + \sin\theta_W \frac{1}{e^2} \frac{\partial S}{\partial \theta_\mu} \right) \exp(\beta S), \qquad (2)$$

where $\delta_\mu(\boldsymbol{m}) = \pm 1$ if $\boldsymbol{m}$ labels a $\pm$ directed link and $= 0$ otherwise. This is the static current generated by the Wilson loop.

$$\delta_\mu(\boldsymbol{m}) = \widehat{J}^e_\mu \qquad (3)$$

Next evaluate the derivative of S. Isolate the six plaquettes affected by the shift

$$S = \sum_{\nu \neq \mu} [\cos\{\theta_\mu(\boldsymbol{m}) + \theta_\nu(\boldsymbol{m}+\mu) - \theta_\mu(\boldsymbol{m}+\nu) - \theta_\nu(\boldsymbol{m})\}$$
$$+ \cos\{\theta_\mu(\boldsymbol{m}) - \theta_\nu(\boldsymbol{m}+\mu-\nu) - \theta_\mu(\boldsymbol{m}-\nu) + \theta_\nu(\boldsymbol{m}-\nu)\}] + \cdots,$$

$$\frac{\partial S}{\partial \epsilon_\mu(\boldsymbol{m})} = \sum_{\nu \neq \mu} [-\sin\{\theta_\mu(\boldsymbol{m}) + \theta_\nu(\boldsymbol{m}+\mu) - \theta_\mu(\boldsymbol{m}+\nu) - \theta_\nu(\boldsymbol{m})\}$$
$$- \sin\{\theta_\mu(\boldsymbol{m}) - \theta_\nu(\boldsymbol{m}+\mu-\nu) - \theta_\mu(\boldsymbol{m}-\nu) + \theta_\nu(\boldsymbol{m}-\nu)\}],$$

$$\frac{\partial S}{\partial \epsilon_\mu(\boldsymbol{m})} = -\Delta^-_\nu \sin\theta_{\mu\nu}(\boldsymbol{m}). \qquad (4)$$

Using Eqn.(1) and Eqn.(4) we can see that Eqn.(2) is the form of Maxwell equations for averages.

$$\frac{1}{e^2} \Delta^-_\nu \langle \widehat{F}^{(2)}_{\mu\nu} \rangle_W = \widehat{J}^e_\mu(\boldsymbol{m}), \qquad (5)$$

where

$$\langle \cdots \rangle_W = \frac{\langle \sin\theta_W \cdots \rangle}{\langle \cos\theta_W \rangle}.$$

Since the charged line in a Wilson loop is closed the electric current is conserved. The local statement of conservation is

$$0 = \Delta^-_\mu \Delta^-_\nu \langle \widehat{F}^{(2)}_{\mu\nu} \rangle_W = \Delta^-_\mu \widehat{J}^e_\mu.$$

It is straightforward to verify on the lattice that the LHS of Eqn.(5) gives $-1, 0, 1$ depending on its position with respect to the Wilson loop. An alternative definition such as $\widehat{F}^{(1)}_{\mu\nu}$ need not vanish off the Wilson loop nor give ± 1 on the Wilson loop and hence would introduce an error. In a $U(1)$ theory there is no dynamical charge density, all charge resides on the Wilson loop.

2.2. *U(1) flux in the SU(2) theory in the maximal Abelian gauge*

We restrict our attention to the maximal Abelian gauge defined as a local maximum of

$$R = \sum_{n,\mu} \mathrm{tr}\left\{\sigma_3 U_\mu(n)\sigma_3 U_\mu^\dagger(n)\right\},$$

over the set of gauge transformations $\{g(m) = e^{i\alpha_i(m)\sigma_i}\}$, $U \longrightarrow U^g$. Taking U to be the stationary value, the stationary condition is given by

$$F_{jn}[U] = \left.\frac{\partial R[U^g]}{\partial \alpha_j(n)}\right|_{\alpha=0} = 0.$$

The second derivatives entering in the Jacobian are given by

$$M_{jn;im}(U) = \left.\frac{\partial^2 R[U^g]}{\partial \alpha_j(n)\partial \alpha_i(m)}\right|_{\alpha=0}.$$

The partition function is

$$Z_W^{g.f.}(\epsilon_\mu^3(\boldsymbol{m})) = \int [dU]\, \frac{1}{2} Tr[i\sigma_3 U_W(\boldsymbol{n})]\ \exp(\beta S) \prod_{jn} \delta(F_{jn}[U])\ \Delta_{FP}, \quad (6)$$

where the Faddeev-Popov Jacobian is

$$\Delta_{FP} = \det|M_{jn;im}(U)|.$$

An infinitesimal shift in this partition function has the added complication that it violates the gauge condition. This can be corrected by an infinitesimal accompanying gauge transformation. Thus the shift in one link affects all links. However experience has shown that the effect drops off rapidly with distance from the shifted link.

The derivative of the partition function Eqn.(6)with respect to $\epsilon_\mu^3(\boldsymbol{m})$ gives

$$\begin{aligned} 0 = &\int [dU] \left\{\delta_\mu(\boldsymbol{m}) \frac{1}{2} Tr[U_W(\boldsymbol{n})]\right\} \exp(\beta S) \prod_{jn} \delta(F_{jn}[U])\Delta_{FP} \\ &+ \int [dU] \left\{\beta \frac{1}{2} Tr[i\sigma_3 U_W(\boldsymbol{n})] \frac{\partial S}{\partial \epsilon_\mu^3(\boldsymbol{m})}\right\} \exp(\beta S) \prod_{jn} \delta(F_{jn}[U])\Delta_{FP} \\ &+ \int [dU] \frac{1}{2} Tr[i\sigma_3 U_W(\boldsymbol{n})] \exp(\beta S) \frac{\partial}{\partial \epsilon_\mu^3(\boldsymbol{m})} \left\{\prod_{jn} \delta(F_{jn}[U])\Delta_{FP}\right\}. \quad (7) \end{aligned}$$

The third integral contains terms in the Ward Takahasni identity coming from the gauge fixing including ghost contributions.

We can cast this into the form of the electric Maxwell's equations for averages as in the case of the U(1) theory. However there are now more terms in the current. Consider the standard U(1) parametrization of an SU(2) element:

$$\begin{aligned} U_\mu(\boldsymbol{n}) &= \begin{pmatrix} C_\mu(\boldsymbol{n})e^{i\theta_\mu(\boldsymbol{n})} & S_\mu(\boldsymbol{n})e^{i\{\gamma_\mu(\boldsymbol{n})-\theta_\mu(\boldsymbol{n})\}} \\ -S_\mu(\boldsymbol{n})e^{-i\{\gamma_\mu(\boldsymbol{n})-\theta_\mu(\boldsymbol{n})\}} & C_\mu(\boldsymbol{n})e^{-i\theta_\mu(\boldsymbol{n})} \end{pmatrix}, \\ &= \begin{pmatrix} C_\mu & S_\mu e^{i\gamma_\mu} \\ -S_\mu e^{-i\gamma_\mu} & C_\mu \end{pmatrix} \begin{pmatrix} e^{i\theta_\mu} & 0 \\ 0 & e^{-i\theta_\mu} \end{pmatrix}, \end{aligned}$$

where

$$\begin{aligned} C_\mu(\boldsymbol{n}) &\equiv \cos\phi_\mu(\boldsymbol{n}), \\ S_\mu(\boldsymbol{n}) &\equiv \sin\phi_\mu(\boldsymbol{n}). \end{aligned} \tag{8}$$

In the Abelian projection factored form, the righthand factor contains the $U(1)$ photon, parameterized by θ. The lefthand factor contains the charged coset matter fields, parameterized by ϕ and γ. The transformation properties are well known and reviewed in DiCecio et.al.[5].

We consider an alternative separation into diagonal and off-diagonal parts which is needed in defining the flux.

$$\begin{aligned} U_\mu(\boldsymbol{n}) &= \begin{pmatrix} C_\mu e^{i\theta_\mu} & 0 \\ 0 & C_\mu e^{-i\theta_\mu} \end{pmatrix} + \begin{pmatrix} 0 & S_\mu e^{i(\gamma_\mu-\theta_\mu)} \\ -S_\mu e^{-i(\gamma_\mu-\theta_\mu)} & 0 \end{pmatrix}, \\ &= D_\mu(\boldsymbol{n}) + O_\mu(\boldsymbol{n}). \end{aligned}$$

The off-diagonal part is the charged matter field $\Phi_\mu \equiv S_\mu e^{-i(\gamma_\mu-\theta_\mu)}$. and diagonal part includes the photon, $e^{i\theta_\mu}$, but also a neutral remnant of the matter field $\sqrt{1-|\Phi_\mu|^2}$ which $\to 1$ in the limit $a \to 0$.

To cast Eqn.(7) into the form of a current conservation law, we first consider the terms to zeroth order in O_μ. First the Wilson loop. Isolating the diagonal contributions gives

$$U_W = D_W + \widetilde{U}_W,$$

where D_W is the product of the diagonal parts

$$\begin{aligned} \frac{1}{2}Tr[D_W(\boldsymbol{n})] &= \prod_W C_\mu(\boldsymbol{n}) \cdot \boldsymbol{\cos\theta_W}, \\ \frac{1}{2}Tr[i\sigma_3 D_W(\boldsymbol{n})] &= -\prod_W C_\mu(\boldsymbol{n}) \cdot \boldsymbol{\sin\theta_W}. \end{aligned}$$

We adhere to the standard choice of an Abelian Wilson loop in which we drop any contributions due to off diagonal elements O_μ and further take the factors $C_\mu(\boldsymbol{n}) = 1$ giving

$$\frac{1}{2} Tr[U_W^{\text{Abelian}}(\boldsymbol{n})] = \cos\theta_W,$$
$$\frac{1}{2} Tr[i\sigma_3 U_W^{\text{Abelian}}(\boldsymbol{n})] = -\sin\theta_W.$$

Second, consider the action. Write

$$S = \sum_{n,\ \mu>\nu} \frac{1}{2}\mathrm{Tr}[D_{\mu\nu}(\boldsymbol{n})] + \widetilde{S},$$

where $\widetilde{S}$ contains terms involving $O_\mu(\boldsymbol{n})$.

$$\frac{\partial(S-\widetilde{S})}{\partial\epsilon_\mu^3(\boldsymbol{m})} = \sum_{\nu\neq\mu}\left[\frac{1}{2}\mathrm{Tr}\left\{-i\sigma_3(\boldsymbol{m})D_\mu(\boldsymbol{m})D_\nu(\boldsymbol{m}+\mu)D_\mu^\dagger(\boldsymbol{m}+\nu)D_\nu^\dagger(\boldsymbol{m})\right\}\right.$$
$$\left. + \frac{1}{2}\mathrm{Tr}\left\{-i\sigma_3(\boldsymbol{m})D_\mu(\boldsymbol{m})D_\nu^\dagger(\boldsymbol{m}+\mu-\nu)D_\mu^\dagger(\boldsymbol{m}-\nu)D_\nu(\boldsymbol{m}-\nu)\right\}\right].$$

Since all matrices are diagonal we can simplify:

$$\frac{\partial(S-\widetilde{S})}{\partial\epsilon_\mu^3(\boldsymbol{m})} = \sum_{\nu\neq\mu}\Delta_\nu^-\left[C_\mu(\boldsymbol{m})C_\nu(\boldsymbol{m}+\mu)C_\mu(\boldsymbol{m}+\nu)C_\nu(\boldsymbol{m})\sin\theta_{\mu\nu}(\boldsymbol{m})\right].$$

The quantity in square brackets is antisymmetric in $\mu\nu$ and we identify this as proportional to the field tensor.

$$\widehat{F}_{\mu\nu}^{(3)} \equiv C_\mu(\boldsymbol{m})C_\nu(\boldsymbol{m}+\mu)C_\mu(\boldsymbol{m}+\nu)C_\nu(\boldsymbol{m})\sin\theta_{\mu\nu}(\boldsymbol{m})$$

Returning to the identity, Eqn.(7), and using the notation

$$\int[dU]\{\cdots\}\exp(\beta S)\prod_{jn}\delta(F_{jn}[U])\Delta_{FP} = \langle\cdots\rangle_{g.f.}$$

we obtain

$$0 = \delta_\mu(\boldsymbol{m})\langle\cos\theta_W\rangle_{g.f.} - \beta\langle\sin\theta_W\Delta_\nu^-\widehat{F}_{\mu\nu}^{(3)}\rangle_{g.f.} - \beta\langle\sin\theta_W\frac{\partial\widetilde{S}}{\partial\epsilon_\mu^3(\boldsymbol{m})}\rangle_{g.f.}$$
$$+ \text{ gauge fixing terms } + \text{ ghost terms}.$$

Rearranging terms as in the $U(1)$ case, we get

$$\beta\frac{\langle\sin\theta_W\Delta_\nu^-\widehat{F}_{\mu\nu}^{(3)}\rangle_{g.f.}}{\langle\cos\theta_W\rangle_{g.f.}} = \delta_\mu(\boldsymbol{m}) - \beta\frac{\langle\sin\theta_W\frac{\partial\widetilde{S}}{\partial\epsilon_\mu^3(\boldsymbol{m})}\rangle_{g.f.}}{\langle\cos\theta_W\rangle_{g.f.}} + \frac{\text{g.f. \& ghosts}}{\langle\cos\theta_W\rangle_{g.f.}},$$

$$\Delta_\nu^- \langle F_{\mu\nu}^{(3)} \rangle_{W,g.f.} = \widehat{J}_\mu^{(e)\text{ Abelian Wilson loop}} + \widehat{J}_\mu^{(e)\text{ matter fields}} + \widehat{J}_\mu^{(e)\text{ gauge fixing}} + \widehat{J}_\mu^{(e)\text{ ghosts}}. \tag{9}$$

The 'Abelian Wilson loop' term is analogous to the above $U(1)$ case. The 'charged matter field' term arises from the off-diagonal elements of links in the action expression and would contribute without gauge fixing. The 'gauge fixing' term arises from the corrective gauge tranformation that accompanies the shift of a link. The 'ghost' term arises from the shift and accompanying gauge transformation on the Faddeev-Popov determinant. See DiCecio et.al.[5] for a complete derivaton of all terms.

It is important here to note that we are not interested in distinguishing the various contributions to the current in the present work. We are only interested in the total current and that can be obtained from the LHS of Eqn.(9).

3. Consistency with the magnetic Maxwell equation

Having defined a unique flux $\widehat{F}_{\mu\nu}^{(i)}$ through the electric Maxwell equation, the magnetic Maxwell equation is

$$\frac{1}{2}\epsilon_{\mu\nu\rho\sigma}\Delta_\nu^+ \widehat{F}_{\rho\sigma}^{(i)} = \widehat{J}_\mu^{m(i)} \qquad i = 2, 3.$$

However the standard DT definition of current is

$$\widehat{J}_\mu^{m(1)} = \frac{1}{2}\epsilon_{\mu\nu\rho\sigma}\Delta_\nu^+ \widehat{F}_{\rho\sigma}^{(1)}.$$

and hence if we use the conventional $\widehat{F}^{(1)}$ to define the monopole current, and $\widehat{F}^{(2)}$ or $\widehat{F}^{(3)}$ respectively for $U(1)$ and $SU(2)$ theories to get an exact expression for flux in the confining string, then the magnetic Maxwell equation is violated.

The solution is to relax the requirement that we use the DT monopole definition and use $\widehat{F}^{(2)}$ or $\widehat{F}^{(3)}$ instead when defining monopoles. A simple configuration for the $U(1)$ case ($\widehat{F}^{(2)}$) will illustrate the effect. Consider a single DT monopole with equal flux out of the six faces of the cube. Then the ratio of the $\widehat{F}^{(2)}$ flux out of this cube compared to the DT $\widehat{F}^{(1)}$ flux gives

$$\frac{6\sin(2\pi/6)}{6(2\pi/6)} \approx 0.83.$$

Since the current is conserved, the balance is made up by magnetic charge in the neighboring cells. On a large surface the total flux is the same for the two definitions.

In Fig.(1) we plot the $\widehat{F}^{(1)}_{\mu\nu}$ as a function of $\theta_{\mu\nu}$, giving a "sawtooth" shape. Monopoles occur as a consequence of $\widehat{F}^{(1)}$ crossing the sawtooth edge, giving a mismatch of 2π in the flux out of a cube. The sine function has no such discreteness and so the notion of discrete Dirac strings and Dirac monopoles is modified. However as one approaches the continuum limit, the action will drive the plaquette to zero and then the regions where the sawtooth differs from the sine are suppressed. Hence we expect both forms to give the standard Dirac picture in the continuum limit.

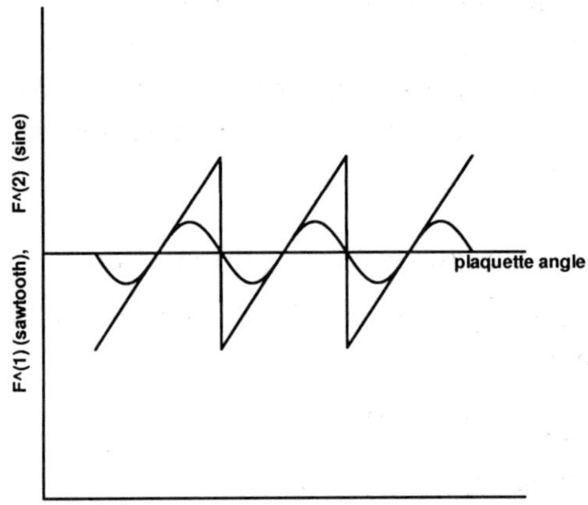

Figure 1. $\widehat{F}^{(1)}_{\mu\nu}$ (sawtooth) and $\widehat{F}^{(2)}_{\mu\nu}$ (sine) as a function of the plaquette angle $\theta_{\mu\nu}$

The electric Maxwell equation determines the total flux in the confining string and the magnetic Maxwell equation determines the transverse shape. Further the latter enters directly in the determination of the London penetration length, λ_d. To see this let us consider the classical Higgs theory which we use to model the simulation data. The dual field is given by

$$\widehat{G}_{\mu\nu}(\boldsymbol{m}) = \Delta^-_\mu \theta^{(d)}_\nu(\boldsymbol{m}) - \Delta^-_\nu \theta^{(d)}_\mu(\boldsymbol{m}),$$

where $\theta^{(d)}_\mu(\boldsymbol{m})$ is a dual link variable. Let us choose to break the gauge symmetry spontaneously through a constrained Higgs field.

$$\Phi(\boldsymbol{m}) = v\rho(\boldsymbol{m})e^{i\chi(\boldsymbol{m})}, \qquad \rho(\boldsymbol{m}) = 1.$$

Under these conditions the magnetic current simplifies to

$$\widehat{J}^m_\mu(\boldsymbol{m}) = 2e^2_m v^2 \sin\left\{\theta_\mu(\boldsymbol{m}) + \chi_\mu(\boldsymbol{m}+\mu) - \chi_\mu(\boldsymbol{m})\right\}.$$

where e_m is the magnetic coupling. The magnetic Maxwell equation is

$$\Delta_\mu^+ \widehat{G}_{\mu\nu} = \widehat{J}_\nu^m. \tag{10}$$

For small $\theta^{(d)}$ it is easy to see that there is a London relation of the form

$$\widehat{G}_{\mu\nu}(m) = \frac{1}{2e_m^2 v^2}\left(\Delta_\mu^- \widehat{J}_\nu^m(m) - \Delta_\nu^- \widehat{J}_\mu^m(m)\right). \tag{11}$$

Taking the confining string along the 3 axis and choosing $\mu = 1$ and $\nu = 2$ we see that the profile of the third component of curl of the magnetic current must match the third component of the electric flux profile. This assumes an infinite Higgs mass M_H. With a finite mass there is a transition region of size $\sim 1/M_H$ in the core of the vortex but the above London relation must hold sufficiently far outside the core.

Combining Eqns.(10) and (11) we get the relation

$$\left(1 - \lambda_d^2 \Delta_\mu^+ \Delta_\mu^-\right) \left\langle \widehat{E}_3(m) \right\rangle_W = 0, \lambda_d^{-1} = e_m v\sqrt{2}$$

The corresponding equations in the simulation of the $SU(2)$ theory must also be satisfied in order to arrive at this correct expression for λ_d and hence the importance of our definitions.

4. Numerical Results

We generated 208 configurations on a 32^4 lattice at $\beta = 2.5115$. The maximal Abelian gauge fixing used over-relaxation. Fig.(2) shows the profile of the electric flux corresponding to $\widehat{F}^{(2)}$ and $\widehat{F}^{(3)}$. Fig.(3) shows the profile of the theta component of the magnetic current corresponding to $\widehat{F}^{(1)}$, $\widehat{F}^{(2)}$ and $\widehat{F}^{(3)}$.

In summary, we showed that consistency requires one use the same definition of flux throughout. If, for example, one uses $\widehat{F}^{(3)}$ definition of electric field (bottom graph in Fig.(2)) in order to account correctly for the total flux but then uses the DT definition of current (top graph in Fig.(3)) we would then incur errors of $\sim 40\%$. For simulations in the scaling window, as long as we use the same definition consistently we do not foresee large changes in the standard analysis of the dual Abrikosov vortex in the maximal Abelian gauge because the order a^2 corrections have small fluctuations and tend to cancel out[8]. However in other gauges, the consequences of our definitions could lead to large effects which may help in understanding the choice of gauge.

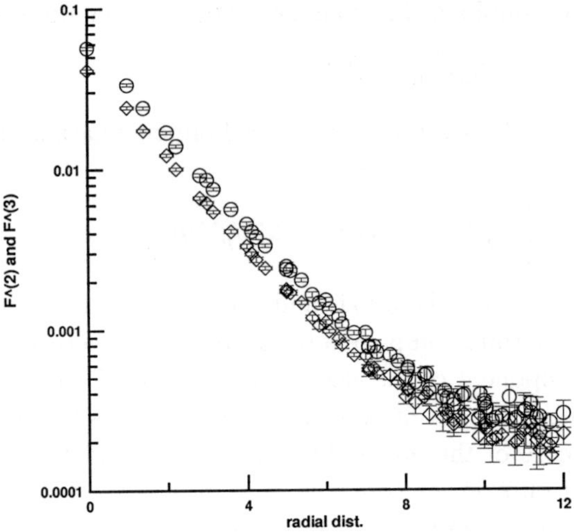

Figure 2. Profile of the electric field (highest to lowest) $\widehat{E}_3^{(2)}$ (circles) and $\widehat{E}_3^{(3)}$ (diamonds) on the mid-plane between q and $\bar{q}$ separated by $13a$.

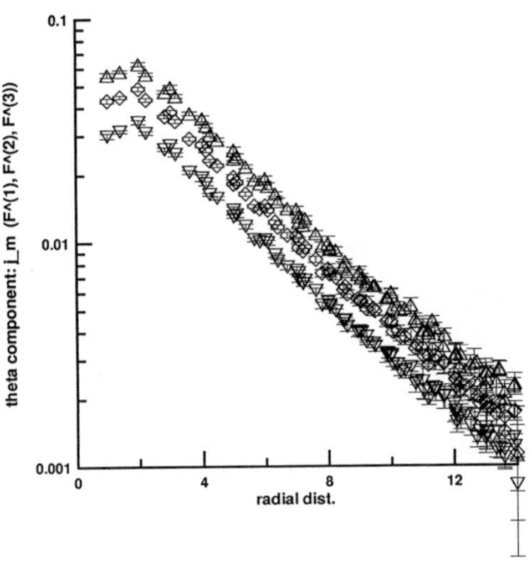

Figure 3. Profile of the theta component of the magnetic current on the mid-plane between q and $\bar{q}$ separated by $13a$ based on (highest to lowest) $\widehat{F}_{\mu\nu}^{(1)}$ (triangles), $\widehat{F}_{\mu\nu}^{(2)}$ (diamonds) and $\widehat{F}_{\mu\nu}^{(3)}$ (inverted triangles).

Finally we report on the effect of truncating DT monopole loops, keeping only the one large connected cluster[9,10]. As mentioned in the Introduction this truncation is expected to have no effect on the confinement signal. This should manifest itself here in that the tail of the profile of the electric field and magnetic current of the vortex are unaffected by the truncation. This is born out as expected. Fig.(4) shows that radial profile of the magnetic current is indistiguishable except for a small deviation in the core of the vortex.

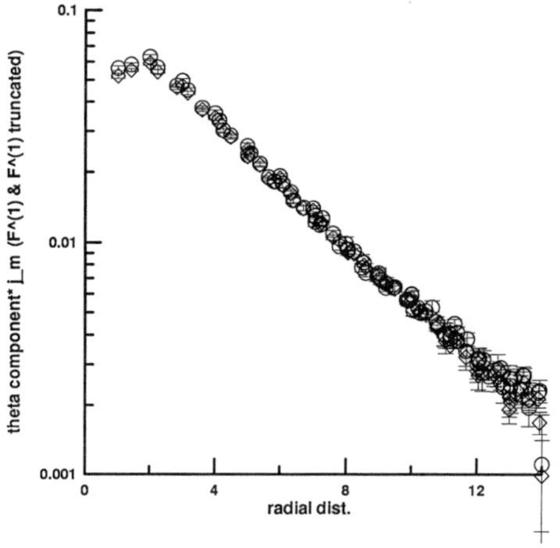

Figure 4. Profile of the theta component of the magnetic current on the mid-plane between q and $\bar{q}$ separated by $13a$ based on (highest to lowest) $\widehat{F}^{(1)}_{\mu\nu}$ (circles), $\widehat{F}^{(1)}_{\mu\nu}$ truncated (diamonds).

Acknowledgments

This work is supported in part by the U. S. Department of Energy under grant no. De-FG05-01 ER 40617.

References

1. V. Singh, D. A. Browne and R. W. Haymaker, *Phys. Lett.* **B306** (1993) 115.
2. G. S. Bali, C. Schlicher and K. Schilling, *Prog. Theor. Phys. Suppl.* **131** (1998) 645.
3. F. V. Gubarev, E.-M Ilgenfritz, M. I. Polikarpov, and T. Suzuki, *Phys. Lett.* **B468** (1999) 134.
4. Y. Koma, M. Koma, E.-M. Ilgenfritz, T. Suzuki, and M.I. Polikarpov, hep-lat/0302006.
5. G. DiCecio, A. Hart and R. Haymaker, *Phys. Lett.* **B441** (1998) 319.
6. T. A. DeGrand and D. Toussaint, *Phys. Rev.* **D22** (1980) 2478.
7. M. Zach, M. Faber, W. Kainz and P. Skala *Phys. Lett.* **B358** (1995) 325.
8. G. Poulis, *Phys. Rev.* **D54** (1996) 6974.
9. A. Hart and M. Teper, *Phys. Rev.* **D58** (1998) 014504.
10. F.V.Gubarev, A.V.Kovalenko, M.I.Polikarpov, S.N.Syritsyn, V.I.Zakharov, hep-lat/0212003; V.G. Bornyakov, P.Yu. Boyko, M.I. Polikarpov, V.I. Zakharov,hep-lat/0305021.

CONNECTING SHORT TO LONG SCALES IN THE CONFINING VACUUM

E. T. TOMBOULIS*

Department of Physics and Astronomy
University of California, Los Angeles
CA 90095-1547, USA
E-mail: tombouli@physics.ucla.edu

We study approximate decimations in SU(N) LGT that connect the short to long distance regimes. Simple 'bond-moving' decimations turn out to provide both upper and lower bounds on the exact partition function. This leads to a representation of the exact partition function in terms of successive decimations whose effective couplings flows are related to those of the easily computable bond-moving decimations. The implications for a derivation of confinement from first principles are discussed.

1. Introduction

Over the last several years an enormous amount of work has been performed by lattice workers on the physics of the QCD vacuum. In particular, isolating the types of configurations in the functional measure that are responsible for maintaining confinement at (arbitrarily) weak coupling has been a central issue. A great deal of information concerning the confinement mechanism has been obtained from these investigations (for recent review, see[1]). However, the goal of a direct derivation of confinement from first principles has remained elusive for the last thirty years.

The origin of the difficulty is clear. One is faced with a multi-scale problem involving the passage from the short-distance weakly coupled, ordered regime to the long distance strongly coupled, disordered, confining regime. Such variable multi-scale behavior can only be addressed by some nonperturbative block-spinning or decimation procedure capable of bridging these different regimes. Exact decimation schemes appear analytically hopeless, and numerically very difficult. It is not even clear what a good definition

*Work partially supported by grant nsf-phy-0309362.

of block-spin variables would be. There is, however, a class of approximate simple decimation procedures which are known in many cases to give qualitatively correct results. They are generally known as 'bond moving' decimations. Here we will consider such decimations in a somewhat more general form and show that they can provide bounds on the exact theory. This leads to a representation of the partition function of the exact theory which allows a connection to be made to the behavior of the approximate, but easily computable, decimations at successive length scales. The implications for the question of an actual derivation of confinement in LGT will be discussed below.

The framework applies to general $SU(N)$, though explicit numerical or analytical calculations supporting the considerations below have for the most part been carried out only for $SU(2)$.

2. Bond moving decimations

Starting with some plaquette action, e.g the Wilson action $A_p(U) = \frac{\beta}{N}\,\mathrm{Re}\,\mathrm{tr}U_p$, at lattice spacing a, we consider the character expansion of the exponential of the action:

$$F(U,a) = e^{A_p(U)} = \sum_j F_j(\beta,a)\, d_j\, \chi_j(U) \tag{1}$$

with Fourier coefficients:

$$F_j = \int dU\; F(U,a)\, \frac{1}{d_j}\, \chi_j(U)\; . \tag{2}$$

Here χ_j denotes the character of the j-th representation of dimension d_j. $j = 0$ will always denote the trivial representation. E.g, for SU(2), $j = 0, \frac{1}{2}, 1, \frac{3}{2}, \ldots$, and $d_j = (2j+1)$. In terms of normalized coefficients:

$$c_j = \frac{F_j}{F_0}\;, \tag{3}$$

one then has

$$\begin{aligned} F(U,a) &= F_0\left[1 + \sum_{j\neq 0} c_j(\beta)\, d_j\, \chi_j(U)\right] \\ &\equiv F_0\; f(U,a) \end{aligned} \tag{4}$$

For a reflection positive action one necessarily has:

$$F_j \geq 0 \qquad \text{hence} \quad 1 \geq c_j \geq 0 \qquad \text{all} \quad j\; . \tag{5}$$

The partition function on lattice Λ is then

$$Z_\Lambda(\beta) = F_0^{|\Lambda|} \int dU_\Lambda \prod_p f_p(U,a) \,. \tag{6}$$

We now consider RG decimation transformations $a \to \lambda a$ in, say, the x^1-direction (Figure 1). Simple approximate transformations of the 'bond moving' type are implemented by 'weakening', i.e. decreasing the c_j's of interior plaquettes (shaded), and 'strengthening', i.e. increasing c_j's of boundary plaquettes (bold) in every decimation cell of side length λ. The

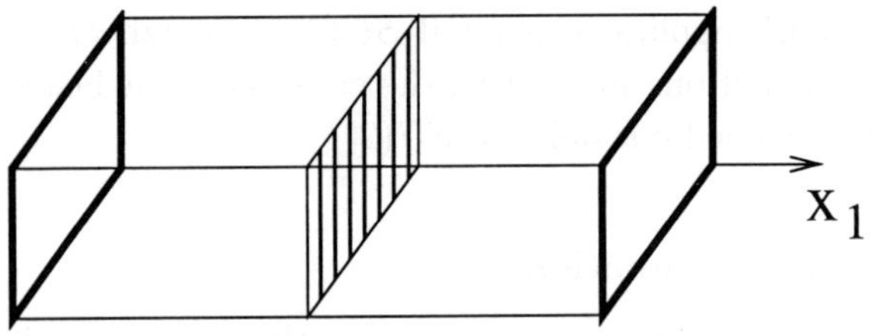

$\lambda = 2$

Figure 1. Basic bond (plaquette) moving operation.

simplest scheme[2], which is adopted in the following, implements complete removal, $c_j = 0$, of interior plaquettes. This is performed simultaneously in all directions (Figure 2).

Under successive decimations

$$a \to \lambda a \to \lambda^2 a \to \cdots \to \lambda^n a$$
$$\Lambda \to \Lambda^{(1)} \to \Lambda^{(2)} \to \cdots \to \Lambda^{(n)}$$

the RG transformation rule is then:

$$f(U, n-1) \to f(U,n) = \Big[1 + \sum_{j \neq 0} c_j(n)\, d_j\, \chi_j(U)\Big] \tag{7}$$

with:

$$c_j(n) = F_j(n)/F_0(n) \,, \quad F_j(n) = \Big[\hat{F}_j(n)\Big]^{\lambda^2} , \tag{8}$$

$$\hat{F}_j(n) = \int dU \, \Big[f(U, n-1) \Big]^{\nu} \frac{1}{d_j} \chi_j(U) \,. \tag{9}$$

The parameter ν controls by how much the remaining plaquettes have been strengthened to compensate for the removed plaquettes. What has been

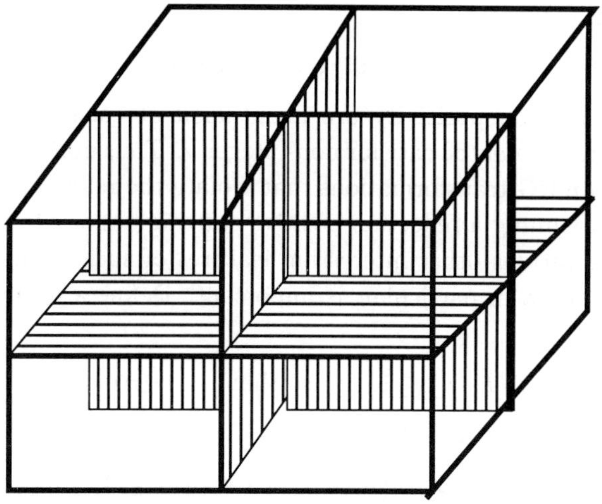

Figure 2. Isotropic interior plaquette moving operation in a decimation cell (hypercube).

considered in the literature before is $\nu = \lambda^{(d-2)}$, where d is the spacetime dimension. This choice of ν defines the MK recursions[2]. Here we generalize to consider ν an arbitrary parameter.

The resulting partition function after n decimation steps is:

$$Z_\Lambda(\beta, n) = \prod_{m=0}^{n} F_0(m)^{|\Lambda|/\lambda^{md}} \int dU_{\Lambda^{(n)}} \prod_p f_p(U, n) . \qquad (10)$$

It is important to note that after each decimation step the resulting action retains the original *one-plaquette* form but will, in general, contain all representations:

$$A_p(U, n) = \sum_j \tilde{\beta}_j(\beta)\, \chi_j(U) . \qquad (11)$$

Furthermore, among the effective couplings $\tilde{\beta}_j$ some negative ones may in general occur. These features are present even after a single decimation step $a \to \lambda a$ starting with the usual single representation (fundamental) Wilson action.

Preservation of the one-plaquette form of the action is of course what makes these decimations simple to explore. The rule specified by (7)- (9) is meaningful for any real (positive) ν. Here, however, a basic distinction can be made. For *integer* ν, the important property of positivity of the Fourier coefficients in (1), (4):

$$F_0(n) \geq 0 \quad , \qquad c_j(n) \geq 0 \,, \tag{12}$$

and hence reflection positivity are maintained at each decimation step. This, in general, is *not* the case for non-integer ν. Thus non-integer ν results in approximate RG transformations that violate the reflection positivity of the theory (assuming a reflection positive starting action).[a]

There are various other interesting features of such decimations. The following property, in particular, is important. Define a normalized $\hat{F}_j(n)$ (cf. (9)):

$$\hat{c}_j(n) \equiv \hat{F}_j(n)/\hat{F}_0(n) \leq 1, \qquad \text{so that} \qquad c_j(n) = \hat{c}_j(n)^{\lambda^2} \,. \tag{13}$$

Then it is possible to prove that

$$\sum_j c_j(n) \left(\hat{c}_j(n+1) - c_j(n) \right) \geq 0 \,. \tag{14}$$

It follows from (14) that the norm (l_2 norm) of the vector formed from the $\hat{c}_j(n+1)$ coefficients is bigger than that of the vector of the $c_j(n)$. In fact one finds in explicit numerical evaluations that (14) holds component-wise, i.e. $\hat{c}_j(n+1) \geq c_j(n)$.

As can be seen from the relation between $\hat{c}_j(n+1)$ and $c_j(n+1)$ in (13), (9), however, it can still be that the norm of the $c_j(n+1)$'s is smaller than that of the $c_j(n)$'s. i.e. the norm of the normalized coefficients $c_j(n)$ in (7) decreases under successive decimations. Note, in particular, that when ν is taken to depend on λ, d, the resulting highly nonlinear dependence can give very nontrivial behavior. This is in fact what happens in the case of the MK recursions where $\nu = \lambda^{(d-2)}$: the normalized coefficients $c_j(n)$ do decrease under successive decimations in the approach to a fixed point in lower dimensions. But an upper critical dimension arises above which the $\hat{c}_j(n+1)$'s become sufficiently large compared to the $c_j(n)$'s so that this is no longer the case, and triviality ensues (for the RG flow on the weak coupling side).

3. The exact partition function

Since our decimations are not exact decimation transformations, the partition function does not in general remain invariant under them. The subse-

[a]It is worth noting in this context that numerical investigations of the standard MK recursions, at least for gauge theories, appear to have been carried out for the most part for fractional λ, $(1 < \lambda < 2)$, which corresponds to non-integer ν; e.g. see[3,4].

quent development hinges on the following two basic statements that can now be proved:

(I) With $\nu = \lambda^{d-2}$:

$$Z_\Lambda(\beta, n) \leq Z_\Lambda(\beta, n+1) . \tag{15}$$

(II) With $\nu = 1$:

$$Z_\Lambda(\beta, n+1) \leq Z_\Lambda(\beta, n) . \tag{16}$$

Note that for $d = 2$ (15) - (16) express the well-known fact that the decimations become exact. For $d > 2$, in both (I), (II) one in fact has strict inequality.

(I) says that modifying the couplings of the remaining plaquettes after decimation by taking $\nu = \lambda^{d-2}$ (standard MK choice[2]) results into overcompensation (upper bound on the partition function). Translation invariance and convexity of the free energy as a function of the couplings in the action underlie (15).

(II) says that decimating plaquettes while leaving the couplings of the remaining plaquettes unaffected results in a lower bound on the partition function. Reflection positivity (positivity of Fourier coefficients) is crucial for this to hold.

Consider now the, say, $(n-1)$-th decimation step with Fourier coefficients $c_j(n-1)$, which we relabel $c_j(n-1) = \tilde{c}_j(n-1)$. Given these $\tilde{c}_j(n-1)$, we proceed to compute the coefficients $F_0(n)$, $c_j(n)$ of the next decimation step according to (7)-(9) above with $\nu = \lambda^{d-2}$.

Then introducing a parameter α, $(0 \leq \alpha)$, define the *interpolating coefficients*:

$$\tilde{c}_j(n, \alpha) = \tilde{c}_j(n-1)^{\lambda^2(1-\alpha)}\, c_j(n)^\alpha . \tag{17}$$

Then,

$$\tilde{c}_j(n, \alpha) = \begin{cases} c_j(n) & : \alpha = 1 \\ \tilde{c}_j(n-1)^{\lambda^2} & : \alpha = 0 \end{cases} \tag{18}$$

The $\alpha = 0$ value is that of the n-th step coefficients resulting from (7)-(9) with $\nu = 1$.

Thus defining the corresponding partition function

$$Z_\Lambda(\beta, \alpha, n) = \Big(\prod_{m=0}^{n-1} F_0(m)^{|\Lambda|/\lambda^{md}} \Big)\, F_0(n)^\alpha \cdot \int dU_{\Lambda^{(n)}} \prod_p f_p(U, n, \alpha) \tag{19}$$

where

$$f_p(U,n,\alpha) = \left[1 + \sum_{j\neq 0} \tilde{c}_j(n,\alpha)\, d_j\, \chi_j(U)\right], \tag{20}$$

we have from (15), (16)), and (18) above:

$$Z_\Lambda(\beta,0,n) \le Z_\Lambda(\beta,n-1) \le Z_\Lambda(\beta,1,n)\,. \tag{21}$$

Now the partition function (19) is a continuous, in fact analytic, in α. So (21) implies that, by continuity, there exist a value of α:

$$\alpha = \alpha^{(n)}(\beta,\lambda,\Lambda)\,, \qquad 0 < \alpha^{(n)}(\beta,\lambda,\Lambda) < 1$$

such that

$$Z_\Lambda(\beta,\alpha^{(n)},n) = Z_\Lambda(\beta,n-1)\,. \tag{22}$$

In other words there is an α at which the n-th decimation step partition function equals that obtained at the previous decimation step; the partition function does not change its value under the decimation step $\lambda^{n-1}a \to \lambda^n a$.

So starting at original spacing a, at every decimation step m, $(m = 0,1,\cdots,n)$, there exist a value $0 < \alpha^{(m)} < 1$ such that

$$Z_\Lambda(\beta,\alpha^{(m+1)},m+1) = Z_\Lambda(\beta,\alpha^{(m)},m)\,. \tag{23}$$

This then gives, after n successive decimations, an *exact representation* of the original partition function in the form:

$$\begin{aligned} Z_\Lambda(\beta) &= F_0^{|\Lambda|} \int dU_\Lambda \prod_p f_p(U,a) \\ &= \prod_{m=0}^{n} F_0(m)^{\alpha^{(m)}|\Lambda|/\lambda^{md}} \\ &\quad \cdot \int dU_{\Lambda^{(n)}} \prod_p f_p(U,n,\alpha^{(n)}), \end{aligned} \tag{24}$$

i.e. in terms of the successive bulk free energy contributions from the $a \to \lambda \to \cdots \to \lambda^n a$ decimations and a one-plaquette effective action on the resulting lattice $\Lambda^{(n)}$.

The $\alpha^{(m)}$'s in (24) may be viewed as effective couplings which, in addition to the $\{\tilde{\beta}_j^{(m)}\}$, enter in the specification of the effective action and bulk free energy at each decimation step m. Thus the flow from scale a to scale $\lambda^n a$ is now specified by $\{\tilde{\boldsymbol{\beta}}(n,\beta,\lambda),\ \boldsymbol{\alpha}(n,\beta,\lambda,\Lambda)\} \equiv \{\,\{\tilde{\beta}_j^{(m)}(\beta,\lambda)\}\,,\alpha^{(m)}(\beta,\lambda,\Lambda) \mid m = 0,\ldots,n\,\}$. This dependence on the additional couplings $\boldsymbol{\alpha}$ may be considered as compensating for the absence in

(24) of additional terms, beyond the one-plaquette interaction, that would normally be expected in an effective action.

At weak and strong coupling $\alpha^{(m)}$ may be estimated analytically. At large β, where the decimations approximate the free energy rather accurately, the appropriate α values are very close to unity. At strong coupling they may be estimated by comparison with the strong coupling expansion. On any finite lattice there is also a weak volume dependence as a correction which goes away as an inverse power of the lattice size.

For most purposes the exact values of the $\alpha^{(m)}$'s, beyond the fact that are fixed between 0 and 1, are not immediately relevant. The main point of the representation (24) is that it can in principle relate the behavior of the exact theory to that of (modifications of) the easily computable approximate decimations.

Indeed, starting from the $\tilde{c}_j(n-1, \alpha^{(n-1)})$'s at the $(n-1)$-th step, consider the coefficients at the next step, and compare those evaluated at $\alpha = \alpha^{(n)}$, i.e. $\tilde{c}_j(n, \alpha = \alpha^{(n)})$, to those evaluated at $\alpha = 1$, i.e $\tilde{c}_j(n, \alpha = 1) \equiv c_j(n)$. The latter will be referred to as the MK coefficients. (Recall that $\alpha = 1 \Longleftrightarrow \nu = \lambda^{d-2}$, the standard MK choice. The absence of a tilde on a coefficient in the following always means that it is computed at $\alpha = 1$.) According to (I), the MK coefficients give an upper bound.

To facilitate the comparison let us rewrite (17) in the form

$$\tilde{c}_j(n, \alpha) = \left(\frac{\tilde{c}_j(n-1)}{\hat{c}_j(n)} \right)^{\lambda^2(1-\alpha)} c_j(n) \, . \tag{25}$$

Now property (14) and the remark following it imply that the ratio in the brackets in (25) is less or equal to unity. It follows that

$$\tilde{c}_j(n, \alpha) \leq c_j(n) \qquad \text{for any} \quad 0 \leq \alpha \leq 1 \, . \tag{26}$$

This has the following important consequence.

Assume we are in a dimension d such that under successive decimations the MK coefficients ($\alpha = 1$) are non-increasing. Then (26) implies:

$$\begin{aligned} \tilde{c}_j(n, \alpha^{(n)}) &\geq c_j(n+1) \geq \tilde{c}_j(n+1, \alpha^{(n+1)}) \\ &\geq c_j(n+2) \geq \tilde{c}_j(n+2, \alpha^{(n+2)}) \\ &\geq \cdots \end{aligned}$$

Thus, if the $c_j(n)$'s are non-increasing, so are the $\tilde{c}_j(n, \alpha)$. The $c_j(n)$'s must then approach a fixed point, and hence so must the $\tilde{c}_j(n, \alpha)$'s, since $c_j(n), \tilde{c}_j(n, \alpha) \geq 0$. Note the fact that this conclusion is independent of the specific value of the α's at every decimation step.

In particular, if the $c_j(n)$'s approach the strong coupling fixed point, i.e. $F_0 \to 1$, $c_j(n) \to 0$ as $n \to \infty$, so must the $\tilde{c}_j(n,\alpha)$'s of the exact representation. If the MK decimations confine, so do those in the exact representation (24). As it is well-known by explicit numerical evaluation, the MK decimations for $SU(2)$ and $SU(3)$ indeed confine for all $\beta < \infty$ and $d \leq 4$. Above the critical dimension $d = 4$, the decimations result in free spin wave behavior.

4. Discussion and outlook

What do the above results say about the question of confinement in the exact theory? They are clearly strongly suggestive of confinement for all β in the exact theory. They cannot, however, as yet be taken to constitute an actual proof. The statement at the end of the previous section concerns the behavior of the long distance action part in the representation (24). Now (24) also includes the large free energy bulk contributions from integration over all scales from a to $\lambda^n a$. It is the complete representation involving both contributions that provides an equality to the value of the exact partition function. This, just by itself, does not suffice to rigorously isolate, at least in any direct way, the actual behavior of the corresponding long distance part in the exact theory. To do this one needs to consider order parameters which can directly couple to the corresponding long distance parts of the effective action in the exact theory and any representation of it like that given by (24). In other words, one would need to carry through the above derivation given for the partition function also for the case of appropriate order parameters.

The derivation of the basic two statements (I) and (II) above (eqs. (15), (16)) assumes translation invariance and reflection positivity. In the presence of observables such as a Wilson loop, translation invariance is broken and reflection positivity is reduced to hold only in the plane bisecting the loop. This does not allow the above derivations to be carried through in any obvious way. Fortunately, there are other order parameters that can characterize the possible phases of the theory while maintaining translational invariance. They are the well-known vortex free energy, and its $Z(N)$ Fourier transform (electric flux free energy). They are in fact the natural order parameters in the present context since they are constructed out of partition functions. Recall that the vortex free energy is defined by

$$e^{-F_v(\tau)} = Z_\Lambda(\tau)/Z_\Lambda \, . \tag{27}$$

Here $Z_\Lambda(\tau)$ denotes the partition function with action modified by the 'twist' $\tau \in Z(N)$ for every plaquette on a coclosed set of plaquettes V winding through the periodic lattice in $(d-2)$ directions; e.g. for the Wilson action one has the replacement $\frac{\beta}{N}\,\mathrm{Re}\,\mathrm{tr}U_p \to \frac{\beta}{N}\,\mathrm{Re}\,\mathrm{tr}U_p\tau$ for every $p \in V$. The twist represents a discontinuous gauge transformation on the set V which introduces $\pi_1(SU(N)/Z(N))$ vortex flux rendered topologically stable by being wrapped around the lattice torus. As indicated by the notation, $Z_\Lambda(\tau)$ depends only on the presence of the flux, and is invariant under changes in the exact location of V. The vortex free energy is then the ratio of the partition function in the presence of this external flux to the partition function in the absence of the flux (the latter is what was considered above). The above development, in particular the derivation of (24), should then be repeated also for $Z_\Lambda(\tau)$. Bulk (local) free energy contributions resulting from integrating over successive scales are insensitive to the presence of the flux. Thus in the analog to (24) for $Z_\Lambda(\tau)$ only the long distance effective action part would be affected by its presence, and the bulk contributions would cancel in (27). Statements, as the ones obtained in the previous section, concerning the behavior of the long distance parts in such representations of the two factors in (27) would then directly constrain the corresponding behavior in the exact theory.

There is, however, an immediate technical complication in obtaining the analog to (24) for $Z_\Lambda(\tau)$. The presence of the flux reduces reflection positivity to hold only in planes perpendicular to the directions in which the flux winds through the lattice. The simple nature of the decimations, however, makes it plausible that this still suffices to allow a generalization of the previous derivation for Z_Λ to go through also in the case of $Z_\Lambda(\tau)$. Further investigation of this and related questions will be reported elsewhere.

Acknowledgments

I would like to thank the participants of Confinement 2003 for many discussions. I am grateful to Prof. H. Suganuma and the organizing committee for the invitation and for organizing such a stimulating and enjoyable conference.

References

1. J. Greensite, Progr. Part. Nucl. Phys. **51**, 1 (2003), (hep-lat/0301023).
2. A. A. Migdal, Sov. Phys. JETP **42**, 413, 743 (1976); L. Kadanoff, Ann. Phys. (N.Y.) **100**, 369 (1976).
3. M. Nauenberg and D. Toussaint, Nucl. Phys. **B190** [FS3], 217 (1981).
4. K. Bitar, S. Gottlieb and C. Zachos, Phys. Rev. **D26**, 2853 (1982).

SO(3) VS. SU(2) YANG-MILLS THEORY ON THE LATTICE: AN INVESTIGATION AT NON-ZERO TEMPERATURE *

A. BARRESI

Dipartimento di Fisica, Universitá di Pisa e
I.N.F.N. Sezione di Pisa,
Via Buonarroti 2, 56127 Pisa, Italy
E-mail: barresi@df.unipi.it

G. BURGIO

School of Mathematics, Trinity College,
Dublin 2, Ireland
E-mail: burgio@maths.tcd.ie

M. MÜLLER-PREUSSKER

Humboldt-Universität zu Berlin, Institut für Physik,
Newtonstr. 15, 12489 Berlin, Germany
E-mail: mmp@physik.hu-berlin.de

The adjoint $SU(2)$ lattice gauge theory in 3+1 dimensions with the Wilson plaquette action modified by a $\mathbb{Z}_2$ monopole suppression term is reinvestigated with special emphasis on the existence of a finite-temperature phase transition decoupling from the well-known bulk transitions.

1. Introduction and motivation

The evidence and our detailed understanding of the deconfinement phase transition in $SU(N)$ gauge theories at finite temperature mainly comes from lattice gauge theories (LGT) formulated in the fundamental representation [1,2]. For pure LGT the transition is associated with the spontaneous breaking of the global center $\mathbb{Z}_N$ symmetry [3,4]:

$$U_4(\vec{x}, x_4) \longrightarrow z \cdot U_4(\vec{x}, x_4), \quad z \in \mathbb{Z}_N \quad \text{for all } \vec{x} \text{ at } x_4 = \text{fixed}.$$

*Contribution based on a plenary talk given by M. Müller-Preussker and a poster presented by G. Burgio at CONFINEMENT 2003, RIKEN, Tokyo 2003 which leaves the lattice gauge action invariant but flips the Polyakov loop variables

$$L_F(\vec{x}) = \frac{1}{N_c}\mathrm{Tr}_F \prod_{x_4=1}^{N_\tau} U_4(\vec{x}, x_4) \tag{1}$$

as $L_F \leftrightarrow zL_F$. As a consequence the standard order parameter for the deconfinement transition is defined as

$$< |L_F| >= \left\langle \left| \frac{1}{N_\sigma^3} \sum_{\vec{x}} L_F(\vec{x}) \right| \right\rangle , \tag{2}$$

where the ensemble average is taken with the Boltzmann distribution represented by the lattice-discretized path integral with periodic boundary conditions for the gauge fields in the imaginary time direction x_4. The above mentioned global symmetry breaking mechanism provides a close analogy to spin models. In particular, the universality class of $SU(2)$ LGT is that of the 3d Ising model [5]. On the other hand the origin of quark and gluon confinement as well as of the occurence of the finite-temperature phase transition has been seen in the condensation of topological excitations like Abelian monopoles [6] and center vortices [7].

Lattice gauge theories can be formulated in different group representations of the gauge fields, e.g. in the *center blind* adjoint representation. In this case (extended) vortices and Abelian monopoles are still present, but the mechanism of spontaneous $\mathbb{Z}(N)$ breaking is obviously not realized. Moreover, the adjoint representation LGT's at strong coupling are strongly affected by bulk phase transitions [8,9] driven by lattice artifacts [10]. A finite temperature transition - if it exists - seems to be completely overshadowed by these bulk transitions.

Therefore, the question of universality in particular of the existence of the finite temperature phase transition remains an important issue. If the existence of this transition turns out to be independent of the group representation, then the question remains whether the driving mechanism related to the condensation of topological excitations is the same.

In the past this principally important question has been studied by several groups mainly in the case of the mixed $SU(2)-SO(3) = SU(2)/\mathbb{Z}_N$ theory realized with the Villain action[11–17]. Still we have not yet reached a completely satisfying answer. Nevertheless, over the last years there has been an interesting progress[18–21] worth to be reviewed at this conference.

In the following Section 2 we shall shortly review $SU(2)$ lattice gauge theories with different mixed fundamental-adjoint actions. In Section 3 we introduce the center-blind model we have further investigated, i.e. the adjoint representation Wilson lattice action with a $\mathbb{Z}_2$ monopole suppression term. In Section 4 we discuss the results of our investigations based on twist variables, the fundamental Polyakov loop distributions as well as on the Pisa disorder operator providing evidence for the existence of a distinct finite-temperature transition in the center-blind theory. Our conclusions are drawn and an outlook is given in Section 5.

2. SU(2) lattice gauge theories with mixed fundamental-adjoint action

Among the "first day" lattice gauge theory models were also those with a mixture of different group representations for the plaquette contribution, e.g. for $SU(2)$

– the Wilson-type mixed action [8]

$$S = \beta_A \sum_P \left(1 - \frac{1}{3}\mathrm{Tr}_A U_P\right) + \beta_F \sum_P \left(1 - \frac{1}{2}\mathrm{Tr}_F U_P\right), \tag{3}$$

$$\frac{1}{g^2} = \frac{\beta_F}{4} + 2\frac{\beta_A}{3}$$

– the Villain-type mixed action [10]

$$S = \beta_V \sum_P \left(1 - \frac{1}{2}\sigma_P \mathrm{Tr}_F U_P\right) + \beta_F \sum_P \left(1 - \frac{1}{2}\mathrm{Tr}_F U_P\right), \tag{4}$$

where $\sigma_P = \pm 1$ is an auxiliary dynamical $\mathbb{Z}_2$ plaquette variable.

The non-trivial phase structure with first order bulk transitions (see Fig. 1) is governed by lattice artifacts: $\mathbb{Z}_2$ magnetic monopoles and electric vortices the densities of which can be defined as follows (N_c and N_l being the number of 3-cubes and lattice links, respectively) [11]

$$M = 1 - \langle \frac{1}{N_c} \sum_c \rho_c \rangle, \qquad \rho_c = \prod_{P \in \partial c} \sigma_P \quad \text{or} \quad \prod_{P \in \partial c} \mathrm{sign}(\mathrm{Tr}_F U_P) \tag{5}$$

$$E = 1 - \langle \frac{1}{N_l} \sum_l \rho_l \rangle, \qquad \rho_l = \prod_{P \in \hat{\partial} l} \sigma_P \quad \text{or} \quad \prod_{P \in \hat{\partial} l} \mathrm{sign}(\mathrm{Tr}_F U_P)\,. \tag{6}$$

These lattice excitations can be suppressed by modifying the action with suppression terms like [11,16]

$$\lambda_V \sum_c (1 - \rho_c)\,, \qquad \gamma_V \sum_l (1 - \rho_l)\,. \tag{7}$$

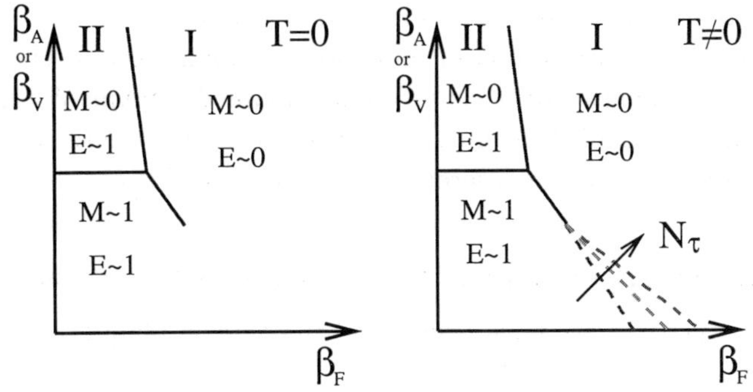

Figure 1. Schematic phase structure at $T = 0$ and $T > 0$.

For the Villain-type action the equivalence between $SO(3)$ and $SU(2)$ has been proven in the limit of complete $\mathbb{Z}(2)$ monopole supression $\lambda_V \to \infty$ for ($\beta_F = \gamma_V = 0$) [22–25,18] in the following form

$$\sum_{\text{twist sectors}} Z_{SU(2)} \equiv A \sum_{\sigma_P=\pm1} \int DU e^{\beta_V \sum_P \sigma_P \frac{1}{2}\mathrm{Tr}_F U_P} \prod_c \delta\Big(\prod_{P\in\partial c} \sigma_P - 1\Big)$$

where on the l.h.s. the twist sectors are imposed by twisted boundary conditions $U_\nu(x + L_\mu) = z_{\mu\nu} U_\nu(x)$, $z_{\mu\nu} \in \pi_1[SU(2)/\mathbb{Z}_2] = \mathbb{Z}(2)$. On the r.h.s. the twist sectors are dynamically encountered, under circumstances separated by large barriers.

The case $T \neq 0$ has been mostly studied with the modified Villain action but always with a non-vanishing admixture of the fundamental representation ($\beta_F \neq 0$). Lines of a finite-T phase transition presumely of second order have been found in the $\beta_V - \beta_F$ plane for $\lambda_V \geq 1$ and $\gamma_V \geq 5$ [16]. Above the finite-T transition the adjoint Polyakov line $\langle L_A \rangle$ has been seen trapped into metastable states [14,15]

$$\langle L_A \rangle \longrightarrow \begin{cases} 1 \\ -\frac{1}{3} \end{cases} \qquad \text{as} \quad \beta_V \to \infty \tag{8}$$

Jahn and de Forcrand [18] related the negative L_A states to non-trivial twists. For demonstrating this they introduced $SO(3)$ – i.e. center-blind – twist variables

$$z_{\mu\nu} \equiv \frac{1}{N_\rho N_\sigma} \sum_{\rho\sigma} \prod_{P\in\mu,\nu-\text{plane}} \text{sign}\,(\mathrm{Tr}_F\, U_P) \in [-1, +1]\,, \quad \epsilon_{\mu\nu\rho\sigma} = 1. \tag{9}$$

The $z_{\mu\nu}$'s measure the $\mathbb{Z}(2)$ fluxes through $\mu\nu$-planes. Then the state

$$L_F = 0 \iff L_A \equiv \frac{1}{3}\mathrm{Tr}_A L = -\frac{1}{3}, \qquad \mathrm{Tr}_A L = (\mathrm{Tr}_F L)^2 - 1 \tag{10}$$

is related to electric twist $z_{i,4} = -1, \qquad i = 1, 2, 3\,.$

Having these observations in mind we are going now to check and to illustrate this scenario for a center-blind modified adjoint Wilson action. We ask how to establish a finite T transition for the center-blind theory and what rôle do play the different twist sectors in this case.

3. Adjoint SU(2) model with $\mathbb{Z}_2$ monopole suppression

In our investigations we have considered the Wilson plaquette action with link variables $U_\mu(x) \in SU(2)$

$$S = \beta_F \sum_P \left(1 - \frac{1}{2}\mathrm{Tr}_F\, U_P\right) + \beta_A \sum_P \frac{4}{3}\left(1 - \frac{1}{4}(\mathrm{Tr}_F\, U_P)^2\right) + \lambda \sum_c (1 - \rho_c) \tag{11}$$

where $\rho_c = \prod_{P \in \partial c} \mathrm{sign}\, \mathrm{Tr}_F\, U_P$. For $\beta_F = 0$ the action S becomes center-blind

$$U_\mu(x) \to -1 \cdot U_\mu(x) \quad \Longrightarrow \quad \rho_c \to \rho_c\,.$$

Fig. 2 shows the phase diagram in the $\beta_F - \beta_A$-plane for varying chemical potential λ for $T = 0$. Obviously, the suppression of $\mathbb{Z}_2$ monopoles ($\lambda > 0$)

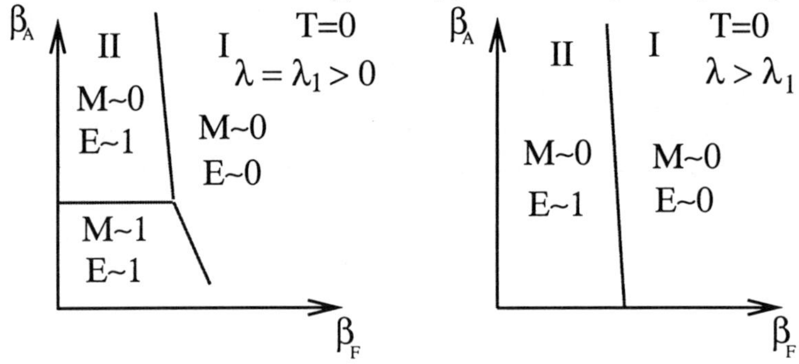

Figure 2. Schematic $\beta_F - \beta_A$ phase diagram for varying λ at $T = 0$.

shifts the horizontal line down to smaller β_A-values. At a first glance the phase II seems to be disconnected from phase I (the ordinary confinement phase) in the range $0 \le \lambda \le 1$. But see the discussion further below.

If we put $\beta_F = 0$ the emerging β_A-λ diagram looks as shown in Fig. 3. Phase I – which at $\beta_F \neq 0$ is connected with the ordinary confinement phase – is characterized by a non-zero $\mathbb{Z}(2)$ monopole density and by twist variables (9) fluctuating close to zero. On the contrary in phase II the

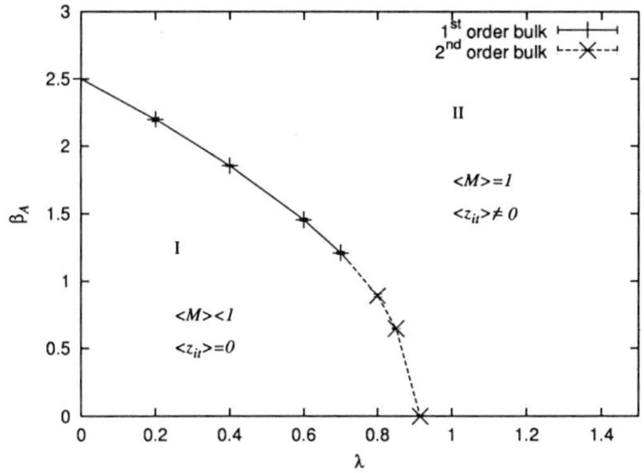

Figure 3. The bulk transition line in the $\beta_A - \lambda$ plane as seen for the lattice size $12^3 \times 4$.

monopoles become suppressed and the twist variables (meta)stable at ± 1. The phase transition line has been established by studying the average plaquette, the adjoint Polyakov loop variable $< L_A >$, the average density M of $\mathbb{Z}(2)$ monopoles and additionally the twist variables (9) as well as their 'susceptibility'

$$\chi_{\text{twist}} = N_\sigma^3 \cdot (\langle \tilde{z}^2 \rangle - \langle z \rangle^2) \tag{12}$$

with $\tilde{z} \equiv \frac{1}{3}(|z_{xt}| + |z_{yt}| + |z_{zt}|)$.

We found out that the bulk transition line in a certain range $0 < \lambda < \lambda_c$ (with $\lambda_c \simeq 0.7$ for the lattice size $12^3 \times 4$) shows a discontinuous behaviour of the monopole density M, of the average plaquette as well as of the adjoint Polyakov loop across the transition. At the same time tunneling between different twist sectors becomes strongly suppressed when passing the border from phase I to phase II. Along this line as long as $\lambda < \lambda_c$ the twist sectors are clearly related to metastable states of the adjoint Polyakov loop. We interpret the transition likely to be a first order transition in this range.

On the contrary, for $\lambda_c < \lambda < 0.95$ the monopole density M and the average plaquette turn out to behave smoothly across the bulk transition line. For the average adjoint Polyakov loop one finds $< L_A > \simeq 0$ on both sides of the line, i.e. phase II in this range seems to be also confining. This already is an indication that for larger λ-values and increasing β_A there should be a further (more or less horizontal) transition line hopefully behaving as a finite temperature transition. The 'tricritical' value λ_c seems to indicate the position, where this additional line might join the bulk transition. We come back to this question in the next section. The bulk transition itself in the range $\lambda_c < \lambda < 0.95$ is visible because of the enhanced tunneling between different twist sectors. This we have observed for all lattice sizes (so far up to $V = 12^4$). The twist susceptibility (12) has a strong peak which increases with increasing lattice size (see Fig. 4). A rough finite-size scaling test shows the transition to resemble the 4D Ising one and, therefore, seems to be of second order.

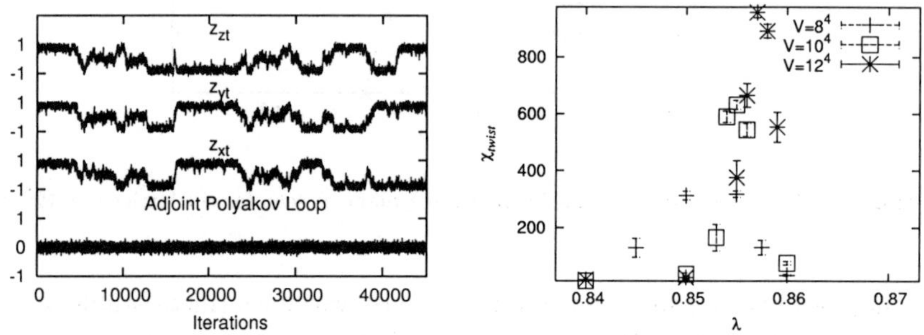

Figure 4. Electric twist histories at $\beta_A = 0.65, \lambda = 0.858$ for 12^4 (left) and twist susceptibility for $\beta_A = 0.65$ at varying λ and lattice sizes $8^4, 10^4$ and 12^4 (right).

4. Evidence for a finite-temperature transition

We observed that for sufficiently large chemical potential $\lambda \geq 1.0$ tunneling between twist sectors becomes completely suppressed. We decided to run the simulations in this range within fixed twist sectors (mostly the trivial one). That is, we suppress the generation or annihilation of extended vortices (up to numbers modulo two).

For fixed λ we carried out measurements with varying β_A. The adjoint Polyakov loop average $\langle L_A \rangle$ as a function of β_A in the zero twist sector is drawn in Fig. 5. Its rise clearly indicates a transition close to $\beta_A \simeq 1$. Frequency distributions of the local fundamental Polyakov loop variable $L_F(\vec{x})$ are plotted for two temporal lattice extensions in Fig. 6. One sees that the shape of the distributions clearly changes from a phase, where they peak at zero, to a phase, where they have two symmetric maxima away from zero. The critical β_A, where two maxima just occur, changes to larger values as N_t is increasing from 4 to 6 in agreement with scaling. Therefore, we can conclude that a finite-temperature transition is really seen. In order to check the existence of the transition at finite T with an independent measurement we have computed also the Pisa disorder parameter[26,27] which on

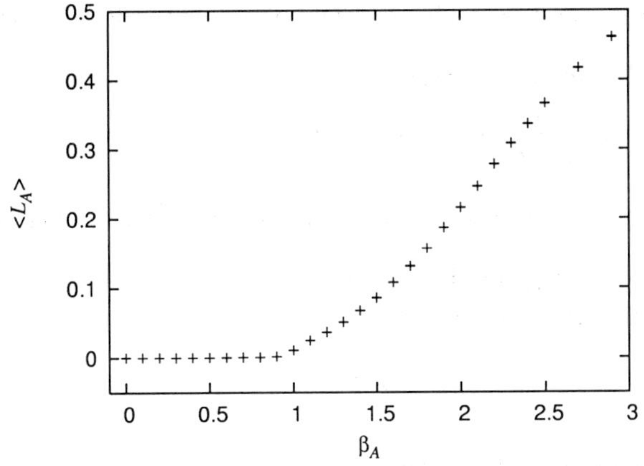

Figure 5. Adjoint Polyakov loop average as a function of β_A at $\lambda = 1.0$ for lattice size $V = 4 \times 12^3$.

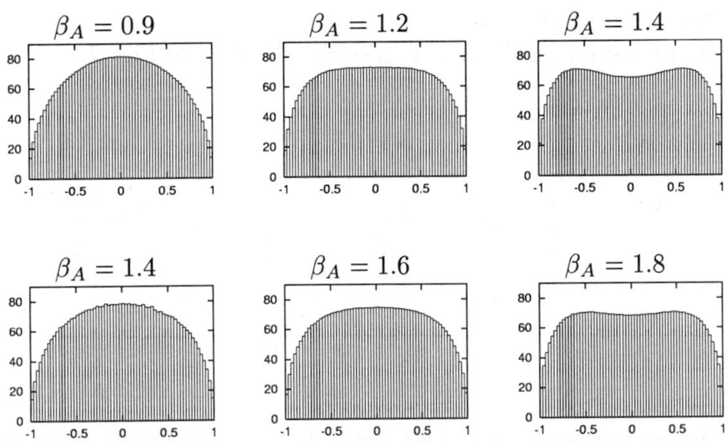

Figure 6. Frequency distributions of the local fundamental Polyakov loop variable $L_F(\vec{x})$. Upper row for lattice size $V = 4 \times 16^3$, lower row for $V = 6 \times 16^3$, all for $\lambda = 1.0$ and zero twist.

the basis of the dual superconductor model [6] allows to test for a condensate of Abelian monopoles in the confinement phase. The order operator for the condensation of magnetic charges $\mu(t)$ is defined by modifying the action at a given time-slice t by a classical Dirac monopole field insertion Φ [26,27]

$$\mu(t) = \exp(-\beta \Delta S(t)) \tag{13}$$

$$\Delta S(t) = \frac{1}{2} \sum_{i,\vec{x}} \mathrm{Tr}\,[U_{i4}(\vec{x},t) - U'_{i4}(\vec{x},t)] \tag{14}$$

$$U'_{i4}(\vec{x},t) = U_i(\vec{x},t)\,\Phi_i(\vec{x}+\hat{\imath},\vec{y})\,U_4(\vec{x}+\hat{\imath},t)\,U_i^\dagger(\vec{x},t+1)\,U_4^\dagger(\vec{x},t) \tag{15}$$

The operator can be generalized to the adjoint ($SO(3)$) action case. In the thermodynamic limit one expects

$$\langle\mu\rangle \quad \begin{cases} \neq 0 & T < T_c \\ = 0 & T > T_c \end{cases}$$

In practice, one measures instead

$$\rho = \frac{d}{d\beta} \log\langle\mu\rangle = -\langle S + \Delta S\rangle|_{S+\Delta S} + \langle S\rangle|_S. \tag{16}$$

The experience for the $SU(2)$ and $SU(3)$ cases tells that the latter quantity exhibits a sharp dip signalling the deconfinement phase transition to be driven by the breaking of a dual magnetic symmetry.

Some of our results for the $SO(3)$ case are collected in Fig. 7.

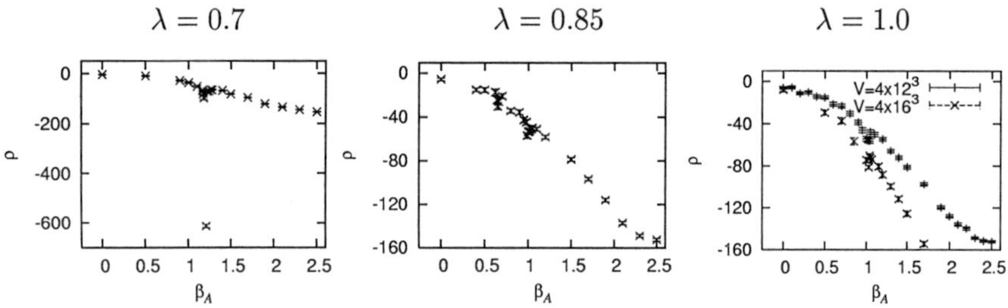

Figure 7. Pisa disorder parameter as a function of β_A for various λ values (lattice size $V = 4 \times 12^3$). The right most figure shows additionally the case with a larger volume $V = 4 \times 16^3$.

The computations show that for $\lambda = 0.7$ the bulk phase transition I $\leftrightarrow$ II is correctly localized. For increasing λ, e.g. at $\lambda = 0.85$, we see simultaneous, distinct signals for the bulk and the finite-T transitions. At even larger chemical potential ($\lambda = 1.$) we find only a signal for the finite-T transition. The localizations of the bulk and finite temperature transitions agree reasonably with those obtained with the other methods mentioned before. The universality class of the transition to be determined e.g. via the critical indices has still to be investigated.

Having convinced ourselves that there is a finite temperature transition also in the completely center-blind adjoint $SU(2)$ LGT we are tempted to ask whether there is a spontaneous breaking of an appropriate global symmetry. If yes, what is the corresponding order parameter?

For $SO(3)$ the question means, whether there is a globally defined operator flipping the adjoint Polyakov loop $L_A = 1 \Leftrightarrow L_A = -\frac{1}{3}$ while leaving the action invariant. There is an approximate solution based on the dual superconductor picture assuming that the long-distance properties are carried by Abelian degrees of freedom.

Let us consider flip operators with $P \in SU(2)$ or $SO(3)$ satisfying the conditions $P^2 = \pm\mathbb{I},\ P^\dagger = \pm P$. The only ones fulfilling these conditions are for

$$\begin{aligned} SU(2): &\quad P = \pm\mathbb{I}_2,\ \ \hat{P} = \ \pm i\vec{n}\cdot\vec{\sigma}\,, \\ SO(3): &\quad P = +\mathbb{I}_3,\ \ \hat{P} = \ \mathbb{I}_3 + 2(\vec{n}\cdot\vec{T})^2, \qquad (\ n^2 = 1\ ). \end{aligned}$$

Let us assume the $SU(2)$ or $SO(3)$ fields to be Abelian (diagonal) with respect to a fixed 'isospin' $\vec{n}$-direction, then $\hat{P}$ applied to a given time sheet, indeed, leaves the action invariant, and the Polyakov loop is flipped to the other state (if its value is different from zero). In practice, this is realized only approximately by fixing the 3D maximally Abelian gauge (MAG) on each time slice x_4 separately. This is achieved by maximizing the gauge functional with respect to gauge transformations g, e.g. for $SU(2)$:

$$F_{x_4}(g) = \sum_{\vec{x}} \sum_{i=1,2,3} \mathrm{Tr}\left((\vec{n}\cdot\vec{\sigma})\ U_i^g(\vec{x},x_4)\ (\vec{n}\cdot\vec{\sigma})\ (U_i^g(\vec{x},x_4))^\dagger \right). \tag{17}$$

The 'order parameter' is then defined as

$$\langle \Delta_{F,A} \rangle = c_{F,A}\ \langle\ |L_{F,A} - \tilde{L}_{F,A}|\ \rangle\,, \tag{18}$$

where

$$\tilde{L}_{F,A} = \frac{1}{N_\sigma^3} \sum_{\vec{x}} \mathrm{Tr}_{F,A} \left(\hat{P} \prod_{x_4=1}^{N_\tau} U_4(\vec{x},x_4) \right). \tag{19}$$

The 'order parameter' should work as for the standard $SU(2)$ case as well as for $SO(3)$. We have checked this. Indeed, for $SU(2)$ we obtained the results drawn in Fig. 8 for the order parameter itself as well as for its susceptibility. The susceptibility develops a peak around $\beta_F = \beta_c \simeq 2.3$ as it should be. The corresponding Binder cumulant computed for varying 3-volume has been seen to provide intersecting lines at the same β_F value. If

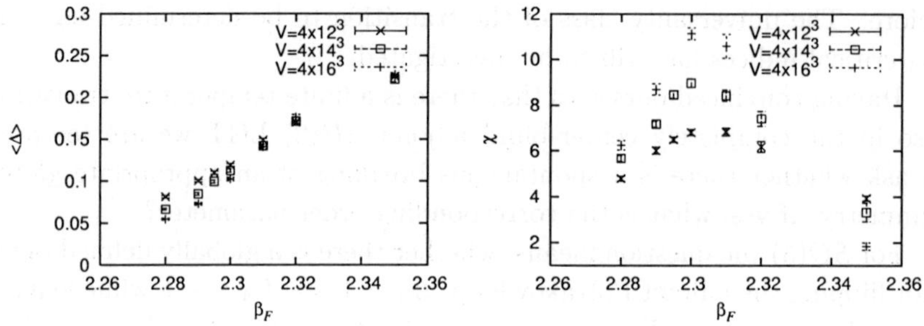

Figure 8. 'Order parameter' $\langle\Delta\rangle$ (left) and its susceptibility χ (right) versus β_F for $\beta_A = 0.0,\ \lambda = 0.0$.

one computes the corresponding 'order parameter' Δ for the adjoint $SO(3)$ action case at $\lambda = 1.$ and in the zero-twist sector one gets a similar picture as the left one in Fig. 8.

5. Conclusions and outlook

We summarize our studies by drawing the phase diagram for the modified adjoint $SU(2)$ theory at $T > 0$ as shown in Fig. 9. We see the existence

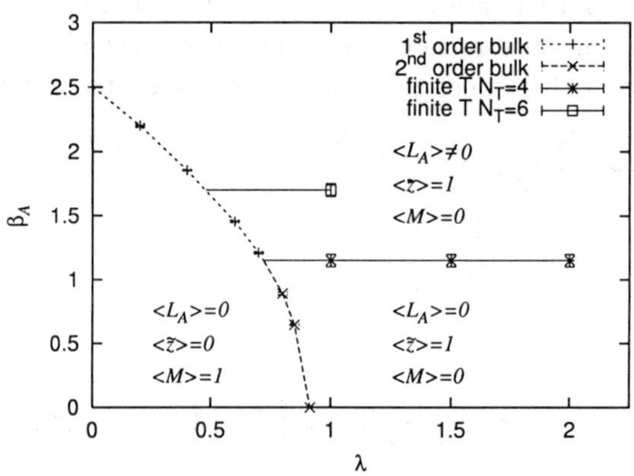

Figure 9. Phase diagram of the center-blind $SO(3)$ model with $\mathbb{Z}(2)$ monopole suppression term. The horizontal lines indicate the position of the finite temperature transition for varying time extension $N_t = 4, 6$.

of a finite temperature transition decoupling from the bulk transition at a position which scales in β_A as one would expect. The finite temperature transition was also localized with the help of the Pisa disorder parameter indicating that the dual superconductor scenario seems to work also for $SO(3)$. Whether also the determination of the free energy of an extended center vortex (see references [28,29]) would also point to the same result remains to be seen. Our numerical study presented here is in many respects still preliminary. Larger lattices, the use of an ergodic algorithm allowing to enhance tunneling between different twist sectors and the determination of critical indices in order to determine the universality class require more extensive computations, until we can draw final conclusions.

References

1. L. D. McLerran and B. Svetitsky, *Phys. Lett.* **B98**, 195 (1981).
2. J. Kuti, J. Polonyi and K. Szlachanyi, *Phys. Lett.* **B98**, 199 (1981).
3. A.M. Polyakov, *Phys. Lett.* **B72**, 477 (1978).
4. L. Susskind, *Phys. Rev.* **D20**, 2610 (1979).
5. B. Svetitsky and L. G. Yaffe, *Nucl. Phys.* **B210**, 423 (1982).
6. Y. Nambu, *Phys. Rev.* **D10**, 4262 (1974);
   G. 't Hooft, in 'High Energy Physics', Proceedings of the EPS International Conference, Palermo 1975, ed. A. Zichichi, Editrice Compositori, Bologna 1976;
   S. Mandelstam, *Phys. Rep.* **23**, 245 (1976);
   A.S. Kronfeld, M.L. Laursen, G. Schierholz and U.J. Wiese, *Phys. Lett.* **B198**, 516 (1987);
   V. Bornyakov, E.-M. Ilgenfritz, M.L. Laursen, V.K. Mitrjushkin, M. Müller-Preussker, A.J. van der Sijs, A.M. Zadorozhny, *Phys. Lett.* **B261**, 116 (1991);
   G. Bali, V. Bornyakov, M. Müller-Preussker and K. Schilling, *Phys. Rev.* **D54**, 2863 (1996).
7. G. 't Hooft, *Nucl. Phys.* **B138**, 1 (1978);
   L. Del Debbio, M. Faber, J. Greensite and S. Olejnik, *Phys. Rev.* **D55**, 2298 (1997);
   L. Del Debbio, M. Faber, J. Giedt, J. Greensite and S. Olejnik, *Phys. Rev.* **D58**, 094501 (1998);
   K. Langfeld, H. Reinhardt and O. Tennert, *Phys. Lett.* **B419**, 317 (1998);
   M. Engelhardt, K. Langfeld, H. Reinhardt and O. Tennert, *Phys. Lett.* **B452**, 301 (1999); *Phys. Rev.* **D61**, 054504 (2000)
8. G. Bhanot and M. Creutz, *Phys. Rev.* **D24**, 3212 (1981).
9. J. Greensite and B. Lautrup, *Phys. Rev. Lett.* **47**, 9 (1981).
10. I.G. Halliday and A. Schwimmer, *Phys. Lett.* **B101**, 327 (1981).
11. I.G. Halliday and A. Schwimmer, *Phys. Lett.* **B102**, 337 (1981).
12. R.V. Gavai, M. Grady and M. Mathur, *Nucl. Phys.* **B423**, 123 (1994).
13. R.V. Gavai, *Nucl. Phys.* **B474**, 446 (1996).
14. S. Cheluvaraja and H. S. Sharathchandra, hep-lat/9611001 (1996).

15. S. Datta and R.V. Gavai, *Phys. Rev.* **D57**, 6618 (1998).
16. S. Datta and R.V. Gavai, *Phys. Rev.* **D60**, 034505 (1999).
17. R.V. Gavai and M. Mathur, *Phys. Lett.* **B458**, 331 (1999).
18. Ph. de Forcrand and O. Jahn, *Nucl. Phys.* **B651**, 125 (2003).
19. A. Barresi, G. Burgio and M. Müller-Preussker, *Nucl. Phys. B (Proc. Suppl.)* **106&107**, 495 (2002).
20. A. Barresi, G. Burgio and M. Müller-Preussker, *Nucl. Phys. B (Proc. Suppl.)* **119**, 571 (2003).
21. A. Barresi, G. Burgio and M. Müller-Preussker, hep-lat/0309010 (2003).
22. E. Tomboulis, *Phys. Rev.* **D23**, 2371 (1981).
23. G. Mack and V.B. Petkova, *Z. Phys.* **C12**, 177 (1982).
24. T. Kovacs and E. Tomboulis, *Phys. Rev.* **D57**, 4054 (1998).
25. A. Alexandru and R.W. Haymaker, *Phys. Rev.* **D62**, 074509 (2000).
26. A. Di Giacomo, B. Lucini, L. Montesi and G. Paffuti, *Phys. Rev.* **D61**, 034503 (2000).
27. A. Di Giacomo, B. Lucini, L. Montesi and G. Paffuti, *Phys. Rev.* **D61**, 034504 (2000).
28. T. Kovacs and E. Tomboulis, *Phys. Rev. Lett.* **85**, 704 (2000).
29. Ph. de Forcrand and L. Von Smekal, *Phys. Rev.* **D66**, 011504 (2002).

ON THE VACUUM OF FULL QCD*

G. SCHIERHOLZ†

John von Neumann-Institut für Computing, Deutsches Elektronen-Sychrotron DESY, 15738 Zeuthen, Germany
Deutsches Elektronen-Sychrotron DESY, 22603 Hamburg, Germany
E-mail: Gerrit.Schierholz@desy.de

In this talk I shall discuss recent results on the s tructure of the vacuum of full QCD, at zero and at finite temperature. Particular attention will be paid on how the vacuum reacts to the introduction of dynamical quarks.

1. Objective

To understand the structure of the QCD vacuum, and in particular the mechanism of color confinement, is one of the biggest problems in theoretical particle physics.

Any intuitive understanding of the QCD vacuum is obscured by the fact that the configuration space $\mathcal{C}$ of the theory is larger than the physical configuration space $\mathcal{C}_{\rm phys}$, because the gauge group $SU(3)$ generates equivalent configurations along the gauge orbit: $\mathcal{C}_{\rm phys} = \mathcal{C}/SU(3)$. $\mathcal{C}_{\rm phys}$ may be obtained by fixing the gauge, provided the gauge condition is unique, i.e. its hypersurface intersects each gauge orbit only once. The gauge fixing must be done in such a way, that no unphysical modes can propagate over macroscopic distances.

A suitable gauge is the maximally abelian gauge (MAG) [1,2]. In this gauge the degrees of freedom are color electric charges, color magnetic monopoles, 'photons' and 'gluons'. The latter appear to become massive [3] due to a yet unresolved mechanism, resulting in an abelian effective theory at large distances. There is evidence that the monopoles condense [2,4] in the low temperature phase of the theory, causing a dual Meissner effect which constricts color electric fields into narrow tubes (strings) of flux, in

*Talk presented at *Confinement 2003*, Riken, Waco, Japan.
† For the *QCDSF* and *DIK* collaborations.

accord with the dual superconductor picture of confinement. So far our understanding is based on simulations of the pure $SU(2)$ and $SU(3)$ gauge theory. It will be interesting now to see if and how this picture changes if we pass to the full theory. It is conceivable that a new class of color-flavor locked quark–anti-quark condensates [5] emerges, which would put the flavor-blind abelian effective theory into question.

In any case, to substantiate our understanding of the QCD vacuum, we need to derive an effective action at the long-distance scale. This has been attempted for the quenched, abelian projected theory with some success [6]. In this approach color electric charges have been integrated out though, which is inappropriate in full QCD. (Actually, color electric currents should be taken into account even in the quenched theory, as they are induced by topological excitations [7,8] and are inextricably linked with the θ vacuum.) Sometimes the long-distance physics is being associated with a string model. But that is no serious option, because in full QCD the string breaks already at distances of about the confinement radius, as we shall see.

As a first step towards a quantitative understanding of the vacuum of full QCD, I shall discuss how the vacuum reacts to the introduction of dynamical color electric charges. The talk is based on recent simulations with $N_f = 2$ flavors of dynamical Wilson fermions, at zero and at finite temperature, performed by the *QCDSF* and *DIK* collaborations. Part of the material presented here has been published elsewhere [9,10,11,12,13].

2. Preliminaries

Throughout this talk we employ the Wilson gauge field action and non-perturbatively $O(a)$ improved Wilson fermions,

$$S_F = S_F^{(0)} - \frac{\mathrm{i}}{2}\,\kappa\, g\, c_{SW}\, a^5 \sum_s \bar{\psi}(s)\sigma_{\mu\nu}F_{\mu\nu}\psi(s)\,, \tag{1}$$

with $N_f = 2$ flavors of dynamical quarks. Here a is the lattice spacing, κ is the hopping parameter which fixes the quark mass, c_{SW} is a known parameter, and $S_F^{(0)}$ is the original Wilson fermion action. Altogether, this action reduces cut-off corrections to $O(a^2)$. Systematic restoration of chiral symmetry to $O(a^2)$ is also possible. I like to stress that the action (1) is local, in contrast to the $N_f = 2$ staggered fermion action, so that we should not have to worry about unphysical singularities, in particular in the course of gauge fixing.

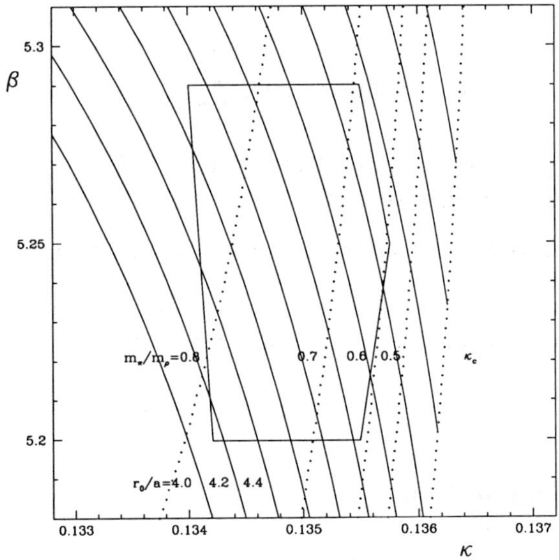

Figure 1. The bare parameter space, together with lines of constant r_0/a (solid curves) and lines of constant m_π/m_ρ (dotted curves). The inner region indicates the area of bare parameters covered in our zero temperature calculations, while the calculations at finite temperature are limited to $\beta = 5.2$.

The calculations are done on $16^3\,32$ and $24^3\,48$ lattices at zero temperature, and on $16^3\,8$ lattices at finite temperature. In Fig. 1 I show the region of bare parameters that is covered presently in our dynamical simulations, together with the corresponding lattice spacings and m_π/m_ρ ratios. I frequently use $r_0 = 0.5$ fm to set the scale.

The lattice gauge field $U(s,\mu)$ is brought into the MAG by maximizing [14]

$$R = \sum_{s,\mu} \sum_{i=1}^{3} |U_{ii}(s,\mu)|^2 \tag{2}$$

with respect to gauge transformations g:

$$U(s,\mu) \longrightarrow g(s) U(s,\mu) g^{-1}(s+\hat{\mu})\,. \tag{3}$$

The abelian link variables

$$u(s,\mu) = \mathrm{diag}\big(u_1(s,\mu), u_2(s,\mu), u_3(s,\mu)\big) \tag{4}$$

with

$$u_i(s,\mu) = \exp(\mathrm{i}\theta_i(s,\mu))\,, \ \sum_{i=1}^{3} \theta_i(s,\mu) = 0\,, \tag{5}$$

are obtained by means of a coset decomposition [15] $U(s,\mu) = c(s,\mu)u(s,\mu)$. From (5) we define plaquette angles

$$\theta_i(s,\mu,\nu) = \partial_\mu \theta_i(s,\nu) - \partial_\nu \theta_i(s,\mu) \equiv \bar{\theta}_i(s,\mu,\nu) + 2\pi m_i(s,\mu,\nu)\,, \tag{6}$$

where $\bar{\theta}_i(s,\mu,\nu) \in (-\pi,\pi]$ and $m_i(s,\mu,\nu) \in \mathbb{Z}$. The monopole currents reside on links of the dual lattice and are defined by

$$k_i(s,\mu) = \frac{1}{4\pi}\epsilon_{\mu\nu\rho\sigma}\partial_\nu \bar{\theta}_i(s+\hat{\mu},\rho,\sigma) = -\frac{1}{2}\epsilon_{\mu\nu\rho\sigma}\partial_\nu m_i(s,\rho,\sigma)\,. \tag{7}$$

3. The Vacuum at T=0

Let me first address the vacuum at zero temperature. Though our present quark masses are much heavier than the physical masses of the u and d quarks, the vacuum shows already significant changes, compared with the quenched theory, as we shall see.

Gross Properties

The first quantity that comes to mind if one thinks of the nonperturbative nature of the QCD vacuum is the topological susceptibility

$$\chi = \frac{\langle Q^2 \rangle}{V}\,, \tag{8}$$

where Q is the topological charge and V the (lattice) volume. In Fig. 2 I show χ as a function of $m_\pi^2 \propto (m_u + m_d)$. [a] The topological charge has been computed by means of a cooling algorithm [16]. Chiral perturbation theory predicts [17]

$$\chi = \frac{1}{N_f} f_\pi^2 m_\pi^2\,. \tag{9}$$

We see that the susceptibility gets gradually suppressed as the pion mass is decreased. The lattice data are consistent with the predictions of chiral perturbation theory. For a quantitative comparison we need, however, to explore smaller quark masses.

---

[a] I like to thank A. Hart for providing this figure.

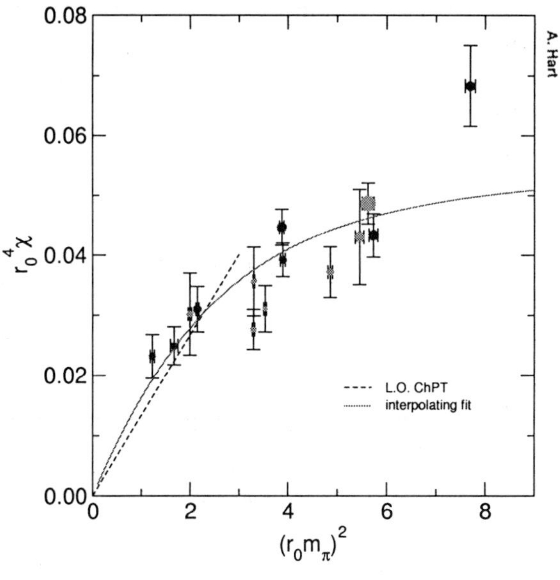

Figure 2. The topological susceptibility χ as a function of m_π^2, together with the prediction of low-order chiral perturbation theory (dotted line).

A related quantity is the monopole density. In Fig. 3 I show

$$\rho = \frac{1}{12\,V}\sum_i \sum_{s,\mu} |k_i(s,\mu)| \tag{10}$$

as a function of m_π/m_ρ. We see that the monopole density increases with decreasing pion mass. We have distinguished between small (ultraviolet) and large (infrared) clusters. The latter consist of monopole loops which wrap around the boundary of the lattice at least once. Later on we will see that the increase of ρ is also reflected in the monopole effective action.

Both effects, the decrease of χ and the increase of ρ with decreasing quark mass, are related by the fact that (anti-)instantons are sources of monopoles [7], and vice versa. In fact, they both are associated with the same gauge singularities. The explanation for this behavior is that the fermion determinant suppresses fermionic zero modes and, by means of the index theorem, isolated (anti-)instantons. This causes an attraction of neighboring instantons and anti-instantons [18], as well as an attraction of monopoles and (anti-)monopoles. As a result, the monopole density increases. If the vacuum was a dilute gas of instantons in the quenched

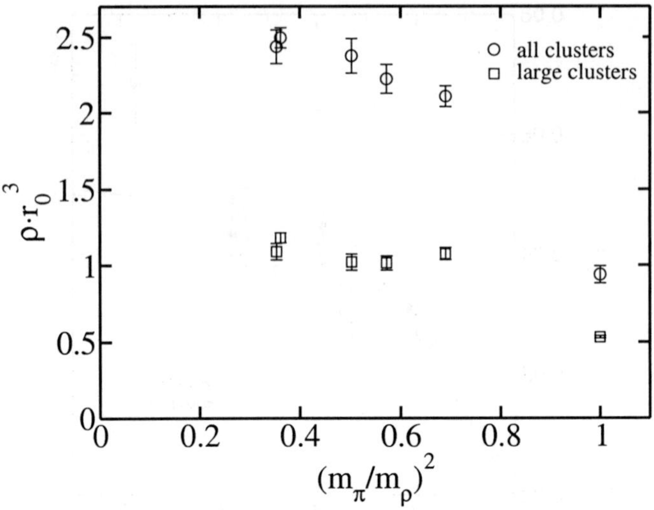

Figure 3. The monopole density ρ as a function of $(m_\pi/m_\rho)^2$, separately for large clusters and for all monopoles. The entries at $m_\pi/m_\rho = 1$ refer to the quenched result.

case [19], it gradually will become an instanton liquid or crystal, as the dynamical quark mass is decreased.

A screened plasma of static monopoles is known to generate a string tension [20]

$$\sigma \propto \rho\xi\,, \tag{11}$$

where ξ is the magnetic screening length. This is defined by the exponential decay of the magnetic flux $\Phi(r)$ through a sphere of radius r around the monopole. On the periodic lattice

$$\Phi(r) = \Phi_0 \exp\big(-\frac{L}{2\xi}\big) \sinh\big(\frac{L-2r}{2\xi}\big)\,, \tag{12}$$

where L is the effective length of the box, which is taken to be a free parameter. The string tension σ does not change significantly between the quenched and unquenched theory [9]. Thus, any increase of the monopole density is expected to be followed by a decrease of the magnetic screening length. In Fig. 4 I show a typical quenched and unquenched flux distribution, together with a fit of the form (12). The results for ξ are given in Table 1. We see that at the smallest quark mass the screening length is reduced by more than 30%, though not as much as suggested by (11).

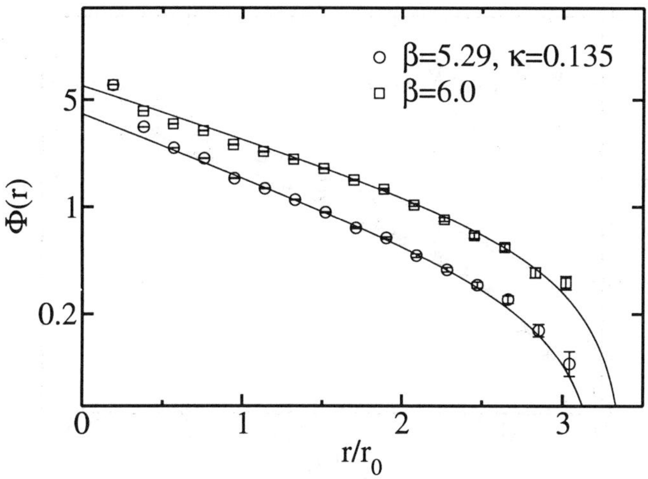

Figure 4. The magnetic flux $\Phi(r)$ as a function of r on the $16^3\,32$ lattice in the quenched ($\beta = 6.0$) and unquenched theory.

m_π/m_ρ	ξ/r_0	$\sigma_{\rm ab}/\sigma$
0.60	0.48(2)	0.90(4)
0.70	0.47(3)	0.96(3)
0.76	0.52(2)	0.99(6)
0.83	0.48(2)	0.99(6)
1	0.66(3)	0.83(3)

Table 1. The monopole screening length and the ratio of abelian to non-abelian string tension in full and quenched QCD.

Abelian Dominance

The dynamical origin of abelian dominance is obscure. It is surprising that most of the community takes it for granted, or is satisfied with the mere demonstration of it. Very probably it is due to a dynamical Higgs mechanism, which is brought forward by a yet unknown color octet condensate. In full QCD there are many more possibilities to form such a condensate, including the recently proposed color-flavor locked composite Higgs field [5], which might lead to a different scenario. For example, in the case of a color-flavor locked condensate also the 'photons' would become massive, leaving

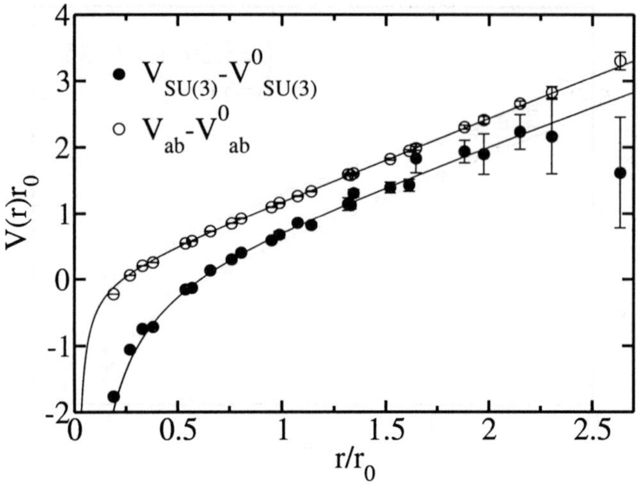

Figure 5. Comparison of the abelian and nonabelian potential in full QCD on the $16^3\,32$ lattice at $\beta = 5.29$, $\kappa = 0.135$. The curves are fits of the form (14).

us with a $\mathbb{Z}_3$ symmetry only.

A measure of abelian dominance is the string tension [21]. In Fig. 5 I compare abelian and nonabelian potentials. They are obtained from Wilson loops:

$$V(r) = \lim_{T\to`\infty'} \log\left(\frac{\langle W(r,T)\rangle}{\langle W(r,T+1)\rangle}\right),\; W(r,T) = \frac{1}{3}\,\mathrm{Re}\,\mathrm{Tr}\,\mathcal{W}_C\,, \tag{13}$$

with $\mathcal{W}_C = \Pi_{s,\mu\in C}\, U(s,\mu)$ and $\mathcal{W}_C = \Pi_{s,\mu\in C}\, u(s,\mu)$, respectively. The Wilson loop projects mainly onto the state of the (unbroken) string. Note that long-range forces between quarks remain to exist in the dynamical theory as well, because the color charge of quarks cannot be screened locally by Higgs scalars. We fit both the abelian and the nonabelian potential by the ansatz

$$V(r) = -\frac{\alpha}{r} + \sigma\, r + V_0\,. \tag{14}$$

The resulting string tensions are listed in Table 1. The abelian string tension $\sigma_{\rm ab}$ turns out to be very close to the nonabelian one, much closer even than in the quenched approximation.

Flux Tube

The chromoelectric flux tube between static color electric charges forms an

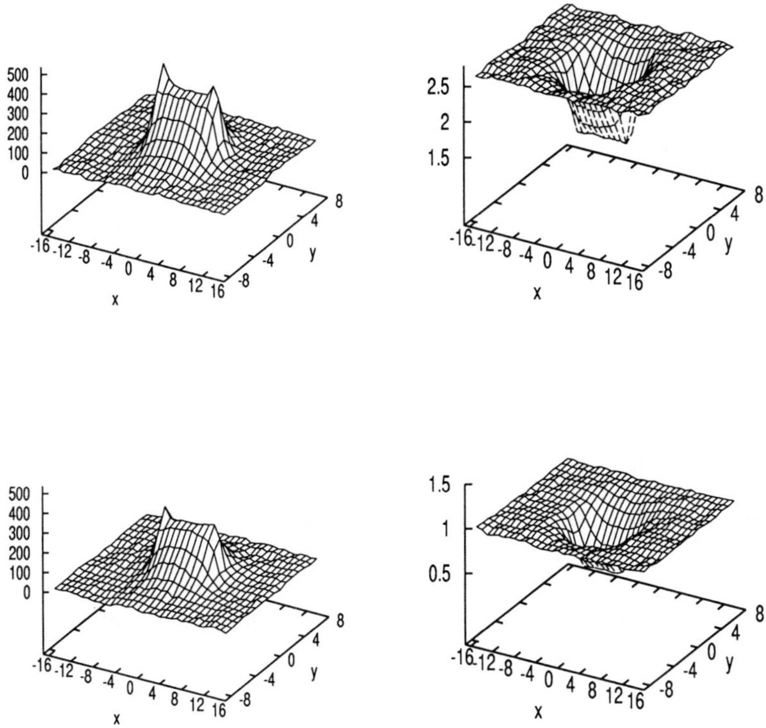

Figure 6. The action density (left figures) and monopole density (right figures) of the abelian flux tube as a function of $x = s_1$, $y = s_2$ at $Z = s_3 = 0$ on the $16^3\,32$ lattice in full (top figures) and quenched QCD (bottom figures) at $\beta = 5.2, \kappa = 0.1355$ and $\beta = 6.0$, respectively.

interface between the superconducting (confining) phase and the conducting phase, and in this capacity bears valuable information on the properties of the QCD vacuum.

The basic idea of the dual superconductor picture of confinement is that the color electric field is squeezed into flux tubes by the monopole supercurrent. In Fig. 6 I show the action density of the abelian flux tube

$$\frac{\beta}{6} \sum_{i,\mu>\nu} \Big(\frac{\langle \cos\theta_i(s,\mu,\nu)\, W_{\rm ab}(R,T)\rangle}{\langle W_{\rm ab}(R,T)\rangle} - \langle \cos\theta_i(s,\mu,\nu)\rangle \Big) \tag{15}$$

and the corresponding monopole density

$$\frac{1}{12}\sum_{i,\mu}\Big(\frac{\langle|k_i(s,\mu)|\,W_{\mathrm{ab}}(R,T)\rangle}{\langle W_{\mathrm{ab}}(R,T)\rangle}-\langle|k_i(s,\mu)|\rangle\Big)\,, \tag{16}$$

for $R = 10, T = 6$. We see that in full QCD the action density is about 40% larger than in the quenched theory, while the width of the flux tube remains unchanged. Outside the flux tube the monopole density is more than two times larger in full QCD as compared to the quenched case. We have looked at the spatial distribution of the color electric field and the monopole current and found that the solenoidal monopole current satisfies dual Ampère's law in full QCD as well. Altogether, we do not see any departure from the dual superconductor picture of confinement.

Monopole Effective Action

The monopole effective action [6] is, by construction, an effective action of the abelian projected theory. In the quenched theory it has been quite successful [22] in reproducing, among others, the string tension and the low-lying glueball masses. We shall discuss now how this action will be affected by the introduction of dynamical quarks.

We assume that the form of the action is the same as in the quenched theory:

$$S(k)=\sum_{i=1}^{29}G_i\,S_i(k) \tag{17}$$

with

$$\begin{aligned} S_1(k)&=\sum_s\sum_\mu k(s,\mu)\,k(s,\mu)\,,\\ S_2(k)&=\sum_s\sum_\mu k(s,\mu)\,k(s+\hat\mu,\mu)\,,\\ &\;\vdots\\ S_{28}&=\sum_s\big(\sum_\mu k(s,\mu)^2\big)^2\,,\\ S_{29}&=\sum_s\big(\sum_\mu k(s,\mu)^2\big)^3\,. \end{aligned} \tag{18}$$

We employ a type-II block spin transformation with up to $n = 4$ blocking steps. The final outcome is an action at the physical length scale $b = na$. We

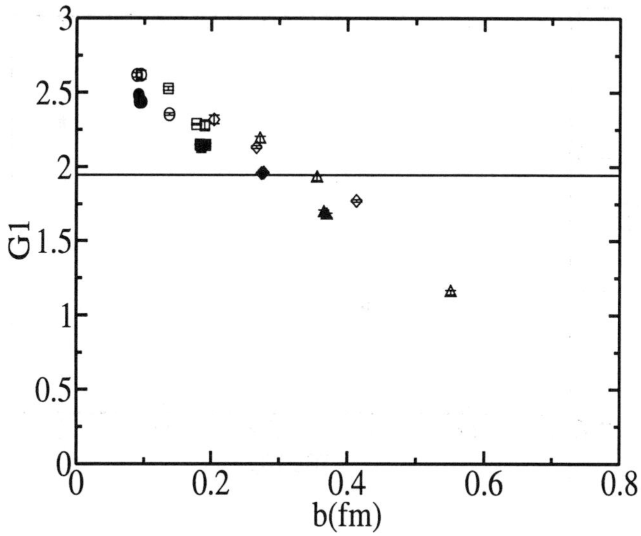

Figure 7. The monopole self-coupling G_1 as a function of b. The symbols are: $n = 1$ (●), $n = 2$ (■), $n = 3$ (♦), $n = 4$ (▲) in full QCD, and $n = 1$ (○), $n = 2$ (□), $n = 3$ (◊), $n = 4$ (△) in quenched QCD. The horizontal line corresponds to $G_1 = \ln 7$.

believe that the monopole action becomes effective at scales $0.4 \lesssim b \lesssim 0.8$ fm.

In Fig. 7 I show the monopole self-coupling G_1 as a function of the physical length scale b for both quenched and full QCD. We see that in full QCD G_1 is systematically smaller than in the quenched theory for all values of b. In Fig. 8 I plot G_1 as a function of m_π/m_ρ for the smallest value of b. This shows clearly that G_1 decreases with decreasing quark mass. A similar behavior is found at larger values of b.

A necessary condition for monopole condensation is $G_1 \leq \ln 7$. This is achieved for $b \gtrsim 0.29$ fm (at $m_\pi/m_\rho \geq 0.6$) in full QCD and for $b \gtrsim 0.35$ fm in the quenched approximation.

The higher coupling constants are found to be smaller by one (G_2) to three orders of magnitude (G_{29}). Considering that the charge of the monopole is ± 1 in almost all cases, we can write

$$\begin{aligned} S(k) &\approx G_1\, S_1(k) \equiv S_{\rm self} \\ &\approx G_1\, L\,, \end{aligned} \tag{19}$$

where L is the length of the monopole loop. As a result, a smaller self-coupling G_1 will give rise to a larger value of L, and hence to a larger

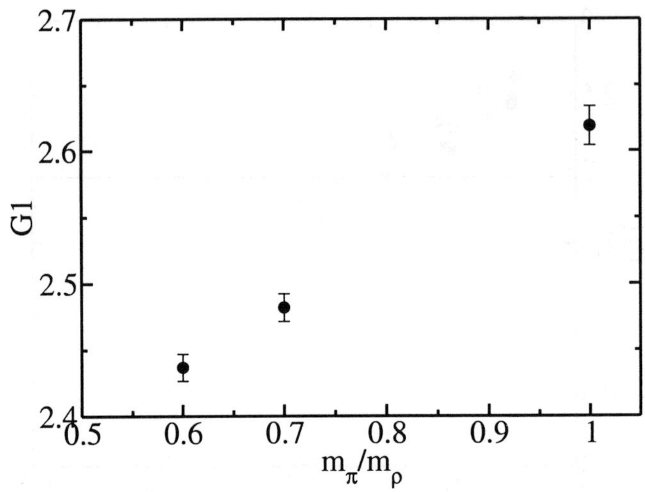

Figure 8. The monopole self-coupling G_1 as a function of m_π/m_ρ.

monopole density, and vice versa. So everything appears to be consistent.

From the dynamical point of view the self-energy contribution is not very interesting. It merely gives a (bare) mass to the monopoles. The higher terms of the action can be cast into the follwing form:

$$S(k) = S_{\text{self}} + S_{\text{Coulomb}} + S_{4-\text{point}} + S_{6-\text{point}} + \cdots \,, \tag{20}$$

where

$$S_{\text{Coulomb}} = \frac{g_{\text{m}}^2}{2} \sum_{s,s'} \sum_{\mu} k(s,\mu)\Delta^{-1}(s-s')k(s',\mu) \,, \tag{21}$$

and $S_{4-\text{point}}, \cdots$ are small in the infrared. This action has the same form as that of Cardy and Rabinovici [23], except that the electric currents are missing.

4. The Vacuum at T>0

Let us now turn to QCD at finite temperature T. The critical temperature T_c of the finite temperature phase transition changes significantly upon the introduction of dynamical quarks, in contrast to the majority of hadronic observables at zero temperature. In Fig. 9 I show T_c as a function of m_π/m_ρ [12]. The critical temperature changes from $\approx$ 280 MeV in the

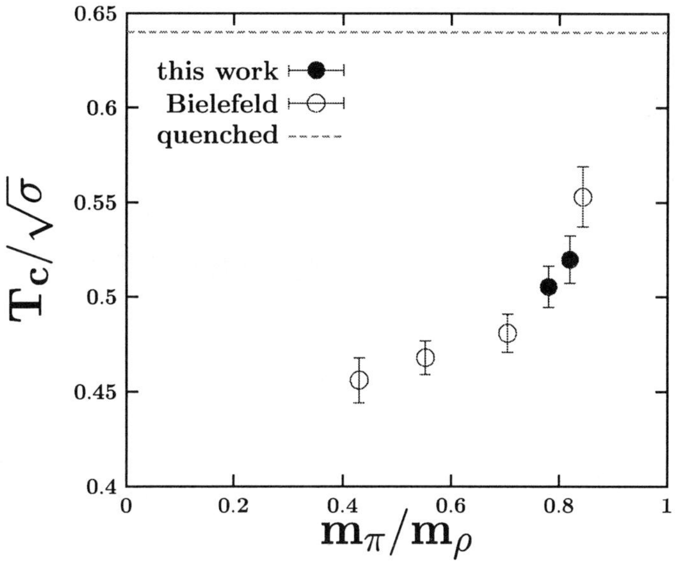

Figure 9. The critical temperature T_c as a function of m_π/m_ρ [12], together with the results of the Bielefeld group [24]. The latter employed improved staggered fermions.

quenched approximation to ≈ 180 MeV in the limit of $N_f = 2$ light quarks. Thus, finite temperature QCD appears to be an ideal place where to search for effects of dynamical fermions.

The most sought after phenomenon is string breaking. Wilson loops are known to have a poor overlap with the broken string. Not so Polyakov loop correlators, which we may expect to give a useful signal at finite temperature. In Fig. 10 I show the effective static potential

$$V_{\text{eff}}(r,T) = -T\ln\langle L(\vec{x})L^\dagger(\vec{y})\rangle \tag{22}$$

at finite temperature $T/T_c \approx 0.8$, where L is the Polyakov loop and $r = |\vec{x}-\vec{y}|$. String breaking is a level crossing phenomenon. At smaller distances the transfer matrix is dominated by the unbroken string state [25],

$$\begin{aligned} V_{\text{eff}}(r,T) = V_{\text{string}}(r,T) &= -\frac{1}{r}\big(\alpha - \frac{1}{6}\arctan(2rT)\big) \\ &+ \big(\sigma(T) + \frac{2T^2}{3}\text{arccot}(2rT)\big)r + \frac{T}{2}\ln(1+4r^2T^2) + V_0(T)\,, \end{aligned} \tag{23}$$

while at large distances the state of two static-light mesons takes over

$$V_{\text{eff}}(r,T) = 2m(T) + V_0(T)\,, \tag{24}$$

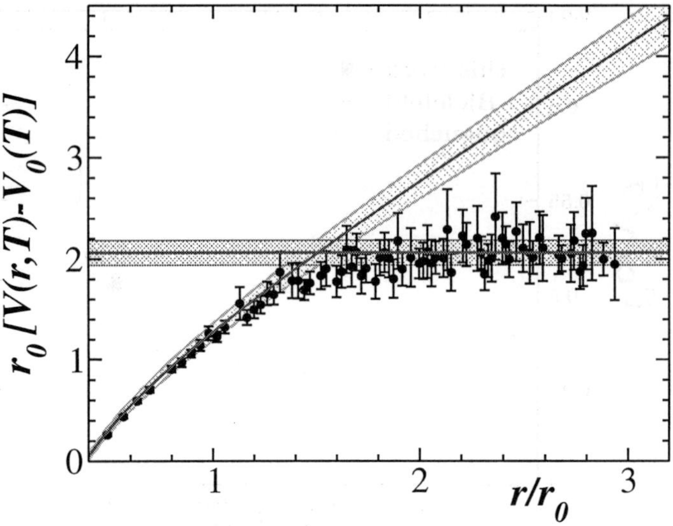

Figure 10. The effective static potential at $T/T_c \approx 0.8$.

where $m(T)$ is the so-called constituent quark mass and V_0 is the self-energy. The string breaking distance was found to decrease with increasing temperature, which has helped to observe string breaking.

By crossing the finite temperature phase transition we may switch the confinement mechanism off and on, which makes finite temperature QCD interesting for studies of the confinement mechanism. In Fig. 11 I show the abelian action density

$$\frac{\beta}{6}\sum_{i,\mu>\nu}\left(\frac{\langle\cos\theta_i(s,\mu,\nu)L(0)L^\dagger(R)\rangle}{\langle L(0)L^\dagger(R)\rangle}-\langle\cos\theta_i(s,\mu,\nu)\rangle\right) \tag{25}$$

below and above the phase transition, and for three different separations of the quarks each: 0.37, 0.64 and 0.97 fm (from top to bottom) in the deconfined phase, and 0.48, 0.82 and 1.24 fm in the confined phase, respectively. In the confined phase at $T/T_c = 0.98$ (right part of the figure) we see a flux tube at small separations, in accord with Fig. 10, while at separations $\gtrsim 0.5$ fm the string appears to be broken. At zero temperature string breaking occurs at ≈ 1.1 fm. One of the reasons for this difference is that the constituent quark mass decreases with increasing temperature, approaching zero at $T/T_c = 1$. In the deconfined phase at $T/T_c = 1.3$ (left part of the figure) we see no flux tube at all, not even at the smallest distance.

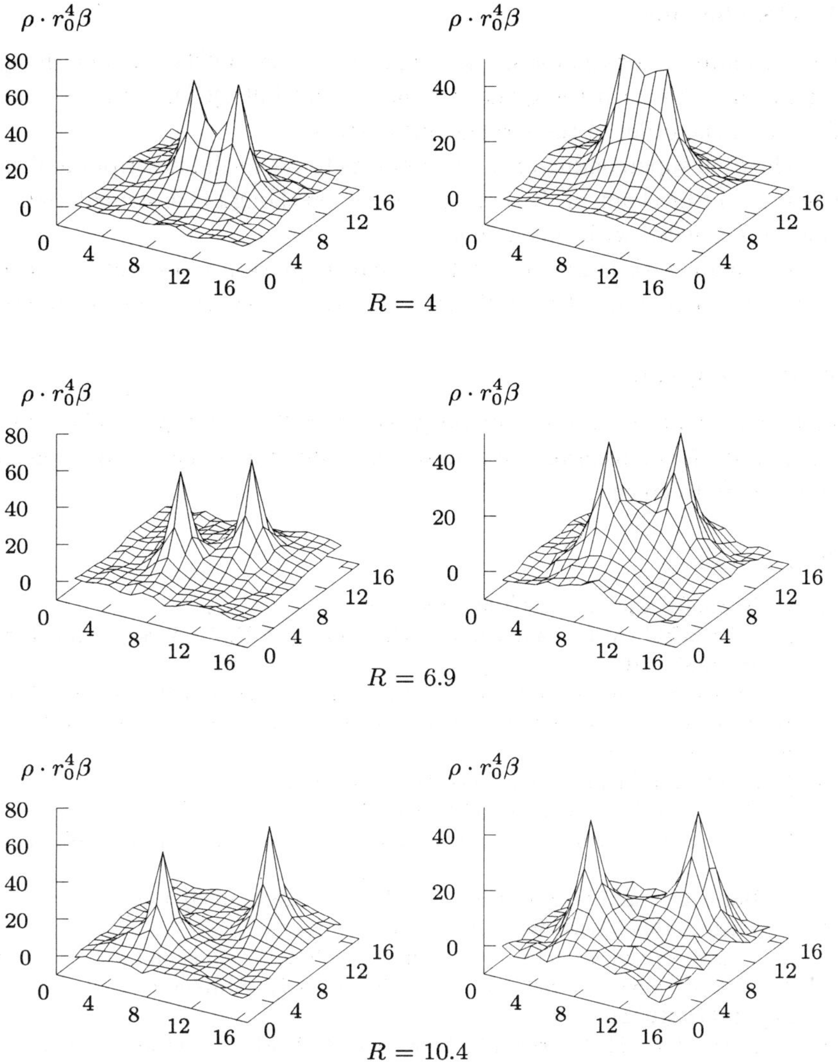

Figure 11. The action density in the deconfined phase at $T/T_c \approx 1.3$ (left) and the confined phase at $T/T_c = 0.98$ (right), respectively.

5. Conclusions

Unlike hadronic observables, the properties of the QCD vacuum change significantly between the quenched theory and full QCD, even for quark masses of the order of the strange quark mass.

The dual superconductor model remains to be a valid description of the confinement mechanism. It is an open question though how string breaking can be reconciled with this model.

The main task is now to explore smaller quark masses and attend to fermionic observables. Hopefully this will give us new clues and surprises.

Acknowledgement

I like to thank the organizers of *Confinement 2003*, and in particular Hideo Suganuma, for a pleasant and most successful conference, and hope for many more to come.

References

1. G. 't Hooft, Nucl. Phys. B190 (1981) 455.
2. A.S. Kronfeld, M.L. Laursen, G. Schierholz and U.-J. Wiese, Phys. Lett. B198 (1987) 516.
3. K. Amemiya and H. Suganuma, Phys. Rev. D60 (1999) 114509; V.G. Bornyakov, M.N. Chernodub, F.V. Gubarev, S.M. Morozov and M.I. Polikarpov 2072 (1995).
4. H. Shiba and Suzuki, Phys. Lett. B351 (1995) 519.
5. C. Wetterich, Phys. Lett. B462 (1999) 164.
6. S. Kato, S. Kitahara, N. Nakamura and T. Suzuki, Nucl. Phys. B520 (1998) 323.
7. V. Bornyakov and G. Schierholz, Phys. Lett. B384 (1996) 190.
8. E. Witten, Phys. Lett. B86 (1979) 283.
9. V.G. Bornyakov, H. Ichie, Y. Koma, Y. Mori, Y. Nakamura, D. Pleiter, M.I. Polikarpov, G. Schierholz, T. Streuer, H. Stüben and T. Suzuki, `hep-lat/0310011`.
10. V. Bornyakov, M. Chernodub, Y. Koma, Y. Mori, Y. Nakamura, M.I. Polikarpov, G. Schierholz, A. Slavnov, H. Stüben, T. Suzuki, P. Uvarov and A. Veselov, `hep-lat/0209157`.
11. V. Bornyakov, H. Ichie, Y. Koma, Y. Mori, Y. Nakamura, M.I. Polikarpov, G. Schierholz, T. Streuer and T. Suzuki, `hep-lat/0212023`.
12. V. Bornyakov, M. Chernodub, Y. Koma, Y. Mori, Y. Nakamura, M.I. Polikarpov, G. Schierholz, D. Sigaev, A. Slavnov, H. Stüben, T. Suzuki, P. Uvarov and A. Veselov, Nucl. Phys. A721 (2003) 930.
13. V. Bornyakov, M. Chernodub, Y. Koma, Y. Mori, Y. Nakamura, M.I. Polikarpov, G. Schierholz, A. Slavnov, H. Stüben, T. Suzuki, P. Uvarov and A. Veselov, `hep-lat/0309144`.

14. F. Brandstaeter, G. Schierholz and U.-J. Wiese, Phys. Lett. B272 (1991) 319.
15. A.S. Kronfeld, G. Schierholz and U.-J. Wiese, Nucl. Phys. B293 (1987) 461.
16. A. Hart and M. Teper, `hep-ph/0004180`.
17. H. Leutwyler and A. Smilga, Phys. Rev. D46 (1992) 5607.
18. A. Hasenfratz, Phys. Lett. B476 (2000) 188.
19. E.-M. Ilgenfritz, B.V. Martemyanov and M. Müller-Preussker, Phys. Rev. D62 (2000) 096004.
20. A. Hart and M. Teper, Phys. Lett. B371 (1966) 261.
21. T. Suzuki and I. Yotsuyanagi, Phys. Rev. D42 (1990) 4257.
22. T. Suzuki, in *Confinement 2000*, p. 120, eds. H. Suganuma, M. Fukushima and H. Toki (World Scientific, Singapore, 2001).
23. J.L. Cardy and E. Rabinovici, Nucl. Phys. B025 (1982) 1.
24. F. Karsch, E. Laermann and A. Peikert, Nucl. Phys. B605 (2001) 579.
25. M. Gao, Phys. Rev. D40 (1989) 2708; see also: P. de Forcrand, G. Schierholz, H. Schneider and M. Teper, Phys. Lett. B160 (1985) 137.

BRANES IN LATTICE SU(2) GLUODYNAMICS

V. I. ZAKHAROV
Max-Planck Institut für Physik
Werner-Heisenberg Institut
Föhringer Ring 6, 80805, Munich
E-mail: xxz@mppmu.mpg.de

In many cases, topological excitations become fundamental variables in a dual formulation of the theory. Assuming this is true in case of $SU(2)$ gluodynamics, we look for hints on the dual variables from the lattice simulations. Two-dimensional vortices, or branes with recently established properties seem to be most natural candidates. The total area of the branes scales in the physical units while the non-Abelian action is ultraviolet divergent. The branes are populated with magnetic monopoles, or tachyonic mode.

1. Introduction

Duality proves a powerful tool to get insight into dynamics of a system, for a review see, e.g., [1]. A recent landmark is a proof of the confinement using duality within certain supersymmetric Abelian field theories [2]. There exist also other examples of Abelian gauge theories which allow for dual formulations in terms of electric and magnetic variables [1]. As for non-Abelian theories, it has become clear that search for dualities goes in this case beyond field theories, see, in particular, [3]. This makes the search for dualities even a more profound problem.

Search for dual formulations is a theoretical challenge, first of all. Here we focus on a much less ambitious aspect of the problem, trying to understand what kind of hints on the dual variables can be extracted from the lattice simulations of $SU(2)$ gluodynamics. The point is that topological excitations found within the original formulation can become dual variables, see, e.g., [1]. Let us explain this on the Seiberg-Witten example. In $N = 2$ SUSY Abelian theory there exist classical monopole solutions with mass equal to

$$M_{mon} = \langle\phi\rangle \cdot Q_M \quad , \tag{1}$$

where ϕ is a scalar filed and Q_M is the magnetic charge which is inverse proportional to the electric charge, $Q_M = const/e$. Moreover, the solution saturates the BPS bound. The $N = 2$ supersymmetry ensures absence of quantum corrections to the mass. Thus, one can prove that the monopole is becoming light for large original couplings e. This observation can be considered as a hint that there exists a dual formulation in terms of the monopoles.

In case of gluodynamics, effective degrees of freedom responsible for the confinement are natural candidates for future dual variables. Phenomenologically, magnetic monopoles and P-vortices are most promising, for a recent review see [4]. Note that both monopoles and P-vortices condense which is a common sign of the dual variables.

In this paper we concentrate on field theoretical properties of the monopoles and vortices. Trying to appreciate the anatomy of the monopoles in terms of the Yang-Mills fields, we note first that apriori it would be easy to argue that the monopoles might not play any dynamical role. Indeed, there are two known types of Abelian monopoles, corresponding to singular and non-singular fields at the origin. We argue that generically both types of the Abelian monopoles cannot be generalized to the non-Abelian case. The only possibility left is a self tuning of the suppression due to the action and of enhancement due to the entropy. Both should be ultraviolet divergent but cancel each other to the order Λ_{QCD}.

There is no known mechanism for the self-tuning in the original Yang-Mills theory and the possibility looks pure academic. Amusingly enough, lattice data do provide strong evidence for such self-tuning. Theoretically, one can then go further and argue that self-tuning of the monopoles is not sufficient, even if it is granted by the data, and monopoles should belong to self-tuned two-dimensional surfaces. The data turn again to support this conclusion and the surfaces in point are the P-vortices, for review see [4]. The P-vortices are indeed closely associated with monopoles. What makes the self-tuning manifest is a recent observation that the P-vortices and monopoles have non-Abelian action which is divergent in the ultraviolet [5,6,7]. In other words, the lattice spacing a is seen in distribution of the non-Abelian action on and around the vortices. On the other hand, Λ_{QCD} controls the monopole mass and vortex string tension.

We will call these thin two-dimensional surfaces populated with tachyonic mode (monopoles) branes. (The term vortex is somehow reserved, in the studies of confinement, for thick, or bulky objects, see, e.g., [4,8].) In Sect. 2 we discuss self-tuning of the magnetic monopoles in the lattice $SU(2)$. In Sect. 3 we describe evidence in favor of self-tuning of the vortices. Conclusions are in Sect. 4. Note that the argumentation outlined above was already presented partly in Refs. [9,10].

2. Self-tuning of the lattice monopoles

2.1. *Fine tuning of the Abelian monopoles*

The size of the BPS monopole (1) is large, that is of order $\langle\phi\rangle^{-1}$. The softening of the fields is due to a non-trivial scalar field configuration. Now we turn to another well understood case of the monopole condensation, that is compact $U(1)$, for review see, e.g., [11]. Generically, the idea is the same as above: condensation of the monopoles is signaled by vanishing of the monopole mass. There are two important changes, however. First, there is no scalar field so that magnetic field is very strong at short distances. Second, lattice theories are considered in the Euclidean space.

In more detail, we consider electromagnetic field on the lattice. In the continuum limit the Lagrangian is that of free field:

$$L \;=\; \frac{1}{4e^2}F_{\mu\nu}F_{\mu\nu} \quad . \tag{2}$$

where $F_{\mu\nu}$ is the field strength tensor. Since there is no softening Higgs field the radiative mass of the monopole is divergent in the ultraviolet:

$$M_{rad} \;=\; \frac{1}{8\pi}\int_a^\infty d^3r|\mathbf{H}|^2 \;\approx\; \frac{const}{e^2}\frac{1}{a} \quad , \tag{3}$$

where a is an ultraviolet cut off, $\mathbf{H}$ is the magnetic field, $|\mathbf{H}| \sim 1/(er^2)$ and e is the electric charge appearing because of the Dirac quantization condition.

Because of the ultraviolet divergence (3), the monopoles can be considered point-like. The probability to observe a monopole trajectory of the length L is suppressed then by the action as

$$exp(\;-S) \;\sim\; \exp\Big(\;-\frac{const}{e^2}\frac{L}{a}\Big) \quad , \tag{4}$$

and at first sight the monopoles are removed from the physical spectrum in the limit $a \to 0$. Remarkably, this conclusion is actually not true. Indeed, the probability $W(L)$ to observe a trajectory of length L is a product of (4) and the entropy factor N_L which is the number of ways the length L can be realized on the lattice. The latter is also ultraviolet divergent and:

$$W(L) \;\sim\; \exp(-M_{ren}\cdot L) \quad , \tag{5}$$

where the renormalized mass is approximately:

$$M_{ren} \approx \frac{1}{a}(\frac{const}{e^2} - \ln 7) \ , \tag{6}$$

where the $\ln 7$ term is of pure geometrical origin, due to the entropy.

Monopole condensation occurs when any trajectory length is not suppressed by (5). In other words, choosing $e^2 = e^2_{crit}$, where

$$\frac{const}{e^2_{crit}} - \ln 7 = 0 \ , \tag{7}$$

ensures condensation of the monopoles. Validity of this condition was checked numerically, see [12] and references therein.

More generally, starting with the classical action for a point-like particle:

$$S_{cl} = M(a) \cdot L \ , \tag{8}$$

one can develop field theory in the Euclidean space-time in the polymer representation, see, e.g. [13]. In particular, evaluating the propagator as a path integral a la Feynman one arrives at the following expression for the propagating, or physical mass:

$$m^2_{phys} \approx \frac{8}{a}(M(a) - \ln 7/a) \ , \tag{9}$$

where the factor $\sim \ln 7$ is specific for the hyper-cubic lattice in $d = 4$ and for charged particles (closed trajectories).

In other words, the condensation occurs again at the point where the monopole mass vanishes, $m^2_{phys} = 0$. Moreover, the mass can be fine tuned to zero by choosing the corresponding value of the electric charge (7).

2.2. *No-go arguments for the $SU(2)$ case*

Imagine now that we would try to extract lessons from the Abelian cases and suggest a mechanism for having a light scalar particle (monopole) in the non-Abelian case. We would fail completely and come actually to a kind of a no-go theorem for the non-Abelian case.

Mild fields. It is reasonable to consider separately mild and hard gluonic fields. By mild fields we understand an analog of the Polyakov- 't Hooft monopole while by hard fields we understand an analog of the monopoles of the compact (lattice) $U(1)$.

There is no Higgs field in our case. Monopoles can still be defined following Ref. [14]. Namely, pick up any scalar field in the triplet representation, H^a. Then vanishing of this field,

$$H^a = 0 \ , \ (a = 1, 2, 3) \tag{10}$$

represent three conditions specifying a world line in $d = 4$ space. The central point is that there exists such a definition of the monopole charge that this trajectory is nothing else but the monopole trajectory [14].

The problem is to identify, in absence of matter, a suitable H^a field. Indeed, locally the field H^a is to reduce to a product of non-Abelian field strength tensors $G^a_{\mu\nu}$. The simplest possibility is to choose the Higgs field proportional to a particular component of the field strength tensor, say,

$$H^a \; = \; G^a_{12} \; .$$

This choice would violate Lorentz invariance, however. If we try a Lorentz scalar then the simplest combination vanishes identically:

$$\epsilon^{abc} G^b_{\mu\nu} G^c_{\mu\nu} \; \equiv \; 0 \; .$$

Increasing the number of the $G's$ in the product makes, first of all, the construction unrealistically complicated. Moreover, a more detailed consideration shows that it does not help at all [15].

Thus, there is no way to construct a light monopole in terms of mild, i.e. nonsingular fields.

Hard fields. Try now to construct a fine tuned monopole, in analogy with the lattice $U(1)$ theory. In the $U(1)$ case the price paid for the absence of the Higgs field is singularities of the gauge field. In particular, the magnetic current is defined in terms of violations of the Bianchi identities,

$$j^{mon}_{\mu} \; \sim \; \epsilon_{\mu\nu\rho\sigma} \partial_{\nu} F_{\rho\sigma} \; . \tag{11}$$

Violation of the Bianchi identities assumes the fields to be singular. However, all the expressions are regularized and well defined on the lattice [16].

In the $SU(2)$ case definition (11) can be generalized to a covariant form in an obvious way. Violation of the Bianchi identities means that the non-Abelian fields are singular in the continuum limit $a \to 0$. Singular fields imply a singular action. Singular action means that the corresponding fluctuation is removed from the physical spectrum unless the radiative mass is fine tuned to the entropy factor. So far, everything is similar to the $U(1)$ pure gauge theory. At this point, the analogy breaks down, however. A non-Abelian theory is an interacting theory even in the absence of matter. As a result the coupling is running and cannot be used as a parameter to tune the propagating mass (9).

Thus, we come to the conclusion that neither finite nor singular non-Abelian fields can ensure appearance of a light scalar particle (monopole).

2.3. *Lattice monopoles*

The no-go arguments above have been realized rather recently. Moreover, the phenomenological studies of the lattice monopoles flourish, for a review see, e.g., [17]. Now, however, we can look backwards and ask, why the no-go arguments did not work. This question can be answered quantitatively, by measuring directly the entropy and *non-Abelian* action of the monopoles (we always mean monopoles of the maximal Abelian projection, see, e.g., [17]). For the monopole action the result of measurements is [5]:

$$S_{mon} \approx \ln 7 \cdot \frac{L}{a} , \tag{12}$$

and for the entropy one gets [18]:

$$N_L \approx (L/a)^7 . \tag{13}$$

We do not discuss here error bars in Eqs (12), (13): details can be found in the original papers. However, the answer to our question is quite clear, compare (12), (13) and (9). Namely, the monopoles exist because of a self-tuning. Both the action and entropy are ultraviolet divergent but cancel each other so that the propagating monopole mass is controlled by Λ_{QCD}^{-1} [10]. Note that the action is tuned to a geometrical factor. On the other hand, the action of a pure Abelian monopole (embedded into the $SU(2)$) would be proportional to $g^{-2}(a)$ and could not be tuned to $\ln 7$ for varying a. The mechanism of the self-tuning in terms of the original Yang-Mills fields is not known. However, it follows from the arguments above that this self-tuning was the only "chance" for the monopoles to exist.

The self-tuning of the monopoles can be tested further, through studies of properties of the monopole clusters. In particular, one can predict the spectrum of finite clusters, $P(L)$ and their radii as functions of their length [19]:

$$P(L) \sim L^{-3} , \quad R(L) \sim \sqrt{L \cdot a} . \tag{14}$$

The prediction is known to be true on the lattice [20,21].

The main idea behind (14) is to use relations common to theory of percolating systems. For further applications of the percolation theory to monopoles see [21,22]. Physicswise, what is most intriguing about the success of the percolation theory is that it treats monopoles as point like at the scale of a. Thus, the monopoles emerge as a new probe of short distances.

3. Self-tuning of P-vortices

3.1. *No-go arguments, second round*

Although the self-tuning of the monopoles allows to describe a lot of data on the monopole clusters [19,21,22], it brings us to a new question which looks even more devastating than the original puzzles. Namely, if relations (12) and (13) is the ultimate truth about the lattice monopoles, then according to (9) we have a light scalar particle. But because of the asymptotic freedom we may have only three original gluons at short distances.

To scrutinize the argument we should translate the problem into the language of the observables, that is monopole trajectories. Note, first of all, that point-like particles in field theory are counted through ultraviolet divergences. Usually, in the continuum theory one deals with logarithmic divergences. In particular, the fact that there are only gluons at short distances can be checked by studying the running of the coupling.

However, the logarithmic running is a relatively small effect, difficult to resolve. Luckily enough, on the lattice one can measure also power-like divergences since the ultraviolet regularization is known explicitly. In particular, one can count particles by measuring the average plaquette action. The standard prediction is:

$$\langle (G^a_{\mu\nu})^2 \rangle \approx \frac{(N_c^2 - 1)}{a^4} \tag{15}$$

Indeed, in the limit $a \to 0$ the plaquette action is to count zero-point fluctuations of gluons. Prediction (15) is easy to check. Moreover, the relation (15) has perturbative corrections calculable, again, in terms of gluons alone. In fact, many perturbative terms are explicitly known [23].

The first non-perturbative contribution to the plaquette action which is allowed by the continuum theory is that of the ultraviolet renormalon, for a recent review and further references see [24]. This observation severely constrains the monopole contribution to the plaquette action:

$$\langle (G^a_{\mu\nu})^2 \rangle_{mon} \sim \Lambda^2_{QCD} \cdot a^{-2} . \tag{16}$$

To appreciate implications of (16) for the monopole trajectories we need to recollect some elements of the geometrical picture. Monopoles occupy centers of elementary cubes on the lattice. The cubes add up to trajectories. The action, associated with the monopole is measured as excess of action density on the plaquettes forming the cubes times the volume of the cube, a^3. In particular, (12) means that the excess of the action density on the plaquettes belonging to the monopoles is comparable to action density of the zero-point fluctuations (15).

Now, for the purpose of orientation consider the standard model of percolation, see, e.g., [25]. One introduces probability p for a cube to belong to a monopole trajectory. At some value $p = p_{cr}$ the trajectories begin to percolate, i.e. there appears a single cluster which spreads over the whole lattice. Such a phenomenon is observed for the lattice monopoles indeed. However, in this model, the total number of monopole cubes would be a finite fraction, p of all the cubes on the lattice. As a result, the excess of the action due to the monopoles, averaged over the lattice would be comparable to (15). and the monopoles would interfere with the counting degrees of freedom. We uncover again the danger that observations (12), (13) signify presence of new particles at short distances, in contradiction with the asymptotic freedom.

Thus, ordinary percolation is not allowed for the monopoles. And what is allowed? Introduce probability $\theta(cube)$ for an arbitrary cube on the lattice to contain a monopole. Then (16) implies:

$$\theta\ (cube)\ \sim\ \left(a \cdot \Lambda_{QCD}\right)^2\ . \tag{17}$$

That is, the probability for a cube to contain a monopole with ultraviolet divergent action should itself depend on a. The prediction might look even absurd at first sight but the drama is that the data at presently available lattices [21,26] do comply with (17).

3.2. *Branes*

Although observation of (17) on the lattice settles our problem with asymptotic freedom, the overall picture might look still very puzzling. Namely, the lattice monopoles share the properties (12), (13) with a free particle in $d = 4$ space. On the other hand, the probability (17) depends on the lattice spacing a, in striking distinction from the free particle in $d = 4$. The geometrical meaning of (17) is actually transparent: monopoles live on a $d = 2$ subspace of the whole $d = 4$ space. Indeed, the factor $(a \cdot \Lambda_{QCD})^2$ gives then the weight of the cubes which belong to the surface provided that the area of the surface is in physical units.

The conclusion might sound bizarre but actually phenomenologically this picture, at least partially, has been known. Namely, the surfaces are nothing else but the P-vortices, for review see [4]. Their total area scales in physical units [27,4]. According to the latest data [6]:

$$A_{vort} \approx 24\,(fm)^{-2} \cdot V_4\,, \tag{18}$$

where V_4 is the lattice volume. Moreover it is known from the numerical simulations that monopoles are strongly correlated with the P-vortices. This was observed first for a single value of the lattice spacing a [28] and later confirmed for all the values of a available [7].

Although scaling (18) and conspiracy of the P-vortices and monopoles were known since some time, no self-tuning was discussed until very recently [6]. The missing point was the measurement of the vortex action as function of the lattice spacing a. The (non-Abelian) action turned to be ultraviolet divergent [6]:

$$S_{vort} \approx 0.54 \frac{A}{a^2}\,. \tag{19}$$

Observation (19) means that the self-tuning is generalized from the monopoles to P-vortices [6]. Moreover, conspiracy of the monopoles and P-vortices is elevated from a simple phenomenological observation to a consequence of the asymptotic freedom.

3.3. *Interplay of the monopoles and vortices*

Let us emphasize that it would be an oversimplification to say that the vortices are primary and monopoles are secondary. The interplay between the monopoles and vortices is subtler. Namely, at short distances monopoles 'know' about $d = 4$. Indeed, the spectrum of finite clusters (14) is sensitive to the number of dimensions:

$$N(L) \sim L^{-1-d/2}\,, \tag{20}$$

and the data indicate $d = 4$, [19,21]. Also, it is the monopole action which is tuned to the entropy of a point-like particle in $d = 4$, compare (12) and (13). Note that both the action and entropy are decided at short distances. There is no such match for the vortices if one would assume that the vortices are structureless [6].

On the other hand, in the infrared the monopoles are associated with a $d = 2$ subspace of the full $d = 4$ space. Thus, it is condensation of the branes that is observed, not of monopoles.

4. Conclusions

The lattice data indicate that there exist branes in the vacuum state of the lattice $SU(2)$ gluodynamics which possess remarkable properties:

(a) total area scales in physical units;
(b) non-Abelian action is ultraviolet divergent;
(c) branes are populated with scalar particles, monopoles;
(d) monopoles are particles living on a $d = 2$ subspace;
(e) the branes percolate and condense.

The branes are natural candidates for dual variables.

An important caveat is that in all the cases the evidence is entirely numerical and is, therefore, true for a limited range of the lattice spacings available now. The only theoretical support for the picture observed we could find is that if existence of a dual description were granted then the basic features of the branes listed above would be uniquely fixed.

Appearance of the branes might be related to a very special status of dimension two condensates in gauge theories , see, in particular, [29]. Naively, the first gauge-invariant condensate, $< (G^2_{\mu\nu} >$ has dimension four. The only Lorentz invariant operator of dimension two, $(A^a_\mu)^2$ is seemingly gauge dependent. However, one can argue that the minimal value of $< (A^a_\mu)^2 >$, over the gauge orbit is gauge invariant. This minimization condition implies non-locality. In case of pure electromagnetic field

$$< A^2_\mu >_{min} \;=\; (const) \int \frac{d^4x d^4x'}{(x - x')^2} F_{\mu\nu}(x) F_{\mu\nu}(x') \;. \tag{21}$$

In the non-Abelian case there is no such closed relation but an explicit calculation in the Hamiltonian formalizm [30] demonstrates sensitivity of the dimension-two condensate to the Gribov horizon.

Moreover, one can readily argue that the ultraviolet divergent part of $< (A^a_\mu)^2 >_{min}$ cannot be relevant to physical observables. Thus, we are left with the non-perturbative part of $< (A^a_\mu)^2 >_{min}$ which is of order Λ^2_{QCD}. As we argued in [10] the monopoles realize

$$< |\phi_M|^2 > \;\sim\; \Lambda^2_{QCD} \;, \tag{22}$$

where ϕ_M is magnetically charged field. Properties of the branes seem to reflect the balance of locality and non-locality inherent to the gauge invariant operator of dimension 2 in gauge theories.

Acknowledgments

This note grew out of discussions of the results of the lattice measurements presented in [5,6,7,18,21]. I am thankful to V.G. Bornyakov, P.Yu. Boyko, M.N. Chernodub, F.V. Gubarev, A.V. Kovalenko, M.I. Polikarpov, T. Suzuki and S.N. Syritsyn for collaboration and thorough discussions.

I am thankful to A. Bartl, M. Faber, A. Di Giacomo, J. Greensite, J. Kubo, P. van Nieuwenhuizen, A.A. Slavnov, L. Stodolsky, H. Terao and E.T. Tomboulis for interesting discussions.

The paper was completed while the author was visiting the Yukawa Institute for Theoretical Physics of the Kyoto University. I am thankful to Prof. T. Kugo and other members of the group for the hospitality.

The material of this note was presented as a talk at the Conference "Confinement 2003" at RIKEN (Japan), July 21-24, (2003). I am thankful to the organizers, and especially to Prof. H. Suganuma, for their efforts and invitation .

References

1. R. Savit, *Rev. Mod. Phys.* **52**, 453 (1980);
   A. Di Giacomo, *"Color confinement and dual superconductivity: an update"*, hep-lat/0204032.
2. N. Seiberg and E. Witten, *Nucl. Phys.* **B426**, 19 (1994), (hep-th/9407087); *Nucl. Phys.* **B431**, 484 (1994), (hep-th/9408099).
3. J.M. Maldacena, *Adv. Theor. Math.* **2**, 231 (1998), (hep-th/9711200);
   A.M. Polyakov, *Int. J. Mod. Phys.* **A14**, 645 (1999), (hep-th/9809057).
4. J. Greensite, *Prog. Part. Nucl. Phys.* **51**, 1 (2003), (hep-lat/0301023).
5. V.G. Bornyakov *et al.*, *Phys. Lett.* **B537** 291 (2002).
6. F.V. Gubarev, A.V. Kovalenko, M.I. Polikarpov, S.N. Syritsyn and V.I. Zakharov, *" Fine tuned vortices in lattice SU(2) gluodynamics"*, hep-lat/0212003; *"Self-tuning of the P-vortices"*, hep-lat/0309034.
7. A.V. Kovalenko, M.I. Polikarpov, S.N. Syritsyn and V.I. Zakharov, *"Interplay of monopoles and P-vortices"*, hep-lat/0309032.
8. T.G. Kovacs and E.T. Tomboulis, *"Vortices and confinement"*, hep-lat/9912051.
9. F.V. Gubarev and V.I. Zakharov, *"Interpreting the lattice monopoles in the continuum limit"*, hep-lat/0211033.
10. V.I. Zakharov, *"Hidden mass hierarchy in QCD"*, hep-ph/0204040; *" 'Branes' in the lattice SU(2) gluodynamics"*; hep-ph/0306361, *"Fine tuning in lattice SU(2) gluodynamics vs continuum-theory constraints"*, hep-ph/0306362.
11. M.E. Peskin, *Ann. Phys.* **113**, 122 (1978).
12. H. Shiba and T. Suzuki, *Phys. Lett.* **B333**, 461 (1994).
13. K. Symanzik, *in* "Local Quantum Theory", (1969) Varenna International

School of Physics, Course XLV, p 152;
A.M. Polyakov, "Gauge Fields and Strings", (1987), Ch. 9;
J. Ambjorn, *"Quantization of geometry"*, hep-th/9411179.

14. G. 't Hooft, *Nucl.Phys.* **B190**, 455 (1981).
15. A. Kovner, M. Lavelle and D. McMullan, *JHEP* **0212**, 045 (2002).
16. T.A. DeGrand and D. Toussaint, *Phys. Rev.* **D22**, 2478 (1980).
17. H. Ichie and H. Suganuma, *"Dual Higgs theory for coloe confinement in quantum chromodynamics"*, hep-lat/9906005.
18. P.Yu. Boyko, M.I. Polikarpov and V.I. Zakharov, *"Geometry of percolating monopole clusters"*, hep-lat/0209075.
19. M.N. Chernodub and V.I. Zakharov, *Nucl. Phys.* **B669**, 233 (2003) (hep-th/0211267).
20. A. Hart and M. Teper, *Phys. Rev.* **D58**, 014504 (1998).
21. V.G. Bornyakov, P.Yu. Boyko, M.I. Polikarpov and V.I. Zakharov, *"Monopole clusters at short and large distance"*, hep-lat/0305021; *"Geometry of the monopole clusters on different scales"*, hep-lat/0309021.
22. M.N. Chernodub, K. Ishiguro, K. Kobayashi and T. Suzuki, *"Numerical determination of monopole entropy in pure $SU(2)$ QCD*, hep-lat/0306001;
T. Suzuki, M.N. Chernodub and K. Ishiguro, *"Energy-entropy study of projected space-like monopoles in finite-T quenched QCD*, hep-lat/0309018.
23. G. Burgio, F. Di Renzo, G. Marchesini and E. Onofri, *Phys. Lett.* **B422**, 219 (1998), (hep-ph/9706209);
R. Horsley, P.E.L. Rakow and G. Schierholz, *Nucl. Phys. Proc. Suppl.* **106**, 870 (2002), (hep-lat/0110210).
24. V.I. Zakharov, *"Non-perturbative match of ultraviolet renormalon"*, hep-ph/0309178.
25. G. Grimmett, *"Percolation"*, A series of comprehensive studies in mathematics, vol. 321, Springer (1999).
26. V.G. Bornyakov and M. Müller-Preussker, *Nucl. Phys.* **B106**, 646 (2002).
27. L. Del Debbio, M. Faber, J. Greensite and S. Olejnik, *Phys. Rev.* **D55**, 2298 (1997), (hep-lat/9610005);
K. Langfeld, H. Reinhardt and O. Tennert, *Phys. Lett.* **B419**, 317 (1998).
28. J. Ambjorn, J. Giedt and J. Greensite, *JHEP* **0002**, 033 (2000) (hep-lat/9907021).
29. F.V. Gubarev, L. Stodolsky, and V.I. Zakharov, *Phys. Rev. Lett.* **86**, 2220 (2001) (hep-ph/0010057);
F.V. Gubarev and V.I. Zakharov, *Phys. Lett.* **B501**, 28 (2001) (hep-ph/0010096);
K.-I. Kondo, *Phys.Lett.* **B514**, 335 (2001) (hep-th/0105299);
K.-I. Kondo, T. Murakami, T. Shinohara and T. Imai, *Phys. Rev* **D65**, 085034 (2002) (hep-th/ 011256);
H. Verschelde, K. Knecht, K. Van Acoleyen and M. Vanderkelen, *Phys. Lett.* **B516**, 307 (2001) (hep-th/0105018);
D. Dudal et al., *Phys. Lett.* **B569**, 57 (2003) (hep-th/0306116).
30. L. Stodolsky, P. van Baal and V.I. Zakharov, *Phys. Lett.* **B552**, 214 (2003).

Nonperturbative Aspects in QCD

ARE GLUEBALLS KNOTTED CLOSED STRINGS ?

ANTTI J. NIEMI
Department of Theoretical Physics, Uppsala University,
Box 803, S-75 108 Uppsala, Sweden
e-mail:Antti.Niemi@teorfys.uu.se

Glueballs have a natural interpretation as closed strings in Yang-Mills theory. Their stability requires that the string carries a nontrivial twist, or then it is knotted. Since a twist can be either left-handed or right-handed, this implies that the glueball spectrum must be degenerate. This degeneracy becomes consistent with experimental observations, when we identify the $\eta_L(1410)$ component of the $\eta(1440)$ pseudoscalar as a 0^{-+} glueball, degenerate in mass with the widely accepted 0^{++} glueball $f_0(1500)$. In addition of qualitative similarities, we find that these two states also share quantitative similarity in terms of equal production ratios, which we view as further evidence that their structures must be very similar. We explain how our string picture of glueballs can be obtained from Yang-Mills theory, by employing a decomposed gauge field. We also consider various experimental consequences of our proposal, including the interactions between glueballs and quarks and the possibility to employ glueballs as probes for extra dimensions: The coupling of strong interactions to higher dimensions seems to imply that absolute color confinement becomes lost.

Lord Kelvin was first to suggest that knotted strings could have fundamental importance. In Ref.[1] he proposed that atoms, which at the time were considered as elementary particles, could be interpreted as knotted vortex tubes in aether. Subsequently he also conjectured[2] that vortex filaments in the shape of torus knots should be stable. Kelvin's theory of vortex atoms has long ago subsided. However, at the time it was taken seriously which led to an extensive study and classification of knots. In particular the results obtained by Tait[3] remain a classic contribution to mathematical knot theory.

Presently, it is commonly accepted that fundamental interactions are described by string theories,[4] with different elementary particles corresponding to the vibrational excitations of a primary string. But the relevance of nontrivial string topology such as an appropriate extension of the concept of a knot in a higher dimensional string (membrane) has until now been studied sparsely. However, there are some obvious connections between string theory and knot theory, for example the Chern-Simons action is related both to conformal field theories and three dimensional knot invariants.[5] Hopefully one day an intimate relationship between knots and nonperturbative strings will be established.

Even though the relevance of knots at the level of a fundamental string theory remains open, in strong interaction physics we have good reasons to believe that knotted strings are present. Indeed, it is widely accepted, and also confirmed by detailed lattice simulations, that in QCD the collective excitations of gluons lead to the formation of a confining string between two widely separated quarks.[6] Consequently it is only natural to expect that *glueballs* could be interpreted as closed configurations of this string, emitted *e.g.* by a fluctuation in a (relatively) long open string that connects two widely separated quarks. Since glueballs relate directly to such fundamental issues in strong interaction physics as confinement and the formation of mass gap in Yang-Mills theory, their theoretical and experimental study is centrally important for our understanding of the properties of hadronic particles, the source of most known mass in the Universe.

A homogeneous and structureless linear string has an energy which is directly proportional to its length. In the absense of other contributions to the energy besides the linear tension, a closed string that has for example a toroidal shape (unknot) is then unstable, it shrinks away by minimizing the length. This is what happens to closed vortex loops for example in the abelian Higgs model with a single complex field, an isolated closed toroidal vortex ring in type-II superconductor is unstable against shrinkage.

But if a glueball corresponds to a closed string of gluonic flux, it must be stable within the Yang-Mills part of QCD. This is a direct consequence of a mass gap, which prevents free massless gluons from appearing in the spectrum. Consequently in the absence of quarks, or in the limit of heavy quark masses, there is nothing to which a closed gluonic fluxtube could decay into. This means that in the case of Yang-Mills theories there must be additional contributions to the energy of a closed gluonic string besides the linear string tension. Since lattice simulations indicate that a straight, linear gluonic string can only be subject to a (classically) linear tension,[6] any additional force which stabilizes *e.g.* a closed toroidal string against shrinkage must have a geometric origin, it is present in a toroidal configuration but absent from a straight linear string. The natural source for such a stabilizing force is in the three dimensional extrinsic geometry of the fluxtube, the way how it twists and bends. In fact, suppose that we bend a finite length linear piece of a fluxtube so that it forms a toroidal ring *i.e.* an unknot. If we twist the fluxtube once around its core before joining the ends, we can expect that there is a centrifugal twist contribution to the energy which ensures the stability of the toroidal fluxtube against shrinkage.[7] Indeed, from general considerations[8] which can be interpreted as amounting to universality arguments in the sense of renormalization groups, we expect that the leading order contribution to the energy from the bending and twisting of a linear fluxtube scales inversely to the length L of the fluxtube. Together with the linear tension we then expect in the limit of long closed fluxtubes[9]

$$E = \sigma L + \frac{a}{L}$$

where σ is the (linear) string tension, and $a \geq 0$ is a parameter characterising the (average) twisting and bending contributions to the energy. From this we find that the energy of the fluxtube attains its minimum at a non-vanishing value of L,

$$L_{min} = \sqrt{\frac{a}{\sigma}}$$

This means that a closed, possibly knotted string becomes stabilized against shrinkage, with a finite value of length L_{min}.

But the present consideration also reveals an obvious degeneracy in the energy spectrum of stable closed fluxtubes. Namely, before we join the ends of an open fluxtube to form the closed configuration, the twist that we introduce can be either left-handed (L) or right-handed (R). In the absence of any parity symmetry breaking terms, we then have two different but stable configurations corresponding to the left-twisted and the right-twisted toroidal string (unknot configuration). If Yang-Mills theory is indeed capable of describing left-handed and right-handed twist, in the absence of any parity violating term in the Hamiltonian we expect that in the quantum theory we have a twofold mass-denegeracy in the eigenstates of the Yang-Mills Hamiltonian, the left-handed twisted closed fluxtube $|L>$ and the right-handed twisted closed fluxtube $|R>$. Specifically, a parity operator P relates these two quantum states,

$$P|L>=|R> \qquad \text{and} \qquad P|R>=|L>$$

and we can diagonalize P by setting

$$|\pm> = \frac{1}{\sqrt{2}}(|L> \pm |R>)$$

with

$$P|\pm> = \pm|\pm>$$

and if

$$[P, H_{ym}] \; = \; 0$$

the energy spectrum of the Yang-Mills hamiltonian H_{ym} is then $P-$degenerate. In particular, for each parity-even closed string there must be a parity-odd closed string. This means that *if glueballs are closed, knotted fluxtubes which are stable within the pure Yang-Mills part of QCD, they must reflect the two-fold twist-parity degeneracy.*[10]

A parity degenerate glueball spectrum in a Yang-Mills theory can actually be considered natural in the following sense: At very high energy the Yang-Mills theory describes asymptotically free massless gluons. These gluons can be polarized in two different ways, they can be chosen to be either left-handed polarized or right-handed polarized. But this implies a doubling in the spectrum which is not akin the doubling we obtain by twisting the string in either left-handed or right-handed manner. Indeed, when we move towards the low energy phase, if the polarization doubling survives the ensuing phase transition it is only natural that it is reflected in a left-right doubling of the glueball spectrum.

Unfortunately, such a parity degeneracy in the glueball spectrum is not consistent with lattice computations, which predict a highly non-degenerate spectrum.[11] Specifically, lattice computations suggest that the lowest energy glueball is a $J^{PC} = 0^{++}$ state with a mass of 1611 ± 163 MeV, while the parity transformed $J^{PC} = 0^{-+}$ state has a mass which is in the vicinity of 2500 MeV. This is even above the lattice prediction of the mass of 2^{++} glueball which is about 2232 ± 310 MeV.

In lattice simulations there does not appear to be any direct evidence why glueballs should be interpreted as stable closed strings. Consequently there is no intrinsic reason for a parity doubling in the spectrum. But the interpretation of a glueball as a closed string is not inconsistent with lattice computations, maybe one just does not know how to describe a twisted closed gluonic string on a lattice. Thus it could simply be, that this aspect of glueballs has not yet emerged. But if lattice computations will eventually identify glueballs as closed gluonic strings which is quite possible, we still have the puzzle to explain the stability of a glueball in the infinite quark mass limit in a manner which is consistent with the computed, apparently highly asymmetric mass spectrum with respect to twisting.

In the light of these lattice computations, at the moment it appears to be very brave to conclude that our proposal to interpret glueballs as twisted closed strings is theoretically viable. But when we make comparisons with experimental results, we arrive at a conclusion that there is nothing apparently inconsistent with our proposal.[10] For this we note that there is a wide experimental agreement on the reliability of lattice computations to the extent, that the lowest mass glueball 0^{++} should have a mass which is about the same as the mass of isoscalar $q\bar{q}$ mesons. Since these can be grouped into SU(3) flavor nonets any candidate for a glueball should appear as a supernumerous state. In addition, a glueball candidate should be consistent with the following experimental signatures:[12]

i) Glueballs should be produced in $p\bar{p}$ annihilation processes, since a $q\bar{q}$ annihilation leads to a gluon-rich environment which strongly favors the formation of gluonic degrees of freedom. In fact, due to the LEAR experiment the nucleon-antinucleon annihilation processes are now the dominant source of data for studying the glueball formation.

ii) The central region of various other high-energy hadron-hadron scattering processes should similarly favor the production of glueballs since it is neither populated by target nor by beam quarks.

iii) The radiative decay of quarkonium states and especially the radiative J/ψ decay should be a prime source of glueballs.

iv) The decay branching fractions of a glueball should be incompatible with SU(3) predictions for $q\bar{q}$ states.

v) Since glueballs have no direct coupling to electromagnetism, they should be absent in any $\gamma\gamma$ collision processes.

The relatively narrow state $f_0(1500)$ is widely considered as the best available candidate for the lowest mass 0^{++} glueball.[12] It does fullfill all of the experimental criteria *i)-v)*, and it also has a mass which fits well the lattice prediction for the 0^{++} glueball state.

The lattice predicts further, that the next glueball should be a 2^{++} state with a mass of $2232 \pm\ 310$MeV.[11] This prediction seems to fit in a wonderful manner the observation of $f_J(2220)$, which has been for this reason proposed as a candidate for the 2^{++} glueball.[12] But with new data from $p\bar{p}$ annihilation processes the existence of a $f_J(2220)$ state is in serious doubt, it is fading away.[12] Consequently at the present there does not seem to be any widely accepted candidate for the 2^{++} glueball, with a mass which is consistent with the computed lattice glueball spectrum.

While the lattice prediction of the 0^{++} glueball appears to be almost perfect, the apparent difficulty of lattice to predict the mass of a 2^{++} glueball could indicate that a full understanding of glueball spectrum remains to be achieved. Indeed, at the moment our experimental observations on the glueball spectrum are also quite lacking and marred with puzzling observations. Notorious in this respect is the $\eta(1440)$, a $J^{PC} = 0^{-+}$ pseudoscalar which is most likely a mixture of two particles, the $\eta_H(1480)$ and the $\eta_L(1410)$.[12]

The $\eta_H(1480)$ couples strongly to kaons. It seems to be an almost ideal mix of $s\bar{s}$ which makes it a likely candidate for the $s\bar{s}$ member of the pseudoscalar nonet. But this implies that the $\eta_L(1410)$ becomes supernumerous in this nonet. Consequently it is a candidate for exotics: In a full analogy with the nearby $f_0(1500)$, the $\eta_L(1410)$ is a relatively narrow state that fullfills *all* of the above criteria *i)-v)* expected from a glueball. But since lattice computations predict that the mass of a 0^{-+} glueball should be in the vicinity of 2500 MeV, in the literature[13] this has been used as an argument to suggest that $\eta_L(1410)$ could be a new degree of freedom which is beyond the Standard Model QCD. According to Ref.[13] it could even relate to a bound state of light gluinos.

Here we consider a much more conventional possibility, with no need to go beyond the Standard Model, by interpreting the $\eta_L(1410)$ as the $J^{PC} = 0^{-+}$ glueball state.[10] In fact, we propose that in the picture where we view glueballs as stable closed strings, the $\eta_L(1410)$ has a natural interpretation as the parity-odd partner of the 0^{++} glueball corresponding to $f_0(1500)$. Consequently the experimental investigation of the $\eta_L(1410)$ state becomes a direct test of our theoretical understanding of fluxtube structures in QCD.

We note that the slight 90 MeV mass difference between the $\eta_L(1410)$ and the $f_0(1500)$ is quite insignificant. Theoretically, this mass difference could signal the presence of a small CP-violating θ-term in the QCD action. Experimentally, it is also well known that interferences with nearby resonances can easily shift masses by as much as 5%. Consequently the quoted values for the $f_0(1500)$, which are in the range of 1445 – 1560 MeV dependending on the reaction channels, are not necessarily inconsistent with a mass degeneracy with $\eta_L(1410)$. In fact, since $f_0(1500)$ is known to be heavily mixed with nearby meson states a precise experimental mass comparison becomes quite delicate.

Besides similarity in mass and in the qualitative criteria *i)-v)* which are also all satisfied by $\eta_L(1410)$, the $f_0(1500)$ and $\eta_L(1410)$ have also additional quantitative similarities, such as production ratios. These are clear indicatives that the two states must have very similar internal structures. For example, in $p\bar{p}$ annihilations the production of $f_0(1500)$ has the following measured branching ratios:[14]

$$\text{BF}\left(\frac{p\bar{p} \to f_0(1500)}{p\bar{p} \to 3\pi^0}\right) = (13 \pm 4)\%$$

$$\text{BF}(p\bar{p} \to 3\pi^0) = (5.5 \pm 1.0) \times 10^{-3}$$

When we take into account that $2\pi^0$ decays represent $(9.3 \pm 2.5)\%$ of all $f_0(1500)$ decays[15] we conclude that

$$\text{BF}(p\bar{p} \to f_0(1500)) = (7.7 \pm 3.8) \times 10^{-3} \tag{1}$$

In our annihilation rate into $3\pi^0$ we have accounted for the latest value for the annihilation into $2\pi^0$ from the Crystal Barrel Collaboration.[16] But we also note that the OBELIX collaboration quotes a value which is more than a factor of 2 smaller. There is also a disagreement that influences the normalization of our production ratio.[17,18]

For $\eta_L(1410)$, the $K\bar{K}\pi$ and $\eta\pi\pi$ decay modes are expected to be the dominant.[13] Both have been measured in $p\bar{p}$ annihilations, with results[19]

$$\text{BF}\left(\frac{p\bar{p} \to \eta_L(1410)\pi\pi}{\eta_L(1410) \to \eta\pi\pi}\right) = (3.3 \pm 1.0) \times 10^{-3}$$

and[20]

$$\text{BF}\left(\frac{p\bar{p} \to \eta_L(1410)\pi\pi}{\eta_L(1410) \to K\bar{K}\pi}\right) = (2.0 \pm 0.2) \times 10^{-3}$$

When we add these, we find

$$\text{BF}(p\bar{p} \to \eta_L(1410)\pi\pi) = (5.3 \pm 1.7) \times 10^{-3} \tag{2}$$

In comparing with (1) we conclude that the $p\bar{p}$ annihilation production rates for our two glueball candidates $f_0(1500)$ and $\eta_L(1410)$ in $p\bar{p}$ are remarkably equal, when accounting for experimental uncertainties.

We have also compared the observed production rates in radiative J/ψ decays. The branching fraction for the production of the $f_0(1500)$ and its subsequent decay into 4π has been measured to be $(8.2 \pm 1.7) \times 10^{-4}$, with 4π decays accounting for $61.7 \pm 9.6\%$ of all $f_0(1500)$ decays.[21] Therefore, we expect a branching fraction of

$$\text{BF}(J/\psi \to \gamma f_0(1500)) = (1.3 \pm 0.3) \times 10^{-3}$$

The branching fraction for the production of the $\eta_L(1410)$ can be determined again from its decays into $K\bar{K}\pi$ [22] and $\eta\pi\pi$.[23] Adding the results leads to a branching fraction

$$\text{BF}(J/\psi \to \gamma\eta_L(1410)) = (1.0 \pm 0.46) \times 10^{-3}$$

Again, we find that in the case of radiative J/ψ decays the production rates of $f_0(1500)$ and $\eta_L(1410)$ are remarkably similar within experimental uncertainties.

We find that these similarities in the rates, the relative narrowness of both $f_0(1500)$ and $\eta_L(1410)$, and the fact that both satisfy all of the overall criteria $i) - v)$ are quite remarkable. To us this strongly suggests that these two states must have very similar internal structures. Since there is wide agreement to interpret $f_0(1500)$ as a 0^{++} glueball, the natural interpretation of $\eta_L(1410)$ is then in terms of a 0^{-+} glueball. As it is furthermore (essentially) degenerate in mass with $f_0(1500)$, the natural explanation is that these two states are the parity even and odd linear combinations of closed toroidal strings with a single left-handed and right-handed twist.[10] Indeed, the present proposal to identify $\eta_L(1410)$ as a 0^{-+} glueball which is denegerate in mass with the 0^{++} glueball $f_0(1500)$ is clearly very consistent with the picture that glueballs correspond to closed strings. However since our proposal to interpret $\eta_L(1410)$ within the Standard Model is somewhat unorthodox and our conclusion that it is a glueball is not consistent with lattice predictions, there is an obvious need for some theoretical backing. For this we shall now explain how our model of glueballs could be supported by the structure of a Yang-Mills theory, and in particular how the presence of the twisting-degree of freedom and the formation of closed fluxtubes can be realized.

In order to identify the twisting degree of freedom directly in the Yang-Mills theory, we employ a decomposition of the gauge field A_μ^a.[24,25,26] The decomposition of vectors and tensors in terms of their irreducible components is a common problem in Physics. For example, it is widely employed in fluid dynamics where the velocity three vector decomposes into its gradient and vorticity components, and in classical electrodynamics the four dimensional Maxwellian field strength tensor $F_{\mu\nu}$ becomes dissected into its electric and magnetic components. For notational simplicity but without any loss of generality we consider here a decomposition of A_μ^a in a SU(2) Yang-Mills theory. This decomposition[26] interprets the Cartan component A_μ^3 as a $U(1) \in SU(2)$ gauge field. Then

$$A_\mu^+ = A_\mu^1 + iA_\mu^2$$

together with its complex conjugate transforms as charged vector field under the corresponding U(1) gauge transformation, corresponding to diagonal SU(2) gauge rotations. The two vectors A_μ^1 and A_μ^2 are in a plane in a four dimensional space. We parametrize this plane using a *twobein* ${e^a}_\mu$

$(a = 1, 2)$ with

$$e^a{}_\mu e^b{}_\nu = \delta^{ab}$$

We define

$$\mathbf{e}_\mu = 1/\sqrt{2}(e^1{}_\mu + ie^2{}_\mu)$$

With these, we can then represent the most general A_μ^+ as

$$A_\mu^1 + iA_\mu^2 \;=\; i\psi_1 \mathbf{e}_\mu + i\psi_2 \mathbf{e}_\mu^\star$$

Here ψ_1 and ψ_2 are two complex fields; see Ref.[26] for details. We define

$$\rho^2 = |\psi_1|^2 + |\psi_2|^2$$

and introduce a three-component unit vector $\vec{n}$ by

$$\vec{n} \;=\; \frac{1}{\rho^2}(\psi_1^\star \; \psi_2^\star)\vec{\sigma}\begin{pmatrix}\psi_1 \\ \psi_2\end{pmatrix} \tag{3}$$

Here $\vec{\sigma}$ are the standard Pauli matrices. We substitute the ensuing decomposed fields in the Yang-Mills action, using the background gauge condition

$$D_\mu^{ab}(A^3)A_\mu^b = 0 \tag{4}$$

w.r.t. the Cartan component A_μ^3. When we keep *only* terms which involve ρ and $\vec{n}$, we get[26,27,28]

$$-\frac{1}{4}\int d^4x F_{\mu\nu}^2 \tag{5}$$
$$= \int d^4x \left\{ (\partial_\mu \rho)^2 + \rho^2(\partial_\mu n^a)^2 + \frac{1}{4}(\epsilon_{abc} n^a \partial_\mu n^b \partial_\nu n^c)^2 + V(\rho^2, \vec{h}\cdot\vec{n}) + \ldots \right\}$$

Here V is a potential term that involves some of the additional terms that we have deleted. This potential term breaks the global $O(3)$ invariance of the first three terms on the *r.h.s.* under rigid rotations of the vector $\vec{n}$. This is necessary, as it gives a mass to the two Goldstone bosons corresponding to the breaking of the global $O(3)$ by boundary condition to $\vec{n}$ at spatial infinity.[26]

The function ρ can be interpreted as the average density of the two scalars ψ_1 and ψ_2,

$$\int d^4x \; (A_\mu^1 A_\mu^1 + A_\mu^2 A_\mu^2) \;=\; \int d^4x \; (|\psi_1|^2 + |\psi_2|^2) \;=\; \int d^4x \; \rho^2 \tag{6}$$

This is *a priori* a gauge dependent quantity, but when we minimize (6) along gauge orbits the minimization selects *exactly* the background gauge condition (4). This means in particular, that the minimum value ρ_{min} of

(6) is a gauge invariant quantity.[29] A one-loop computation suggests that this minimum value $<\rho_{min}>$ is nonvanishing.[30]

For $<\rho^2>\neq 0$ the action (6) is known to support stable, knotted solitons.[7,32] The simplest soliton solution is either a left-handed or a right-handed unknot *i.e.* a twisted, doubly-degenerate state as we need for our picture of glueballs.

The knottedness and in particular the twisting of the soliton relates to the topology of the unit vector field $\vec{n}$. For a finite energy configuration, $\vec{n}$ must go to a constant vector $\vec{n} \to \vec{n}_0$ at large distances, with $\vec{n}_0$ a minimum of the potential $V(\rho^2, \vec{h}\cdot\vec{n})$ at spatial infinity. Consequently $\vec{n}$ defines a mapping from the compactified $R^3 \sim S^3 \to S^2$. These mappings fall into nontrivial homotopy classes $\pi_3(S^2) \sim Z$ and can be characterized by the Hopf invariant.[7] For this we introduce the two-form

$$F \ = < d\vec{n} \wedge d\vec{n}\ ,\ \vec{n} >$$

Since the cohomology group $H_2(S^3) = 0$, the preimage of F on S^3 is exact and we can write

$$F = dA$$

The Hopf invariant Q_H is then given by

$$Q_H \ = \ \int F \wedge A$$

It computes the self-linking number of a knotted stringlike soliton, and in particular describes its self-twisting. Provided the knotted solitons of (6) indeed survive in the quantum Yang-Mills theory, we then have natural candidates for the glueballs as knotted as twisted, $L-R$ degenerate fluxtubes.

Obviously it would be too *naive* to expect that the solitons of (6) actually provide a *quantitatively* accurate description of glueballs. For this we must account for the roughening which is due to quantum fluctuations in the additional fields that appear in the pertinent decomposition of A^a_μ in the full SU(3) Yang-Mills theory. However, it is interesting to consider some of the predictions of these solitons. Maybe some of their properties are sufficiently universal to survive a more comprehensive analysis in the full SU(3) theory.

The energy spectrum of the solitons in (6) follows a rational curve in their self-linking number (Hopf invariant) Q_H [31]:

$$E_Q \ \geq \ c \cdot |Q_H|^{3/4}$$

If $f_0(1500)$ and $\eta_L(1410)$ are indeed the lowest mass states, we then have $c \approx 1500$ MeV which suggests the mass spectrum

$$M_Q \approx 1500 \cdot |Q_H|^{3/4} \text{ MeV} \tag{7}$$

which predicts that the next ($|Q_H| = 2$) glueball has a mass in the vicinity of $2500 - 2600$ MeV. This value is well within the range of the planned radiative charmonium decay experiments at CLEO-II and BES. Even though we expect this estimate to be very crude, there are general model independent, topological arguments[33] which suggest that the 3/4-scaling law should describe a universal behaviour, in the limit of large values $|Q_H|$. If this universality extends to smaller values of $|Q_H|$, there could be very interesting physics around $6.4 - 6.5$ GeV: According to (7) this energy corresponds to $Q_H = \pm 7$, and numerical simulations[32] suggest that the ensuing soliton is a trefoil which is a *chiral* knot. Consequently, besides the twisting we have an additional discrete symmetry with ensuing mass degeneracy, corresponding to the two-fold mirror symmetry of the trefoil. Curiously, we note that this mass value is near the upper reach of energy at the recently approved antiproton facility at GSI.

The 3/4 scaling-law of energy also suggests that the interaction between two knots is attractive. In particular, two singly-twisted unknots can form a bound state which is a doubly-twisted unknot. This could eventually develop into a probe on the properties of glueballs, if indeed described by knotted fluxtubes.

In three spatial dimensions the self-linking number of a knot is a topological invariant. But if a knot becomes embedded in a space with more than three spatial dimensions its self-linking number ceases to be a topological invariant, and the knot can disentangle. This suggests that the stability and decay properties of glueballs could be developed to an experimental test, to explore the dimensionality of space-time as seen by strong interactions. In particular, in three space dimensions the lowest energy glueball can only decay by the formation of quark-antiquark pairs *i.e.* into mesons. But the presence of extra dimensions suggests that there are additional, purely gluonic decay channels where the glueball could decay by unknotting itself thru higher dimensions, thus not necessarily decaying into quarks. On the other hand, the presence of a mass gap in four dimensional Yang-Mills theory excludes the presence of massless gluons. Consequently the existence of extra dimensions could mean that absolute confinement of gluons is lost.

Indeed, in full four dimensional QCD we expect that (virtual) $q\bar{q}$ pairs instabilize a closed and knotted gluonic string: A closed string opens itself and becomes disentangled into another closed string but with a different self-linking number, through the formation and subsequent annihilation of a quark-antiquark pair. This leads to an intuitively very attractive picture of interactions between guarks and glueballs, where quarks act much like certain enzymes act in the process of DNA replication by allowing one strand of the gluonic fluxtube to pass through another, thus changing its self-linking number and eventually leading to its decay into mesons. In particular, when quark loops are supressed like in the limit of large quark masses or large-N, we expect the knotted fluxtubes to be stable provided the space-time is four dimensional.

In conclusion, we have studied the possibility that glueballs could correspond to closed fluxtubes. Very general universality arguments then imply, that the stability of glueballs is possible only if the string is twisted or knotted. Since the twist can be either left-handed or right-handed, this implies that the glueball spectrum must be degenerate, and in particular the lowest mass glueball is a 0^{++} and 0^{-+} pair. We have inspected the known meson spectrum up to energies around 2.3 GeV, which is the upper limit that has been experimentally studied in $p\bar{p}$ annihilation processes. Our proposal that glueballs are closed gluonic fluxtubes then indicates, that the mysterious $\eta_L(1410)$ should be interpreted as the 0^{-+} glueball, (essentially) degenerate in mass with the widely accepted 0^{++} glueball $f_0(1500)$. We have explained how this interpretation of glueballs as twisted and knotted strings arises in Yang-Mills theory, where such configurations appear as knotted solitons in an effective low-energy theory. But our arguments are quite model-independent, suggesting that the interpretation of $\eta_L(1410)$ as the 0^{-+} glueball provides a test of various model independent qualitative aspects of strong interactions. These include the general properties of QCD string and confinement, the formation of a mass gap in Yang-Mills theory, and the possibility of space-time to have more than four dimensions. Our estimate of the glueball mass spectrum suggests that these could all be experimentally studied by the recently approved antiproton facility at GSI in Darmstadt.

This article presents my talk at Quarks 2003, and I thank the organizers for an excellent meeting. The article is largely based on original articles written together with Ludvig Faddeev and Ulrich Wiedner, and I thank them both for wonderful collaborations. The research has been partially supported by STINT IG2001-062.

References

1. W.H. Thomson, *Trans. Royal Soc. Edinburgh* **25**, 217 (1869).
2. W.H. Thomson, in *Mathematical and Physical Papers IV* (Cambridge, 1911).
3. P.G. Tait, *On Knots I, II, III* (Cambridge University Press, Cambridge, 1898).
4. M.B. Green, J.H. Schwarz and E. Witten, *Superstring Theory I, II* (Cambridge University Press, Cambridge, 1987).
5. E. Witten, *Commun. Math. Phys.* **121**, 351 (1989).
6. K. J. Juge, J. Kuti and C. Morningstar, *Phys. Rev. Lett.* **90**, 161601 (2003); and hep-lat/0310039.
7. L.D. Faddeev and A.J. Niemi, *Nature* **387**, 58 (1997).
8. For instance, L.D. Landau and E.M. Lifshitz, *Theory of Elasticity, 3rd edition* (Pergamon Press, Oxford, 1986).
9. M. Miettinen, A.J. Niemi and Y. Stroganov, *Phys. Lett.* **B474**, 303 (2000).
10. L. Faddeev, A.J. Niemi and U. Wiedner, hep-ph/0308240.
11. C.J. Morningstar and M. Peardon, *Phys. Rev.* **D60**, 034509 (1999).
12. Particle Data Group, *Phys. Rev.* **D66**, 010001 (2002).
13. F.E. Close, G.R. Farrar, Z. Li, *Phys. Rev.* **D55**, 5749 (1997).
14. Crystal Barrel Collaboration (C. Amsler *et al.*), *Phys. Lett.* **B342**, 433 (1995).
15. C. Amsler, *Rev. Mod. Phys.* **70**, 1293 (1998).
16. Crystal Barrel Collaboration (A. Abele *et al.*), *Nucl. Phys.* **A679**, 563 (2001).
17. OBELIX Collaboration (M. Bargiotti *et al.*), *Phys. Rev.* **D65**, 012001 (2002).
18. Crystal Barrel Collaboration (C. Amsler *et al.*), *Phys. Rev.* **D66**, 058101 (2002).
19. Crystal Barrel Collaboration (C. Amsler *et al.*), *Phys. Lett.* **B358**, 389 (1995).
20. P. Baillon *et al.*, *Nuovo Ciemento* **50A**, 393 (1967).
21. D. V. Bugg *et al.*, *Phys. Lett.* **B353**, 378 (1995).
22. MARK III Collaboration (Z. Bai *et al.*), *Phys. Rev. Lett.* **65**, 2507 (1990).
23. MARK III Collaboration (T. Bolton *et al.*), *Phys. Rev. Lett.* **69**, 1328 (1992).
24. L.D. Faddeev and A.J. Niemi, *Phys. Rev. Lett.* **82**, 1624 (1999).
25. L.D. Faddeev and A.J. Niemi, *Phys. Lett.* **B464**, 90 (1999); *ibid.* **B449**, 214 (1999).
26. L.D. Faddeev and A.J. Niemi, *Phys. Lett.* **B525**, 195 (2002).
27. E. Langmann and A.J. Niemi, *Phys. Lett.* **B463**, 252 (1999).
28. L. Faddeev, "Quantisation of Solitons", preprint IAS Print-75-QS70, (1975); "Einstein and Several Contemporary Tendencies in the Field Theory of Elementary Particles" in *Relativity, Quanta and Cosmology Vol. 1*, M. Pantaleo and F. De Finis (eds.), (Johnson Reprint, 1979).
29. F.V. Gubarev, L. Stodolsky, V.I. Zakharov, *Phys. Rev. Lett.* **86**, 2220 (2001).
30. L. Freyhult, *Int. J. Mod. Phys.* **A17**, 3681 (2002).
31. A.F. Vakulenko and L.V. Kapitanski, *Dokl. Akad. Nauk USSR* **248**, 810 (1979).
32. R. Battye and P. Sutcliffe, *Phys. Rev. Lett.* **81**, 4798 (1998); *Proc. R. Soc. London* **A455**, 4305 (1999); J. Hietarinta and P. Salo, *Phys. Lett.* **B451**, 60 (1999); *Phys. Rev.* **D62**, 081701 (2000).
33. T.A. Bolokhov, unpublished.

VACUUM CONDENSATES, EFFECTIVE GLUON MASS AND COLOR CONFINEMENT

K.-I. KONDO*

Department of Physics, Faculty of Science, Chiba University,
Chiba 263-8522, Japan
E-mail: kondok@faculty.chiba-u.jp

We propose a new reformulation of Yang-Mills theory in which three- and four-gluon self-interactions are eliminated at the price of introducing a sufficient number of auxiliary fields. We discuss the validity of this reformulation in the possible applications such as dynamical gluon mass generation, color confinement and glueball mass calculation. We emphasize the transverse-gluon pair condensation as the basic mechanism for dynamical mass generation. The confinement is realized as a consequence of a fact that the auxiliary fields become dynamical in the sense that they acquire the kinetic term due to quantum corrections.

1. Introduction

In the classical level, Yang-Mills theory [1] is a scale-invariant (actually, conformal invariant) field theory describing massless gauge boson. This is because the existence of the mass term is prohibited in the classical Yang-Mills theory, once we require *gauge invariance.*

In the quantum level, on the other hand, this is not necessarily the case. To quantize the Yang-Mills theory, the procedure of gauge fixing (GF) is necessary. The gauge-fixed Yang-Mills theory does not prohibit the mass term, since the gauge invariance no longer hold. Moreover, we need to introduce the Faddeev–Popov (FP) ghost field in the non-Abelian gauge group. Consequently, the BRST invariance of the total Yang-Mills action is expected to hold in the quantum Yang-Mills theory instead of the gauge invariance. In fact, it is known [2] that a mass term of the gauge boson

*Work supported by Sumitomo foundations, Grant-in-aid for Scientific Research (c)14540243 from Japan Society for the Promotion of Science (JSPS), and in part by Grant-in-aid for Scientific Research on Priority Areas (b)13135203 from the Ministry of Education, Culture, Sports, Science and Technology (MEXT).

can be introduced in the Lagrangian by adding simultaneously the (gauge-parameter dependent) mass term of the ghost. In this case, the BRST transformation should be understood as the *on-shell* BRST transformation. Here on-shell means that the Nakanishi-Lautrup field B is eliminated by using the equation of motion for B. However, the on-shell BRST transformation does not have the nilpotency, unless we use the equation of motion for the (anti)ghost. There is an option [3] to modify the off-shell BRST transformation so that the mass term can be made off-shell invariant under the modified transformation by adopting an appropriate GF+FP term. However, the off-shell nilpotency is lost in the modified off-shell BRST transformation. No one has found the mass term which is invariant under the off-shell BRST transformation with nilpotency. Remember that the nilpotency plays the key role in the proof of the unitarity of the physical S-matrices.[4] Therefore, the failure of nilpotency endangers maintaining the unitarity of the theory. In fact, it is shown that the physical space specified by the subsidiary condition $Q'_B|phys\rangle = 0$ using the modified BRST charge Q'_B contains the negative norm state. These facts prevent us from pursuing the massive Yang-Mills theory at the tree level, i.e., Yang-Mills theory with a mass term in the Lagrangian level.

In view of these, we are urged to consider the Yang-Mills theory whose mass is purely generated through quantum effects. Thus we are lead to examine the dynamical mass generation of gauge boson in Yang-Mills theory. In fact, such a possibility and its implications were investigated long ago in the following papers: [Eichten and Feinberg, 1974], [Smit, 1974], [Tye, Tomboulis and Poggio, 1975], [Fukuda and Kugo, 1978], [Gusynin and Miransky, 1978], [Cornwall and Papavassiliou, 1982], by using non-perturbative methods: Schwinger-Dyson equation, Bethe-Salpeter equation, Effective potential, Variational technique, Cooper equation, Bogoliubov transformation, etc. However, I have no time to explain the details of these works.

Here we mention recent studies of the coupled SD equations for the gluon and ghost propagators (in the Landau gauge) initiated by Smekal, Hauck and Alkofer.[5,6] They revealed the surprising results: Euclidean propagators exhibit *infrared (IR) asymptotic power-law behaviors characterized by the critical exponent* κ $(1/2 < \kappa < 1)$:

· *IR suppression of the gluon propagator*

$$D_T(Q^2) \cong \frac{A}{Q^2}(Q^2)^{2\kappa} \downarrow 0 \quad \text{as} \quad Q^2 \downarrow 0. \tag{1}$$

· *IR enhancement of the ghost propagator*

$$\Delta_{FP}(Q^2) \cong \frac{B}{Q^2}(Q^2)^{-\kappa} \uparrow \infty \quad \text{as} \quad Q^2 \downarrow 0. \tag{2}$$

These approximate solutions lead to
· *Existence of IR fixed point*

$$\beta(g^2(\mu)) \uparrow 0 \quad \text{as} \quad \mu \downarrow 0. \tag{3}$$

· *Fulfillment of a color confinement criterion due to Kugo-Ojima*

$$\lim_{Q^2 \to 0} [Q^2 \Delta_{FP}(Q^2)]^{-1} \equiv 1 + u(0) = 0. \tag{4}$$

However, it should be remarked that these solutions are obtained under the specific truncation of neglecting the two-loop diagrams in the gluon SD equation. Nevertheless, recent lattice simulations support some of the above results: [Bonnet, Bowman, Leinweber and Williams, hep-lat/0002020], [Langfeld, Reinhardt and Gattnar, hep-lat/0107141], [Alexandrou, de Forcrand and Follana, hep-lat/0112043], [Furui and Nakajima, hep-lat/0305010].

It is obvious that the above solutions of the SD equations do not correspond to the dynamical masss generation. However, the latest results of SD equations seem to suggest that $\kappa = 0.5$ is allowed as a solution if the two-loop diagrams and hence all the diagrams are included in the SD equations. [7] On the other hand, it was argued by the author [8] that the axiomatic consideration yields $\kappa = 1/2$:

$$D_T(Q^2) \cong \text{const.} + \mathcal{O}(Q^2) \downarrow \text{const.} \quad \text{as} \quad Q^2 \downarrow 0, \tag{5}$$

$$\Delta_{FP}(Q^2) \cong \frac{1}{Q^2}(Q^2)^{-1/2} \cong \frac{1}{(Q^2)^{3/2}} \uparrow \infty \quad \text{as} \quad Q^2 \downarrow 0. \tag{6}$$

These solutions also satisfy (3) and (4). See Fig. 1.

To find a non-perturbative solution with dynamical mass generation is an interesting question in itself. However, more important issue is to answer the question: What is the mechanism of generating the gluon mass? This is because the equation can not be solved without approximation in almost all cases, whereas the mechanism does not depend on the approximation if it is true. In fact, the above solutions of the SD equations are gauge dependent and not systematically improvable.

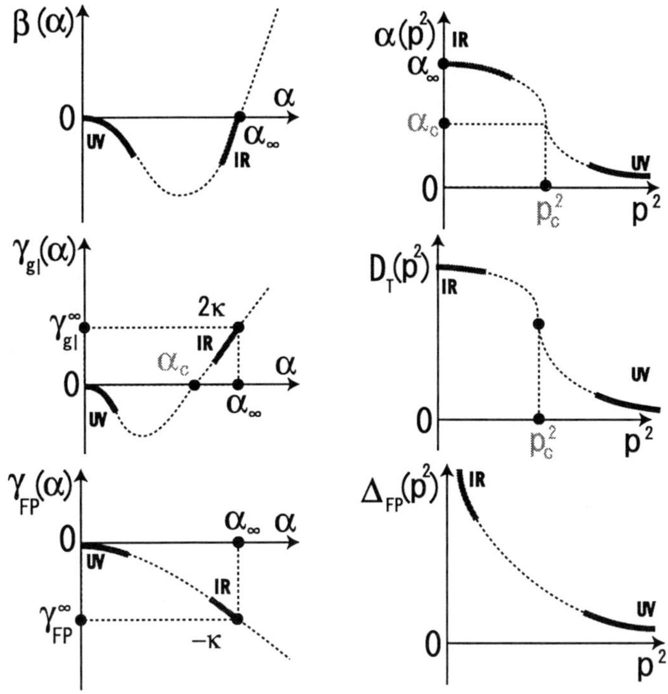

Figure 1. The schematic behavior of RG functions, running coupling constant and propagators for gluons and ghosts for Yang–Mills theory in the Lorentz–Landau gauge. (a) Beta function vs. Running coupling, (b) Anomalous dim. of gluon vs. Gluon propagator, (c) Anomalous dim. of ghost vs. FP Ghost propagator.

In this talk, we propose the *gluon pair condensation* [9] as a mechanism of *gluon mass generation and color confinement* in $SU(N)$ Yang-Mills theory. See Fig. 2. A purpose of this work is to construct an approximate *solution of QCD which is blessed with both dynamical gluon mass generation and color confinement.*

For such Bose-Einstein (BE) condensation to take place, we need the net attractive force between gauge boson pair. In Yang-Mills theory, gluon self-interactions exist. See Fig. 3.

In the perturbation theory, short-distance gluon-gluon potential leads to the net attractive force [10], although the details depend on the total spin of the gluon pair. See Fig. 4. In order to see the long-distance effect realizing the gluon confinement, we must go beyond perturbation theory.

For this purpose, we set up a new *large N expansion* based on a new

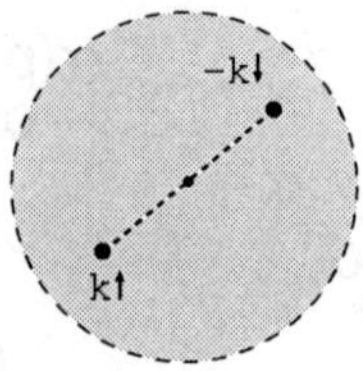

Figure 2. Gluon pair condensation in spin singlet and color singlet channel (center of mass frame).

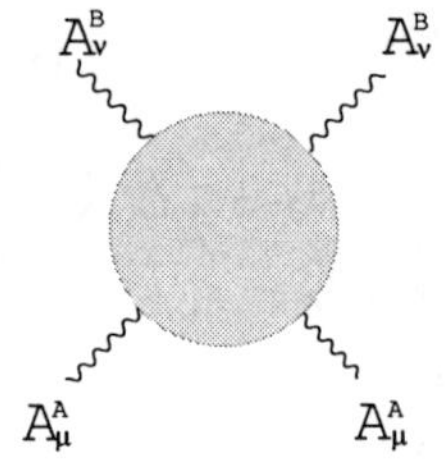

Figure 3. Interacting pair of gauge bosons.

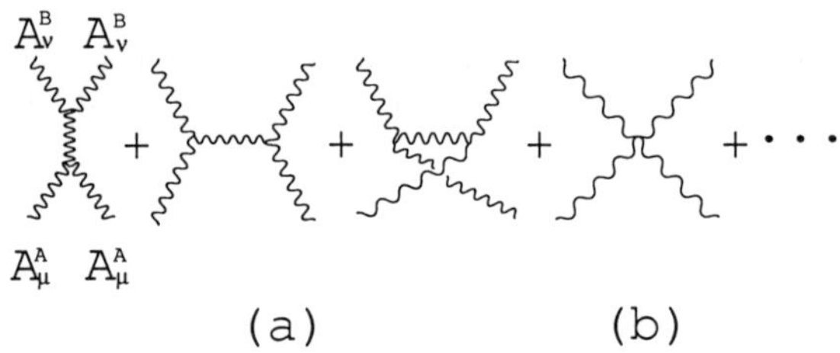

Figure 4. Perturbative gluon–gluon interaction to the lowest order (at tree level): (a) attractive interaction for $s-$,$t-$,$u-$channel gluon exchange, (b) attractive or repulsive force depending on the total spin $\boldsymbol{S} = \boldsymbol{S}_1 + \boldsymbol{S}_2$ of two-gluons .

formulation of Yang-Mills theory. It is said that N is only one free parameter in $SU(N)$ Yang-Mills theory. In general, the large N (or the $1/N$) expansion is

· non-perturbative
· systematically improvable
· gauge symmetry preserving

· related to string theory ['t Hooft,1974]
We also suggest a new scheme of the coupled SD equation to include all the diagrams. For details, see [11].

2. Yang-Mills theory without the gluon self-interactions

The Euclidean $SU(N_c)$ Yang-Mills Lagrangian is given by

$$\mathcal{L}_{\rm YM} = \frac{1}{4}\mathcal{F}_{\mu\nu}^2 = \frac{1}{4}(\partial_\mu \mathcal{A}_\nu - \partial_\nu \mathcal{A}_\mu + g\mathcal{A}_\mu \times \mathcal{A}_\nu)^2. \tag{7}$$

In what follows, we use a notation $F^2 := F \cdot F$, $(F \cdot G) := F^A G^A = 2\mathrm{tr}(FG)$, $(F \times G)^A := f^{ABC} F^B G^C = -2i\mathrm{tr}(T^A[F,G])$, and $(F \star G)^A := d^{ABC} F^B G^C = 2\mathrm{tr}(T^A\{F,G\})$.

We rewrite the Yang-Mills theory into an equivalent theory in which the gluon self-interactions are eliminated by introducing a number of auxiliary fields $\phi, B_{\mu\nu}^A, G_{\mu\nu}, \varphi^A, V_{\mu\nu}^A$ and by choosing the parameters appropriately.

$$\begin{aligned}\mathcal{L}_{\rm YM} =& \frac{1-\rho^2}{4}(\partial_\mu \mathcal{A}_\nu - \partial_\nu \mathcal{A}_\mu)^2 + \frac{1}{4}B_{\mu\nu}\cdot B^{\mu\nu} \\ & - \frac{i}{2}B^{\mu\nu}\cdot {}^*[\rho(\partial_\mu \mathcal{A}_\nu - \partial_\nu \mathcal{A}_\mu) + \sigma g(\mathcal{A}_\mu \times \mathcal{A}_\nu)] \\ & + \frac{\sigma_\phi}{2}\phi^2 + \frac{\sigma_\phi}{2}\phi(\mathcal{A}_\mu \cdot \mathcal{A}^\mu) + \frac{\sigma_G}{2}G_{\mu\nu}G^{\mu\nu} + \frac{\sigma_G}{2}G_{\mu\nu}S^{\mu\nu} \\ & + \frac{\sigma_\varphi}{2}\varphi\cdot\varphi + \frac{\sigma_\varphi}{2}\varphi\cdot(\mathcal{A}_\mu \star \mathcal{A}^\mu) + \frac{\sigma_V}{2}V_{\mu\nu}\cdot V^{\mu\nu} + \frac{\sigma_V}{2}V_{\mu\nu}\cdot T^{\mu\nu}, \end{aligned} \tag{8}$$

where the three-gluon interactions have been eliminated by choosing

$$\rho = \sigma^{-1}, \tag{9}$$

and the four-gluon interactions have been eliminated by choosing

$$\sigma_\phi = \frac{3(\sigma^2-1)g^2}{N_c},\ \sigma_G = \frac{4(1-\sigma^2)g^2}{N_c},\ \sigma_\varphi = (3/2)(\sigma^2-1)g^2,\ \sigma_V = 2(1-\sigma^2)g^2. \tag{10}$$

Here $*$ denotes the Hodge star (duality) operation * which is defined for the second-rank tensor as ${}^*H_{\mu\nu} := \frac{1}{2}\epsilon_{\mu\nu\rho\sigma}H^{\rho\sigma}$.

In addition, the GF+FP term is given by

$$\mathcal{L}_{\rm GF+FP} = \boldsymbol{\delta}_{\rm B}\bar{\boldsymbol{\delta}}_{\rm B}\left(\frac{1}{2}\mathcal{A}_\mu\cdot\mathcal{A}_\mu + \frac{\alpha'}{2}\mathcal{C}\cdot\bar{\mathcal{C}}\right) - \frac{\alpha}{2}\mathcal{B}\cdot\mathcal{B}. \tag{11}$$

By introducing further auxiliary fields ϕ_{FP}, φ_{FP}, the four-ghost self-interactions are eliminated as

$$\begin{aligned}\mathcal{L}_{\text{GF+FP}} =& \frac{1}{2\lambda}(\partial_\mu \mathcal{A}_\mu)^2 + \bar{\mathcal{C}} \cdot \partial_\mu \partial_\mu \mathcal{C} + (1-\xi) g \mathcal{A}_\mu \cdot (\partial_\mu \bar{\mathcal{C}} \times \mathcal{C}) \\ & - \xi g \mathcal{A}_\mu \cdot (\bar{\mathcal{C}} \times \partial_\mu \mathcal{C}) - \frac{1}{2}\lambda \xi (1-\xi) g^2 (\mathcal{C} \times \bar{\mathcal{C}}) \cdot (\mathcal{C} \times \bar{\mathcal{C}}) \\ & + \frac{\sigma_{FP}}{2}\phi_{FP}^2 + \frac{\sigma'_{FP}}{2}\varphi_{FP} \cdot \varphi_{FP} + \sigma_{FP}\phi_{FP}\bar{\mathcal{C}} \cdot \mathcal{C} + \sigma'_{FP}\varphi_{FP} \cdot (\bar{\mathcal{C}} \star \mathcal{C}),\end{aligned} \tag{12}$$

where

$$\sigma_{FP} = \lambda \xi (1-\xi) g^2, \quad \sigma'_{FP} = \lambda \xi (1-\xi) g^2, \tag{13}$$

and

$$\lambda := \alpha + \alpha', \quad \xi := \frac{\alpha'/2}{\alpha + \alpha'} = \frac{\alpha'}{2\lambda}. \tag{14}$$

Finally, we obtain the total Lagrangin $\mathcal{L}_{\text{YM}}^{\text{tot}} = \mathcal{L}_{\text{YM}} + \mathcal{L}_{\text{GF+FP}}$ which is at most quadratic in the gluon field $\mathcal{A}_\mu$:

$$\begin{aligned}\mathcal{L}_{\text{YM}}^{\text{tot}} =& \frac{1}{4}B_{\mu\nu} \cdot B^{\mu\nu} + \frac{\sigma_\phi}{2}\phi^2 + \frac{\sigma_G}{2}G_{\mu\nu}G^{\mu\nu} + \frac{\sigma_\varphi}{2}\varphi \cdot \varphi + \frac{\sigma_V}{2}V_{\mu\nu} \cdot V^{\mu\nu} \\ & + \frac{\sigma_{FP}}{2}\phi_{FP}^2 + \frac{\sigma'_{FP}}{2}\varphi_{FP} \cdot \varphi_{FP} \\ & + \bar{\mathcal{C}} \cdot \partial_\mu \partial^\mu \mathcal{C} + \sigma_{FP}\phi_{FP}\bar{\mathcal{C}} \cdot \mathcal{C} + \sigma'_{FP}\varphi_{FP} \cdot (\bar{\mathcal{C}} \star \mathcal{C}) \\ & + \frac{1}{2}\mathcal{A}_\mu^A \mathcal{K}^{AB\mu\nu} \mathcal{A}_\nu^B + \mathcal{A}_\mu \cdot \mathcal{J}^\mu,\end{aligned} \tag{15}$$

where we have defined $\mathcal{K}_{\mu\nu}^{AB}$ and $\mathcal{J}_\mu^A$ by

$$\begin{aligned}\mathcal{K}_{\mu\nu}^{AB} :=& \delta^{AB}\left[-(1-\rho^2)(\partial^2\delta_{\mu\nu} - \partial_\mu\partial_\nu) - \lambda^{-1}\partial_\mu\partial_\nu\right] \\ & - ig\sigma f^{ABC} {}^* B_{\mu\nu}^C + \sigma_\phi \delta^{AB}\delta_{\mu\nu}\phi + \sigma_\varphi d^{ABC}\delta_{\mu\nu}\varphi^C \\ & + \sigma_G \delta^{AB}\left(G_{\mu\nu} - \frac{1}{4}\delta_{\mu\nu}G_\rho{}^\rho\right) + \sigma_V d^{ABC}\left(V_{\mu\nu}^C - \frac{1}{4}\delta_{\mu\nu}V_\rho^{C\rho}\right), \\ \mathcal{J}_\mu^A :=& J_\mu^A - i\rho\partial_\nu {}^* B_{\mu\nu}^A + (1-\xi)g(\partial_\mu\bar{\mathcal{C}} \times \mathcal{C})^A - \xi g(\bar{\mathcal{C}} \times \partial_\mu \mathcal{C})^A.\end{aligned} \tag{16}$$

Here J_μ^A is the source for the gluon field $\mathcal{A}_\mu^A$. Note that $\mathcal{K}_{\mu\nu}^{AB}$ is at most linear in the auxiliary field and $\mathcal{J}_\mu^A$ contains the bilinear term in the ghost and antighost fields. [The source term for the auxiliary field will be introduced later to avoid the complication.]

An advantage of this formulation is that we can integrate out the gluon field $\mathcal{A}_\mu$ exactly:

$$S_{\rm EYM} = \int_x \Big\{\frac{1}{4}B_{\mu\nu}\cdot B^{\mu\nu} + \frac{\sigma_\phi}{2}\phi^2 + \frac{\sigma_G}{2}G_{\mu\nu}G^{\mu\nu} + \frac{\sigma_\varphi}{2}\varphi\cdot\varphi + \frac{\sigma_V}{2}V_{\mu\nu}\cdot V^{\mu\nu}$$
$$+ \frac{\sigma_{FP}}{2}\phi_{FP}^2 + \frac{\sigma'_{FP}}{2}\varphi_{FP}\cdot\varphi_{FP}$$
$$+ \bar{\mathcal{C}}\cdot\partial_\mu\partial^\mu\mathcal{C} + \sigma_{FP}\phi_{FP}\bar{\mathcal{C}}\cdot\mathcal{C} + \sigma'_{FP}\varphi_{FP}\cdot(\bar{\mathcal{C}}\star\mathcal{C})$$
$$- \frac{1}{2}\mathcal{J}_\mu^A[\mathcal{K}^{AB\mu\nu}]^{-1}\mathcal{J}_\nu^B\Big\} + \frac{1}{2}\ln {\rm Det}[\mathcal{K}^{AB\mu\nu}]. \tag{17}$$

When $\lambda = 0$, $\sigma_F = 0 = \sigma'_{FP}$ and the ghost self-interaction term disappears. Therefore, we do not need to introduce ϕ_{FP} and φ_{FP}^A in (17) which decouple from the theory.

Some remarks are in order.

1. The resulting theory has one undetermined parameter σ. In the classical level, any value of σ reproduces the original Yang-Mills theory, as is trivial from the above construction. However, in the quantum level, this is not the case. The renormalization urges the parameter σ to run according to the renormalization scale.

2. For a special choice $\sigma^2 = 1$, we have $\rho^2 = \sigma^2 = 1$ and hence the four coefficients vanish $\sigma_\Phi = 0$, which implies that all the auxiliary fields $\phi, G_{\mu\nu}, \varphi, V_{\mu\nu}$ except for $B_{\mu\nu}$ decouple from the theory. Therefore, this case reproduces the field strength formulation [12] of the Yang-Mills theory, which is identified with the deformation of the topological BF theory [13,14] with the Lagrangian:

$$\mathcal{L}_{\rm BF} = \frac{1}{4}B_{\mu\nu}\cdot B^{\mu\nu} - \frac{i}{2}B^{\mu\nu}\cdot {}^*\mathcal{F}_{\mu\nu}. \tag{18}$$

(The signature of ρ and σ can be absorbed by the redefinition of the fields $\mathcal{A}_\mu$ and $B_{\mu\nu}$.) When $\sigma^2 \neq 1$, therefore, we have the scalar field ϕ, φ and the symmetric tensor fields $G_{\mu\nu}, V_{\mu\nu}$ in addition to the antisymmetric tensor field $B_{\mu\nu}$.

3. Dynamical gluon mass generation, vacuum condensate and glueball

If the auxiliary (composite) field ϕ develops the non-vanishing vacuum expectation value (VEV),

$$\langle\phi\rangle := \phi_0 \neq 0, \tag{19}$$

the gluons acquire the common mass M_A:

$$M_A^2 = g\phi_0. \tag{20}$$

We argue that the mechanism of dynamical gluon mass generation is the pair condensation of transverse gluons. The VEV is obtained as a nontrivial solution of the gap equation:

$$\frac{1}{3(\sigma^2-1)}\phi + \frac{1}{2}\hat{g}^2\mathrm{tr}(\mathcal{K}_{\mu\mu}^{-1}[\phi]) = 0, \tag{21}$$

where $\hat{g}^2$ is the 't Hooft coupling $\hat{g}^2 := g^2 N_c$ and $\mathcal{K}_{\mu\nu}^{-1}$ is the gluon full propagator in the condensed vacuum. We can in principle calculate the effective action $S_E[\phi]$ from which the gap equation is obtained as the saddle-point equation. The fluctuation mode $\tilde{\phi}$ of $S_E[\phi]$ around ϕ_0 gives the glueball. For a constant ϕ, the effective action reduces to the effective potential, i.e., $S_E[\phi] = V(\phi)$. The ϕ_0 gives the absolute minimum of $V(\phi)$. See Fig. 5.

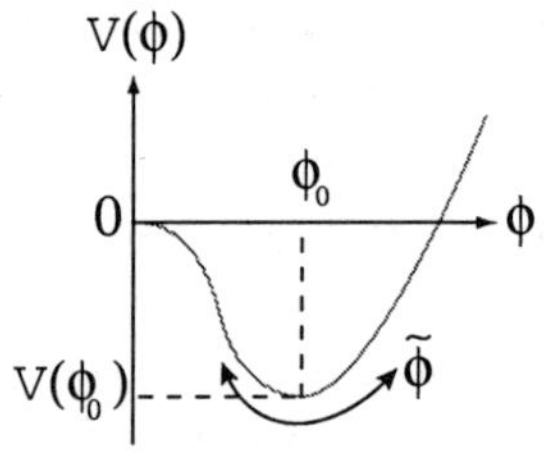

Figure 5. The effective potential for the auxiliary field ϕ.

The ratio of the glueball mass 0^{++} to the gluon mass is given by

$$M(0^{++})/M_A = \sqrt{6} \cong 2.45, \tag{22}$$

in the leading order.

Among a number of the auxiliary fields introduced above, it is the anti-symmetric tensor field $B_{\mu\nu}$ that is most important from the viewpoint of confinement. This is because the anti-symmetric tensor field can couple in a natural way to a surface element of the two-dimensional surface whose boundary is the Wilson loop and this property is desirable for deriving the area law decay of the Wilson loop average, i.e., quark confinement. Although the auxiliary fields do not have the kinetic term in the beginning by definition, the kinetic terms may be generated due to quantum effects. Then $B_{\mu\nu}$ mediates the confining force. For this scenario to be successful, the existence of the condensation ϕ_0, i.e., non-vanishing vacuum expectation value (VEV) is indispensable, since it provides with the mass scale which is compensated for the dimension introduced through the kinetic term generated. (Note that ∂_μ^2/ϕ_0 is dimensionless.) Without the condensation $\phi_0 = 0$, $B_{\mu\nu}$ can not acquire the kinetic term.

4. Vacuum condensate of mass dimension 2

The existence of *vacuum condensate of mass dimension two, i.e.,* $\langle \mathcal{A}_\mu^2 \rangle \neq 0$, was argued to exist recently by several groups. The composite operator $\mathcal{A}_\mu^2$ is not gauge invariant, while $\bar{\psi}\psi$ and $\mathcal{F}_{\mu\nu}^2$ are gauge invariant. Does such a vacuum condensate have any physical meaning?

· The minimum of $\mathcal{A}_\mu^2$ along the gauge orbit can have a definite physical meaning, as argued by [Zakharov et al.,2001]

$$\delta_\omega \int d^4x \frac{1}{2}\mathcal{A}_\mu^2 = \int d^4x \mathcal{A}_\mu \cdot \delta_\omega \mathcal{A}_\mu = \int d^4x \mathcal{A}_\mu \cdot \partial_\mu \omega(x)$$
$$= -\int d^4x \partial_\mu \mathcal{A}_\mu \cdot \omega(x). \qquad (23)$$

· Lattice simulations + OPE fit performed by [Boucaud et al., 2000,2001,2002] yield the sizable value $\langle \mathcal{A}_\mu^2 \rangle \cong (1.4\text{GeV})^2$ and the operator product expansion of the running coupling reads

$$\alpha(p) = \alpha_{pert}(p) \left\{ 1 + R \left(\ln \frac{p}{\Lambda_{QCD}} \right)^{-9/44} \langle \mathcal{A}_\mu \mathcal{A}_\mu \rangle / p^2 \right\}. \qquad (24)$$

This implies the confining potential at *short* distance. (It is argued how the instanton or monopole contributes to this condensate.)

Thus the existence of vacuum condensate with mass dimension two means the existence of non-perturbative power correction even in the high-energy region.

We have proposed a on-shell BRST-invariant composite operator of mass dimension two. It is given in the partial gauge fixing $G \to H$ by[15]:

$$\mathcal{O} := \Omega^{-1} \int d^4x \, \mathrm{tr}_{G/H} \left[\frac{1}{2} \mathcal{A}_\mu \mathcal{A}^\mu + \lambda i \bar{\mathcal{C}} \mathcal{C} \right]. \qquad (25)$$

Therefore, $\mathcal{O}$ is invariant under the on-shell BRST (and anti-BRST) transformation when $\xi = \frac{1}{2}$ or $\lambda = 0$. In particular, in the Landau gauge $\lambda = 0$, $\partial^\mu \mathcal{A}_\mu(x) = 0$. Hence $\mathcal{O}$ is off-shell BRST invariant.

We have argued in [16]

1. The on-shell BRST closed operator $\mathcal{O}$ ($\boldsymbol{\delta}_{\rm os}\mathcal{O} = 0$) is written as a sum of the gauge-invariant (but nonlocal) operator $\mathcal{O}'$ ($\boldsymbol{\delta}_\omega \mathcal{O}' = 0$) and the on-shell BRST exact part:

$$\mathcal{O} = \mathcal{O}' + \boldsymbol{\delta}_{\rm os}\mathcal{O}'', \tag{26}$$

where

$$\mathcal{O}' = \Omega^{-1} \int \frac{d^3\boldsymbol{k}}{(2\pi)^3 2|\boldsymbol{k}|} \sum_{\sigma=\pm} a^A(\boldsymbol{k},\sigma) a^{A\dagger}(\boldsymbol{k},\sigma),$$

$\leftarrow$ gauge-invariant transverse (but nonlocal)

$$\mathcal{O}'' = \Omega^{-1} \int \frac{d^3\boldsymbol{k}}{(2\pi)^3 2|\boldsymbol{k}|} \lambda [i\bar{c}^A(\boldsymbol{k}) a^{A\dagger}(\boldsymbol{k},L) + \text{h.c.}],$$

$\leftarrow$ on-shell BRST exact (27)

and

$$\boldsymbol{\delta}_{\rm os}\mathcal{O}'' = \Omega^{-1} \int \frac{d^3\boldsymbol{k}}{(2\pi)^3 2|\boldsymbol{k}|} [a^A(\boldsymbol{k},S) a^{A\dagger}(\boldsymbol{k},L) + \lambda i\bar{c}^A(\boldsymbol{k}) c^{A\dagger}(\boldsymbol{k}) + \text{h.c.}].$$

$\leftarrow$ longitudinal and scalar modes cancel the ghost and antighost

2. The VEV of $\mathcal{O}$ is equivalent to the VEV of the gauge invariant operator $\mathcal{O}'$ which has zero ghost number and is written in terms of the transverse gauge boson alone,

$$\langle \mathcal{O} \rangle = \langle \mathcal{O}' \rangle, \quad \mathcal{O}' = \Omega^{-1} \int d^4x \sum_{\sigma=\pm} \frac{1}{2} [A_\mu^{(\sigma)}(x)]^2. \tag{28}$$

Therefore, $\langle \mathcal{O} \rangle \neq 0$ is a gauge invariant and gauge independent statement.

3. The gauge invariant operator $\mathcal{O}'$ is nonlocal and can be non-linear in the non-Abelian gauge theory. The transverse gauge boson is both BRST-invariant and gauge invariant, i.e., $\boldsymbol{\delta}_{\rm B}\mathcal{A}^i_{\rm phys}(x) = 0$, $\boldsymbol{\delta}_\omega \mathcal{A}^i_{\rm phys}(x) = 0$.

5. Conclusion and discussion

We have proposed an equivalent formulation of the Yang-Mills theory in which a sufficient number of auxiliary fields are introduced to eliminate all the gluon self-interactions.

We have set up a new large N_c expansion of $SU(N_c)$ Yang-Mills theory on a non-perturbative vacuum which is stabilized by transverse gluon pair condensation. Then we have given the Feynman rule of the large N_c

expansion. This large N_c expansion is defined on a non-trivial vacuum in which the vacuum condensate of mass dimension two take place. A physical meaning of the on-shell BRST invariant composite operator of mass dimension two is discussed.

We determined the mass ratio of the lowest scalar glueball to the gluon: $M(0^{++})/M_A = \sqrt{6} \cong 2.45$ to the leading order. This should be compared with the potential model[10] $M(0^{++})/M_A \cong 2.3$. For $M_A = 600 \sim 700(MeV)$, $M(0^{++}) = 1.47 \sim 1.72(GeV)$.

To the leading order, the static potential between a pair of color charges is calculated. It is given by the sum of the Yukawa-type potential and the linear potential. The non-zero string tension is provided by the condensation. The resulting expression is very similar to that [17] obtained in the MA gauge assuming the Abelian dominance. The string representation is also derived within the same approximation. The corresponding confining string is nothing but the rigid string due to Polyakov.

References

1. C.N. Yang and R.L. Mills, *Phys. Rev.* **96**, 191 (1954).
2. G. Curci and R. Ferrari, *Nuovo Cimento* **32A**, 151 (1976).
   G. Curci and R. Ferrari, *Nuovo Cimento* **35A**, 1 (1976).
3. I. Ojima, *Z. Phys.* **C 13**, 173 (1982).
4. T. Kugo and I. Ojima, *Prog. Theor. Phys. Suppl.* **No. 66**, 1 (1979).
5. L. von Smekal, A. Hauck and R. Alkofer, *Phys. Rev. Lett.* **79**, 3591 (1997).
   L. von Smekal, A. Hauck and R. Alkofer, *Ann. Phys. (N.Y.)* **267**, 1 (1998).
6. R. Alkofer and L. von Smekal, *Phys. Rept.* **353**, 281 (2001).
7. J.C.R. Bloch, hep-ph/0303125, Few Body Syst., to be published.
8. K.-I. Kondo, hep-th/0303251.
9. R. Fukuda and T. Kugo, *Prog. Theor. Phys.* **60**, 565 (1978).
10. J.M. Cornwall and A. Soni, *Phys. Lett.* **B 120**, 431 (1983).
    W.S. Hou and A. Soni, *Phys. Rev.* **D 29**, 101 (1984).
11. K.-I. Kondo, hep-th/0307279.
12. M.B. Halpern, *Phys. Rev.* **D 16**, 1798 (1977).
13. F. Fucito, M. Martellini and M. Zeni, *Nucl. Phys.* **B496**, 259 (1997).
    M. Martellini and M. Zeni, theory, *Phys. Lett.* **B401**, 62 (1997).
14. K.-I. Kondo, *Phys. Rev.* **D 58**, 105019 (1998).
15. K.-I. Kondo, *Phys. Lett.* **B 514**, 335 (2001).
16. K.-I. Kondo, *Phys. Lett.* **B 572**, 210 (2003).
17. T. Suzuki, *Prog. Theor. Phys.* **80**, 929 (1988).
    S. Maedan and T. Suzuki, *Prog. Theor. Phys.* **81**, 229 (1989).
    T. Suzuki, *Prog. Theor. Phys.* **81**, 752 (1989).
    H. Suganuma, S. Sasaki and H. Toki, *Nucl. Phys.* **B 435**, 207 (1995).

ADVENTURES IN COULOMB GAUGE *†

JEFF GREENSITE

Physics and Astronomy Dept., San Francisco State University, San Francisco, CA 94117. E-mail: greensit@stars.sfsu.edu

ŠTEFAN OLEJNÍK

Institute of Physics, Slovak Academy of Sciences, SK-845 11 Bratislava, Slovakia. E-mail: fyziolej@savba.sk

We study the phase structure of SU(2) gauge theories at zero and high temperature, with and without scalar matter fields, in terms of the symmetric/broken realization of the remnant gauge symmetry which exists after fixing to Coulomb gauge. The symmetric realization is associated with a linearly rising color Coulomb potential (which we compute numerically), and is a necessary but not sufficient condition for confinement.

There are several reasons why Coulomb gauge may be interesting and/or useful in the study of the confining force. First of all there is the speculation by Gribov [1] and Zwanziger [2] that confinement in Coulomb gauge is due to instantaneous (dressed) one-gluon exchange. Secondly, the behavior of the color Coulomb potential, defined in Coulomb gauge, is an important element in the gluon-chain model of QCD string formation [3]. Finally, as we will see below, the confining property of the color Coulomb potential is associated with the unbroken realization of a remnant gauge symmetry, and this suggests a new order parameter for studying the phase structure of lattice gauge theories.

We begin with the idea that confinement arises from one-gluon exchange in Coulomb gauge; specifically, from the instantaneous piece of the $\langle A_0 A_0 \rangle$ propagator

$$\langle A_0^a(x) A_0^b(y) \rangle = P(\vec{x} - \vec{y})\delta^{ab}\delta(x_0 - y_0) + \text{non-instantaneous} \qquad (1)$$

*Talk presented by J. Greensite at *Confinement 2003*, Tokyo, July 21-24.

†Work supported by the US Dept. of Energy, Grant No. DE-FG03-92ER40711 (J.G.), and the Slovak Grant Agency for Science, Grant No. 2/3106/2003 (Š.O.)

where

$$P(\vec{x}-\vec{y})\delta^{ab} = \left\langle \left[\frac{1}{\nabla \cdot D(A)}(-\nabla^2)\frac{1}{\nabla \cdot D(A)} \right]^{a,b}_{x,y} \right\rangle \tag{2}$$

where $D_i(A)$ is the covariant derivative. This quantity is directly related to the Coulomb interaction energy in Coulomb gauge.

Recall that the classical Hamiltonian in Coulomb gauge, $H = H_{glue} + H_{coul}$, has the form

$$\begin{aligned} H_{glue} &= \frac{1}{2}\int d^3x \; (E^{tr,a} \cdot E^{tr,a} + B^a \cdot B^a) \\ H_{coul} &= \frac{1}{2}\int d^3x d^3y \; \rho^a(x) K^{ab}(x,y;A)\rho^b(y) \\ K^{ab}(x,y;A) &= \left[\frac{1}{\nabla \cdot D(A)}(-\nabla^2)\frac{1}{\nabla \cdot D(A)} \right]^{ab}_{xy} \\ \rho^a &= \rho^a_{matter} - g f^{abc} A^b_k E^c_k \end{aligned} \tag{3}$$

Note that $\langle K \rangle$ is the instantaneous piece of the $\langle A_0 A_0 \rangle$ propagator. Gribov and Zwanziger argue that this propagator is enhanced by configurations at the Gribov horizon, defined as a boundary in function space where the operator $\nabla \cdot D(A)$ aquires a zero eigenvalue. The conjecture is that this enhancement leads to a confining Coulomb potential, and therefore confinement by one-gluon exchange.

One objection to this idea is that it is difficult to see how the string-like properties of the QCD flux tube, namely, the logarithmic growth of the flux tube cross-section (roughening), and the universal $-\pi/12R$ contribution to the static quark potential (the Lüscher term), could arise from one-gluon exchange. On the other hand, as we will see below, the color Coulomb potential is an upper bound on the confining static quark potential. This means that confinement by one-gluon exchange is a necessary condition for confinement.

Let

$$|\Psi_{qq}\rangle = \overline{q}^a(0) q^a(R) |\Psi_0\rangle \tag{4}$$

be a physical state in Coulomb gauge containing two static charges; Ψ_0 is the ground state. Then

$$\begin{aligned} \Delta E &= \langle \Psi_{qq}|H|\Psi_{qq}\rangle - \langle \Psi_0|H|\Psi_0\rangle \\ &= V_{coul}(R) + E_{se} \end{aligned} \tag{5}$$

is the expectation value of the excitation energy, where the R-dependent Coulomb potential $V_{coul}(R)$ can only arise from the expectation value of the non-local $\rho K \rho$ piece of the Hamiltonian. We want to address the following questions: Is $V_{coul}(R)$ confining? If so, is it asymptotically linear? If it is linear, then is the associated string tension σ_{coul} equal to string tension σ of the static quark potential? Finally, we would like to study the effect, on the Coulomb string tension, of removing center vortices.

We begin by defining the correlator, for SU(N) gauge theory in Coulomb gauge, of two timelike Wilson lines

$$\begin{aligned} G(R,T) &= \langle \frac{1}{N} \mathrm{Tr}[L^\dagger(0,T) L(R,T)] \rangle \\ &= \langle \Psi_{qq} | e^{-(H-E_0)T} | \Psi_{qq} \rangle \end{aligned} \tag{6}$$

where

$$L(\vec{x},T) = P \exp \left[i \int_0^T dt A(\vec{x},t) \right] \tag{7}$$

Note that L is a timelike Wilson line (*not* a Polyakov line) of time extent T. The existence of a transfer matrix implies

$$G(R,T) = \sum_n |\langle \Psi_n | \Psi_{qq} \rangle|^2 \, e^{-\Delta E_n T} \tag{8}$$

where the sum is over energy eigenstates, and ΔE_n is the energy above the ground state. Denote

$$V(R,T) = -\frac{d}{dT} \log[G(R,T)] \tag{9}$$

Then its not hard to see that

$$\begin{aligned} \Delta E &= V_{coul}(R) + E_{se} \\ &= V(R,0) \end{aligned} \tag{10}$$

while

$$\begin{aligned} \Delta E_{min} &= V(R) + E'_{se} \\ &= \lim_{T \to \infty} V(R,T) \end{aligned} \tag{11}$$

where ΔE_{min} is the minimum energy of the $q\bar{q}$ system, and $V(R)$ is the static quark potential. The use of Wilson line correlators, in Coulomb gauge, to compute the static potential was first suggested by Marinari et al. [4].

With a lattice regularization, the self-energies E_{se} and E'_{se} become negligible, at large R, compared to the confining static potential $V(R)$. Then, since $\Delta E_{min} \leq \Delta E$, it follows that asymptotically

$$V(R) \leq V_{coul}(R) \tag{12}$$

This inequality was originally derived by Zwanziger [5]. It implies that if confinement exists, it exists already at the level of dressed one-gluon exchange.

With a lattice regularization, we have

$$\begin{aligned} L(x,T) &= U_0(x,a)U_0(x,2a)\cdots U_0(x,T) \\ V(R,T) &= \frac{1}{a}\log\left[\frac{G(R,T)}{G(R,T+a)}\right] \end{aligned} \tag{13}$$

so that

$$\lim_{\beta\to\infty} V(R,0) = V_{coul}(R) \ + \ \text{const.}$$

$$\lim_{T\to\infty} V(R,T) = V(R) \ + \ \text{const.} \tag{14}$$

where, in lattice units $a = 1$,

$$V(R,0) = -\log[G(R,1)] \tag{15}$$

Via lattice Monte Carlo we can then arrive at an estimate, exact in the continuum limit, of $V_{coul}(R)$ from $V(R,0)$, and compare this to the static quark potential $V(R)$.

This procedure was carried out in ref. [6] for SU(2) lattice gauge theory. The result for $V(R,0)$ at $\beta = 2.5$ is shown in Fig. 1 (upper set of data points). The two lines shown are best fits to the data by a linear, and by a linear + Lüscher, potential. The data immediately answers three of the questions posed above: The Coulomb potential is confining, and it is linear (see also [7]). However, it turns out that the associated string tension σ_{coul} is substantially greater than σ, by almost a factor of three. This means that the QCD flux tube is not simply the static charges and their associated Coulomb (longitudinal color electric) field. The minimum energy string state is more complicated than (4); it must also contain some constituent gluons, as in the gluon chain picture of string formation advocated in [3].

The same figure shows the effect, on $V(R,0)$, of center vortex removal. The center vortex theory of confinement has been studied very actively in recent years; the theory and the numerical evidence in its favor are reviewed in ref. [8]. Center vortices are identified by first fixing to an adjoint gauge,

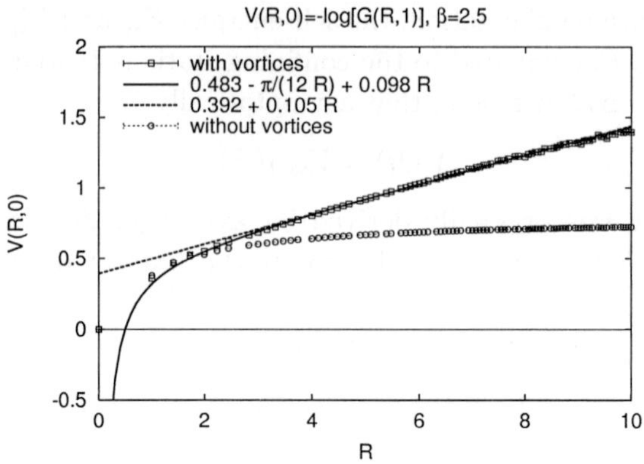

Figure 1. Lattice approximation $V(R,0)$ (upper data set, open squares) to the color Coulomb potential $V_{coul}(R)$ + constant. The lines are best fits to linear, and linear + Lüscher, potentials. The lower data set (open circles) for $V(R,0)$ is obtained after removing center vortices from the lattice configurations, as described in the text.

and then projecting link variables to the center subgroup of the gauge group. An example is the direct maximal center gauge (= Landau gauge in the adjoint representation), where the procedure is to gauge fix to a local maximum of

$$R = \sum_{x,\mu} \Big| \mathrm{Tr}[U_\mu(x)] \Big|^2 \tag{16}$$

and then to project each link to the closest center element, e.g. for SU(2)

$$U_\mu(x) \to Z_\mu(x) = \mathrm{signTr}[U_\mu(x)] \tag{17}$$

Vortices are removed from a given lattice configuration by multiplying the adjoint gauge-fixed configuration by the projected configuration, i.e.

$$U_\mu(x) \to U'_\mu(x) = Z_\mu(x) U_\mu(x) \tag{18}$$

In ref. [9] it was shown that after vortex removal the string tension vanishes, chiral symmetry breaking is eliminated, and each vortex-removed configuration has zero topological charge.

One can then ask what vortex removal does to the color Coulomb potential. In this connection, a relevant fact is that thin center vortices can be shown [10] to lie on the Gribov horizon, which is thought to play an important role in the enhancement of the Coulomb energy. In our numerical study, the modified configuration $U'_\mu(x)$ is gauge-fixed to Coulomb gauge, the timelike link correlators are calculated, and $V(R,0)$ is extracted. The result at $\beta = 2.5$ is shown in the lower data set (open circles) in Fig. 1. It is clear that vortex removal completely removes the confining property of the Coulomb potential. Further details can be found in ref. [6].

Of course, pure gauge theory at zero temperature is a special, albeit very important, case. In general gauge theories can exist in various phases, and it is useful to characterize these in terms of the distribution of the electric field emanating from a static isolated source.

- *Massless Phase:* The electric field is spherically symmetric, and falls off like $1/R^2$. This is the case for compact QED_4, and for lattice SU(N) gauge theory in $D > 4$ dimensions, at weak couplings.
- *Confined Phase:* The color electric field is collimated into a flux tube; global center symmetry is unbroken. Examples include SU(N) pure gauge theories at low temperature, and SU(N) gauge theories with matter in the adjoint representation of the gauge group.
- *Screened Phases:* There is a Yukawa-like falloff of the color electric field. This happens in SU(N) gauge theories when the Z_N center symmetry is broken spontaneously, as in high temperature gauge theory and gauge-Higgs theories with the Higgs field in the adjoint representation. Gauge theories with only a trivial center symmetry (consisting only of the identity element) are also in the screened phase; these theories include SU(N) gauge theories with matter in the fundamental representation, and G_2 gauge theory with or without matter fields.

For the purpose of studying Coulomb energy in these various phases, we find it useful to introduce a new order parameter, related to the realization of a remnant gauge symmetry in Coulomb gauge.

Let $U_\mu(x)$ be a lattice configuration fixed to Coulomb gauge. Note that the Coulomb gauge condition is preserved by the gauge transformation

$$\begin{aligned} U_k(x,t) &\to g(t)U_k(x,t)g^\dagger(t) \\ U_0(x,t) &\to g(t)U_0(x,t)g^\dagger(t+1) \end{aligned} \tag{19}$$

On any time slice, this is a global transformation, and therefore can be spontaneously broken in the following sense: At any fixed time t, in the infinite volume limit, the average of timelike link variables $U_0(x,t)$ is non-zero in any thermalized configuration. This means that

$$\begin{aligned}\lim_{R\to\infty} G(R,1) &> 0 \\ \lim_{R\to\infty} V(R,0) &= \text{finite const.} \\ \sigma_{coul} &= 0 \end{aligned} \tag{20}$$

in the broken phase. Therefore Coulomb confinement or non-confinement can be understood as the symmetric or spontaneously broken realization, respectively, of the remnant gauge symmetry in Coulomb gauge.

We now introduce the order parameter Q, as the modulus of the spatial average of timelike links, i.e.

$$\begin{aligned}U_0^{av}(t) &= \frac{1}{L^3}\sum_{\vec{x}} U_0(\vec{x},t) \\ Q &= \left\langle \sqrt{\mathrm{Tr}[U_0^{av}(t)U_0^{av\dagger}(t)]} \right\rangle \end{aligned} \tag{21}$$

On general grounds

$$Q = c + \frac{b}{L^{3/2}} \quad \text{with} \quad \begin{cases} c = 0 \text{ symmetric phase} \\ c > 0 \text{ broken phase} \end{cases} \tag{22}$$

Thus $Q > 0$ in the infinite volume limit implies that $V_{coul}(R)$ is non-confining. Since $V_{coul}(R)$ is an upper bound on $V(R)$, this implies that $Q = 0$ is a necessary (but not sufficient) condition for confinement.

It is useful to try out this order parameter in compact QED_4, where there is a transition from the confining to the massless phase at $\beta \approx 1$. Figure 2 shows our results for Q vs. the root inverse 3-volume $L^{-3/2}$ at $\beta = 0.7$ (confining phase) and $\beta = 1.3$ (massless phase). In this case the Q parameter seems to nicely distinguish between the two phases, extrapolating to zero only in the confined phase.

Next, we consider SU(2) gauge-Higgs theory with a "frozen" Higgs field in the adjoint representation. The lattice Lagrangian is

$$\begin{aligned}S &= \beta \sum_{plaq} \frac{1}{2}\mathrm{Tr}[UUU^\dagger U^\dagger] \\ &+ \frac{\gamma}{4}\sum_{x,\mu} \phi^a(x)\phi^b(x+\widehat{\mu})\mathrm{Tr}[\sigma^a U_\mu(x)\sigma^b U_\mu^\dagger(x)] \end{aligned} \tag{23}$$

where ϕ is a real 3-component field satisfying the constraint $\sum_a (\phi^a)^2 = 1$. This is a theory with a confining, center symmetric phase, and a non-confining phase with spontaneously broken center symmetry. Our finding

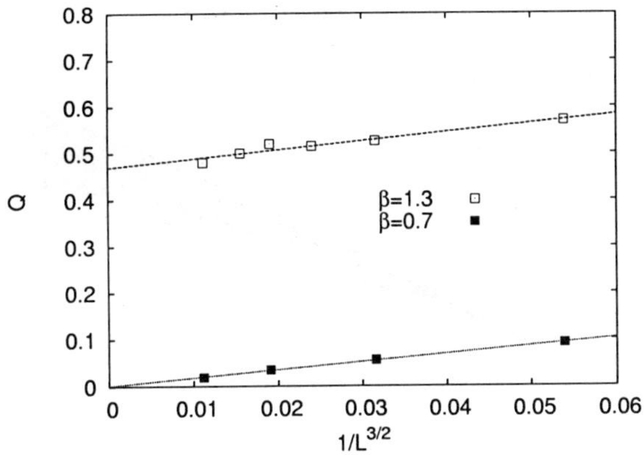

Figure 2. Extrapolation of Q to infinite volume in QED_4, for $\beta = 0.7$ (confining phase) and $\beta = 1.3$ (massless phase).

is that the transition line in the $\beta - \gamma$ phase diagram corresponding to the remnant symmetry-breaking transition is identical to the transition line for confinement-deconfinement, mapped out long ago by Brower et al. [11] from measurements of the plaquette energy. In the confined phase we find $Q = 0$ (when extrapolated to infinite volume), and $Q > 0$ in the Higgs phase, as indicated schematically in Figure 3.

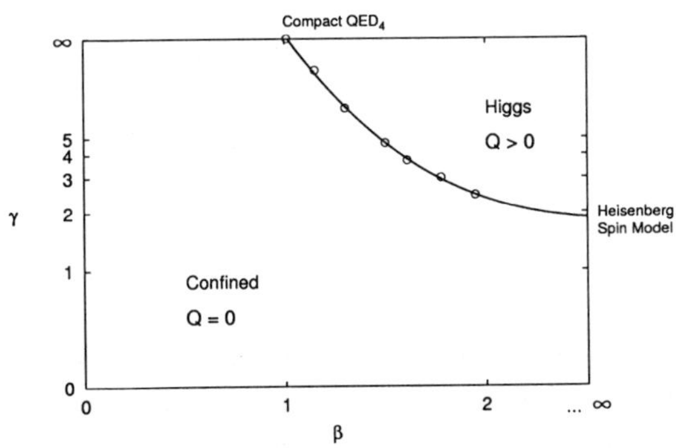

Figure 3. Phase diagram of the SU(2) adjoint Higgs model. The plaquette energy and the remnant symmetry order parameter Q locate the same transition line between the confined and Higgs phases.

One might guess that the transition from the confined to the deconfined phase is always accompanied by remnant symmetry breaking. Surprisingly,

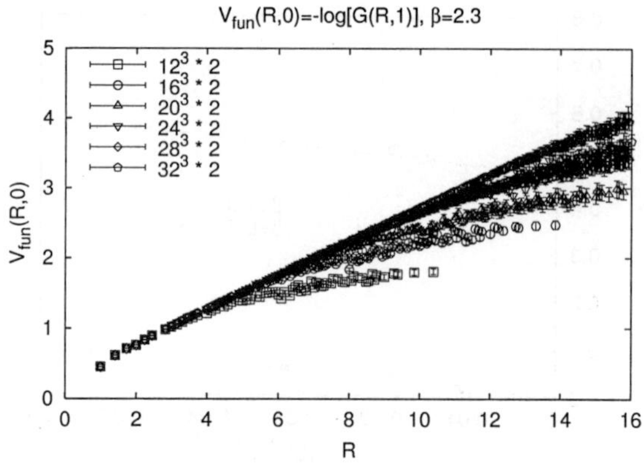

Figure 4. Coulomb potential in the deconfined phase, at $\beta = 2.3$ and $L_t = 2$ lattice spacings in the time direction, with spatial volumes from 12^3 to 32^3.

this turns out not to be true. We have also computed $V(R,0)$ and Q in the deconfined phase of pure SU(2) lattice gauge theory, with the results shown in Figs. 4 and 5. This data was taken at $\beta = 2.3$ on lattices with time extension of two lattice spacings, well within the deconfined phase. Yet the Coulomb potential is clearly linear and confining at large lattice volume, while the extrapolation of Q to infinite volume seems compatible with zero. A possible reason for this behavior is the fact that $K(x,y;A)$, whose expectation value gives the instantaneous Coulomb propagator, depends only on the spacelike components A_k at a fixed time. On the lattice, this translates to dependence only on spacelike links on a time slice. But we know that spacelike links on a time slice are a confining ensemble even in the deconfined phase, since spacelike Wilson loops are known to have an area law falloff at any temperature. If the Coulomb propagator depends only on the confining properties of spacelike links, then it is not so surprising that the Coulomb potential is confining in the deconfined regime (nor is this a paradox: the Coulomb potential is only an upper limit on the static potential). A test of this explanation is to remove the confining properties of the spacelike links by removing center vortices, via the de Forcrand/D'Elia procedure explained above. Then one expects the Coulomb potential to be non-confining, and this is, in fact, what is observed.

Finally, we study a gauge-Higgs system with the radially frozen Higgs field in the fundamental representation. For the SU(2) gauge group, the

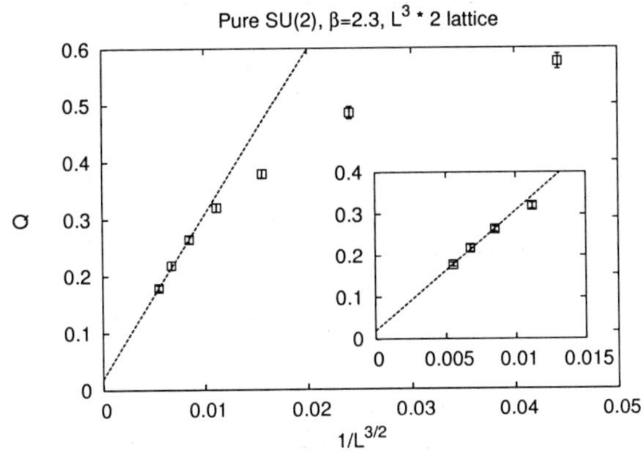

Figure 5. The Q parameter vs. root inverse 3-volume in the deconfined phase, $\beta = 2.3$ and $L_t = 2$ lattice spacings.

lattice Lagrangian can be expressed as [12]

$$S = \beta \sum_{plaq} \frac{1}{2}\mathrm{Tr}[UUU^{\dagger}U^{\dagger}] + \gamma \sum_{x,\mu} \frac{1}{2}\mathrm{Tr}[\phi^{\dagger}(x)U_{\mu}(x)\phi(x+\widehat{\mu})] \tag{24}$$

with ϕ an SU(2) group-valued field. This is a theory with only a screened phase; it can be proven that no transition to a confined phase is possible [13]. There is a first-order phase transition line in the $\beta - \gamma$ phase diagram, but this line has an endpoint, and does not divide the diagram into thermodynamically separate phases.

The remnant symmetry transition line coincides with the (thermodynamic) line of first-order transitions found by Lang et al. [12], but it then extends beyond the thermodynamic line all the way to $\beta = 0$ and $\gamma = 2$. This line divides the phase diagram into $Q = 0$ and $Q > 0$ regions, as indicated schematically in Fig. 6. In Fig. 7 we plot Q vs. γ at $\beta = 0$. If Q were the magnetization of an Ising spin system, this would surely be a second order phase transition, with the solid line in the figure representing the infinite volume limit. Nevertheless, there is no thermodynamic transition. At $\beta = 0$ one can easily compute the free energy exactly, which is found to be

$$F(\gamma) = 4V \log\left[\frac{2I_1(\gamma)}{\gamma}\right] \tag{25}$$

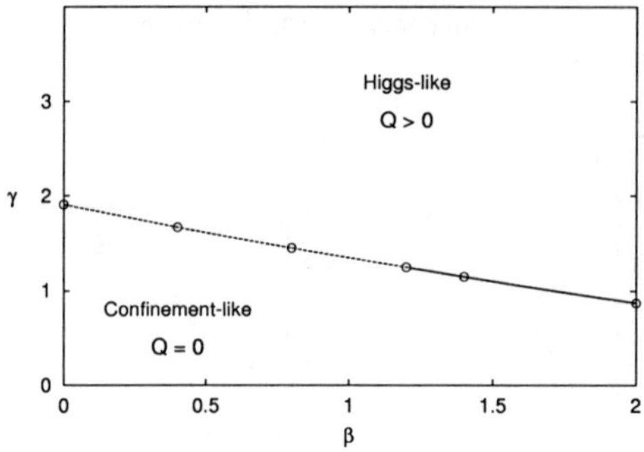

Figure 6. Phase diagram of the fundamental Higgs model. There is a thermodynamic transition and a Q transition along the solid line, but a non-thermodynamic transition (Kertész line) in Q along the dashed line.

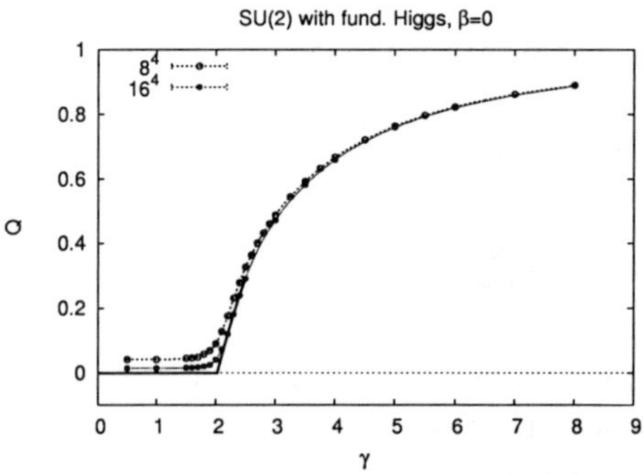

Figure 7. Q vs. γ at $\beta = 0$ in the SU(2) fundamental Higgs model, on 8^4 and 16^4 lattices. The solid line is the presumed extrapolation to infinite volume.

This expression is perfectly analytic at all $\gamma > 0$. On the other hand, a strong-coupling analysis of $G(R, 1)$ at fixed $\beta \ll 1$ [10] arrives at an exponential decay to zero as $R \to \infty$ at small γ, but a non-zero large-R limit at large γ. This implies a symmetry-breaking transition at some critical value $\gamma_{cr}(\beta)$, which motivated our numerical study of Q in this model.

The remnant symmetry breaking transition in the gauge-Higgs system, in the absence of a thermodynamic transition, is probably an example of a *Kertész line* [14] in statistical mechanics. This possibility was first suggested by Langfeld [15], who discovered remnant gauge symmetry breaking in Landau gauge in a closely related model (see also Satz [16]). Kertész lines are associated with percolation transitions, and it is natural to ask, in this case, what is percolating. Recent investigations [17] indicate that in the gauge-Higgs system, the Kertész line locates center vortex percolation transitions.

This concludes a summary of our investigations of phase structure in lattice gauge theory, as seen by Coulomb energy and remnant symmetry. Our study has uncovered Coulomb "over-confinement" ($\sigma_{coul} \approx 3\sigma$) in the low temperature confined phase, the persistence of Coulomb confinement in the deconfined phase, connections between vortex and Coulomb confinement, and (in accord with Langfeld [15]) symmetry breaking in the absence of a thermodynamic phase transition. These aspects of non-perturbative gauge theory are somewhat surprising, and merit further study.

Acknowledgments

Results reported in the second half of this talk were obtained in collaboration with Daniel Zwanziger. These results will be presented in more detail in a later publication [10].

J.G. thanks Hideo Suganuma for the opportunity to speak at this enjoyable and stimulating meeting.

References

1. V. Gribov, *Nucl. Phys.* **B139**, 1 (1978).
2. D. Zwanziger, *Nucl. Phys.* **B518**, 237 (1998).
3. J. Greensite and C. Thorn, *Journal of High Energy Phys.* **0202**, 014 (2002).
4. E. Marinari, M. Paciello, G. Parisi, and B. Taglienti, *Phys. Lett.* **B298**, 400 (1993).
5. D. Zwanziger, *Phys. Rev. Lett.* **90**, 0209105 (2003).
6. J. Greensite and Š. Olejník, *Phys. Rev.* **D67**, 094503 (2003).
7. A. Cucchieri and D. Zwanziger, *Nucl. Phys. Proc. Suppl.* **119**, 727 (2003).
8. J. Greensite, *Prog. Part. Nucl. Phys.* **51**, 1 (2003); arXiv: hep-lat/0301023.
9. P. de Forcrand and M. D'Elia, *Phys. Rev. Lett.* **85**, 4582 (1999).
10. J. Greensite, Š. Olejník, and D. Zwanziger, in preparation.
11. R. Brower et al., *Phys. Rev.* **D25**, 3319 (1982).
12. C. Lang, C. Rebbi, and M. Virasoro, *Phys. Lett.* **104B**, 294 (1981).
13. E. Fradkin and S. Shenker, *Phys. Rev.* **D19**, 3682 (1979).
14. J. Kertész, *Physica* **A 161**, 58 (1989).
15. K. Langfeld, arXiv:hep-lat/0212032 and hep-lat/0109033.
16. H. Satz, *Comput. Phys. Commun.* **147**, 46 (2002).
17. R. Bertle and M. Faber, arXiv:hep-lat/0212027;
R. Bertle, M. Faber, J. Greensite, and Š. Olejník, in preparation.

CONFORMAL ASPECTS OF QCD *

S. J. BRODSKY
Stanford Linear Accelerator Center
2575 Sand Hill Road
Menlo Park, CA 94025, USA
E-mail: sjbth@slac.stanford.edu

Theoretical and phenomenological evidence is now accumulating that the QCD coupling becomes constant at small virtuality; *i.e.*, $\alpha_s(Q^2)$ develops an infrared fixed point in contradiction to the usual assumption of singular growth in the infrared. For example, the hadronic decays of the τ lepton can be used to determine the effective charge $\alpha_\tau(m^2_{\tau'})$ for a hypothetical τ-lepton with mass in the range $0 < m_{\tau'} < m_\tau$. The τ decay data at low mass scales indicates that the effective charge freezes at a value of $s = m^2_{\tau'}$ of order $1\,\mathrm{GeV}^2$ with a magnitude $\alpha_\tau \sim 0.9 \pm 0.1$. The near-constant behavior of effective couplings suggests that QCD can be approximated as a conformal theory even at relatively small momentum transfer and why there are no significant running coupling corrections to quark counting rules for exclusive processes. The AdS/CFT correspondence of large N_C supergravity theory in higher-dimensional anti-de Sitter space with supersymmetric QCD in 4-dimensional space-time also has interesting implications for hadron phenomenology in the conformal limit, including an all-orders demonstration of counting rules for exclusive processes and light-front wavefunctions. The utility of light-front quantization and light-front Fock wavefunctions for analyzing non-perturbative QCD and representing the dynamics of QCD bound states is also discussed.

1. The Infrared Behavior of the Effective QCD Couplings $\alpha_\tau(s)$

It is often assumed that color confinement in QCD can be traced to the singular behavior of the running coupling in the infrared, *i.e.* "infrared slavery." For example if $\alpha_s(q^2) \to \frac{1}{q^2}$ at $q^2 \to 0$, then one-gluon exchange leads to a linear potential at large distances. However, theoretical [1,2,3,4,5,6] and phenomenological [7,8,9] evidence is now accumulating that the QCD

*This work is supported by the department of energy under contract number DE–AC03–76SF00515.

coupling becomes constant [10,11,12,13] at small virtuality; *i.e.*, $\alpha_s(Q^2)$ develops an infrared fixed point in contradiction to the usual assumption of singular growth in the infrared. If QCD running couplings are bounded, the integration over the running coupling is finite and renormalon resummations are not required. If the QCD coupling becomes scale-invariant in the infrared, then elements of conformal theory [14] become relevant even at relatively small momentum transfers.

One can define the fundamental coupling of QCD from virtually any physical observable [15,16]. Such couplings, called effective charges, are all-order resummations of perturbation theory, so they correspond to the complete theory of QCD; it is thus guaranteed that they are analytic and non-singular. For example, it has been shown that unlike the $\overline{\rm MS}$ coupling, a physical coupling is analytic across quark flavor thresholds [17,18]. Furthermore, a physical coupling must stay finite in the infrared when the momentum scale goes to zero. In turn, this means that integrals over the running coupling are well defined for physical couplings. Once such a physical coupling $\alpha_{\rm phys}(k^2)$ is chosen, other physical quantities can be expressed as expansions in $\alpha_{\rm phys}$ by eliminating the $\overline{\rm MS}$ coupling which now becomes only an intermediary [19]. In such a procedure there are in principle no further renormalization scale (μ) or scheme ambiguities. The physical couplings satisfy the standard renormalization group equation for its logarithmic derivative, $d\alpha_{\rm phys}/d\ln k^2 = \widehat{\beta}_{\rm phys}[\alpha_{\rm phys}(k^2)]$, where the first two terms in the perturbative expansion of the Gell-Mann Low function $\widehat{\beta}_{\rm phys}$ are scheme-independent at leading twist, whereas the higher order terms have to be calculated for each observable separately using perturbation theory.

In a recent paper, Menke, Merino, and Rathsman [8,20] and I have presented a definition of a physical coupling for QCD which has a direct relation to high precision measurements of the hadronic decay channels of the $\tau^- \to \nu_\tau h^-$ A related effective charge has also been considered by Neubert [21]. Let R_τ be the ratio of the hadronic decay rate to the leptonic one. Then $R_\tau \equiv R_\tau^0 \left[1 + \frac{\alpha_\tau}{\pi}\right]$, where R_τ^0 is the zeroth order QCD prediction, defines the effective charge α_τ. The data for τ decays is well-understood channel by channel, thus allowing the calculation of the hadronic decay rate and the effective charge as a function of the τ mass below the physical mass. The vector and axial-vector decay modes which can be studied separately.

Using an analysis of the τ data from the OPAL collaboration [22], we have found that the experimental value of the coupling $\alpha_\tau(s) = 0.621 \pm 0.008$ at $s = m_\tau^2$ corresponds to a value of $\alpha_{\overline{\rm MS}}(M_Z^2) = (0.117\text{-}0.122) \pm 0.002$, where the range corresponds to three different perturbative methods used in analyzing the data. This result is in good agreement with the world average $\alpha_{\overline{\rm MS}}(M_Z^2) = 0.117 \pm 0.002$. However, from the figure we also see that the effective charge only reaches $\alpha_\tau(s) \sim 0.9 \pm 0.1$ at $s = 1\,\text{GeV}^2$, and it even stays within the same range down to $s \sim 0.5\,\text{GeV}^2$. This result is in good agreement with the estimate of Mattingly and Stevenson [7] for the effective coupling $\alpha_R(s) \sim 0.85$ for $\sqrt{s} < 0.3\,\text{GeV}$ determined from e^+e^- annihilation, especially if one takes into account the perturbative commensurate scale relation, $\alpha_\tau(m_{\tau'}^2) = \alpha_R(s^*)$, where $s^* \simeq 0.10\, m_{\tau'}^2$. This behavior is not consistent with the coupling having a Landau pole, but rather shows that the physical coupling is close to constant at low scales, suggesting that physical QCD couplings are effectively constant or "frozen" at low scales.

Figure 1 shows a comparison of the experimentally determined effective charge $\alpha_\tau(s)$ with solutions to the evolution equation for α_τ at two-, three-, and four-loop order normalized at m_τ. At three loops the behavior of the perturbative solution drastically changes, and instead of diverging, it freezes to a value $\alpha_\tau \simeq 2$ in the infrared. The reason for this fundamental change is, the negative sign of $\beta_{\tau,2}$. This result is not perturbatively stable since the evolution of the coupling is governed by the highest order term. This is illustrated by the widely different results obtained for three different values of the unknown four loop term $\beta_{\tau,3}$ which are also shown[a] It is interesting to note that the central four-loop solution is in good agreement with the data all the way down to $s \simeq 1\,\text{GeV}^2$.

It has also been argued that $\alpha_R(s)$ freezes perturbatively to all orders [3]. In fact since all observables are related by commensurate scale relations, they all should have an IR fixed point [5]. This result is also consistent with Dyson-Schwinger equation studies of the physical gluon propagator in Landau gauge [1]. In contrast, Cucchieri and Zwanziger [24,25,26,2,27] have shown that the QCD coupling defined from the D_{44} term in the Coulomb gauge propagator in quenched lattice gauge theory exhibits a confining $\frac{1}{k^2}$ behavior. For a discussion on how to reconcile these disparate results, see ref. [28].

The results for α_τ resemble the behavior of the one-loop "time-like"

[a]The values of $\beta_{\tau,3}$ used are obtained from the estimate of the four loop term in the perturbative series of R_τ, $K_4^{\overline{\rm MS}} = 25 \pm 50$ [23].

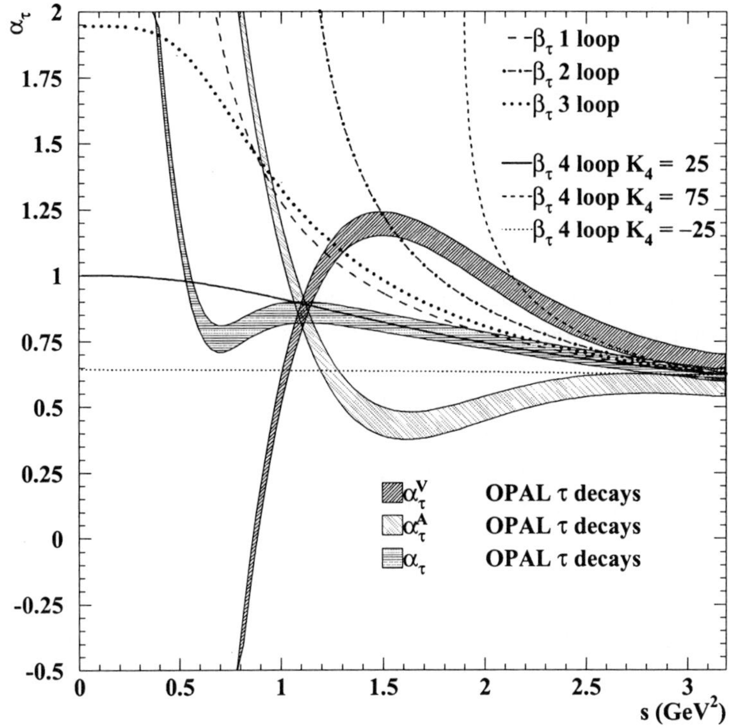

Figure 1. The effective charge α_τ for non-strange hadronic decays of a hypothetical τ lepton with $m_{\tau'}^2 = s$ compared to solutions of the fixed order evolution equation for α_τ at two-, three-, and four-loop order. The error bands include statistical and systematic errors.

effective coupling [29,30,31]

$$\alpha_{\text{eff}}(s) = \frac{4\pi}{\beta_0}\left\{\frac{1}{2} - \frac{1}{\pi}\arctan\left[\frac{1}{\pi}\ln\frac{s}{\Lambda^2}\right]\right\} \tag{1}$$

which is finite in the infrared and freezes to the value $\alpha_{\text{eff}}(s) = 4\pi/\beta_0$ as $s \to 0$. It is instructive to expand the "time-like" effective coupling for large s,

$$\alpha_{\rm eff}(s) = \frac{4\pi}{\beta_0 \ln(s/\Lambda^2)} \left\{ 1 - \frac{1}{3}\frac{\pi^2}{\ln^2(s/\Lambda^2)} + \frac{1}{5}\frac{\pi^4}{\ln^4(s/\Lambda^2)} + \dots \right\}$$
$$= \alpha_s(s) \left\{ 1 - \frac{\pi^2\beta_0^2}{3}\left(\frac{\alpha_s(s)}{4\pi}\right)^2 + \frac{\pi^4\beta_0^4}{5}\left(\frac{\alpha_s(s)}{4\pi}\right)^4 + \dots \right\}.$$

This shows that the "time-like" effective coupling is a resummation of $(\pi^2\beta_0^2\alpha_s^2)^n$-corrections to the usual running couplings. The finite coupling $\alpha_{\rm eff}$ given in Eq. (1) obeys standard PQCD evolution at LO. Thus one can have a solution for the perturbative running of the QCD coupling which obeys asymptotic freedom but does not have a Landau singularity.

The near constancy of the effective QCD coupling at small scales helps explain the empirical success of dimensional counting rules for the power law fall-off of form factors and fixed angle scaling. As shown in the references [32,33], one can calculate the hard scattering amplitude T_H for such processes [34] without scale ambiguity in terms of the effective charge α_τ or α_R using commensurate scale relations. The effective coupling is evaluated in the regime where the coupling is approximately constant, in contrast to the rapidly varying behavior from powers of α_s predicted by perturbation theory (the universal two-loop coupling). For example, the nucleon form factors are proportional at leading order to two powers of α_s evaluated at low scales in addition to two powers of $1/q^2$; The pion photoproduction amplitude at fixed angles is proportional at leading order to three powers of the QCD coupling. The essential variation from leading-twist counting-rule behavior then only arises from the anomalous dimensions of the hadron distribution amplitudes.

Parisi [35] has shown that perturbative QCD becomes a conformal theory for $\beta \to 0$ and zero quark mass. There are a number of useful phenomenological consequences of near conformal behavior: the conformal approximation with zero β function can be used as template for QCD analyses [36,37] such as the form of the expansion polynomials for distribution amplitudes [14,38]. The near-conformal behavior of QCD is also the basis for commensurate scale relations [19] which relate observables to each other without renormalization scale or scheme ambiguities [39]. An important example is the generalized Crewther relation [40]. In this method the effective charges of observables are related to each other in conformal gauge theory; the effects of the nonzero QCD $\beta-$ function are then taken into account using the BLM method [41] to set the scales of the respective couplings. Since the QCD running couplings are bounded, integration over the running coupling is finite, and the arguments leading to renormalon resummations do not result. The magnitude of the corresponding effective charge [32] $\alpha_s^{\text{exclusive}}(Q^2) = F_\pi(Q^2)/4\pi Q^2 F^2_{\gamma\pi^0}(Q^2)$ for exclusive amplitudes is connected to α_τ by a commensurate scale relation. Its magnitude: $\alpha_s^{\text{exclusive}}(Q^2) \sim 0.8$ at small Q^2, is sufficiently large as to explain the observed magnitude of exclusive amplitudes such as the pion form factor using the asymptotic distribution amplitude.

2. AdS/CFT and Near-Conformal Field Theory

As shown by Maldacena [42], there is a remarkable correspondence between large N_C supergravity theory in a higher dimensional anti-de Sitter space and supersymmetric QCD in 4-dimensional space-time. String/gauge duality provides a framework for predicting QCD phenomena based on the conformal properties of the ADS/CFT correspondence. For example, Polchinski and Strassler [43] have shown that the power-law fall-off of hard exclusive hadron-hadron scattering amplitudes at large momentum transfer can be derived without the use of perturbation theory by using the scaling properties of the hadronic interpolating fields in the large-r region of AdS space. Thus one can use the Maldacena correspondence to compute the leading power-law falloff of exclusive processes such as high-energy fixed-angle scattering of gluonium-gluonium scattering in supersymmetric QCD. The resulting predictions for hadron physics effectively coincide [43,44,45] with QCD dimensional counting rules:[46,47,48]

$$\frac{d\sigma_{H_1 H_2 \to H_3 H_4}}{dt} = \frac{F(t/s)}{s^{n-2}} \tag{2}$$

where n is the sum of the minimal number of interpolating fields. (For a recent review of hard fixed θ_{CM} angle exclusive processes in QCD see the references [49].) As shown by Brower and Tan [44], the non-conformal dimensional scale which appears in the QCD analysis is set by the string constant, the slope of the primary Regge trajectory $\Lambda^2 = \alpha'_R(0)$ of the supergravity theory. Polchinski and Strassler [43] have also derived counting rules for deep inelastic structure functions at $x \to 1$ in agreement with perturbative QCD predictions [50] as well as Bloom-Gilman exclusive-inclusive duality.

The supergravity analysis is based on an extension of classical gravity theory in higher dimensions and is nonperturbative. Thus the usual analyses of exclusive processes, which were derived in perturbation theory can be extended by the Maldacena correspondence to all orders. An interesting point is that the hard scattering amplitudes which are normally or order α_s^p

in PQCD appear as order $\alpha_s^{p/2}$ in the supergravity predictions. This can be understood as an all-orders resummation of the effective potential [42,51].

The superstring theory results are derived in the limit of a large N_C [52]. For gluon-gluon scattering, the amplitude scales as $1/N_C^2$. Frampton has shown how to extend the analysis to the fundamental representation [53]. For color-singlet bound states of quarks, the amplitude scales as $1/N_C$. This large N_C-counting in fact corresponds to the quark interchange mechanism [54]. For example, for $K^+p \to K^+p$ scattering, the u-quark exchange amplitude scales approximately as $\frac{1}{u}\frac{1}{t^2}$, which agrees remarkably well with the measured large θ_{CM} dependence of the K^+p differential cross section [55]. This implies that the nonsinglet Reggeon trajectory asymptotes to a negative integer [56], in this case, $\lim_{-t\to\infty}\alpha_R(t) \to -1$.

Pinch contributions corresponding to the independent scattering mechanism of Landshoff [57] are absent in the superstring derivation. This can be understood by the fact that amplitudes based on gluon exchange between color-singlet hadrons is suppressed at large N_C [58]. Furthermore, the independent scattering amplitudes are suppressed by Sudakov form factors which fall faster than any power in a theory with a fixed-point coupling such as conformal QCD [48,59].

The leading-twist results for hard exclusive processes correspond to the suppression of hadron wave functions with non-zero orbital angular momentum, which is the principle underlying the selection rules corresponding to hadron helicity conservation [60]. The suppression can be understood as follows: the LF wave function with nonzero angular momentum in the constituent rest frame $\sum \vec{k}_i = 0$ can be determined by iterating the one gluon exchange kernel. They then have the structure [61,62]

$$\psi_{L_z=1} = \frac{\vec{S}\cdot\widehat{n}\times\vec{k}_\perp}{D(k_\perp^2,x)}\psi_{L_z=0} \tag{3}$$

or

$$\psi_{L_z=1} = \frac{\widehat{\epsilon}\cdot\widehat{n}\times\vec{k}_\perp}{D(k_\perp^2,x)}\psi_{L_z=0} \tag{4}$$

where the light-front energy denominator $D(k_\perp^2,x) \sim k_\perp^2$ at high transverse momentum, $\widehat{n}$ is the light-front quantization direction, and $\widehat{\epsilon}$ is a spin-one polarization vector. This leads to the Λ/Q suppression of spin-flip amplitudes in QCD. For example, such wave functions lead to the large momentum transfer prediction $A_{LL} \sim 1/3$ for $pp \to pp$ elastic scattering [58] at large angles and momentum transfer and the asymptotic prediction $F_2(t)/F_1(t) \propto t^{-2}$ modulo powers of $\log t$ [63].

3. Light-Front Quantization

The concept of a wave function of a hadron as a composite of relativistic quarks and gluons is naturally formulated in terms of the light-front Fock expansion at fixed light-front time, $\tau = x \cdot \omega$. The four-vector ω, with $\omega^2 = 0$, determines the orientation of the light-front plane; the freedom to choose ω provides an explicitly covariant formulation of light-front quantization [64]. Although LFWFs depend on the choice of the light-front quantization direction, all observables such as matrix elements of local current operators, form factors, and cross sections are light-front invariants – they must be independent of ω_μ.

The light-front wave functions (LFWFs) $\psi_n(x_i, k_{\perp_i}, \lambda_i)$, with $x_i = \frac{k_i \cdot \omega}{P \cdot \omega}$, $\sum_{i=1}^{n} x_i = 1$, $\sum_{i=1}^{n} k_{\perp_i} = 0_\perp$, are the coefficient functions for n partons in the Fock expansion, providing a general frame-independent representation of the hadron state. Matrix elements of local operators such as spacelike proton form factors can be computed simply from the overlap integrals of light front wave functions in analogy to nonrelativistic Schrödinger theory. In principle, one can solve for the LFWFs directly from the fundamental theory using methods such as discretized light-front quantization, the transverse lattice, lattice gauge theory moments, or Bethe–Salpeter techniques. The determination of the hadron LFWFs from phenomenological constraints and from QCD itself is a central goal of hadron and nuclear physics. Reviews of nonperturbative light-front methods may be found in the references [65,64,66,67]. One can also project the known solutions of the Bethe–Salpeter equation to equal light-front time, thus producing hadronic light-front Fock wave functions. A potentially important method is to construct the $q\bar{q}$ Green's function using light-front Hamiltonian theory, with DLCQ boundary conditions and Lippmann-Schwinger resummation. The zeros of the resulting resolvent projected on states of specific angular momentum J_z can then generate the meson spectrum and their light-front Fock wavefunctions. The DLCQ properties and boundary conditions allow a truncation of the Fock space while retaining the kinematic boost and Lorentz invariance of light-front quantization.

One of the central issues in the analysis of fundamental hadron structure is the presence of non-zero orbital angular momentum in the bound-state wave functions. The evidence for a "spin crisis" in the Ellis-Jaffe sum rule signals a significant orbital contribution in the proton wave function [68,69]. The Pauli form factor of nucleons is computed from the overlap of LFWFs differing by one unit of orbital angular momentum $\Delta L_z = \pm 1$. Thus the fact that the anomalous moment of the proton is non-zero requires nonzero orbital angular momentum in the proton wavefunction [70]. In the light-front method, orbital angular momentum is treated explicitly; it includes the orbital contributions induced by relativistic effects, such as the spin-orbit effects normally associated with the conventional Dirac spinors.

In recent work, Dae Sung Hwang, John Hiller, Volodya Karmonov [62], and I have studied the analytic structure of LFWFs using the explicitly Lorentz-invariant formulation of the front form. Eigensolutions of the Bethe-Salpeter equation have specific angular momentum as specified by the Pauli-Lubanski vector. The corresponding LFWF for an n-particle Fock state evaluated at equal light-front time $\tau = \omega \cdot x$ can be obtained by integrating the Bethe-Salpeter solutions over the corresponding relative light-front energies. The resulting LFWFs $\psi^I_n(x_i, k_{\perp i})$ are functions of the light-cone momentum fractions $x_i = k_i \cdot \omega / p \cdot \omega$ and the invariant mass squared of the constituents $M_0^2 = (\sum_{i=1}^n k_i^\mu)^2 = \sum_{i=1}^n \left[\frac{k_\perp^2 + m^2}{x}\right]_i$ and the light-cone momentum fractions $x_i = k \cdot \omega / p \cdot \omega$ each multiplying spin-vector and polarization tensor invariants which can involve ω^μ. The resulting LFWFs for bound states are eigenstates of the Karmanov–Smirnov kinematic angular momentum operator [71]. Thus LFWFs satisfy all Lorentz symmetries of the front form, including boost invariance, and they are proper eigenstates of angular momentum.

4. AFS/CFT Correspondence and Light-Front Wavefunctions

One can also use the scaling properties of the hadronic interpolating operator in the extended AdS/CFT space-time theory to determine the scaling of light-front hadronic wavefunctions at high relative transverse momentum. De Teramond and I [58] have also shown how the angular momentum dependence of the light-front wavefunctions also follow from the conformal properties of the AdS/CFT correspondence. The scaling predictions agree with the perturbative QCD analysis given in the references [61], but the AdS/CFT analysis is performed at strong coupling without the use of perturbation theory. The near-conformal scaling properties of light-front wavefunctions lead to a number of other predictions for QCD which are normally discussed in the context of perturbation theory, such as constituent counting scaling laws for the leading power fall-off of form factors and hard exclusive scattering amplitudes for QCD processes. The ratio of Pauli to Dirac baryon form factor have the nominal asymptotic form $F_2(Q^2)/F_1(Q^2) \sim 1/Q^2$, modulo logarithmic corrections, in agreement with the perturbative results [63]. Our analysis can also be extended to study the spin structure of scattering amplitudes at large transverse momentum and other processes which are dependent on the scaling and orbital angular momentum structure of light-front wavefunctions.

Acknowledgements

I wish to thank Professor H. Suganuma and his colleagues at the Tokyo Institute of Technology, and The Institute of Physical and Chemical Research (RIKEN) for hosting this outstanding meeting. This talk is based on collaborations with Guy de Teramond, John Hiller, Dae Sung Hwang, Volodya Karmanov, Sven Menke, Carlos Merino, and Johan Rathsman. This work was supported by the U.S. Department of Energy, contract DE–AC03–76SF00515.

References

1. L. von Smekal, R. Alkofer and A. Hauck, Phys. Rev. Lett. **79**, 3591 (1997) [arXiv:hep-ph/9705242].
2. D. Zwanziger, arXiv:hep-ph/0303028.
3. D. M. Howe and C. J. Maxwell, Phys. Lett. B **541**, 129 (2002) [arXiv:hep-ph/0204036].
4. A. C. Aguilar, A. A. Natale and P. S. Rodrigues da Silva, Phys. Rev. Lett. **90**, 152001 (2003) [arXiv:hep-ph/0212105].
5. D. M. Howe and C. J. Maxwell, arXiv:hep-ph/0303163.
6. S. Furui and H. Nakajima, arXiv:hep-lat/0309166.
7. A. C. Mattingly and P. M. Stevenson, Phys. Rev. D **49**, 437 (1994) [arXiv:hep-ph/9307266].
8. S. J. Brodsky, S. Menke, C. Merino and J. Rathsman, Phys. Rev. D **67**, 055008 (2003) [arXiv:hep-ph/0212078].
9. M. Baldicchi and G. M. Prosperi, Phys. Rev. D **66**, 074008 (2002) [arXiv:hep-ph/0202172].
10. G. Parisi and R. Petronzio, Phys. Lett. B **94**, 51 (1980).
11. V. Gribov, Eur. Phys. J. C **10**, 71 (1999) [arXiv:hep-ph/9807224].
12. Y. L. Dokshitzer and B. R. Webber, Phys. Lett. B **352**, 451 (1995) [arXiv:hep-ph/9504219].
13. D. V. Shirkov and I. L. Solovtsov, Phys. Rev. Lett. **79**, 1209 (1997) [arXiv:hep-ph/9704333].
14. For an excellent review of the uses of conformal symmetry in QCD, see V. M. Braun, G. P. Korchemsky and D. Muller, arXiv:hep-ph/0306057.
15. G. Grunberg, Phys. Lett. **B95**, 70 (1980) [Erratum-ibid. **B110**, 501 (1982)].
16. G. Grunberg, Phys. Rev. **D29**, 2315 (1984).

17. S. J. Brodsky, M. S. Gill, M. Melles and J. Rathsman, Phys. Rev. D **58**, 116006 (1998) [arXiv:hep-ph/9801330].
18. S. J. Brodsky, M. Melles and J. Rathsman, Phys. Rev. D **60**, 096006 (1999) [arXiv:hep-ph/9906324].
19. S. J. Brodsky and H. J. Lu, Phys. Rev. D **51**, 3652 (1995) [arXiv:hep-ph/9405218].
20. S. J. Brodsky, J. R. Pelaez and N. Toumbas, Phys. Rev. D **60**, 037501 (1999) [arXiv:hep-ph/9810424].
21. M. Neubert, Nucl. Phys. B **463**, 511 (1996) [arXiv:hep-ph/9509432].
22. K. Ackerstaff *et al.* [OPAL Collaboration], Eur. Phys. J. C **7**, 571 (1999) [arXiv:hep-ex/9808019].
23. F. Le Diberder and A. Pich, Phys. Lett. **B289**, 165 (1992).
24. A. Cucchieri and D. Zwanziger, Nucl. Phys. Proc. Suppl. **119**, 727 (2003) [arXiv:hep-lat/0209068].
25. D. Zwanziger, Phys. Rev. Lett. **90**, 102001 (2003) [arXiv:hep-lat/0209105].
26. J. Greensite and S. Olejnik, Phys. Rev. D **67**, 094503 (2003) [arXiv:hep-lat/0302018].
27. E. S. Swanson, arXiv:hep-ph/0310089.
28. A. P. Szczepaniak, arXiv:hep-ph/0306030.
29. M. Beneke and V. M. Braun, Phys. Lett. **B348**, 513 (1995) [arXiv:hep-ph/9411229].
30. P. Ball, M. Beneke and V. M. Braun, Nucl. Phys. **B452**, 563 (1995) [arXiv:hep-ph/9502300].
31. Y. L. Dokshitzer, G. Marchesini and B. R. Webber, Nucl. Phys. **B469**, 93 (1996) [arXiv:hep-ph/9512336].
32. S. J. Brodsky, C. R. Ji, A. Pang and D. G. Robertson, Phys. Rev. D **57**, 245 (1998) [arXiv:hep-ph/9705221].
33. B. Melic, B. Nizic and K. Passek, Phys. Rev. D **65**, 053020 (2002) [arXiv:hep-ph/0107295].
34. G. P. Lepage and S. J. Brodsky, Phys. Rev. D **22**, 2157 (1980).
35. G. Parisi, Phys. Lett. B **39**, 643 (1972).
36. S. J. Brodsky, Y. Frishman and G. P. Lepage, Phys. Lett. B **167**, 347 (1986).
37. S. J. Brodsky, P. Damgaard, Y. Frishman and G. P. Lepage, Phys. Rev. D **33**, 1881 (1986).
38. V. M. Braun, S. E. Derkachov, G. P. Korchemsky and A. N. Manashov, Nucl. Phys. B **553**, 355 (1999) [arXiv:hep-ph/9902375].
39. S. J. Brodsky, E. Gardi, G. Grunberg and J. Rathsman, Phys. Rev. D **63**, 094017 (2001) [arXiv:hep-ph/0002065].
40. S. J. Brodsky, G. T. Gabadadze, A. L. Kataev and H. J. Lu, Phys. Lett. **B372**, 133 (1996) [arXiv:hep-ph/9512367].
41. S. J. Brodsky, G. P. Lepage and P. B. Mackenzie, Phys. Rev. D **28**, 228 (1983).
42. J. M. Maldacena, Adv. Theor. Math. Phys. **2**, 231 (1998) [Int. J. Theor. Phys. **38**, 1113 (1999)] [arXiv:hep-th/9711200].
43. J. Polchinski and M. J. Strassler, Phys. Rev. Lett. **88**, 031601 (2002) [arXiv:hep-th/0109174].

44. R. C. Brower and C. I. Tan, Nucl. Phys. B **662**, 393 (2003) [arXiv:hep-th/0207144].
45. O. Andreev, Phys. Rev. D **67**, 046001 (2003) [arXiv:hep-th/0209256].
46. S. J. Brodsky and G. R. Farrar, Phys. Rev. Lett. **31**, 1153 (1973).
47. V. A. Matveev, R. M. Muradian and A. N. Tavkhelidze, Lett. Nuovo Cim. **7**, 719 (1973).
48. S. J. Brodsky and G. R. Farrar, Phys. Rev. D **11**, 1309 (1975).
49. S. J. Brodsky, arXiv:hep-ph/0208158.
50. S. J. Brodsky, M. Burkardt and I. Schmidt, Nucl. Phys. B **441**, 197 (1995) [arXiv:hep-ph/9401328].
51. S. J. Rey and J. T. Yee, Eur. Phys. J. C **22**, 379 (2001) [arXiv:hep-th/9803001].
52. G. 't Hooft, Nucl. Phys. B **72**, 461 (1974).
53. P. H. Frampton, arXiv:hep-th/0305160.
54. J. F. Gunion, S. J. Brodsky and R. Blankenbecler, Phys. Rev. D **8**, 287 (1973).
55. D. W. Sivers, S. J. Brodsky and R. Blankenbecler, Phys. Rept. **23**, 1 (1976).
56. R. Blankenbecler, S. J. Brodsky, J. F. Gunion and R. Savit, Phys. Rev. D **8**, 4117 (1973).
57. P. V. Landshoff, Phys. Rev. D **10**, 1024 (1974).
58. S. J. Brodsky and G. de Teramond, SLAC-PUB-10020 (in preparation).
59. A. Duncan and A. H. Mueller, Phys. Lett. B **90**, 159 (1980).
60. S. J. Brodsky and G. P. Lepage, Phys. Rev. D **24**, 2848 (1981).
61. X. D. Ji, F. Yuan and J. P. Ma, Phys. Rev. Lett. **90**, 241601 (2003).
62. S. J. Brodsky, J. Hiller, D. S. Hwang, and V. Karmanov (in preparation).
63. A. V. Belitsky, X. d. Ji and F. Yuan, arXiv:hep-ph/0212351.
64. J. Carbonell, B. Desplanques, V. A. Karmanov, and J. F. Mathiot, Phys. Rep. **300**, 215 (1998) [arXiv:nucl-th/9804029].
65. S. J. Brodsky, H. C. Pauli and S. S. Pinsky, Phys. Rept. **301**, 299 (1998) [arXiv:hep-ph/9705477].
66. S. Dalley, Nucl. Phys. B (Proc. Suppl.) **108**, 145 (2002).
67. S. J. Brodsky, arXiv:hep-th/0304106.
68. R. L. Jaffe and A. Manohar, Nucl. Phys. B **337**, 509 (1990).
69. X. Ji, arXiv:hep-lat/0211016.
70. S. J. Brodsky and S. D. Drell, Phys. Rev. D **22**, 2236 (1980).
71. V. A. Karmanov and A.V. Smirnov, Nucl. Phys. A **546**, 691 (1992).

NON-PERTURBATIVE RENORMALIZATION GROUP ANALYSIS FOR DYNAMICAL CHIRAL SYMMETRY BREAKING IN QCD *

KEN-ICHI AOKI

Institute for Theoretical Physics
Kanazawa University
KANAZAWA 9201192, Japan
E-mail: aoki@hep.s.kanazawa-u.ac.jp

We analyze the dynamical chiral symmetry breaking by the non-perturbative renormalization group. We calculate the effective potential for the chiral order parameter and identify the critical behavior of the dynamical symmetry breaking. Using the newly obtained β-functions for the multi-fermion operators, we can go beyond the ladder approximation. Applying this method to QCD, we evaluate the chiral condensates which exhibits improvement of the gauge independence.

1. Introduction

We challenge a beyond-the-ladder calculation of the dynamical chiral symmetry breaking in QCD by using a non-ladder extension in the Non-Perturbative Renormalization Group (NPRG) method. The ladder approximation of the NPRG Local Potential β function had been integrated to give exactly the same results of the (improved) ladder Schwinger-Dyson equation for the chiral condensates and the dynamical mass of quark $\Sigma(0)$[1].

We add non-ladder diagrams to the NPRG ladder β function, restoring the gauge independence of the physical results. We obtain a set of β functions using the effective gluon vertex defined by the sum of the ladder type and the crossed type couplings. We numerically integrate this new β function to get the chiral condensates. It is enhanced compared with the previous ladder results, which are favorable phenomenologically. Also we evaluate the gauge parameter dependence of our results and find it is fairly improved compared to the ladder case[2].

*Collaborated work with Kaoru Takagi, Haruhiko Terao and Masashi Tomoyose

We stress here that our results are the first result in the long history of analyzing the dynamical chiral symmetry breaking in gauge theories, which goes beyond the (improved) ladder equipping with a systematic approximation method. This is realized by quite a new viewpoint of the NPRG method for the dynamical chiral symmetry breaking[3,8].

2. NPRG Equation and its Approximation

The starting point is the Euclidean path integral with the controlled momentum cutoff $\Lambda(t) = e^{-t}\Lambda_0$:

$$Z = \int^{\Lambda(t)} D\phi \exp[-S_{\text{eff}}(\phi; t)], \tag{1}$$

where S_{eff} is called the Wilsonian effective action. The NPRG equation describes how the Wilsonian effective action S_{eff} should change as the higher momentum degrees of freedom are integrated out. It is obtained by reducing $\Lambda(t)$ infinitesimally with fixing the partition function Z. Simultaneously we rescale the momentum variables and the fields by cutoff $\Lambda(t)$, since the change of the dimensionless quantities are of our physical interest. We obtain the following differential equation,

$$\begin{aligned}\frac{dS_{\text{eff}}[\phi;t]}{dt} = DS_{\text{eff}} - \int_{|p|\leq 1} \frac{d^Dp}{(2\pi)^D}\, \phi_i(p) \left(D_\phi - p^\mu \frac{\partial'}{\partial p^\mu}\right) \frac{\delta}{\delta\phi_i(p)} S_{\text{eff}} \\ - \frac{1}{2dt}\int' \frac{d^Dp}{(2\pi)^D} \left\{ \frac{\delta S_{\text{eff}}}{\delta\phi_i(p)} \left(\frac{\delta^2 S_{\text{eff}}}{\delta\phi_i(p)\delta\phi_j(-p)}\right)^{-1} \frac{\delta S_{\text{eff}}}{\delta\phi_j(-p)} \right. \\ \left. - \text{str}\ln\left(\frac{\delta^2 S_{\text{eff}}}{\delta\phi_i(p)\delta\phi_j(-p)}\right)\right\},\end{aligned} \tag{2}$$

where D is the space time dimension, D_ϕ is the dimension of ϕ including its anomalous dimension, the second primed integral denotes integration over the infinitesimal shell modes of momenta $e^{-dt} \leq p \leq 1$, and the prime in the derivative indicates that it does not act on the δ function in $\frac{\delta}{\delta\phi_p}S_{\text{eff}}$. The subscript i represents every Lorentz and internal symmetry indices. This equation is known as a sharp cutoff version of the NPRG, and is called the Wegner-Houghton (W-H) equation.[4]

It is inevitable to approximate the above equation for the practical calculations. We employ the Local Potential Approximation (LPA), in which any derivative couplings are dropped except for the fixed kinetic terms,

$$S_{\text{eff}} = \int \frac{d^Dp}{(2\pi)^D} \left\{ \frac{1}{2}\phi_i K_{ij}(p)\phi_j + V_{\text{eff}}(\phi) \right\}, \tag{3}$$

where $K_{ij}(p)$ is a matrix of the canonical kinetic terms, and $V_{\rm eff}$ is called the Wilsonian effective potential. For example, we take a theory of one scalar φ and one Dirac fermion ψ and its conjugate $\bar{\psi}$. The matrix $K(p)$ in $(\phi, \psi, \bar{\psi})$-space is written as

$$K(p) = \begin{pmatrix} p^2 & 0 & 0 \\ 0 & 0 & -i\not{p}^T \\ 0 & -i\not{p} & 0 \end{pmatrix} . \tag{4}$$

In this approximation the W-H equation is reduced to a nonlinear partial differential equation for the Wilsonian effective potential $V_{\rm eff}(\phi, t)$,[5]

$$\frac{\partial V_{\rm eff}(\phi;t)}{\partial t} = DV_{\rm eff} - D'_\phi \phi_i \frac{\partial V_{\rm eff}}{\partial \phi_i} + \frac{1}{2}\int' \frac{d^D p}{(2\pi)^D} {\rm str}\ln\left(K_{ij} + \frac{\partial^2 V_{\rm eff}}{\partial\phi_i \partial\phi_j}\right), \tag{5}$$

where D'_ϕ denotes the canonical dimension of field ϕ.

3. NPRG for the dynamical chiral symmetry breaking

Now we apply Eq.(5) to QCD with three massless quarks. We take the local potential effective action,

$$S_{\rm eff}[\psi, \bar{\psi}, A_\mu;\ t] = \int d^4x \left[V_{\rm eff}(\psi, \bar{\psi};\ t) + \bar{\psi}(\not{\partial} - g\not{A})\psi \right.$$
$$\left. + \frac{1}{4}(F^a_{\mu\nu})^2 + \frac{1}{2\alpha}(\partial_\mu A^a_\mu)^2\right], \tag{6}$$

where α is the gauge parameter, and ψ denotes massless triplet quarks. Furthermore, we take a sub theory space spanned by polynomials in a scalar operator up to some maximum power $nmax$,

$$V_{\rm eff}(\psi, \bar{\psi};\ t) = \sum_n^{nmax} \frac{G_{2n}(t)}{n}\left[(\bar{\psi}\psi)^2 + (\bar{\psi}i\gamma_5\psi)^2\right]^n . \tag{7}$$

In fact in this subspace we can analyze the critical behavior of the dynamical chiral symmetry breaking. However, to evaluate the physical quantities like the chiral condensates, we need to avoid some singular behaviors intrinsic to the fermionic theory space. Introducing a composite operator ϕ corresponding to the order parameter $\langle\bar{\psi}\psi\rangle$ enables us to carry out the calculation of the Wilsonian effective potential.[1]

Then the bare lagrangian is modified as follows:

$$L = L_{\rm org}\,[\psi, \bar{\psi}, A;\ t] + \frac{1}{2}(\phi - y\bar{\psi}\psi)^2$$
$$= \frac{1}{4}F^a_{\mu\nu}F^a_{\mu\nu} + \frac{1}{2\alpha}(\partial_\mu A^a_\mu)^2 + \bar{\psi}(\not{\partial} - g\not{A} - y\phi)\psi + \frac{1}{2}\phi^2 + \frac{y^2}{2}(\bar{\psi}\psi)^2. \tag{8}$$

and we work with the following Wilsonian effective potential:

$$L_{\text{eff}} = \frac{1}{4} F^a_{\mu\nu} F^a_{\mu\nu} + \bar{\psi}(\partial\!\!\!/ - gA\!\!\!/)\psi + V_{\text{eff}}(\phi, \sigma;\ t), \tag{9}$$

$$\begin{aligned} V_{\text{eff}}(\phi, \sigma;\ t) &= \tilde{G}_0(\phi;\ t) + \sum_{n}^{nmax} \frac{1}{n} \tilde{G}_n(\phi;\ t)\ \sigma^n \\ &= \tilde{G}_0(\phi;\ t) + \tilde{G}_1(\phi;\ t)\sigma + \frac{1}{2}\tilde{G}_2(\phi;\ t)\sigma^2 + \cdots, \end{aligned} \tag{10}$$

where the notation $\sigma = \bar{\psi}\psi$ is introduced. In this formalism, it was shown that the chiral condensate $\langle\bar{\psi}\psi\rangle$ is proportional to the minimum position of the scalar potential $\tilde{G}_0(\phi)$, denoted by $\langle\phi\rangle$,

$$\langle\bar{\psi}\psi\rangle = \frac{1}{y}\langle\phi\rangle. \tag{11}$$

4. Improvement of the gauge independence

As noted before, the ladder part NPRG exactly reproduces the results obtained by the ladder SD equations. Namely the results by the ladder part NPRG depend on the gauge parameter α strongly[6]. In order to improve the gauge dependence, we develop a non-ladder extended approximation in the course of the systematic approximation of NPRG.

First we define the "massive" quark propagator,

$$\frac{1}{i\not{p} - m(\phi, \sigma)}, \tag{12}$$

where

$$\begin{aligned} m(\phi, \sigma) \equiv \frac{\partial V}{\partial \sigma} &= \frac{\partial}{\partial \sigma} \sum_{n=1}^{nmax} \frac{\tilde{G}_n(\phi)}{n} \sigma^n = \sum_{n=1}^{nmax} \tilde{G}_n(\phi)\ \sigma^{n-1} \\ &= \tilde{G}_1(\phi) + \tilde{G}_2(\phi)\sigma + \tilde{G}_3(\phi)\sigma^2 + \tilde{G}_4(\phi)\sigma^3 + \cdots. \end{aligned} \tag{13}$$

In the Feynman diagram language, the "massive" quark propagator is represented as

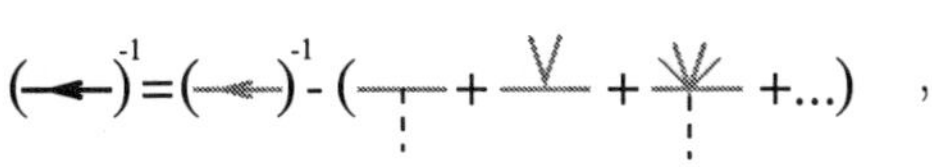

where the deep full line is a "massive" fermion propagator and the pale full lines are massless fermion operators and the dashed lines are the auxiliary fields ϕ. Using this massive propagator, the ladder part of NPRG β function is represented as follows:

$$\frac{dV_{\text{eff}}}{dt} =$$ 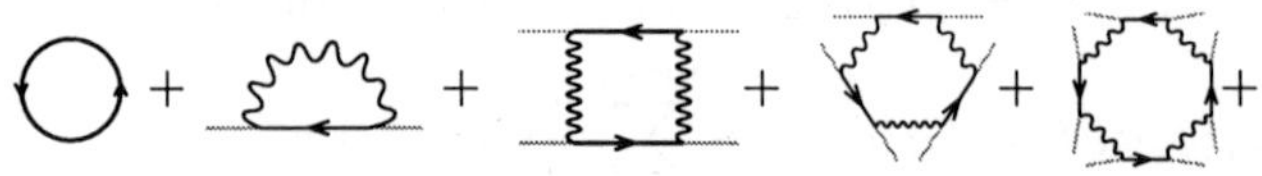 (14)

This β functions are known to reproduce the ladder SD results exactly[1].

Towards constituting a gauge invariant subset, we define a complex vertex, which is composed of "two" diagrams, the ladder type connection and the crossed type connection, using the "massive" quark propagator:

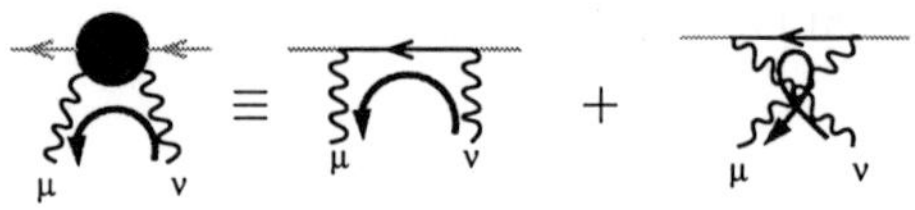

$$= \frac{-2g^2}{p^2+m^2}\left[ip_\alpha\epsilon^{\mu\nu\alpha\beta}\gamma_5\gamma_\beta + mg^{\mu\nu}\right], \tag{15}$$

where the wavy lines are gluons, the deep full lines are "massive" quarks, the pale full lines are external quark operators, and the curved arrows denote the direction of the shell-mode momentum p. Therefore this vertex itself comprises an infinite number of diagrams.

Then we construct a beyond-the-ladder β function as follows. Take the ladder type β function, and replace every pairs of the gluon vertices with the complex vertex defined above,

$$\frac{dV}{dt} = \frac{d}{dt}\left\{\tilde{G}_0 + \tilde{G}_1(\bar{\psi}\psi) + \frac{1}{2}\tilde{G}_2(\bar{\psi}\psi)^2 + \frac{1}{3}\tilde{G}_3(\bar{\psi}\psi)^3 + \frac{1}{4}\tilde{G}_4(\bar{\psi}\psi)^4 + \cdots\right\}$$

$$=$$ 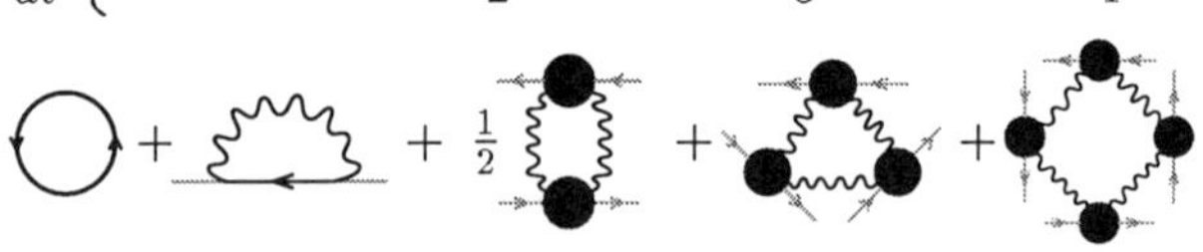 (16)

For example, the box diagram (the third term) contributing to the four-fermi operator now contains the crossed box diagram which is the key to restore the gauge invariance. Each diagram in the above contributes to an infinite number of coupling constants in the polynomial expansion.

We further limit our approximation by restricting the operator projection method to give polinomials in σ so that it corresponds to the ladder part β function when the complex vertex is approximated to be the simple ladder type. Now we write down general formulae of g^{2n} terms in the

NPRG equations as follows:

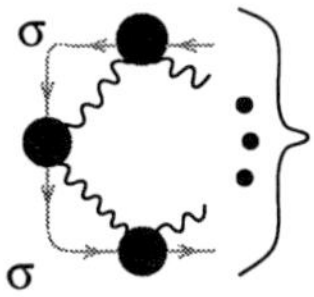

$$\sigma^n = -\frac{1}{8\pi^2}\frac{g^{2n}}{2^{n-2}\,n!}\,\frac{1}{(1+m^2)^n}\,\frac{1}{1+\delta_{n,1}+\delta_{n,2}}$$

$$\times\left[m^n(3+\alpha^n)+\sum_{k=1}^{[\frac{n}{2}]}\frac{(-)^k n!(2+4^k)}{(2k)!\,(n-2k)!}m^{n-2k}\right]\sigma^n . \quad (17)$$

5. Numerical Calculation and Results

Now we describe how to get the chiral condensates in QCD with our method. We work with the Wilsonian effective potential defined in Eq.(10) with some finite highest powers $nmax$, and we numerically integrate its NPRG equation. The NPRG equation is defined by the β function given in Eq.(17), that is, we take only the quantum loops of quarks and gluons and not of the scalar composites. The initial effective potential is taken from Eq.(8). During evolution the scalar field ϕ is kept fixed to be a certain value. The gauge coupling constant is set to follow the one-loop perturbative β function with three flavor quarks. Also the one loop anomalous dimension of quark fields are taken into account. We take the QCD scale parameter $\Lambda_{\rm QCD}$ to be 490 MeV and adopt the same infrared cutoff scheme of the gauge coupling constant divergence as in the previous work, since our results should be first compared with the ladder results.[7]

Integrating the NPRG equation, the effective potential finally stops moving except for the canonical scaling, where the cutoff scale has been lowered well below the quark mass scale. Then we get the scalar potential $\tilde{G}_0(\phi)$ at the fixed ϕ value. Scanning with respect to ϕ, we obtain the scalar potential function $\tilde{G}_0(\phi)$ and its minimum point $\langle\phi\rangle$, which gives the chiral condensates using Eq.(11). The chiral condensates obtained above should be regarded as the bare operator condensation at the initial highest cutoff scale. It should be renormalized through the standard procedure to give the renormalized condensates at 1 GeV scale.

Fig.1 shows the results with number of operators in our subspace. Convergence through increasing the number of operators looks very fine. Compaison with the ladder approximation results, which are equal to the ladder SD results, and the gauge parameter dependence of the results are also depicted. We conclude that the gauge independence of our new results are greatly improved compared to the ladder case. Using the Landau gauge,

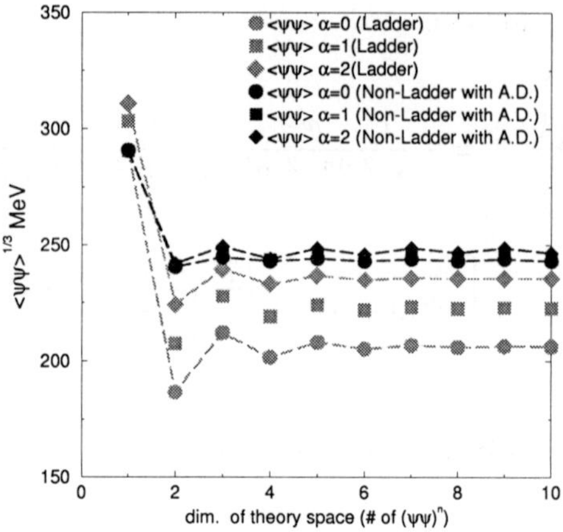

Figure 1. The chiral condensates with various gauge parameters.

our main results should read,

$$\left.\frac{\langle\bar{\psi}\psi\rangle^{1/3}}{\Lambda_{\rm QCD}}\right|_{\text{non-ladder}} = 0.512 \pm 0.014, \tag{18}$$

which is better in the phenomenological fitting compared with the previous ladder results of 0.439.

References

1. K-I. Aoki, K. Morikawa, J.-I. Sumi, H. Terao and M. Tomoyose, Prog. Theor. Phys. **102** (1999), 1151; Phys. Rev. **D61** (2000) 045008.
2. K-I. Aoki, K. Takagi, H. Terao and M. Tomoyose, Prog. Theor. Phys. **103-4** (2000) 815
3. K-I. Aoki, Prog. Theor. Phys. Suppl. **131** (1998), 129; Int. J. Mod. Phys. B **14** (2000) 1249
4. F.J.Wegner and A.Houghton, Phys. Rev. **A8** (1973), 401.
5. A. Hasenfratz and P. Hasenfratz, Nucl. Phys. **B270** (1986), 269.
   T. R. Morris, Phys. Lett, **B334** (1994), 355.
6. K-I. Aoki, M. Bando, T. Kugo, K. Hasebe and H. Nakatani, Prog. Theor. Phys. **81** (1989), 866.
7. K-I. Aoki, T. Kugo and M. G. Mitchard, Phys. Lett. **B266** (1991), 467.
   K-I. Aoki, M. Bando, T. Kugo, M. G. Mitchard and H. Nakatani, Prog. Theor. Phys. **84** (1990), 683.
8. K-I. Aoki, K. Morikawa, W. Souma, J. I. Sumi and H. Terao, Prog. Theor. Phys. **95** (1996), 409; Prog. Theor. Phys. **99** (1998), 451.
   K-I. Aoki, K. Morikawa, J.-I. Sumi, H. Terao and M. Tomoyose, Prog. Theor. Phys. **97** (1997), 479.

LIGHT MESONS AND INFRARED BEHAVIOR OF THE RUNNING COUPLING CONSTANT IN QCD

M. BALDICCHI AND G. M. PROSPERI

Dipartimento di Fisica dell'Università di Milano; I.N.F.N. Sez. di Milano; via Celoria 16, I20133 Milano, Italy; E-mail: prosperi@mi.infn.it

A previous method for handling bound states in QCD is briefly revisited. Taking advantage of the Feynman-Schwinger representation for the iterated quark propagator in an external field, it is possible to give closed representations for certain appropriate (second order) two point and four point Green functions, $H^{(2)}(x-y)$ and $H^{(4)}(x_1,x_2,y_1,y_2)$, as path integrals on quark world lines. Then, starting from reasonable assumptions on the Wilson line correlators, a Bethe-Salpeter equation for $H^{(4)}$ and a Dyson-Schwinger equation for $H^{(2)}$ can be obtained, which are consistent with the Goldstone theorem in the chiral limit. Such equations are too complicate to be solved directly. However, a reduced Salpeter equation can be derived which is tractable and has been applied to a calculation of the meson spectrum. The results are in general good agreement with the data, but with the important exceptions of the light pseudo scalars (that are related to the breaking of the chiral symmetry). In this scenario two important improvements can be introduced: a) the fixed coupling constant can be replaced by a running coupling constant $\alpha_s(Q^2)$ appropriately modified in the infrared region; b) the fixed mass in the reduced equation can be replaced for light quarks by an effective mass depending on the momentum of the particle, as suggested by the form of the DS equation. Then even the light pseudo scalar mesons can be made to agree with to their experimental value.

1. Introduction

In previous papers we have introduced what we have called a second order Bethe-Salpeter formalism, which works in terms of certain appropriate four point and two point Green functions, $H^{(4)}(x_1,x_2,y_1,y_2)$ and $H^{(2)}(x-y)$ [1]. Taking advantage of a Feynman-Schwinger representation for the "second order" quark propagator in an external field, it is possible to write $H^{(2)}$ and $H^{(4)}$ as path integrals on quark or antiquark world lines joining y to x, $y_{1,2}$ to $x_{1,2}$. In such representations the gauge field appears only trough *Wilson line* correlators like $\frac{1}{\sqrt{3}}\mathrm{Tr}_{\mathrm{color}}\langle\exp[i\int_y^x dz^\mu A_\mu(z)]$ or $\frac{1}{3}\mathrm{Tr}_{\mathrm{color}}\langle\exp[i\int_{y_1}^{x_1} dz^\mu A_\mu(z)]\exp[i\int_{x_2}^{y_2} dz^\mu A_\mu(z)]\rangle$ which reduce to ordinary Wilson loops W in the limit $x\to y$ or $x_1\to x_2$ and $y_1\to y_2$. In analogy with a usual assumption on W such correlators are written as the sum of their perturbative expressions and appropriate area terms and it is possible to obtain a confining Bethe-Salpeter equation for the quantity $H^{(4)}$ and a corresponding Dyson-Schwinger equation for $H^{(2)}$.

The above equations are too complicate to be solved directly. However, by a conventional three dimensional reduction, one can derive a tractable squared mass operator that can be applied to an evaluation of the spectrum [2]. With an appropriate choice of the parameters (quark masses, coupling constant, string tension) the results can be made in good agreement with the data in the entire framework of the light-light, light-heavy and heavy-heavy quark-antiquark sectors, with the important exception, however, of the light pseudo scalar mesons [3].

In this paper we want to discuss two important improvements to the above scenario:

a) the fixed coupling constant α_s is replaced by a running coupling constant $\alpha_s(Q^2)$, which amounts to take into account higher order contributions in α_s (and in particular closed quark loops);

b) an effective mass for the light quarks is introduced which depends on the momentum of the particle, as suggested by the form of the DS equation.

As well known, the usual perturbative expressions for $\alpha_s(Q^2)$ have an unphysical singularity for $Q^2 = \Lambda^2_{\rm QCD}$, which would be disastrous for our purpose. However, various modification have been proposed for the infrared region [4,5]. We have considered in particular the Shirkov-Solovtsov prescription, which rests only on general analyticity requirements. We find that with such prescription, and parametrazing the effective squared light quark masses by a simple polynomial, even the light pseudo scalar mesons can be made in agreement with their experimental values. It should be stressed that in the fit only the quark masses are treated as free parameters, while the constant Λ and σ occurring in the BS kernel are a priori fixed on the basis of high energy phenomenology and lattice simulations. Results in this line were already published in ref. [[6]]; here we present a more systematic study.

The remaining part of the paper is organized as the following plan: in Sects. 2 and 3 we briefly revue the second order Bethe-Salpeter formalism, in Sect. 4 we discuss the running coupling constant and effective mass, in sections 5 and 6 we reports the results and make some conclusions.

2. Second order correlators

After integrating out the fermionic fields, the appropriate ordinary (first order) four point function for the quark-antiquark bound state problem can be written

$$G^{(4)}(x_1, x_2; y_1, y_2) = -\frac{1}{3}\mathrm{Tr}_{\mathrm{color}}\langle S_1(x_1, y_1; A)S_2(y_2, x_2; A)\rangle. \qquad (1)$$

Here the subscripts 1 and 2 refer to the quark and the antiquark respectively, projection on the color singlet has been performed, $S(x, y; A)$ denotes the quark propagator in an external field,

$$(i\gamma^\mu D_\mu - m)S(x, y; A) = \delta^4(x - y)\,, \qquad (2)$$

and

$$\langle f[A]\rangle = \int DA\, M_F[A]\, e^{iS_G[A]} f[A]\,, \qquad (3)$$

with $M_F[A] = \mathrm{Det}\Pi_{j=1}^2[1 + g\gamma^\mu A_\mu(i\gamma_j^\nu \partial_{j\nu} - m_j)^{-1}]$.

The *second order* four point function is defined by

$$H^{(4)}(x_1, x_2; y_1, y_2) = -\frac{1}{3}\mathrm{Tr}_{\mathrm{color}}\langle \Delta_1^\sigma(x_1, y_1; A)\Delta_2^\sigma(y_2, x_2; A)\rangle\,, \qquad (4)$$

where $\Delta_1^\sigma(x, y; A)$ is the *second order* propagator that satisfies the second order differential equation

$$(D_\mu D^\mu + m^2 - \frac{1}{2}g\,\sigma^{\mu\nu}F_{\mu\nu})\Delta^\sigma(x, y; A) = -\delta^4(x - y)\,, \qquad (5)$$

($\sigma^{\mu\nu} = \frac{i}{2}[\gamma^\mu, \gamma^\nu]$) and is related to $S(x, y; A)$ by

$$S(x, y; A) = (i\gamma^\nu D_\nu + m)\Delta^\sigma(x, y; A)\,. \qquad (6)$$

The quantity $H^{(4)}$ is related to $G^{(4)}$ by an integro-differential operator, of the type $[(i\gamma_1^\mu\partial_{1\mu}+m_1)(i\gamma_2^\nu\partial_{2\nu}+m_2)+\ldots]$, which we do not need to specify in detail. The important fact is that the two functions are completely equivalent for the determination of the bound states since their Fourier transforms $\hat{H}^{(4)}$ and $\hat{G}^{(4)}$ have related analyticity properties and the same poles.

The advantage in considering second order quantities is that it is possible to write for $\Delta^\sigma(x, y; A)$ a generalized Feynman-Schwinger representation, i. e. to solve eq. (5) in terms of a quark path integral

$$\Delta^\sigma(x, y; A) = -\frac{i}{2}\int_0^\infty ds \int_y^x Dz \exp[-i\int_0^s d\tau \frac{1}{2}(m^2 + \dot{z}^2)]$$
$$S_0^s \mathrm{P}\exp[ig\int_0^s d\tau \dot{z}^\mu A_\mu(z)]\,, \qquad (7)$$

where the world-line $z^\mu = z^\mu(\tau)$ connecting y to x is written in the four-dimensional language in terms of an additional parameter τ, $S_0^s = \mathrm{T}\exp\Big[-\frac{1}{4}\int_0^s d\tau\sigma^{\mu\nu}\frac{\delta}{\delta S^{\mu\nu}(z)}\Big]$ and $\delta S^{\mu\nu} = dz^\mu\delta z^\nu - dz^\nu\delta z^\mu$ (the functional derivative being defined through an arbitrary deformation, $z \to z + \delta z$, of the world-line).

Replacing eq. (7) in eq. (4) a similar representation can be obtained for the 4-point function

$$H^{(4)}(x_1, x_2; y_1, y_2) = (\frac{1}{2})^2 \int_0^\infty ds_1 \int_0^\infty ds_2 \int_{y_1}^{x_1} Dz_1 \int_{y_2}^{x_2} Dz_2$$
$$\exp\Big\{-\frac{i}{2}\int_0^{s_1} d\tau_1(m_1^2 + \dot{z}_1^2) - \frac{i}{2}\int_0^{s_2} d\tau_2(m_2^2 + \dot{z}_2^2)\Big\} \tag{8}$$
$$\frac{1}{3}S_0^{s_1}S_0^{s_2}\mathrm{Tr}_{\mathrm{color}}\langle \mathrm{P}\exp\{ig\int_{y_1}^{x_1} dz_1^\mu\, A_\mu(z_1)\}\mathrm{P}\exp\{ig\int_{x_2}^{y_2} dz_2^\mu\, A_\mu(z_2)\}\rangle\,,$$

The interesting aspect of the above equation is that the gauge field appears in it only through the expectation value of the product of the two Wilson lines.

Similarly the second order two point function (uncolored full quark propagator) $H^{(2)}(x-y) = \frac{i}{\sqrt{3}}\mathrm{Tr}_{\mathrm{color}}\langle S(x, y : A)\rangle$ can be written

$$H^{(2)}(x-y) = \frac{1}{2}\int_0^\infty ds \int_y^x Dz \exp\Big\{-\frac{i}{2}\int_0^s d\tau(m^2 + \dot{z}^2)\Big\}$$
$$\frac{1}{\sqrt{3}}S_0^s\mathrm{Tr}_{\mathrm{color}}\langle \mathrm{P}\exp\{ig\int_y^x dz^\mu\, A_\mu(z)\}\rangle\,. \tag{9}$$

3. Bethe-Salpeter and Dyson-Schwinger equations

In the limit $x_2 \to x_1$, $y_2 \to y_1$ or $y \to x$ the two Wilson lines occurring in (8) or the single line occurring in (9) close in a single Wilson loop Γ,

$$W = \langle\frac{1}{3}\mathrm{Tr}_{\mathrm{color}}\mathrm{P}\exp\Big[\oint_\Gamma dz^\mu A_\mu(z)\Big]\rangle\,. \tag{10}$$

As in the previous papers we assume that in a first approximation $i\ln W$ can be written as the sum of its perturbative expression and an area term $i\ln W = (i\ln W)_{\mathrm{pert}} + \sigma S$. Then, at the lowest order in the coupling constant, we can assume

$$i\ln W = \frac{4}{3}g^2\oint dz^\mu \oint dz^{\nu\prime}D_{\mu\nu}(z-z') + \tag{11}$$
$$\sigma\oint dz^0\oint dz^{0\prime}\delta(z^0 - z^{0\prime})|\mathbf{z}-\mathbf{z}'|\int_0^1 d\lambda\Big\{1 - [\lambda\frac{d\mathbf{z}_{\mathrm{T}}}{dz^0} + (1-\lambda)\frac{d\mathbf{z}'_{\mathrm{T}}}{dz^{0\prime}}]^2\Big\}^{\frac{1}{2}}\,.$$

Notice that the surface term in (11) is written as the algebraic sum of successive equal time straight strips ($d\mathbf{z}_\mathrm{T}$ denotes the transversal component of $d\mathbf{z}$). For a flat loop or for other special geometries (e. g. for two quarks uniformly rotating around their fixed center of mass) this coincides obviously with the plane or the minimum surface delimited by Γ. That is not generally the case; and indeed, the right hand side of (11) usually depends on the reference frame. Since, however, in contrast e. g. with $S_{\min}$, such quantity maintains many of the analytic properties of the original $i \ln W$, we shall assume (11) to be valid for an arbitrary loop in the center of mass reference frame. Actually we shall assume this even for $x_2 \neq x_1$, $y_2 \neq y_1$ or $y \neq x$. In analogy with what happens in the pure perturbative case. In this way single perturbative and confinement contributions are put on the same foot and we may refer to them as a *gluon exchange* and a *string connection* between the two quarks.

Replacing (11) in (8) and (9) we obtain integral representations in terms of the quark World line alone. Then by various manipulation and using an appropriate iterative procedure, a Bethe-Salpeter equation for the function $H^{(4)}(x_1, x_2; y_1, y_2)$ and a Dyson-Schwinger equation for $H^{(2)}(x-y)$ can be derived in a kind of generalized ladder and rainbow approximation.

In momentum representation, the corresponding homogeneous BS-equation can be written

$$\Phi_P(k) = -i \int \frac{d^4u}{(2\pi)^4} \hat{I}_{ab}\big(k-u, \frac{1}{2}P + \frac{k+u}{2}, \frac{1}{2}P - \frac{k+u}{2}\big) \hat{H}_1^{(2)}\big(\frac{1}{2}P + k\big)\sigma^a \Phi_P(u) \sigma^b \hat{H}_2^{(2)}\big(-\frac{1}{2}P + k\big), \quad (12)$$

where we have set $\sigma^0 = 1$ and $a, b = 0, \mu\nu$, the center of mass frame has to be understood $P = (m_B, \mathbf{0})$ and $\Phi_P(k)$ denotes an appropriate *second order* wave function [a]. Similarly, in terms of the irreducible self-energy, defined by $\hat{H}^{(2)}(k) = \frac{i}{k^2-m^2} + \frac{i}{k^2-m^2} i\Gamma(k)\hat{H}^{(2)}(k)$, the DS-equation can be written also

$$\hat{\Gamma}(k) = \int \frac{d^4l}{(2\pi)^4} \hat{I}_{ab}\Big(k-l; \frac{k+l}{2}, \frac{k+l}{2}\Big)\sigma^a \hat{H}^{(2)}(l)\,\sigma^b\,. \quad (13)$$

[a] In terms of the second order field $\phi(x) = (i\gamma^\mu D_\mu + m)^{-1}\psi(x)$ this wave function is defined by

$$\langle 0|\phi(\frac{\xi}{2})\bar{\psi}(-\frac{\xi}{2})|P\rangle = \frac{1}{(2\pi)^2}\Phi_P(k) e^{-ik\xi}\,.$$

The kernels in (12) and (13) are the same in the two equations, consistently with the requirement of chiral symmetry limit [7], and are given by

$$\hat{I}_{0;0}(Q;p,p') = 4\int d^4\zeta\, e^{iQ\zeta} E(\zeta;p,p') = 16\pi\frac{4}{3}\alpha_s p^\alpha p'^\beta \hat{D}_{\alpha\beta}(Q) +$$

$$+4\sigma\int d^3\zeta e^{-i\mathbf{Q}\cdot\zeta}|\zeta|\epsilon(p_0)\epsilon(p_0')\int_0^1 d\lambda\{p_0^2 p_0'^2 - [\lambda p_0'\mathbf{p}_\mathrm{T} + (1-\lambda)p_0\mathbf{p}_\mathrm{T}']^2\}^{\frac{1}{2}}$$

$$\hat{I}_{\mu\nu;0}(Q;p,p') = 4\pi i\frac{4}{3}\alpha_s(\delta_\mu^\alpha Q_\nu - \delta_\nu^\alpha Q_\mu)p'_\beta \hat{D}_{\alpha\beta}(Q) -$$

$$-\sigma\int d^3\zeta\, e^{-i\mathbf{Q}\cdot\zeta}\epsilon(p_0)\frac{\zeta_\mu p_\nu - \zeta_\nu p_\mu}{|\zeta|\sqrt{p_0^2 - \mathbf{p}_\mathrm{T}^2}}p_0'$$

$$\hat{I}_{0;\rho\sigma}(Q;p,p') = -4\pi i\frac{4}{3}\alpha_s p^\alpha(\delta_\rho^\beta Q_\sigma - \delta_\sigma^\beta Q_\rho)\hat{D}_{\alpha\beta}(Q) +$$

$$+\sigma\int d^3\zeta\, e^{-i\mathbf{Q}\cdot\zeta}p_0\frac{\zeta_\rho p'_\sigma - \zeta_\sigma p'_\rho}{|\zeta|\sqrt{p_0'^2 - \mathbf{p}_\mathrm{T}'^2}}\epsilon(p_0')$$

$$\hat{I}_{\mu\nu;\rho\sigma}(Q;p,p') = \pi\frac{4}{3}\alpha_s(\delta_\mu^\alpha Q_\nu - \delta_\nu^\alpha Q_\mu)(\delta_\rho^\alpha Q_\sigma - \delta_\sigma^\alpha Q_\rho)\hat{D}_{\alpha\beta}(Q)\,, \qquad (14)$$

where in the second and in the third equation $\zeta_0 = 0$ has to be understood.

To find the $q\bar{q}$ spectrum, in principle one should solve first (13) and use the resulting propagator in (12). In practice this turns out to be a difficult task and one has to resort to the three dimensional equation which can be obtained from (12) by the so called instantaneous approximation. This consists in replacing $\hat{H}_j^{(2)}(k)$ in (12) with the free quark propagator $\frac{i}{k^2-m_j^2}$ and the kernels $\hat{I}_{ab}$ with its so called instantaneous approximation $\hat{I}_{ab}^{\mathrm{inst}}(\mathbf{k},\mathbf{k}')$ [1].

The reduced equation takes the form of the eigenvalue equation for a squared mass operator, $M^2 = M_0^2 + U$, with $M_0 = \sqrt{m_1^2+\mathbf{k}^2} + \sqrt{m_2^2+\mathbf{k}^2}$ and

$$\langle\mathbf{k}|U|\mathbf{k}'\rangle = \frac{1}{(2\pi)^3}\sqrt{\frac{w_1+w_2}{2w_1w_2}}\,\hat{I}_{ab}^{\mathrm{inst}}(\mathbf{k},\mathbf{k}')\sqrt{\frac{w_1'+w_2'}{2w_1'w_2'}}\sigma_1^a\sigma_2^b \qquad (15)$$

(for an explicit expression see ref. [6,3]). The quadratic form of the above equation obviously derives from the second order formalism we have used.

Alternatively, in more usual terms, one can look for the eigenvalue of the mass operator or center of mass Hamiltonian $H_{\mathrm{CM}} \equiv M = M_0 + V$ with V defined by $M_0V + VM_0 + V^2 = U$. Neglecting the term V^2 the linear form potential V can be obtained from U simply by the kinematic replacement $\sqrt{\frac{(w_1+w_2)(w_1'+w_2')}{w_1w_2w_1'w_2'}} \to \frac{1}{2\sqrt{w_1w_2w_1'w_2'}}$. Such expression is particularly useful for a comparison with models based on potential. In particular in the static limit V reduces to the Cornell potential

$$V = -\frac{4}{3}\frac{\alpha_s}{r} + \sigma r\,. \tag{16}$$

in the semirelativistic limit (up to $\frac{1}{m^2}$ terms after an appropriate Foldy-Wouthuysen transformation) equals the potential discussed in ref. [8], if the spin dependent terms are neglected but full relativistic kinematics is kept, it becomes identical to the potential of the (relativistic) flux tube model [1].

4. Running coupling constant and effective mass

In ref. [3] the spectrum was evaluated for both the operators M^2 and $H_{\rm CM}$ introduced in the preceding section, omitting the spin-orbit terms in the potential but including the hyperfine ones. As we told, with fixed coupling constant and quark masses, a general good fit of the data was obtained over the entire calculable spectrum with, however, the relevant exception of the light pseudoscalar mesons. At the light of the idea that such mesons should be Goldston massless particles in the chiral limit, this is not surprising [7]. The results obtained in ref. [6] suggest, however, that the situation can be greatly improved using a running coupling constant and an effective light quark mass function of the momentum, as implied in (13) for a kernel of the type (14).

At one loop the running coupling constant in QCD is usually written as

$$\alpha_s(Q^2) = \frac{4\pi}{\beta_0 \ln{(Q^2/\Lambda^2)}} \tag{17}$$

Q being the relevant energy scale, $\beta_0 = 11 - \frac{2}{3}N_f$, and N_f the number of flavors with masses smaller than Q. Such expression becomes singular and completely inadequate as Q^2 approaches Λ^2.

The most naive modification of eq. (17) consists in saturating $\alpha_s(Q^2)$ to a certain maximum value $\bar{\alpha}_s$ as Q^2 decrease (fig. 1) and in treating this value as a phenomenological parameter (truncation prescription).

Alternatively Shirkov and Solovtsov [5] replace (17) with

$$\alpha_s(Q^2) = \frac{4\pi}{\beta_0}\left(\frac{1}{\ln{(Q^2/\Lambda^2)}} + \frac{\Lambda^2}{\Lambda^2 - Q^2}\right). \tag{18}$$

This remains regular for $Q^2 = \Lambda^2$ and has a finite Λ independent limit, $\alpha_s(0) = 4\pi/\beta_0$, for $Q^2 \to 0$. Eq. (18) is obtained assuming a dispersion relation with a cut from $-\infty < Q^2 < 0$ and applying (17) to the evaluation of the spectral function alone.

In the quark-antiquark bound state problem the variable Q^2 can be identified with the squared momentum transfer $\mathbf{Q}^2 = (\mathbf{k} - \mathbf{k}')^2$. Then, typically, $\langle \mathbf{Q}^2 \rangle$ ranges between $(0.1\,\text{GeV})^2$ and $(1\,\text{GeV})^2$ for different quark masses and internal excitations and values of $\mathbf{Q}^2$ smaller than Λ^2 can be important. The specific infrared behavior is therefore expected to affect the spectrum and other properties of mesons.

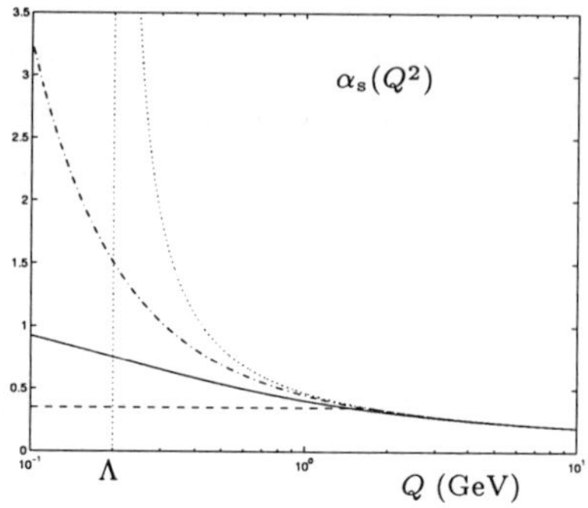

Figure 1. Running coupling constant $\alpha_s(Q^2)$ on logarithmic scale. Perturbative expression (dotted line), truncation prescription (dashed line), Shirkov-Solovtsov prescription (full line), Nesterenko prescription (dot-dashed line).

Coming to the quark masses, notice that, even neglecting the spin dependent part in $\hat{H}^{(2)}(k)$ (consistently with what we do for $\langle \mathbf{k}|U|\mathbf{k}'\rangle$), we could more sensibly write $\hat{H}^{(2)}(k) = i/(k^2 - m^2 - \Gamma_0(k))$ in the reduction procedure, rather than simply identify such quantity with its free expression ($\Gamma_0(k) = 0$). However, since the $I_a(Q, p, p')$ are not formally covariant, but given in terms of C.M. variables, $\Gamma_0(k)$ must depend separately on k_0^2 and $\mathbf{k}^2$. Then the pole of $H_2(k)$, defined by

$$k_0^2 - \mathbf{k}^2 - m^2 - \Gamma(k_0^2, \mathbf{k}^2) = 0\,, \tag{19}$$

could be written

$$k_0^2 = m_{\text{eff}}^2(|\mathbf{k}|) + \mathbf{k}^2\,, \tag{20}$$

where $m_{\text{eff}}^2(|\mathbf{k}|)$ would be a $|\mathbf{k}|$ dependent expression that is expected to approach the current m^2 for $|\mathbf{k}| \to 0$ and to increase toward a kind of constituent m'^2 as $|\mathbf{k}|$ increases [b]. Then, eventually, we obtain the same operator M^2 as given above, but with m_1 and m_2 replaced by expressions like $m_{\text{eff}}^2(|\mathbf{k}|)$.

[b]Take into account that from the virial theorem for a linear potential in the extreme relativistic approximation we have $\langle |\mathbf{k}| \rangle = \sigma \langle r \rangle$ and a large $|\mathbf{k}|$ corresponds to a peripheral interaction or, what is the same, to a small $|\mathbf{Q}|$.

As matter of fact we have not tried in this paper to evaluate $\Gamma(p)$ from first principles by solving eq. (13) but have simply parametrized $m^2_{\text{eff}}(|\mathbf{k}|)$ as a polynomial in $|\mathbf{k}|$ for the light quarks and used a fixed mass as in the preceding papers for strange and heavy quarks.

5. Numerical procedure and results

In the calculation of the eigenvalues of the operator M^2 introduced in sect. 4 we have used the same approximations and the same numerical procedure as in the previous papers. We have retained only the hyperfine part of the complicate spin dependent terms (spin orbit and tensorial ones) occurring in the potential U. We have solved the eigenvalue equation for H_{CM} with V equal its static limit (16) by the Rayleigh-Ritz method, using the three dimensional Harmonic oscillator basis and diagonalizing a 30×30 matrix. Then we have evaluated $\langle \phi_n | M^2 | \phi_n \rangle$ for the eigenfunction ϕ_n obtained in the first step.

In figs. 2 and 3 we have reported the results already obtained in ref. [3] with a truncated running coupling constant (crosses) and the new ones performed with the Shirkov-Solovtsov prescription (circlets), contrasted with the data (small lines) [9].

In both case we have taken $N_{\text{f}} = 4$ and $\Lambda = 200\,\text{MeV}$. With the truncation prescription, the light quark masses were fixed on typical current values, $m_u = m_s = 10\,\text{MeV}$, $m_s = 200\,\text{MeV}$, the maximum value for $\alpha_{\text{s}}(Q^2)$, the string tension and the heavy quark masses were treated as adjustable parameters and chosen as $\bar{\alpha}_{\text{s}} = 0.35$, $\sigma = 0.2\,\text{GeV}^2$ (in order to reproduce the correct $J/\Psi - \eta_c$ separation and Regge trajectory slope) and $m_c = 1.394\,\text{GeV}$, $m_b = 4.763\,\text{GeV}$ (in order to obtain exact masses for J/Ψ and Υ). On the contrary, with the Shirkov-Solovtsov prescription, we have fixed the string tension on the commonly accepted value $\sigma = 0.18\,\text{GeV}^2$ and used all masses as adjustable parameters. For the light quarks we have taken (in GeV^2)

$$m_u^2 = m_d^2 = 0.121\,k - 0.025\,k^2 + 0.25\,k^4\,, \tag{21}$$

(k being the quark momentum in the meson C. M. frame), for the strange and the heavy quarks $m_s = 0.381\,\text{GeV}$, $m_c = 1.545\,\text{GeV}$ and $m_b = 4.898\,\text{GeV}$.

As an example we also report a numerical table for the $u\bar{u}$ case and the ρ Regge trajectories (fig. 4).

As it is apparent the use of a running constant with infrared truncation does not improve essentially the situation in comparison with the fixed

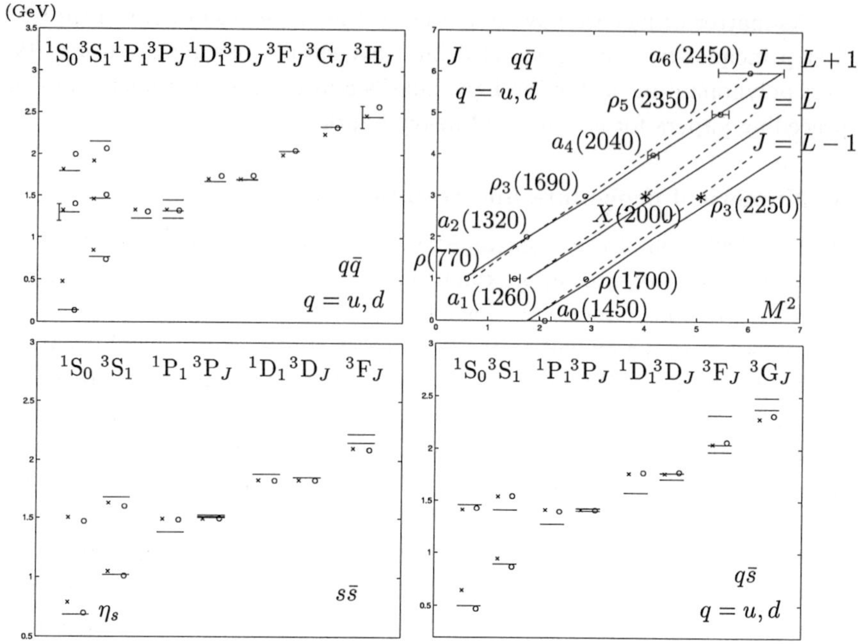

Figure 2. Light-light quarkonium spectrum and Regge trajectories. $n_f = 4$, $\Lambda = 0.2$ GeV. Crosses and dashed line: truncation prescription, $\bar{\alpha}_s = 0.35$, $\sigma = 0.2$ GeV^2, $m_s = 0.2$ GeV, $m_u = m_d = 0.01$ GeV. Circlets and full line: Shirkov-Solovtsov $\alpha_s(Q^2)$, $\sigma = 0.18$ GeV^2, $m_s = 0.381$ GeV, $m_u^2 = m_d^2 = 0.121\,k - 0.025\,k^2 + 0.25\,k^4$.

constant case. The same happen with other prescription like the Dokshitzer et al. prescription [4]. On the contrary with the Shirkov-Solovtsov constant, together with the running u, d mass of eq. (21) the main difficulties seem to be solved. Even the masses of π, η_s and K turn correct without destroying the agreement in the other part of the spectrum. An exception could be represented by the masses of the $u\bar{b}$ channel that for some reason turn out to high. The use of an effective running mass even for s does not seem to produce any substantial modification. With reference to the mass of η_s reported as a data in fig. 2 concerning the $s\bar{s}$ sector, it should be mentioned that the quantity is derived from the experimental values of η and η' masses with conventional assumptions on the origin of this particles as a mixing of $\frac{1}{\sqrt{2}}(u\bar{u} + d\bar{d})$ and $s\bar{s}$ states.

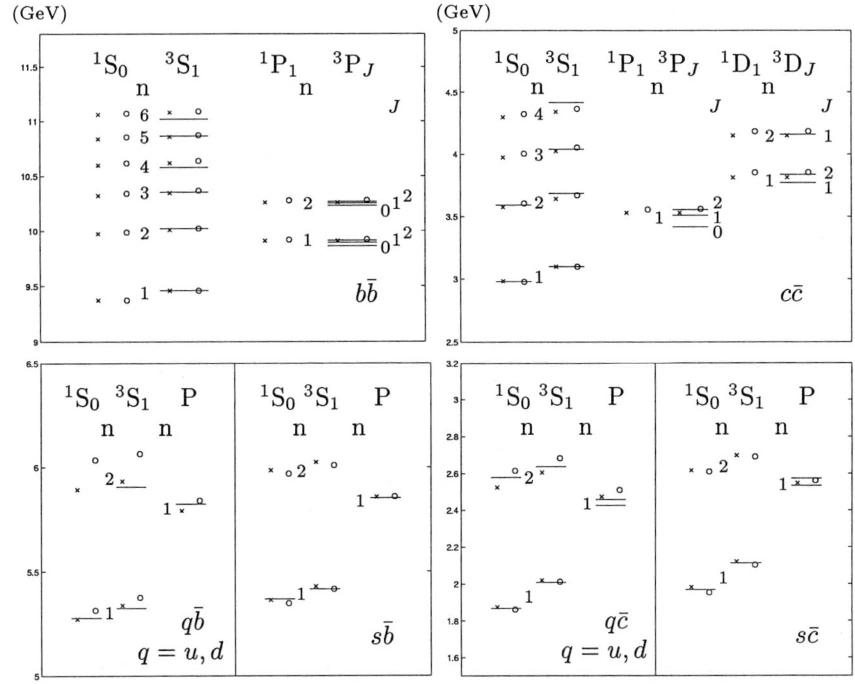

Figure 3. Heavy-heavy and heavy-light quarkonium spectrum. $n_f = 4$, $\Lambda = 0.2$ GeV. Crosses: truncation prescription, $\bar{\alpha}_s = 0.35$, $\sigma = 0.2\ \text{GeV}^2$, $m_b = 4.763$ GeV, $m_c = 1.394$ GeV, $m_s = 0.2$ GeV, $m_u = m_d = 0.01$ GeV. Circlets: Shirkov-Solovtsov $\alpha_s(Q^2)$, $\sigma = 0.18\ \text{GeV}^2$, $m_b = 4.898$, $m_c = 1.545$, $m_s = 0.381$ GeV, $m_u^2 = m_d^2 = 0.121\,k - 0.025\,k^2 + 0.25\,k^4$.

States		experimental values (MeV)		(a) (MeV)	(b) (MeV)
$1\,^1S_0$	π^0	134.9764 ± 0.0006		479	135
	$\pi^\pm$	139.56995 ± 0.00035			
$1\,^3S_1$	$\rho(770)$	768.5 ± 0.6		846	739
$1\Delta SS$		630		367	604
$2\,^1S_0$	$\pi(1300)$	1300 ± 100		1326	1402
$2\,^3S_1$	$\rho(1450)$	1465 ± 25		1461	1509
$2\Delta SS$		165		135	107
$3\,^1S_0$	$\pi(1800)$	1795 ± 10		1815	1994
$3\,^3S_1$	$\rho(2150)$	2149 ± 17		1916	2064
$3\Delta SS$		354		101	70
$1\,^1P_1$	$b_1(1235)$	1231 ± 10		1333	1310
$1\,^3P_2$	$a_2(1320)$	1318.1 ± 0.7			
$1\,^3P_1$	$a_1(1260)$	1230 ± 40	1303	1333	1324
$1\,^3P_0$	$a_0(1450)$	1450 ± 40			

States		experimental values (MeV)	(a) (MeV)	(b) (MeV)
$1\,^1D_2$	$\pi_2(1670)$	1670 ± 20	1701	1739
$1\,^3D_3$	$\rho_3(1690)$	1691 ± 5		
$1\,^3D_2$			1701	1743
$1\,^3D_1$	$\rho(1700)$	1700 ± 20		
$1\,^1F_3$			1990	2043
$1\,^3F_4$	$a_4(2040)$	2037 ± 26		
$1\,^3F_3$	$X(2000)$		1990	2044
$1\,^3F_2$				
$1\,^1G_4$			2238	2318
$1\,^3G_5$	$\rho_5(2350)$	2330 ± 35		
$1\,^3G_4$			2238	2321
$1\,^3G_3$	$\rho_3(2250)$			
$1\,^1H_5$			2460	2570
$1\,^3H_6$	$a_6(2450)$	2450 ± 130		
$1\,^3H_5$			2460	2570
$1\,^3H_4$				

Figure 4. $q\bar{q}$ sector, $n_f = 4$, $\Lambda = 0.2$ GeV. (a) Truncation prescription, $\bar{\alpha}_s = 0.35$, $\sigma = 0.2\ \text{GeV}^2$, $m_u = m_d = 0.01$ GeV, (b) Shirkov-Solovtsov $\alpha_s(Q^2)$, $\sigma = 0.18\ \text{GeV}^2$, $m_u^2 = m_d^2 = 0.121\,k - 0.025\,k^2 + 0.25\,k^4$.

6. Conclusions

In conclusion, starting from a a reasonable ansatz on the Wilson loop it is possible to develop a *second order* Bethe-Salpeter approach to the $q\bar{q}$ bound state problem. From this, by a standard reduction technique, an eigenvalue for a mass squared operator $M^2 = (w_1 + w_2)^2 + U$ is obtained, that can effectively applied to the calculation of the spectrum.

If only the one gluon exchange contribution is included in the perturbative part of the Bethe-Salpeter kernel, but a truncated running constant $\alpha_s(Q^2)$ is used, a reasonable good reproduction of the data is already obtained (as far as calculable) with the relevant exception of the light pseudo scalar mesons. However if the Shirkov-Solovtsov prescription for $\alpha_s(Q^2)$ is adopted and a phenomenological running mass is used for the u and d quarks, even the light pseudo scalar mesons occur in the correct range.

An open problem is a resolution of the complicated Dyson-Schwinger equation in some reasonable approximation (13) and an actual derivation of an equation of the type (21) from first principle.

References

1. N. Brambilla, E. Montaldi, G.M. Prosperi, *Phys. Rev.* **D 54** (1996) 3506; G.M. Prosperi, in *Problems of Quantum Theory of Fields*, Pag. 381, B.M. Barbashov, G.V. Efimov, A.V. Efremov Eds. JINR Dubna 1999, hep-ph/9906237.
2. For a different approach: Yu.S. Kalashnikova, A.V. Nefediev, Yu.A. Simonov, *Phys. Rev.* **D 64** (2001) 014037; and references therein.
3. M. Baldicchi, G.M. Prosperi, *Phys. Rev.* **D 62** (2000) 114024; *Fizika* **B 8** (1999) 2, 251.
4. A.I. Sanda *Phys. Rev. Lett.* **42** (1979) 1658; Yu.L. Dokshitzer, A. Lucenti, G. Marchesini, G.P. Salam, *JHEP* 9805 (1998) 003; Yu.L. Dokshitzer, V.A. Khoze, S.I. Troyan, *Phys. Rev.* **D 53** (1996) 89.
5. D.V. Shirkov, I.L. Solovtsov, *Phys. Rev. Lett.* **79** (1997) 1209; *Theor. Math. Phys.* **120** (1999) 1220; A. V. Nesterenko, *Phys. Rev.* **D 62** (2000) 094028; hep-ph/0210122; S. J. Brodsky, S. Menke, C. Merino, *Phys. Rev.* **D 67** (2003), 055008.
6. M. Baldicchi and G. M. Prosperi *Phys. Rev.* **D 66** (2002) 074008.
7. M.B. Hecht, C.D. Roberts, S.M. Schmidt, *Phys. Rev.* **C 63** (2001) 025213, and references therein.
8. A. Barchielli, E. Montaldi, G.M. Prosperi, *Nucl. Phys.* **B 296** (1988) 625; Erratum-ibid. **B 303** (1988) 752; N. Brambilla, P. Consoli, G.M. Prosperi, *Phys. Rev.* **D 50** (1994) 5878.
9. The Review of Particle Physics, K. Hagiwara *et al.*, *Phys. Rev.* **D 66** 010001 (2002).

RENORMALON CANCELLATION AND PERTURBATIVE QCD POTENTIAL AS A COULOMB+LINEAR POTENTIAL

Y. SUMINO

Department of Physics, Tohoku University
Sendai, 980-8578 Japan

Recently evidence has been found that the perturbative QCD potential agrees well with phenomenological potentials and lattice computations of the QCD potential. We discuss the present status of the perturbative QCD potential and theoretical backgrounds. We also report our recent analysis which shows analytically, on the basis of renormalon dominance picture, that the perturbative QCD potential quickly "converges" to a Coulomb-plus-linear form. The Coulomb-plus-linear potential can be computed systematically as we include more terms of the perturbative series; up to three-loop running (our current best knowledge), it shows a convergence towards lattice results. e.g. At one-loop running, the linear potential is σr with $\sigma = (2\pi C_F/\beta_0)\Lambda^2_{\overline{\rm MS}}$.

1. Introduction

In this article, we review the perturbative QCD predictions of the QCD potential for a static quark-antiquark ($Q\bar{Q}$) pair, in the distance region relevant to the bottomonium and charmonium states, namely, $0.5\ \mathrm{GeV}^{-1}(0.1\ \mathrm{fm}) \lesssim r \lesssim 5\ \mathrm{GeV}^{-1}(1\ \mathrm{fm})$.

Several years ago, the perturbative prediction of the QCD potential became much more accurate in this region. There were two important developments: (1) The complete $O(\alpha_S^3)$ corrections to the QCD potential have been computed [1]; also the relation between the pole quark mass and the $\overline{\rm MS}$ mass has been computed [2] up to $O(\alpha_S^3)$. (2) The renormalon cancellation was discovered [3] in the total energy of a static $Q\bar{Q}$ pair[a], $E_{\rm tot}(r) \equiv 2m_{\rm pole} + V_{\rm QCD}(r)$. Consequently, convergence of the perturbative series of $E_{\rm tot}(r)$ improves drastically, if it is expressed in terms of the $\overline{\rm MS}$ quark mass instead of the pole mass.

Let us demonstrate the improvement of accuracy of the perturbative

[a]By "static", we mean that the kinetic energies of Q and $\bar{Q}$ are neglected.

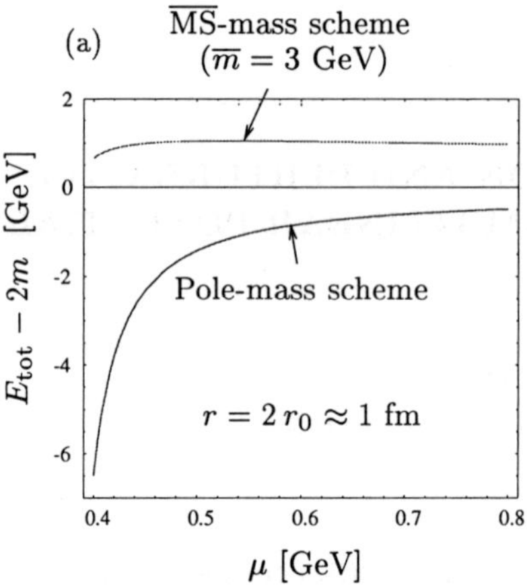

(b)

Scheme	$E_{\text{tot}}(2r_0)$ at $\mu = 0.53\,\text{GeV}$		
	$\mathcal{O}(\alpha_S)$	$\mathcal{O}(\alpha_S^2)$	$\mathcal{O}(\alpha_S^3)$
Pole mass	-0.14	-0.26	-0.75
$\overline{\text{MS}}$ mass	$+1.20$	-0.14	-0.022

Figure 1. (a)Scale dependences of $E_{\text{tot}}(r)$ at $r = 2r_0 \approx 1$ fm, for the pole-mass and $\overline{\text{MS}}$-mass schemes. ($\overline{m} \equiv m_{\overline{\text{MS}}}(m_{\overline{\text{MS}}}) = 3$ GeV) (b) Perturbative series of $E_{\text{tot}}(r)$ at $r = 2r_0$, and at $\mu = 0.53$ GeV determined from $\partial E_{\text{tot}}/\partial\mu = 0$ in (a) ($\overline{\text{MS}}$-mass scheme).

prediction for $E_{\text{tot}}(r)$ up to $O(\alpha_S^3)$, in the case without light quark flavors ($n_l = 0$). In Fig. 1(a), we fix r to be $2r_0 \approx 1$ fm as an extreme case[b] and show the renormalization-scale (μ) dependence of $E_{\text{tot}}(r)$. We see that E_{tot} is much less scale dependent when we use the $\overline{\text{MS}}$ mass instead of the pole mass. Fig. 1(b) shows the convergence behaviors of the perturbative series of E_{tot} for the same r and when μ is fixed to the value where E_{tot} becomes least sensitive to variation of μ (minimal-sensitivity scale). If we use the pole mass, the series is divergent, whereas for the $\overline{\text{MS}}$ mass, the series is convergent. In fact, we observe qualitatively the same features at different r and for different n_l. Generally, the prediction becomes more stable and more convergent at smaller r.

The aim of this paper is to study properties of $E_{\text{tot}}(r)$, given the much more accurate prediction as compared to several years ago. In Sec. 2 we examine $E_{\text{tot}}(r)$ up to $O(\alpha_S^3)$. Sec. 3 provides our present theoretical understanding based on renormalon dominance picture and operator-product-expansion (OPE). We show that the perturbative QCD potential "converges" to a Coulomb+linear potential in Sec. 4. Conclusions are given in Sec. 5.

[b] $r_0 \approx 0.602\,\Lambda_{\overline{\text{MS}}}^{-1}$ represents the Sommer scale [4]. It is translated to physical unit following the convention of lattice calculations in the quenched approximation: $r_0 \approx 0.5$ fm.

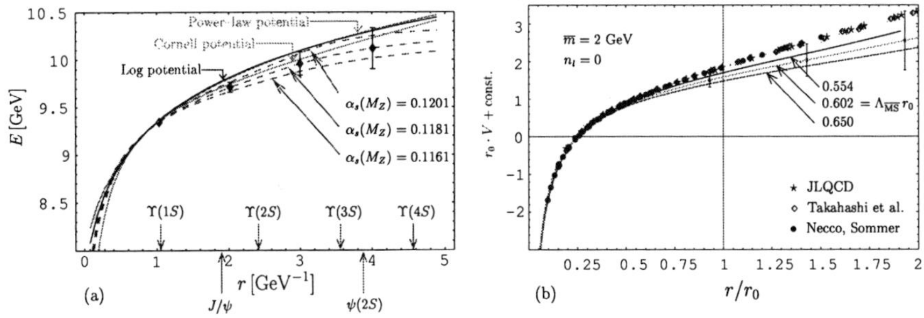

Figure 2. (a)Comparison between $E_{\rm tot}(r)$ and typical phenomenological potentials. Arrows at the bottom show the r.m.s. radii of the heavy quarkonium states. (b) Comparison between $E_{\rm tot}(r)$ and lattice computations of the QCD potential.

2. $E_{\rm tot}(r)$ up to $O(\alpha_S^3)$

The most solid part of our study is an examination of the perturbative predictions of $E_{\rm tot}(r)$ up to $O(\alpha_S^3)$, through comparisons with phenomenological potentials and with lattice computations of the QCD potential.

In Fig. 2(a) $E_{\rm tot}(r)$ is compared with typical phenomenological potentials. Since the latter are determined from the heavy quarkonium spectra, we chose realistic values for the input parameters of $E_{\rm tot}(r)$: $\overline{m}_b = 4.190$ GeV; $n_l = 4$ with $\overline{m}_u = \overline{m}_d = \overline{m}_s = 0$ and $\overline{m}_c = 1.243$ GeV for the light quarks in internal loops; $\mu = \mu(r)$ is fixed by either $\partial E_{\rm tot}/\partial\mu = 0$ or $|E_{\rm tot}^{(3)}| =$ minimum, but both prescriptions lead to almost same values of $E_{\rm tot}(r)$. See [5,6] for details. We see that $E_{\rm tot}(r)$ corresponding to the present values of the strong coupling constant (dashed lines) agree well with the phenomenological potentials within estimated perturbative uncertainties (indicated by error bars). We also note that the agreement is lost quickly if we take $\alpha_S(M_Z)$ outside of the present world-average values, so that the agreement is unlikely to be accidental.

In Fig. 2(b) $E_{\rm tot}(r)$ is compared with the recent lattice computations of the QCD potential in the quenched approximation. Accordingly we set $n_l = 0$ in $E_{\rm tot}(r)$. We take $\overline{m} = 2$ GeV, which stabilizes the perturbative prediction up to largest r. The scale $\mu = \mu(r)$ is fixed in the same way as in Fig. 2(a). See [7] for details. The three solid lines in Fig. 2(b) represent the same perturbative prediction for $E_{\rm tot}(r)$; in plotting these lines, we varied the relation between $\Lambda_{\overline{\rm MS}}$ and the lattice scale within the error band given by [4]. Taking this uncertainty into account and also uncertainties of the perturbative prediction (error bars), we observe a good agreement between

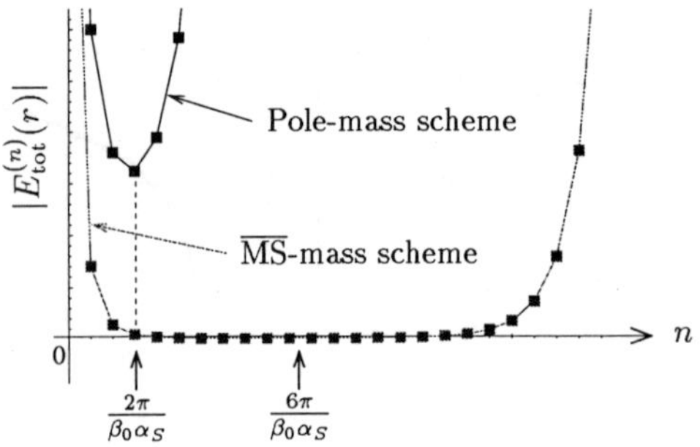

Figure 3. Schematic diagram showing n-dependence of $|V_{\beta_0}^{(n)}(r)|$ (or $|E_{\rm tot}^{(n)}(r)|$ in the pole-mass scheme) [black] and that of $|E_{\rm tot}^{(n)}(r)|$ in the $\overline{\rm MS}$-mass scheme [red].

$E_{\rm tot}(r)$ and lattice results.

In fact, by now several works have confirmed the agreement [5,8,6,9,7]. Although details depend on how the renormalon in the QCD potetial is cancelled, qualitatively the same conclusions were drawn, i.e. perturbative predictions become stable and agree with phenomenological potentials/lattice results up to much larger r than before.

3. Theoretical Backgrounds

We would like to understand why the perturbative predictions of $E_{\rm tot}(r)$ behave in the way we observed in the previous sections. In this section, we review our current theoretical understanding.

3.1. *Perturbative uncertainties: Leading-order renormalon*

The nature of the perturbative series of $V_{\rm QCD}(r)$ and $E_{\rm tot}(r)$, including their uncertainties, can be understood within the argument based on renormalons. This argument gives certain estimates of higher-order corrections in perturbative QCD, and empirically it gives good estimates even at relatively low orders of perturbative series.

According to the renormalon argument, the perturbative series of $V_{\rm QCD}(r)$ (or $E_{\rm tot}(r)$ in the pole-mass scheme) behaves as depicted in Fig. 3 [black points]. (see e.g. [10]) Namely, the series shows apparent convergence up to order $\alpha_S^{n_0}$ with $n_0 \approx 2\pi/(\beta_0\alpha_S)$. Beyond this order, the series diverges. Due to the divergence (the series is an asymptotic series), there is a limitation to the achievable accuracy of the perturbative prediction. It can be estimated by the size of the terms around the minimum, $O(\alpha_S^{n_0})$; this gives an uncertainty for $V_{\rm QCD}(r)$ of order $\Lambda_{\rm QCD} \sim 300$ MeV [11].

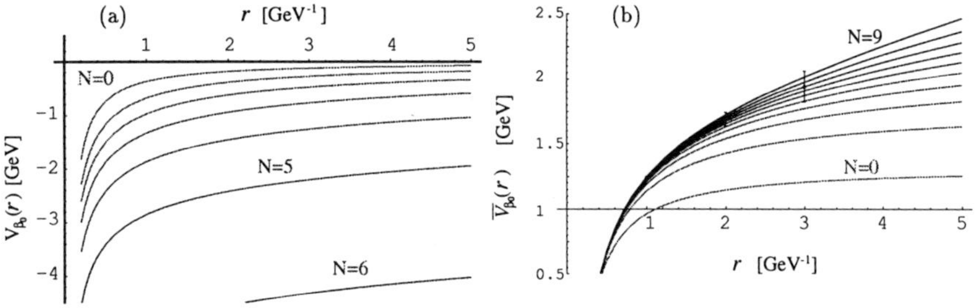

Figure 4. $V_{\beta_0}(r)$ truncated at $O(\alpha_S^{N+1})$ term, $\sum_{n=0}^{N} V_{\beta_0}^{(n)}(r)$. We set $\mu = 2.49$ GeV, $n_l = 4$ and $\alpha_S(\mu) = 0.273$. (a) Before cancellation of the leading-order renormalon. (b) After cancellation of the leading-order renormalon. Note that the vertical sales are different in two figures.

To obtain the above behavior, we may use, for instance, "the large-β_0 approximation" [12] to estimate the higher-order corrections. The order α_S^{n+1} term of $V_{\rm QCD}(r)$ for $n \gg 1$ is estimated as

$$V_{\beta_0}^{(n)}(r) \sim -\frac{2C_F\alpha_S(\mu)}{\pi}\,\mu\, e^{5/6} \times \left(\frac{\beta_0\alpha_S(\mu)}{2\pi}\right)^n \times n!, \tag{1}$$

where $C_F = 4/3$ is a color factor and $\beta_0 = 11 - 2n_l/3$ is the 1-loop coefficient of the beta function. From this expression, it is easy to obtain the behavior of the series as explained above.

It is important to note that Eq. (1) is independent of r. Although $V_{\beta_0}^{(n)}(r)$ is dependent on r, the leading part in the large n limit is independent of r. This feature can be verified in Fig. 4(a): As we include more terms of the perturbative series of $V_{\beta_0}(r)$, there are large corrections to the potential, which shift the potential almost r-independently. Furthermore, one sees no sign of convergence of the perturbative series in the range of r we are interested in.

3.2. *Renormalon cancellation and its implications*

As already stated, the leading-order renormalon uncertainty in the QCD potential gets cancelled in the $Q\bar{Q}$ total energy, $E_{\rm tot}(r) = 2m_{\rm pole} + V_{\rm QCD}(r)$, if we use the $\overline{\rm MS}$ mass. As seen above, the leading-order renormalon is independent of r, hence, it can be absorbed by a redefinition of the quark mass. The pole mass is expressed in terms of the $\overline{\rm MS}$ mass as a perturbative series, $m_{\rm pole} = \overline{m}\,(1 + c_1\alpha_S + c_2\alpha_S^2 + \cdots)$. The leading part of this series at large orders takes the same form as Eq. (1) but in opposite sign [13]. As a result, the leading-order renormalon is cancelled in $E_{\rm tot}(r)$ and its perturbative series behaves as [3]

$$E_{\rm tot}^{(n)}(r) \sim {\rm const.} \times r^2 \times \left(\frac{\beta_0 \alpha_S(\mu)}{6\pi} \right)^n \times n! \qquad (2)$$

at large n. As compared to $V_{\beta_0}(r)$, the series converges faster and up to larger order, $n \approx 6\pi/(\beta_0\alpha_S)$, but beyond this order again the series diverges; see Fig. 3. An uncertainty of this perturbative series can be estimated from the size of the terms around the minimum, which gives $\delta E_{\rm tot}(r) \sim \Lambda_{\rm QCD} \times (r\Lambda_{\rm QCD})^2$. Thus, at distances $r \lesssim \Lambda_{\rm QCD}^{-1}$, the uncertainty becomes smaller than the original uncertainty $\delta V_{\rm QCD}(r) \sim \Lambda_{\rm QCD}$. In fact, the error bars in Figs. 2(a,b) represent $\pm\frac{1}{2}\Lambda^3 r^2$ with $\Lambda = 300$ MeV. The agreement of $E_{\rm tot}(r)$ and phenomenological potentials/lattice results holds within this uncertainty.

The improvement of convergence can be seen by comparing Figs. 4(a) and (b), where higher-order corrections are estimated using the large-β_0 approximation. Moreover, in Fig. 4(b) we see that the higher-order corrections make the potential steeper at large r, as compared to the tree-level ($N = 0$) Coulomb potential, after cancellation of the leading-order renormalon. In fact, due to this very feature, an agreement between $E_{\rm tot}(r)$ and phenomenological potentials/lattice results was observed in Sec. 2. We may understand the reason as follows. Let us define a strong coupling constant $\alpha_F(\mu)$ from the interquark force:

$$F(r) \equiv -\frac{d}{dr} V_{\rm QCD}(r) \equiv -C_F \, \frac{\alpha_F(1/r)}{r^2}. \qquad (3)$$

Since the leading-order renormalon in $V_{\rm QCD}(r)$ is r-independent, it is *killed* upon differentiation by r. $\alpha_F(1/r)$ grows at infrared (IR) due to the running (the first two coefficients of the β-function are universal), which makes $|F(r)|$ stronger than the Coulomb force at large r. This means that $V_{\rm QCD}(r)$, after cancelling the leading-order renormalon, becomes steeper than the Coulomb potential. See [5,8] for details.

Conventionally, theoretical calculations of the energy of a heavy $Q\bar{Q}$ boundstate closely followed that of a QED boundstate such as positronium: it starts from the natural picture that, when an electron and a positron are at rest and far apart from each other, they tend to be free particles and the total energy of the system is given by the sum of the energies of the two particles (pole masses); as the electron and positron approach each other, the energy of the system decreases due to the negative potential energy, so that the total energy of the boundstate is given as the sum of the pole masses minus the binding energy. Applying the same description to the energy of a $Q\bar{Q}$ system, however, is not natural, because when Q and $\bar{Q}$ are far apart from each other, the free quark picture is not good. As a result, the perturbative expansion of the boundstate energy turns out to be poorly convergent, due to the contributions from IR gluons with wave-lengths of order $\Lambda_{\rm QCD}^{-1}$. On the other hand, intuitively we expect that there should be a way to calculate the boundstate energy in which the contributions of IR gluons can be mostly eliminated. This is because when the boundstate size is sufficiently smaller than $\Lambda_{\rm QCD}^{-1}$, IR gluons cannot resolve the color charges of the constituent particles, so that they decouple from this color-singlet system. Such a calculation is possible if we use a quark mass (e.g. $\overline{\rm MS}$ mass), into which only contributions from short wave-lenghth gluons to the quark self-energy are absorbed (renormalized).

In fact, qualitatively the $Q\bar{Q}$ total energy can be expressed as [14]

$$E_{\rm tot}(r) \simeq 2\overline{m} + \int\limits_{r^{-1} \lesssim q \lesssim \overline{m}} \frac{d^3\vec{q}}{(2\pi)^3}\, C_F \frac{4\pi\alpha_S(q)}{q^2}. \tag{4}$$

It shows that the energy is mainly composed of (i) the $\overline{\rm MS}$ masses of Q and $\bar{Q}$, and (ii) the self-energies of Q and $\bar{Q}$ originating from gluons whose wavelengths are shorter than the size of the system and longer than those absorbed into (i), i.e. $1/\overline{m} \lesssim \lambda_g \lesssim r$. Thus, IR gluons ($\lambda_g > r$) have decoupled from Eq. (4). If $\alpha_S(q)$ were a constant (non-running), $E_{\rm tot}(r)$ would be a Coulomb potential. Due to the running of $\alpha_S(q)$, however, there is a large positive contribution in Eq. (4) as r increases and gets closer to $\Lambda_{\rm QCD}^{-1}$. This gives a microscopic description for how $E_{\rm tot}(r)$ becomes steeper than the Coulomb potential at large r: It is the rapid growth of the self-energies (ii), as more IR gluons couple to this system with increasing r. — Again it stems from the *running* of the strong coupling constant.

3.3. *Consistency with operator-product-expansion*

An OPE of the QCD potential can be performed within the effective field theory "potential-NRQCD" [15]. The dynamical degrees of freedom in this effective theory have only very soft scale, where typical momenta are $k \ll 1/r$. Higher momentum scales are integrated out and the effective interactions are generally non-local in space but local in time (potentials). In this framework, IR part of perturbative QCD calculations is factorized and absorbed into matrix elements of operators. A factorization scale $\mu_{\rm fac}$ is set to satisfy $\Lambda_{\rm QCD} \ll \mu_{\rm fac} \ll 1/r$. Then the total energy of a static $Q\bar{Q}$ pair is given by[16,9]

$$E_{\rm tot}(r) = 2m_{\rm pole} + V_{\rm PT}(r;\mu_{\rm fac}) + \delta E_{\rm US}(r;\mu_{\rm fac}), \tag{5}$$

$$\delta E_{\rm US} = r^2 \times \frac{4\pi\alpha_S}{18}\int_0^\infty dt\, e^{-it(V_O-V_S)}\left\langle \vec{E}^a(t)\phi(t,0)_{ab}\vec{E}^b(0)\right\rangle(\mu_{\rm fac}) + \cdots \tag{6}$$

The perturbative potential $V_{PT}(r;\mu_{\rm fac})$ is now free from renormalons. The renormalons are absorbed into $\delta E_{\rm US}(r;\mu_{\rm fac})$. In the leading-order of multipole expansion, $\delta E_{\rm US}$ is proportional to r^2, as shown above. This is consistent with our observation in Sec. 2. Namely, the agreement between $E_{\rm tot}(r)$ and phenomenological potentials/lattice results holds within the $O(\Lambda_{\rm QCD}^3 r^2)$ renormalon uncertainty. This means, a difference of order $\Lambda_{\rm QCD}^2 r$ between them was ruled out (numerically) at distances $r \lesssim \Lambda_{\rm QCD}^{-1}$. According to OPE, indeed such a difference is not allowed theoretically, since there exists no operator that can absorb a linear potential.

4. Pert. QCD Potential as a Coulomb+Linear Potential

The observations and theoretical arguments in Secs. 2,3 indicate that the perturbative prediction of $E_{\rm tot}(r)$ takes a Coulomb+linear form (up to an r-independent constant) in the range $r \lesssim \Lambda_{\rm QCD}^{-1}$, as represented by a typical phenomenological potential. We would like to substantiate this claim analytically. Since the uncertainty of the perturbative prediction is of order r^2, there is a possibility that $E_{\rm tot}(r)$ is predictable up to order r (linear) term at $r \lesssim \Lambda_{\rm QCD}^{-1}$ within perturbative QCD.

Nevertheless, one immediately comes up with counter arguments: (1) A finite-order perturbative prediction of $E_{\rm tot}(r)$ *cannot* be expressed in a Coulomb+linear form. This is because the perturbative QCD potential up to $O(\alpha_S^N)$ has a form

$$V_N(r) = -C_F\frac{\alpha_S(\mu)}{r} \times \sum_{n=0}^{N-1} \alpha_S(\mu)^n\, P_n(\log \mu r), \tag{7}$$

where $P_n(L)$ denotes an n-th degree polynomial of L. Therefore, as $r \to \infty$, $V_N(r) \to 0$, i.e. the tangent is zero, and there is no linear potential. (2) From dimensional analysis, the coefficient of a linear potential should be non-analytic in α_S, i.e. $\Lambda^2 \sim \mu^2 \exp[-4\pi/(\beta_0\alpha_S)]$. Therefore, it should vanish at any order of perturbative expansion.

Recently we have shown analytically [17] that, based on the renormalon dominance picture, $V_N(r)$ quickly "converges" to a Coulomb+linear potential for $N \gg 1$, up to an $O(\Lambda_{\rm QCD}^3 r^2)$ uncertainty. In showing this, (a) the higher-order terms of $V_N(r)$ are estimated using the renormalization group (RG), and (b) we adopt a scale-fixing prescription based on the renormalon dominance picture.

According to the renormalon dominance picture, if we choose a scale μ such that $\alpha_S(\mu) \approx 6\pi/(\beta_0 N)$, around this scale, $V_N(r)$ (after cancelling the leading-order renomalon) becomes least μ-dependent and the perturbative series becomes most convergent; cf. Fig. 3. In view of this argument, we fix μ such that[c]

$$\alpha_S(\mu) = \frac{6\pi}{\beta_0 N} \times \xi \qquad (\xi \sim 1). \tag{8}$$

Then we consider $V_N(r)$ for $N \gg 1$ while keeping $\Lambda_{\overline{\rm MS}}$ finite.

Let us first demonstrate the result in the simplest case. We set $\xi = 1$ and the higher-order terms of $V_N(r)$ are estimated using the 1-loop running coupling constant (leading log approximation):

$$V_N(r) = -\int \frac{d^3\vec{q}}{(2\pi)^3} \frac{e^{i\vec{q}\cdot\vec{r}}}{q^2} \, 4\pi C_F \, [\alpha_{1L}(q)]_N \qquad ; \qquad q = |\vec{q}|, \tag{9}$$

where $[\alpha_{1L}(q)]_N$ denotes the perturbative expansion of the 1-loop running coupling constant truncated at $O(\alpha_S^N)$:

$$[\alpha_{1L}(q)]_N = \left[\frac{\alpha_S(\mu)}{1 - \frac{\beta_0 \alpha_S(\mu)}{2\pi} \log\left(\frac{\mu}{q}\right)} \right]_N = \alpha_S(\mu) \sum_{n=0}^{N-1} \left\{ \frac{\beta_0 \alpha_S(\mu)}{2\pi} \log\left(\frac{\mu}{q}\right) \right\}^n . \tag{10}$$

In the limit $N \gg 1$, $V_N(r)$ can be decomposed into four parts corresponding to $\{r^{-1}, r^0, r^1, r^2\}$ terms (with log corrections in the r^{-1} and r^2 terms):

$$V_N(r) = \frac{4C_F}{\beta_0} \Lambda \, v(\Lambda r, N), \tag{11}$$

$$v(\rho, N) = v_C(\rho) + B(N) + C\rho + D(\rho, N) + (\text{terms that vanish as } N \to \infty), \tag{12}$$

where we rescaled r and V_N by $\Lambda \equiv \mu \exp[-2\pi/(\beta_0 \alpha_S(\mu))]$. $v_C(\rho)$, $C\rho$ and $D(\rho, N)$ are shown in Fig. 5(a); see [17] for the explicit formulas. The constant $[B(N)]$ and quadratic $[D(\rho, N)]$ parts are divergent as $N \to \infty$, while the "Coulomb"[d] $[v_C(\rho)]$ and linear $[C\rho]$ parts are finite in this limit. The constant part diverges rapidly, but since it can be absorbed by a redefinition of quark mass in the total energy $E_{\rm tot}(r)$, we will not be concerned with it.

---

[c]Here, we generalize the prescription of [17] by introducing the parameter ξ, cf. [18].

[d]The "Coulomb" part $v_C(\rho)$ contains a log correction at short-distances and its short-distance behavior is consistent with the 1-loop RG equation for the potential.

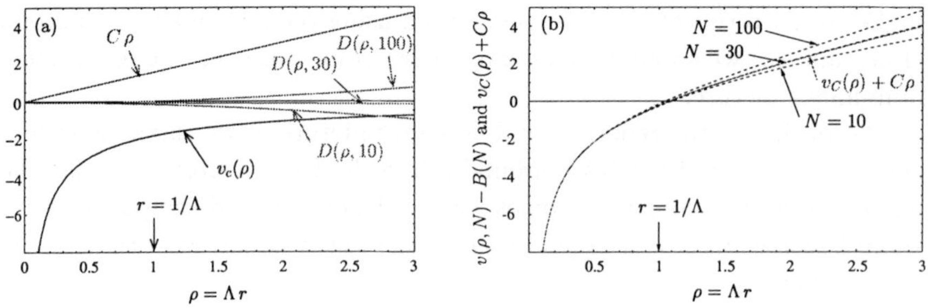

Figure 5. (a) $v_C(\rho)$, $C\rho$ and $D(\rho,N)$ ($N = 10,30,100$) vs. ρ. (b) Comparison of $v(\rho,N) - B(N)$ ($N = 10,30,100$) and the Coulomb+linear potential $v_C(\rho) + C\rho$.

The quadratic part is divergent slowly as $D \sim (\rho^2/12)\log N$, and its size is small for $r \lesssim \Lambda^{-1}$ and $N \lesssim 100$ as compared to the Coulomb+linear part; see Fig. 5(a). As a result, when N is increased up to e.g. 10–30, $v(\rho,N)$ quickly "converges" to $v_C(\rho) + C\rho$ at $r \lesssim \Lambda^{-1}$, while it slowly varies as $(\rho^2/12)\log N$, see Fig. 5(b). Note that, although $v(\rho,N)$ in this figure have the form of Eq. (7), they approximate well $v_C(\rho) + C\rho$ at $r \lesssim \Lambda^{-1}$.

If we vary ξ in the scale-fixing prescription Eq. (8), as long as $\xi > 2/3$, we obtain the same Coulomb+linear potential, $v_C(\rho) + C\rho$, whereas $B(N)$ and $D(\rho,N)$ change: for instance, if $2/3 < \xi < 1$, $D(\rho,N)$ is finite as $N \to \infty$, $D(\rho) = \rho^P \times$ (log-corr.) with $1 < P < 2$.

We may consider that $B(N)$ and $D(\rho,N)$ represent renormalons in $V_{\rm QCD}(r)$. If we perform OPE, these will be (partially) absorbed into $2m_{\rm pole}$ and $\delta E_{\rm US}$ in Eq. (5). On the other hand, $v_C(\rho) + C\rho$ may be regarded as a genuine perturbative part, which cannot be absorbed into $2m_{\rm pole}$ and $\delta E_{\rm US}$.

The above result can be systematically improved by incorporating higher-loop effects into the RG estimate of the higher-order terms of $V_N(r)$. In Fig. 6 we show the Coulomb+linear potentials determined from the 1-loop running, 2-loop running (with 1-loop non-logarithmic term), and 3-loop running (with 2-loop non-logarithmic term). They are compared with the lattice results. Since the 3-loop non-logarithmic term is not yet known, the 3-loop running case represents our current best knowledge. Up to this order, the Coulomb+linear potential agrees with the lattice results up to larger r as we increase the order.

The coefficient of the linear potential σr can be expressed analytically.

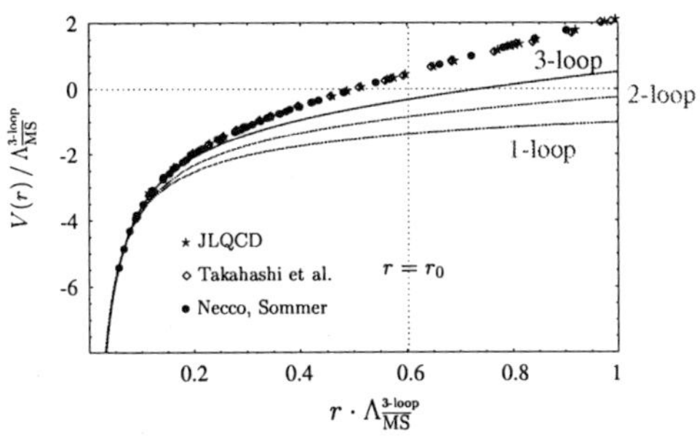

Figure 6. Comparison of the Coulomb+linear potential ($n_l = 0$) with the lattice results in the quenched approximation.

For the 1- and 2-loop running cases, the expressions read

$$\sigma_{1L} = \frac{2\pi C_F}{\beta_0} \left(\Lambda_{\overline{\rm MS}}^{\text{1-loop}}\right)^2, \tag{13}$$

$$\sigma_{2L} = \frac{2\pi C_F}{\beta_0} \left(\Lambda_{\overline{\rm MS}}^{\text{2-loop}}\right)^2 \frac{e^{-\delta}}{\Gamma(1+\delta)} \left[1 + \frac{a_1}{\beta_0}\, \delta^{-1-\delta}\, e^{\delta}\, \gamma(1+\delta, \delta)\right], \tag{14}$$

where $\delta = \beta_1/\beta_0^2$; see [17] for details.

Finally, let us comment on the counter arguments given at the beginning of this section. (1) $V_N(r)$ "converges" to a Coulomb+linear form for $N \gg 1$. In fact, already at relatively low orders and at $r \lesssim \Lambda^{-1}$, $V_N(r)$ approximates the Coulomb+linear potential fairly well. (2) Consider a perturbative term $T = \{\frac{\beta_0 \alpha_S(\mu)}{2\pi} \log(\mu r)\}^N$. If we insert the relation (8) and take the limit $N \to \infty$, it is easy to see that $T \to (\Lambda r)^{3\xi}$. Thus, if we choose a scale μ à la renormalon dominance picture, perturbative terms can converge towards $(\Lambda r)^P$ with some positive power P.

5. Conclusions

After cancellation of the leading-order renormalons by using the $\overline{\rm MS}$ quark mass, perturbative uncertainty of $E_{\rm tot}(r) = 2m_{\rm pole} + V_{\rm QCD}(r)$ reduces from $O(\Lambda_{\rm QCD})$ to $O(\Lambda_{\rm QCD}^3 r^2)$ and much more accurate perturbative prediction is obtained at $r \lesssim \Lambda_{\rm QCD}^{-1} \sim 1$ fm. Consequently we observe the following:

- $E_{\rm tot}(r)$ up to $O(\alpha_S^3)$ agrees well with phenomenological potentials/lattice results within the estimated uncertainty.
- New physical picture on the composition of the energy of a static $Q\bar{Q}$ system or on the interquark force is obtained: The self-energies of Q and $\bar{Q}$ from gluons with $\lambda_g < 1/r$ increase rapidly as r increases, which makes $E_{\rm tot}(r)$ steeper than the Coulomb potential.

- Our observations are consistent with OPE within potential-NRQCD framework.
- Based on the renormalon dominance picture, the perturbative QCD potential "converges" quickly to a Coulomb+linear potential, which can be computed systematically as we include more terms in the estimate of higher-order terms via RG. In particular, the linear potential can be computed analytically [Eqs. (13,14)]. The Coulomb+linear potential agrees with lattice results up to larger r as we include more terms in the higher-order estimate.

Acknowledgements

Many of the analyses were done in collaboration with S. Recksiegel. The author is grateful for fruitful discussion.

References

1. M. Peter, *Phys. Rev. Lett.* **78**, 602 (1997); Y. Schröder, *Phys. Lett.* **B447**, 321 (1999).
2. K. Melnikov and T. v. Ritbergen, *Phys. Lett.* **B482**, 99 (2000).
3. A. Hoang, M. Smith, T. Stelzer and S. Willenbrock, *Phys. Rev.* **D59**, 114014 (1999); M. Beneke, *Phys. Lett.* **B434**, 115 (1998).
4. S. Capitani, M. Lüscher, R. Sommer and H. Wittig [ALPHA Collaboration], *Nucl. Phys.* **B544**, 669 (1999); Erratum ibid. **B582**, 762 (2000).
5. Y. Sumino, *Phys. Rev.* **D65**, 054003 (2002).
6. S. Recksiegel and Y. Sumino, *Phys. Rev.* **D65**, 054018 (2002).
7. S. Recksiegel and Y. Sumino, hep-ph/0212389.
8. S. Necco and R. Sommer, *Phys. Lett.* **B523**, 135 (2001).
9. A. Pineda, *J. Phys.* **G29**, 371 (2003).
10. M. Beneke, *Phys. Rept.* **317** (1999) 1; Y. Sumino, hep-ph/0004087.
11. U. Aglietti and Z. Ligeti, *Phys. Lett.* **B364**, 75 (1995).
12. M. Beneke and V. Braun, *Phys. Lett.* **B348**, 513 (1995).
13. M. Beneke and V. Braun, *Nucl. Phys.* **B426**, 301 (1994); I. Bigi, M. Shifman, N. Uraltsev and A. Vainshtein, *Phys. Rev.* **D50**, 2234 (1994).
14. N. Brambilla, Y. Sumino and A. Vairo, *Phys. Lett.* **B513**, 381 (2001).
15. A. Pineda and J. Soto, *Nucl. Phys. Proc. Suppl.* **64**, 428 (1998).
16. N. Brambilla, A. Pineda, J. Soto and A. Vairo, *Nucl. Phys.* **B566**, 275 (2000).
17. Y. Sumino, *Phys. Lett.* **B571**, 173 (2003).
18. K. Van Acoleyen and H. Verschelde, hep-ph/0307070.

INFRARED LIMIT AND MASS GAP IN QCD

DANIEL ZWANZIGER*

Physics Department, New York University
4 Washington Place
New York, NY 10003, USA
E-mail: daniel.zwanziger@nyu.edu

We show that an exact non-perturbative quantization of continuum gauge theory is provided by the Faddeev-Popov formula in Landau gauge, restricted to the region where the Faddeev-Popov operator is positive $-\partial \cdot D(A) > 0$ (Gribov region). Although there are Gribov copies inside this region, they have no influence on expectation-values. The cut-off of the functional integral on the boundary of the Gribov region does not change the form of the Dyson-Schwinger equations, because $\det[-\partial \cdot D(A)]$ vanishes on the boundary, so there is no boundary contribution. Instead the cut-off provides supplementary conditions that govern the choice of solution. In particular the "horizon condition", though consistent with the perturbative renormalization group, puts QCD into a non-perturbative phase. The asymptotic infrared limit reveals the existence of a mass gap.

1. Introduction

The work that is reported here was motivated by the discovery of non-perturbative solutions of the truncated Dyson-Schwinger (DS) equations in QCD with paradoxical properties.[1,2,3,4,5,6] It was found that in Landau gauge *the gluon propagator is suppressed in the infrared*, compared to the free propagator while *the ghost propagator is enhanced.* Doubt about the reliability of the truncation of the Dyson-Scwhinger equations is allayed by the fact that numerical simulation in lattice gauge theory also yields propagators with the same infrared behavior.[7,8,9,10,11,12,13,14,15,16,17] Moreover these same infrared properties were shown to be a consequence of the cut-off of the functional integral in A-space at the Gribov horizon.[18,19] This agreement of three different methods inspires confidence in the result. The paradox arises because the cut-off at the Gribov horizon was apparently not imposed when the solutions of the DS equation were originally obtained.[1]

*Work partially supported by National Science Foundation, under grant no. phy-0099393

It is this paradox that will be addressed here, with more details provided elsewhere.[20] (There is no paradox about the fact that the numerically determined gluon and ghost propagators manifest these properties because numerical gauge fixing is achieved by a minimization procedure that yields configurations that lie inside the Gribov horizon.)

Since the work of Gribov,[18] a non-perturbative formulation of continuum gauge theory has appeared problematical due to the existence of Gribov copies. These are distinct but gauge-equivalent configurations $A^{(2)} = {}^{g}A^{(1)}$ that both satisfy the gauge condition, $\partial \cdot A^{(1)} = \partial \cdot A^{(2)} = 0$, where ${}^{g}A_\mu = g^{-1}A_\mu g + g^{-1}\partial_\mu g$ is a local gauge transformation. The difficulty arises when one wishes to quantize by *gauge fixing* namely by taking a single representative configuration on each gauge orbit. It has been proven that this cannot be done in a continuous way when space-time is compactified.[21] Geometrically this reflects the intricacy of gauge orbit space, the space of configurations A modulo local gauge transformations g. Fortunately the method of stochastic quantization by-passes the difficulties of Gribov copies.

2. Review of stochastic quantization of gauge fields

2.1. *Time-dependent stochastic quantization*

Historically, stochastic quantization originated[22] with the observation that the formal, unnormalizable, Euclidean proabability distribution $P_0(A) = N\exp[-S_{\rm YM}(A)]$, with 4-dimensional Euclidean Yang-Mills action $S_{\rm YM}(A)$, is the equilibrium distribution of the stochastic process defined by the equation,

$$\frac{\partial P}{\partial t} = \int d^4x \, \frac{\delta}{\delta A^a_\mu(x)} \Big(\frac{\delta P}{\delta A^a_\mu(x)} + \frac{\delta S_{\rm YM}}{\delta A^a_\mu(x)} P \Big) \tag{1}$$

for the time-dependent probability distribution $P(A,t)$. This equation is a continuum analog of the diffusion equation in the presence of the drift force K_i,

$$\frac{\partial P}{\partial t} = \frac{\partial}{\partial A^i} \Big(\frac{\partial P}{\partial A^i} - K_i P \Big) = 0, \tag{2}$$

that is known as the Fokker-Planck equation. If the drift force is conservative, $K_i = -\frac{\partial S_{\rm YM}}{\partial A^i}$, then $\exp[-S_{\rm YM}(A)]$ is a time-independent solution. In Euclidean quantum field theory, t is an artificial 5th time that corresponds to the number of sweeps in a Monte-Carlo simulation. The same stochastic process may equivalently be described by the Langevin equation

$$\frac{\partial A_\mu^a}{\partial t} = -\frac{\delta S_{\rm YM}}{\delta A_\mu^a} + \eta_\mu^a. \tag{3}$$

Here $A_\mu^a = A_\mu^a(x,t)$ depends on the 5th time, and $\eta_\mu^a = \eta_\mu^a(x,t)$ is Gaussian white noise defined by $\langle \eta_\mu^a(x,t) \rangle = 0$ and $\langle \eta_\mu^a(x,t)\eta_\nu^b(y,t')\rangle = 2\delta(x-y)\delta_{\mu\nu}\delta^{ab}\delta(t-t')\rangle$. If $N\exp[-S_{\rm YM}(A)]$ were a normalizable probability distribution — which it is not — every normalized solution to (1) would relax to it as equilibrium distribution. However the process defined by (1) or (3) does not provide a restoring force in gauge orbit directions, so probability escapes to infinity along the gauge orbits. As a result $P(A,t)$ does not relax to a well-defined limiting distribution $\lim_{t\to\infty} P(A,t) \neq N\exp[-S_{\rm YM}(A)]$.

A remedy is provided by the observation[23] that the Langevin equation may be modified by the addition of an infinitesimal gauge transformation, $D_\mu^{ac}v^c$,

$$\frac{\partial A_\mu^a}{\partial t} = -\frac{\delta S}{\delta A_\mu^a} + D_\mu^{ac}v^c + \eta_\mu^a, \tag{4}$$

where v^c is at our disposal. This cannot alter the expectation-value of gauge-invariant quantities, for only a harmless infinitesimal gauge-transformation $K_{{\rm gt},\mu} = D_\mu v$ has been introduced. The additional drift force $K_{{\rm gt},\mu}$ is tangent to the gauge orbit. The modified Langevin equation is equivalent to the modified Fokker-Planck equation

$$\frac{\partial P}{\partial t} = HP \equiv \int d^4x \frac{\delta}{\delta A_\mu^a(x)}\Big(\frac{\delta P}{\delta A_\mu^a(x)} - K_\mu^a(x)P\Big)$$
$$K_\mu^a(x) \equiv -\frac{\delta S_{\rm YM}}{\delta A_\mu^a(x)} + D_\mu^{ac}v^c(x), \tag{5}$$

We call H the "Fokker-Planck hamiltonian". (It is *not* the quantum mechanical hamiltonian!). We will choose $v^c(x)$ to make $D_\mu^{ac}v^c(x)$ globally restoring along gauge orbit directions, so every normalized solution $P(A,t)$ relaxes to a unique equilibrium distribution $\lim_{t\to\infty} P(A,t) = P(A)$.

Stochastic quantization in the time-dependent formulation has been developed by a number of authors who have expressed the solution as a functional integral[24] and demonstrated the renormalizability of this approach[25,26]. A systematic development is presented in,[27,28,29,30,31,32] and reviewed in,[33]. Renormalizability has also been established by an elaboration of BRST techniques[34,35]. Stochastic quantization may be and has been exactly simulated numerically[36] including on rather large lattices.[37,38,39,40]

2.2. *Time-independent stochastic quantization*

When the drift force is globally restoring, $P(A)$ may be calculated directly without reference to the artificial 5th time as the positive normalized solution of the time-independent Fokker-Planck equation

$$HP = \int d^4x \frac{\delta}{\delta A_\mu^a(x)} \Big(- \frac{\delta P}{\delta A_\mu^a(x)} + K_\mu^a P \Big) = 0, \tag{6}$$

and Euclidean expectation values are calculated from $\langle O \rangle = \int dA\, O(A) P(A)$. Equation (6) determines a probability distribution $P(A)$ that is correctly weighted at the non-perturbative level. The Gribov problem of globally correct gauge-fixing by identifying gauge orbits has been by-passed.

To ensure that $K_{\mathrm{gt},\mu} = D_\mu v$ is globally restoring, we introduce a minimizing functional[41,42,43] and choose $K_{\mathrm{gt},\mu}$ to be in direction of steepest descent among gauge-orbit directions. A convenient choice of minimizing functional is the Hilbert norm $||A||^2 = \int d^4x |A|^2$. For an infinitesimal variation in the gauge-orbit direction $\delta A_\mu = \epsilon D_\mu v$, we have

$$\delta ||A||^2 = 2(A_\mu, \delta A_\mu) = 2\epsilon(A_\mu, D_\mu v) = 2\epsilon(A_\mu, \partial_\mu v) = -2\epsilon(\partial_\mu A_\mu, v), \tag{7}$$

so steepest descent among gauge orbit directions is provided by $v = a^{-1}\partial \cdot A$ with $a > 0$, and the time-independent Fokker-Planck equation is now specified to within a single gauge parameter,

$$K_\mu^a(x) \equiv - \frac{\delta S_{\mathrm{YM}}}{\delta A_\mu^a(x)} + a^{-1} D_\mu^{ac} \partial \cdot A^c(x). \tag{8}$$

The Gribov region Ω may be characterized as the set of *local* minima[a] of the minimizing functional $F_A(g) \equiv ||{}^gA||^2$ with respect to local gauge transformations $g(x)$, whereas the fundamental modular region Λ may be characterized as the set of *absolute* minima. The set of absolute minima is free of Gribov copies, apart from the identification of gauge-equivalent points on the boundary $\partial\Lambda$, and may be identified with the gauge orbit space.[44] In a lattice discretization the minimization problem is of spin-glass type, and one expects many nearly degenerate local minima on a typical

[a]At any local minimum, this functional is (a) stationary, and (b) the matrix of second derivatives is non-negative. These two conditions fix the properties that define the Gribov region: (a) transversality, $\partial \cdot A = 0$, and (b) positivity of the Faddeev-Popov operator $-D(A) \cdot \partial$. Property (a) follows from (7), which states that the first variation of the minimizing functional is $\delta ||A||^2 = -2(\omega, \partial \cdot A)$. Property (b) follows because the second variation is $\delta^2 ||A||^2 = -2(\omega, \partial \cdot D(A)\omega)$.

gauge orbit, as is verified by numerical studies. Thus Λ is a proper subset of Ω, $\Lambda \subset \Omega$, but $\Lambda \neq \Omega$.

3. Solution of Fokker-Planck equation in the Landau gauge limit

3.1. *Region of stable equilibrium of $K_{\rm gt}$ is the Gribov region*

The gauge transformation "force" $K_{\rm gt}$ is not conservative, and cannot be written as the gradient of some 4-dimensional gauge-fixing action, $K_{{\rm gt},\mu} = a^{-1}D_\mu^{ac}\partial \cdot A^c(x) \neq -\delta S_{\rm gf}\delta A_\mu^a(x)$, so we cannot write the solution $P(A)$ explicitly in general. However we shall solve (6) exactly for $P(A)$ in the limit $a \to 0$. In this limit $P(A)$ gets concentrated in the region of *stable equilibrium* of the force $K_{{\rm gt},\mu} = a^{-1}D_\mu\partial \cdot A$. It is easily shown[20] that the region of stable equilibrium under the gauge transformation force $K_{{\rm gt},\mu} = D_\mu\partial \cdot A$ is the set of transverse configurations (Landau gauge condition) restricted to the Gribov region Ω.

3.2. *A well-defined change of variable*

In order to solve the time-independent Fokker-Planck equation in the limit $a \to 0$, we only need the solution for small a *in a coordinate patch U in A-space that includes the Gribov region Ω*. In U, we make the change of variable $A \to (B, g)$, defined by the gauge transformation,

$$A_\mu = A_\mu(B,g) = {}^gB_\mu = g^{-1}\partial_\mu g + g^{-1}B_\mu g, \tag{9}$$

where $\partial \cdot B = 0$ and $B \in \Omega$. Local gauge transformations are parametrized by $g(x) = \exp[t^a\theta^a(x)]$ where, for each x, the $\theta^a(x)$ are coordinates for the SU(N) group.[b] The notation $A = A(B,g)$ is understood to stand for $A = A(B,\theta)$, and we have $B = A(B,0)$.

Although this change of variable is not well-defined everywhere, it is well-defined in a coordinate patch U that includes Ω. This is true, even though there are Gribov copies within Ω, because the gauge orbits intersect Ω transversely. The coordinate patch U must be small enough in the θ-directions that the gauge transformations $g(\theta)$ that relate these Gribov copies are not in U. To verify that the gauge-orbits intersect Ω transversely,

[b]Here and below we use the notation $A_\mu \equiv t^aA_\mu^a$ and $B_\mu \equiv t^aB_\mu^a$. The t^a are set of anti-hermitian traceless matrices that form the fundamental representation of the Lie algebra of SU(N), $[t^a, t^b] = f^{abc}t^c$, where the structure constants f^{abc} are completely anti-symmetric.

it is sufficient to show that the change of variables (9) is invertible for infinitesimal angles $\theta^a(x) = \epsilon^a(x)$ for all $B \in \Omega$. It follows that it is also invertible, and thus well-defined, on some finite cordinate patch U that includes Ω.[20]

3.3. *Explicit solution*

In terms of the new variables, the solution *inside* the coordinate patch U has been found explicitly in the limit $a \to 0$, and shown to be the Faddeev-Popov weight in Landau gauge,[20]

$$P(A) = N\ \delta(\partial \cdot A) \det[-\partial \cdot D(A)]\ \exp[-S_{\rm YM}(A)]. \tag{10}$$

Its support lies on the set of transverse configurations, $\partial \cdot A = 0$, and it vanishes with $\det[-\partial \cdot D(A)]$ on the boundary $\partial\Omega$ of the Gribov region. This is the region of stable equilibrium of the drift force. We extend the solution to all of A-space by stipulating that *it vanishes outside* U. For the diffusion equation with a drift force, the equilibrium distribution is unique.[45] Although there are Gribov copies inside Ω,[43] they have no influence on expectation values, which is possible because the number of degrees of freedom is infinite.

4. Dyson-Schwinger equation for partition function

For this solution to be of use, we must be able to implement the cut-off of the functional integral at the boundary $\partial\Omega$ of the Gribov region. In an earlier approach this was rather cumbersome.[46] A very simple way to implement the cut-off [47] rests on the observation that the Gribov horizon $\partial\Omega$ is *a nodal surface* of the integrand because the Faddeev-Popov determinent, $\det[-\partial \cdot D(B)] = \prod_{n=1}^{\infty} \lambda_n(B)$ vanishes with $\lambda_1(B)$, that is to say, on $\partial\Omega$. The DS equations, which are derived by a partial integration, do not pick up a boundary term, and have the same form as they would if the integral were extended to infinity. In this approach we never have to know where the Gribov horizon actually is.

The partition function for the distribution (10) is given by

$$Z(J) = N \int_\Omega dB\ \det M(B)\ \exp[-S_{\rm YM}(B) + (J, B)], \tag{11}$$

where we have written $B \equiv A^{\rm tr}$, $M(B) \equiv -\partial \cdot D(B)$, and $(J, B) \equiv \int d^4x\ J^a_\mu(x) B^a_\mu(x)$. Only the transverse part of J contributes, and we also take J to be identically transverse, $J = J^{\rm tr}$. The Faddeev-Popov determinent $\det M(B)$ vanishes on the boundary $\partial\Omega$, so the identity

$$0 = \int_\Omega dB \, \frac{\delta}{\delta B_\mu^b(x)} \Big(\det M(B) \, \exp[-S_{\rm YM}(B) + (J, B)] \Big) \tag{12}$$

holds, *without any contribution from boundary terms even though the integral is cut-off at the Gribov horizon* $\partial\Omega$. As shown in detail in [20], the standard set of DS equations follows from this identity.

5. Supplementary conditions and resolution of paradox

Although the cut-off of the functional integral at the Gribov horizon does not change the *form* of the DS equations, it does provide *supplementary conditions* that govern the *choice* of solution. These include the conditions that both the gluon propagator $D_{\mu\nu}(x-y) = \langle B_\mu(x) B_\nu(y) \rangle$ and ghost propagator $G(x-y) = \langle M_{xy}^{-1}(B) \rangle$ be positive matrices. The first property follows because the Faddeev-Popov weight is positive inside the Gribov region Ω, and the second because the Fadeev-Popov operator $M(B)$ is positive there. Another supplementary condition results from the fact, that in a space of high-dimension, entropy favors a high concentration of population very near the boundary $\partial\Omega$ of the bounded region Ω. The boundary occurs where the lowest non-trivial eigenvalue of the Faddeev-Popov operator $M(B)$ vanishes. Thus, for typical configurations B, the positive operator $M(B)$ has a very small eigenvalue and, in fact, it has a high density of eigenvalues $\rho(\lambda, B)$ at $\lambda = 0$, per unit Euclidean volume V, as compared to the Laplacian operator.[48] This causes the ghost propagator to be enhanced[18,48] at $p = 0$ compared to $1/p^2$,

$$\lim_{p \to 0} [p^2 \tilde{G}(p)]^{-1} = 0. \tag{13}$$

We call this supplementary condition the "horizon condition".

The resolution of the paradox discussed in the Introduction is now apparent. The DS equations are not modified by the cut-off at the Gribov horizon, and the solutions found in [1] satisfy these supplementary conditions that are a consequence of the cut-off at the Gribov horizon.

6. Horizon condition and renormalization

The DS equation for the ghost propagator reads

$$\tilde{G}^{-1}(p) = p^2 - N g_0^2 (2\pi)^{-4} \, p_\mu p_\nu \int d^4k \tilde{D}_{\mu\nu}(k) \, \tilde{G}(q) \, V(q^2, k^2, p^2), \tag{14}$$

where $\tilde{D}_{\mu\nu}(k)$ and $\tilde{G}(q)$ are the gluon and ghost propagators, and $q \equiv p-k$. The ghost-ghost-gluon vertex function has been written

$$\tilde{\Gamma}_\mu(p,q) = -ig_0 \; P^{\rm tr}_{\mu\nu}(k) \; p_\nu \; V(p^2,k^2,q^2), \tag{15}$$

where the scalar vertex function is symmetric $V(p^2,k^2,q^2) = V(q^2,k^2,p^2)$. This expression for the vertex is possible because the transversality condition is imposed on-shell. All quantities are unrenormalized.

Factorization of the external ghost momentum is explicit in (14). This is a special property of the Landau gauge that makes it less divergent than other gauges. We renormalize according to

$$D_{\mu\nu} = Z_3 D_{R,\mu\nu}; \quad G = \tilde{Z}_3 G_R; \quad V = \tilde{Z}_1^{-1} V_R; \quad g_0 = \tilde{Z}_1 (\tilde{Z}_3 Z_3^{1/2})^{-1} g_R, \tag{16}$$

and in Landau gauge the additional special property

$$\tilde{Z}_1 = 1; \qquad V = V_R; \qquad g_0 = (\tilde{Z}_3 Z_3^{1/2})^{-1} g_R \tag{17}$$

holds. In terms of renormalized quantities, the DS equation for the ghost propagator reads,

$$\begin{aligned} \tilde{G}_R^{-1}(p) = p^2 \tilde{Z}_3 - N g_R^2 (2\pi)^{-4} \; p_\mu p_\nu \\ \times \int d^4k \; \tilde{D}_{R,\mu\nu}(k) \; \tilde{G}_R(p-k) \; V_R(q^2, \; k^2, \; p^2). \end{aligned} \tag{18}$$

To avoid infrared difficulties, the ghost propagator is usually renormalized at some finite renormalization mass μ. However the horizon condition (13) allows us to renormalize at $p = 0$. It requires that in the DS equation (18), the first term, $p^2\tilde{Z}_3$, be cancelled by the term of order p^2 in the second term. This gives a renormalization condition at $p = 0$, in the form of an equation for $\tilde{Z}_3$,

$$\tilde{Z}_3 = N g_R^2 \; (2\pi)^{-4} \int_{|k|<\Lambda} d^4k \; \hat{p}_\mu \hat{p}_\nu \tilde{D}_{R,\mu\nu}(k) \; \tilde{G}_R(k) \; V_R(k^2 \; k^2, \; 0), \tag{19}$$

where Λ is an ultraviolet cut-off. We have set $p = 0$ in the integrand, and the integral is independent of the direction $\hat{p}$. This statement of the horizon condition is flagrantly non-perturbative because, in perturbation theory, the left hand side is of leading order 1, but the right hand side is of leading order g_R^2.

The last equation gives the renormalization-group flow,

$$\Lambda \frac{\partial \tilde{Z}_3}{\partial \Lambda} = N g_R^2 (4\pi)^{-2} (3/2) \; \Lambda^4 \; \tilde{D}_R(\Lambda) \; \tilde{G}_R(\Lambda) \; V_R(\Lambda^2, \Lambda^2, 0). \tag{20}$$

As a check, we note that if we take the tree values $\tilde{D}_R(\Lambda) = \tilde{G}_R(\Lambda) = 1/\Lambda^2$, and $V(p^2, k^2, q^2) = 1$, we obtain

$$\Lambda \frac{\partial \tilde{Z}_3}{\partial \Lambda} = (4\pi)^{-2}(3/2)N g_0^2 + O(g_0^4) \tag{21}$$

which agrees with the standard one-loop expression in Landau gauge. Thus the horizon condition provides a normalization condition for the ghost propagator at $p = 0$ that is in flagrant disagreement with perturbation theory, but nevertheless satisfies the perturbative renormalization-group flow equation.

We substitute (19) into the DS equation (18) for the ghost propagator, and obtain

$$\begin{aligned}\tilde{G}_R^{-1}(p) = N g_R^2\ p_\mu p_\nu\ (2\pi)^{-4} \int d^4k\ \tilde{D}_{R,\mu\nu}(k) \\ \times[\tilde{G}_R(k)\ V_R(k^2, k^2, 0) - \tilde{G}_R(p-k)\ V_R((p-k)^2, k^2, p^2)].\end{aligned} \tag{22}$$

This is a finite, renormalized DS equation for the ghost propagator. It is invariant under the renormalization group in the sense that it is form-invariant under the transformation (16) and (17) of renormalization theory in Landau gauge. This equation, from which the tree term k^2 has been cancelled by the horizon condition, gives the ghost propagator an infrared anomalous dimension a_G, so it behaves like $G(k) \sim (\mu^2)^{a_G}/(k^2)^{1+a_G}$ in the infrared. This puts QCD into a non-perturbative phase.

7. Infrared limit of QCD and mass gap

7.1. *Infrared asymptotic limit of QCD*

Recent solutions of the truncated coupled DS equations yield a ghost propagator that is enhanced in the infrared, and a gluon propagator that is suppressed,[1,2,3,4,5,6] with typical values for the infrared asymptotic gluon and ghost propagators given by,[2,3]

$$\begin{aligned} D^{\mathrm{as}}(k) &= \mu^{2a_D}/(k^2)^{1+a_D} \approx (k^2)^{0.187}/(\mu^2)^{1.187} \\ G^{\mathrm{as}}(k) &= \mu^{2a_G}/(k^2)^{1+a_G} \approx (\mu^2)^{0.595}/(k^2)^{1.595}\,, \end{aligned} \tag{23}$$

where a_D and a_G are the infrared critical exponents of the ghost and gluon, with $a_G = (93 - \sqrt{1201})/98 \approx 0.595$, $a_D = -2a_G$.

One may use these expressions for the asymptotic propagators to estimate the convergence and magnitude of the various terms on the right hand side of the DS equations. The dominant terms in the infrared region are the ones that contain the most ghost propators $G(k)$ in the loop integrals. Keeping only the infrared-dominant terms, one finds: (i) *The infrared limit of the DS equations decouples from the degrees of freedom associated with finite momentum and is free of ultraviolet divergences.* (ii) *The dominant terms in the infrared limit come from the action* $-\ln\det(-\partial\cdot D)$, *whereas the subdominant terms come from Yang-Mills action* $S_{\rm YM}$.

If one extends the non-perturbative formulation to a BRST-invariant theory, one obtains the standard BRST-invariant local action in Landau gauge,

$$S = \int d^4x \; [s(\partial_\mu \bar{c} A_\mu) + S_{\rm YM}(A)], \tag{24}$$

where the BRST operator acts according to

$$sA_\mu = D_\mu c; \qquad sc = -c^2; \qquad s\bar{c} = \lambda; \qquad s\lambda = 0. \tag{25}$$

The asymptotic infrared limit is described by the local BRST-exact action

$$S^{\rm as} \equiv \int d^4x \; s(\partial_\mu \bar{c} A_\mu) = \int d^4x \; (-\;\partial_\mu \bar{c}\; D_\mu c + \partial_\mu \lambda \; A_\mu), \tag{26}$$

and the infrared asymptotic correlators satisfy the Slavnov-Taylor identities. The cut-off at the Gribov horizon makes the functional integral converge, even though we have set $S_{\rm YM} = 0$.

7.2. *Mass gap*

The action $S^{\rm as}$ that describes the infrared asymptotic theory is not only BRST-invariant, it is BRST-exact, $S^{\rm as} = sX$. It defines a topological quantum field theory that corresponds to a pure gauge-fixing term. Consequently the correlators of phyiscal (i.e. BRST-exact) operators are trivial and are characterized by correlation lengths that vanish or, equivalently, by an infinite mass gap. Recall however that we have taken the infrared limit in which all momenta k vanish compared to the single mass scale Λ_{QCD}. So the mass gap that appears infinite in the infrared limit is in fact a mass gap of order Λ_{QCD}.

Acknowledgments

I am grateful to the organizers of Confinement 2003, particularly Professor Suganuma, for their kind hospitality and the opporturnity to speak at this stimulating meeting.

References

1. L. von Smekal, A. Hauck and R. Alkofer, *A Solution to Coupled Dyson-Schwinger Equations in Gluons and Ghosts in Landau Gauge,* Ann. Phys. **267** (1998) 1; L. von Smekal, A. Hauck and R. Alkofer, *The Infrared Behavior of Gluon and Ghost Propagators in Landau Gauge QCD,* Phys. Rev. Lett. **79** (1997) 3591; L. von Smekal *Perspectives for hadronic physics from Dyson-Schwinger equations for the dynamics of quark and glue,* Habilitationsschrift, Friedrich-Alexander Universität, Erlangen-Nürnberg (1998).
2. C. Lerche and L. von Smekal *On the infrared exponent for gluon and ghost propagation in Landau gauge QCD*, hep-ph/0202194
3. D. Zwanziger, *Non-perturbative Landau gauge and infrared critical exponents in QCD*, Phys. Rev. D, **65** 094039 (2002) and hep-th/0109224.
4. C. S. Fischer, R. Alkofer and H. Reinhardt, *The elusiveness of infrared critical exponents in Landau gauge Yang-Mills theories,* Phys. Rev. D **65**, 094008 (2002)
5. C. S. Fischer and R. Alkofer, *Infrared exponents and running coupling of SU(N) Yang-Mills Theories*, Phys. Lett. B **536**, 177 (2002).
6. D. Zwanziger, *Time-independent stochastic quantization, DS equations, and infrared critical exponents in QCD*, Phys. Rev. D**67**, 105001 (2003), and hep-th/0206053.
7. A. Cucchieri, Phys. Rev. D **60** , 034508 (1999).
8. A. Cucchieri, F. Karsch, P Petreczky, Phys. Lett. B **497** , 80 (2001).
9. A. Cucchieri, F. Karsch, P Petreczky, Phys. Rev. D **64** , 036001 (2001).
10. A. Cucchieri, Phys. Lett. **B422** 233, (1998), and hep-lat/9709015.
11. A. Cucchieri, *Gribov copies in the minimal Landau gauge: the influence on gluon and ghost propagators,* Nucl. Phys. B **508**, 353 (1997).
12. K. Langfeld, H. Reinhardt, and J. Gattnar, Nucl. Phys. B **621**, 131 (2002).
13. F. Bonnet, P. O. Bowman, D. B. Leinweber, A. G. Williams and J. M. Zanotti, Phys. Rev. **D64** 034501 (2001).
14. H. Nakajima and S. Furui, Nucl. Phys. Proc. Suppl. **73** 635 (1999).
15. I. L. Bogolubsky and V. K. Mitrjushkin, hep-lat/0204006.
16. A. Cucchieri, and D. Zwanziger, *Fit to gluon propagator and Gribov formula,* Phys. Lett. **B524** 123 (2002), and hep-lat/0012024.
17. A. Cucchieri, T. Mendes, and A. R. Taurines, *SU(2) Landau gluon propagator on a* 140^3 *lattice*, hep-lat/0302022.
18. V.N. Gribov, Nucl. Phys. B **139**, 1 (1978).
19. D. Zwanziger, *Vanishing of zero momentum lattice gluon propagator and color confinement,* Nucl. Phys. B **364**, 127–161 (1991).
20. D. Zwanziger *Non-perturbative Faddeev-Popov formula and the infrared limit of QCD*, Phys. Rev. D, (to be published), and hep-ph/0303028.
21. I.M. Singer, Comm. of Math. Phys. **60** (1978) 7.
22. G. Parisi, Y.S. Wu, Sci. Sinica **24**, 484 (1981).
23. D. Zwanziger, *Covariant Quantization of Gauge Fields without Gribov Ambiguity*, Nucl. Phys. B **192**, 259 (1981).
24. E. Gozzi, *Functional Integral approach to Parisi–Wu Quantization: Scalar*

Theory, Phys. Rev. **D28**, 1922 (1983).
25. J. Zinn-Justin, Nucl. Phys. B **275**, 135, (1986).
26. J. Zinn-Justin, D. Zwanziger, Nucl. Phys. B **295** 297, (1988).
27. Z. Bern, M.B. Halpern, L. Sadun, C. Taubes, Phys. Lett. **165B,** 151, (1985).
28. Z. Bern, M.B. Halpern, L. Sadun, C. Taubes, Nucl. Phys. **B284,** 1, (1987).
29. Z. Bern, M.B. Halpern, L. Sadun, C. Taubes, Nucl. Phys. **B284,** 35, (1987).
30. Z. Bern, M.B. Halpern, L. Sadun, Nucl. Phys. **B284,** 92, (1987).
31. Z. Bern, M.B. Halpern, L. Sadun, Z. Phys. **C35,** 255, (1987).
32. L. Sadun, Z. Phys. **C36,** 467, (1987).
33. M. B. Halpern, Prog. Theor. Phys. Suppl. **111,** 163, (1993).
34. P. A. Grassi, L. Baulieu and D. Zwanziger, *Gauge and Topological Symmetries in the Bulk Quantization of Gauge Theories*, Nucl. Phys. **B597** 583, (2001) and hep-th/0006036.
35. L. Baulieu and D. Zwanziger, *From stochastic quantization to bulk quantization; Schwinger-Dyson equations and the S-matrix*, JHEP 08:016, 2001 hep-th/0012103.
36. A. Nakamura and M. Mizutani, *Numerical study of gauge fixing ambiguity,* Vistas in Astronomy **37** 305 (1993).
37. M. Mizutani and A. Nakamura, *Stochastic gauge fixing for compact lattice theories* Nucl. Phys. B (Proc. Suppl.) **34**, 253 (1994).
38. H. Aiso, M. Fukuda, T. Iwamiya, A. Nakamura, T. Nakamura, and M. Yoshida, *Gauge fixing and gluon propagators,* Prog. Theor. Physics. (Suppl.) **122**, 123 (1996).
39. H. Aiso, J. Fromm, M. Fukuda, T. Iwamiya, A. Nakamura, T. Nakamura, M. Stingl and M. Yoshida, *Towards understanding of confinement of gluons,* Nucl. Phys. B (Proc. Suppl.) **53**, 570 (1997).
40. F. Shoji, T. Suzuki, H. Kodama, and A. Nakamura, Phys. Lett. **B476**, 199 (2000).
41. T. Maskawa and H. Nakajima, Prog. Theor. Phys. **60**, 1526 (1978); **63**, 641 (1980).
42. D. Zwanziger, Phys. Lett. **114B**, 337 (1982).
43. M. Semenov-Tyan-Shanskii and V. Franke, Zap. Nauch. Sem. Leningrad. Otdeleniya Matematicheskogo Instituta im V. A. Steklov, AN SSSR, vol 120, p 159, (1982) (English translation: New York: Plenum Press (1986).
44. P. van Baal, *Gribov ambiguities and the fundamental domain*, hep-th/9711070.
45. S. R. S. Varadhan, *Diffusion problems and partial differential equations,* Tata Institute of Fundamental Research, Bombay, Springer Verlag, Berlin (1980), pp. 249-251.
46. D. Zwanziger, *Critical limit of lattice gauge theory,* Nucl. Phys. B **378** 525-590 (1992).
47. D. Zwanziger, *Non-perturbative Landau gauge and infrared critical exponents in QCD,* Phys. Rev. D **65**, 094309 (2002), and hep-th/0109224.
48. D. Zwanziger, *Renormalizability of the critical limit of lattice gauge theory by BRS invariance,* Nucl. Phys. B **399** , 477 (1993).

Lattice QCD for Quark-Hadron Physics

EXCITATIONS OF THE STATIC QUARK-ANTIQUARK SYSTEM IN SEVERAL GAUGE THEORIES*

K. JIMMY JUGE

School of Mathematics, Trinity College, Dublin 2, Ireland

JULIUS KUTI

Department of Physics, University of California at San Diego, La Jolla, USA 92093-0319

COLIN MORNINGSTAR

Department of Physics, Carnegie Mellon University, Pittsburgh, PA, USA 15213-3890

The spectrum of gluons in the presence of a static quark-antiquark pair is studied using Monte Carlo simulations on anisotropic space-time lattices. For very small quark-antiquark separations R, the level orderings and approximate degeneracies disagree with the expectations from an effective string theory. As the quark-antiquark separation R increases, a dramatic rearrangement of the energies occurs, and above 2 fm, all of the levels studied show behavior consistent with an effective string description. The energy spacings are nearly π/R, but a tantalizing fine structure remains. In addition to 4-dimensional $SU(3)$ gauge theory, results from 3-dimensional $SU(2)$ and compact $U(1)$ gauge theories are also presented.

1. Introduction

An important part of understanding confinement in quantum chromodynamics (QCD) is understanding the low-lying physics of the confining gluon field. The spectrum of gluons in the presence of a static quark-antiquark pair provides valuable clues about the nature of the low-lying stationary states of the gluon field. Innumerable lattice QCD simulations have confirmed that the energy of the ground state rises linearly with the separation R between the quark and antiquark, naively suggesting that the gluon field forms a string-like confining object connecting the quark and the antiquark.

*Talk presented by C. Morningstar.

However, it should be noted that the spherical bag model also predicts a linearly rising energy for moderate R, and hence, the linearly rising ground-state energy is *not* conclusive evidence of string formation. Computations of the gluon action density surrounding a static quark-antiquark pair in $SU(2)$ gauge theory also hint at flux tube formation[1].

Adopting the viewpoint that the nature of the confining gluon field is best revealed in its excitation spectrum, we have embarked on a series of studies[2] employing recent advances in lattice simulation technology, including anisotropic lattices, improved gauge actions, and large sets of creation operators, to investigate the onset of string-like behavior in the gluon field surrounding a quark-antiquark pair for a wide range of separations R from 0.1 to 3 fm. Energy gaps given by multiples of π/R and a well-defined pattern of degeneracies and level orderings among the different symmetry channels form a very distinctive and robust signature of the onset of the Goldstone modes of the effective QCD string. Non-universal details of the underlying string description, such as higher order interactions and their couplings, are encoded in the fine structure of the spectrum at large separations.

In this talk, results from this series of studies clearly demonstrating the onset of string formation for large R are presented. The spectra of three-dimensional $SU(2)$ and compact $U(1)$ gauge theories are also presented to address questions about the dependence of results on the gauge group and the dimensionality of space-time. First, the classification of the states is discussed in Sec. 2. The expected level orderings at large R from an effective string description are detailed in Sec. 3. The spectrum in four-dimensional $SU(3)$ is discussed in Sec. 4, and three-dimensional $SU(2)$ and compact $U(1)$ results are presented in Sec. 5. A complementary study of the spectrum and Casimir energy in three-dimensional $Z(2)$ gauge theory is presented elsewhere[3].

2. Classification of states

The first step in determining the energies of the stationary states of gluons in the presence of a static quark and antiquark, fixed in space some distance R apart, is to classify the levels in terms of the symmetries of the problem. Such a system has cylindrical symmetry about the axis $\hat{\boldsymbol{R}}$ passing through the quark and the antiquark (the molecular axis). The total angular momentum $\vec{\boldsymbol{J}}_g$ of the gluons is not a conserved quantity, but its projection $\vec{\boldsymbol{J}}_g \cdot \hat{\boldsymbol{R}}$ onto the molecular axis is and can be used to label the energy levels of the gluons. Here, we adopt the standard notation from the physics of diatomic molecules and denote the magnitude of the eigenvalue of $\vec{J}_g \cdot \hat{R}$ by Λ. States with $\Lambda = 0, 1, 2, 3, 4, \ldots$ are typically denoted by the capital Greek letters $\Sigma, \Pi, \Delta, \Phi, \Gamma, \ldots$, respectively. The energy of the gluons is unaffected by reflections in a plane containing the molecular axis; since such reflections interchange states of opposite handedness, given by the sign of the eigenvalue of $\vec{J}_g \cdot \hat{R}$, such states must necessarily be degenerate (Λ doubling). However, this doubling does not apply to the Σ states; Σ states which are even (odd) under a reflection in a plane containing the molecular axis are denoted by a superscript $+$ $(-)$. Another symmetry is the combined operation of charge conjugation and spatial inversion about the midpoint between the quark and the antiquark. Here, we denote the eigenvalue of this transformation by η_{CP} which can take values ± 1. States which are even (odd) under this parity–charge-conjugation operation are indicated by subscripts g (u). Thus, the low-lying levels in four space-time dimensions are labeled Σ_g^+, Σ_g^-, Σ_u^+, Σ_u^-, Π_g, Π_u, Δ_g, Δ_u, and so on.

In three space-time dimensions, there is no longer a rotational symmetry about the molecular axis since there are only two spatial dimensions. Instead, the analogous symmetry is a reflection in the molecular axis, and states are either symmetric S or antisymmetric A under this transformation. The combined operation of charge conjugation and spatial inversion about the midpoint between the quark and the antiquark is still a symmetry in three space-time dimensions. Once again, states which are even (odd) under this parity–charge-conjugation operation are indicated by the subscripts g (u). To summarize, the low-lying states in three space-time dimensions are labeled by S_g, A_g, S_u, A_u.

One last note concerning the classification of states should be made. In a gauge theory based on the group $SU(2)$, the subscript g and u refers only to spatial inversion about the midpoint between the static sources, without charge conjugation. This is due to the fact that in $SU(2)$, the complex conjugate representation $\overline{\Gamma}$ is equivalent to the Γ representation.

3. String modes

The ground-state energy of gluons in the presence of a static quark-antiquark pair rises linearly with the quark-antiquark separation R. This fact has led to the general belief that at sufficiently large R, the chromoelectric and chromomagnetic fields become confined to a long tube-like region of space connecting the quark and the antiquark. A treatment of the gluon field in terms of the collective degrees of freedom associated with the position of the long flux might then be sufficient for reproducing the long-wavelength physics. If true, one then hopes that the oscillating flux can be well described in terms of an effective string theory. In such a case, the lowest-lying excitations are expected to be the Goldstone modes associated with the spontaneously broken transverse translational symmetry. These modes are a universal feature of any low-energy description of the effective QCD string and have energy separations above the ground state given by multiples of π/R. For the gluonic excitations at small R, no robust expectations from string theory presently exist. In this section, the pattern of degeneracies and level orderings of the expected string modes for large R in both three and four space-time dimensions are deduced.

The excitations of long flux lines are expected to be described by a local derivative expansion of a massless vector field $\vec{\xi}$ with two transverse components in four-dimensional space-time[4,5]. Assume that the quark is located at $(0,0,0)$ and the antiquark is at $(0,0,R)$, then $\vec{\xi}(x_3,x_4)$ represents the transverse displacement in the x_1 and x_2 directions of the thin flux line from its equilibrium position. We further assume that this displacement field is continuous and single-valued, so that string configurations which double-back on themselves or overhang the ends are not allowed. Symmetry arguments then require that the effective QCD string action in Euclidean space-time should have a leading term given by

$$S_{\rm eff}^{(0)} = \tfrac{1}{2}c_0 \int dx_4 \int_0^R dx_3\ \partial_\mu\vec{\xi}\cdot\partial_\mu\vec{\xi}, \qquad \vec{\xi}(0,x_4)=\vec{\xi}(R,x_4)=0, \quad (1)$$

where the derivatives are taken with respect to the two worldsheet coordinates x_3 and x_4, and c_0 has the dimension of a mass squared and is proportional to the string tension. The stationary states are found by expressing the displacement field $\vec{\xi}$ in terms of normal modes. For fixed ends, the normal modes are standing waves $\sin(m\pi x_3/R)$. These modes have energies $m\omega$ for positive integer m and $\omega=\pi/R$. For two transverse dimensions, one defines right $(+)$ and left $(-)$ circularly polarized ladder operators $a^\dagger_{m\pm}$, then the string eigenmodes are

$$\prod_{m=1}^{\infty}\left(\ (a^\dagger_{m+})^{n_{m+}}\ (a^\dagger_{m-})^{n_{m-}}\right)\ |0\rangle, \quad (2)$$

where $|0\rangle$ denotes the ground state of the string, and n_{m+} and n_{m-} are the occupation numbers which take values $0,1,2,\ldots$. If E_0 denotes the energy of the ground state, then the eigenvalues E (energy), Λ, and η_{CP} associated

Table 1. Low-lying string levels for fixed ends in four space-time dimensions. The $N = 1$ level is two-fold degenerate, and the $N = 2, 3, 4$ levels are 5,10,15-fold degenerate, respectively. The $+(-)$ signs refer to right (left) circular polarizations, and positive integers indicate standing wave normal modes.

$N = 0$:	Σ_g^+	$\|0\rangle$	
$N = 1$:	Π_u	$a_{1+}^\dagger\|0\rangle$	$a_{1-}^\dagger\|0\rangle$
$N = 2$:	$\Sigma_g^{+\prime}$	$a_{1+}^\dagger a_{1-}^\dagger\|0\rangle$	
	Π_g	$a_{2+}^\dagger\|0\rangle$	$a_{2-}^\dagger\|0\rangle$
	Δ_g	$(a_{1+}^\dagger)^2\|0\rangle$	$(a_{1-}^\dagger)^2\|0\rangle$
$N = 3$:	Σ_u^+	$(a_{1+}^\dagger a_{2-}^\dagger + a_{1-}^\dagger a_{2+}^\dagger)\|0\rangle$	
	Σ_u^-	$(a_{1+}^\dagger a_{2-}^\dagger - a_{1-}^\dagger a_{2+}^\dagger)\|0\rangle$	
	Π_u'	$a_{3+}^\dagger\|0\rangle$	$a_{3-}^\dagger\|0\rangle$
	Π_u''	$(a_{1+}^\dagger)^2 a_{1-}^\dagger\|0\rangle$	$a_{1+}^\dagger(a_{1-}^\dagger)^2\|0\rangle$
	Δ_u	$a_{1+}^\dagger a_{2+}^\dagger\|0\rangle$	$a_{1-}^\dagger a_{2-}^\dagger\|0\rangle$
	Φ_u	$(a_{1+}^\dagger)^3\|0\rangle$	$(a_{1-}^\dagger)^3\|0\rangle$
$N = 4$:	$\Sigma_g^{+\prime\prime}$	$a_{2+}^\dagger a_{2-}^\dagger\|0\rangle$	
	$\Sigma_g^{+\prime\prime\prime}$	$(a_{1+}^\dagger)^2(a_{1-}^\dagger)^2\|0\rangle$	
	$\Sigma_g^{+(\mathrm{iv})}$	$(a_{1+}^\dagger a_{3-}^\dagger + a_{1-}^\dagger a_{3+}^\dagger)\|0\rangle$	
	Σ_g^-	$(a_{1+}^\dagger a_{3-}^\dagger - a_{1-}^\dagger a_{3+}^\dagger)\|0\rangle$	
	Π_g'	$a_{4+}^\dagger\|0\rangle$	$a_{4-}^\dagger\|0\rangle$
	Π_g''	$(a_{1+}^\dagger)^2 a_{2-}^\dagger\|0\rangle$	$(a_{1-}^\dagger)^2 a_{2+}^\dagger\|0\rangle$
	Π_g'''	$a_{1+}^\dagger a_{1-}^\dagger a_{2+}^\dagger\|0\rangle$	$a_{1+}^\dagger a_{1-}^\dagger a_{2-}^\dagger\|0\rangle$
	Δ_g'	$a_{1+}^\dagger a_{3+}^\dagger\|0\rangle$	$a_{1-}^\dagger a_{3-}^\dagger\|0\rangle$
	Δ_g''	$(a_{2+}^\dagger)^2\|0\rangle$	$(a_{2-}^\dagger)^2\|0\rangle$
	Δ_g'''	$(a_{1+}^\dagger)^3 a_{1-}^\dagger\|0\rangle$	$a_{1+}^\dagger(a_{1-}^\dagger)^3\|0\rangle$
	Φ_g	$(a_{1+}^\dagger)^2 a_{2+}^\dagger\|0\rangle$	$(a_{1-}^\dagger)^2 a_{2-}^\dagger\|0\rangle$
	Γ_g	$(a_{1+}^\dagger)^4\|0\rangle$	$(a_{1-}^\dagger)^4\|0\rangle$

with the string eigenstates are given by

$$
\begin{aligned}
E &= E_0 + \frac{N\pi}{R}, & N &= \sum_{m=1}^{\infty} m\ (n_{m+} + n_{m-}), \\
\eta_{CP} &= (-1)^N, & \Lambda &= \left|\sum_{m=1}^{\infty} (n_{m+} - n_{m-})\right|.
\end{aligned}
\tag{3}
$$

For the Σ states, the evenness or oddness under exchange $(-) \leftrightarrow (+)$ of the circular polarizations yields a superscript $+$ or $-$, respectively. Using

Table 2. Low-lying string levels for fixed ends in three space-time dimensions. The $N = 1$ level is nondegenerate, and the $N = 2, 3, 4$ levels are 2,3,5-fold degenerate, respectively. The positive integers indicate the standing wave normal modes.

$N = 0$:	S_g	$\vert 0\rangle$
$N = 1$:	A_u	$a_1^\dagger\vert 0\rangle$
$N = 2$:	S_g'	$(a_1^\dagger)^2\vert 0\rangle$
	A_g	$a_2^\dagger\vert 0\rangle$
$N = 3$:	A_u'	$(a_1^\dagger)^3\vert 0\rangle$
	A_u''	$a_3^\dagger\vert 0\rangle$
	S_u	$a_1^\dagger a_2^\dagger\vert 0\rangle$
$N = 4$:	S_g''	$(a_2^\dagger)^2\vert 0\rangle$
	S_g'''	$a_1^\dagger a_3^\dagger\vert 0\rangle$
	$S_g^{(\mathrm{iv})}$	$(a_1^\dagger)^4\vert 0\rangle$
	A_g'	$a_4^\dagger\vert 0\rangle$
	A_g''	$(a_1^\dagger)^2 a_2^\dagger\vert 0\rangle$

these properties, the orderings and degeneracies of the Goldstone string energy levels and their symmetries are as shown in Table 1. One sees that the $N\pi/R$ behavior and a well-defined pattern of degeneracies and level orderings among the different channels form a very distinctive signature of the onset of the Goldstone modes for the effective QCD string.

In three space-time dimensions, there is only one transverse direction for the string, so the ladder operators are written $a_m^\dagger$ since there are no right and left circular polarizations. The orderings and degeneracies of the Goldstone modes are given in Table 2.

4. $SU(3)$ gauge theory in 4 dimensions

The spectrum shown in Fig. 1 provides clear evidence that the gluon field can be well approximated by an effective string theory for large separations R. Energy gaps ΔE above the ground state are compared to asymptotic string gaps for 15 excited states in Fig. 2. The quantity $\Delta E/(N\pi/R) - 1$ is plotted to show percentage deviations from the asymptotic string levels for string quantum number $N = 1, 2, 3, 4$. For small $R < 2$ fm, the energy gaps lie far below the null lines and are strongly split for fixed N. In other words, string formation does not appear to set in until the quark and the antiquark are separated by about 2 fm. For small separations, the level

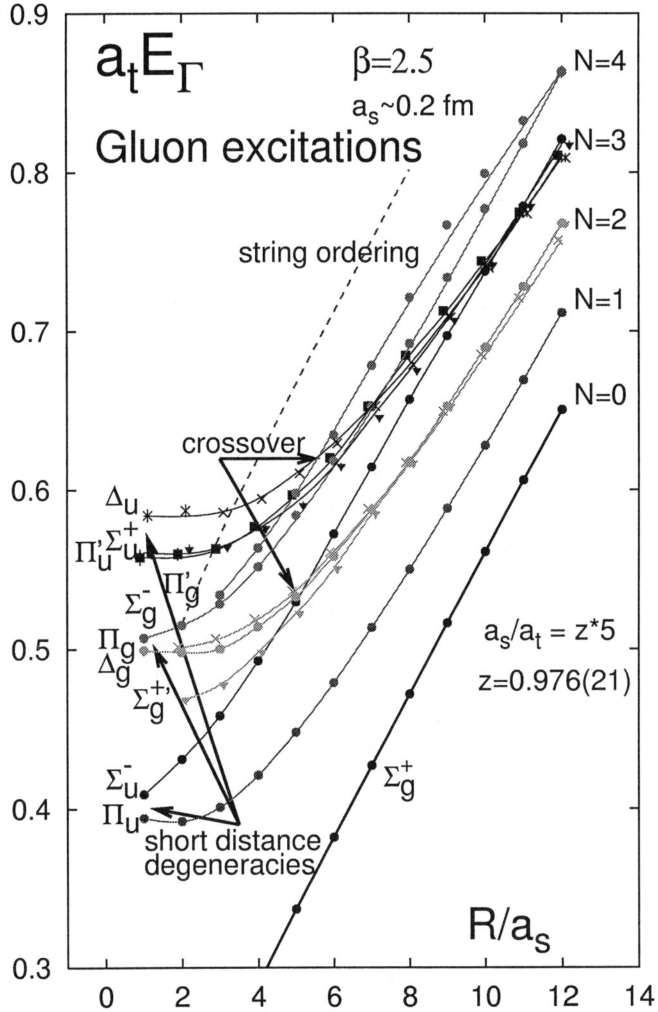

Figure 1. The spectrum of gluonic excitations in the presence of a static quark-antiquark pair separated by a distance R in 4-dimensional $SU(3)$ gauge theory (from Ref. [2]). Results are from one simulation for lattice spacing $a_s \sim 0.2$ fm using an improved action on a $(10^2 \times 30) \times 60$ anisotropic lattice with coupling $\beta = 2.5$ and bare aspect ratio $\xi = 5$. At large distances, *all* levels without exception are consistent with the expectations from an effective string theory description. A dramatic level rearrangement is observed in the crossover region between $0.5 - 2.0$ fm. The dashed line marks a lower bound for the onset of mixing effects with glueball states.

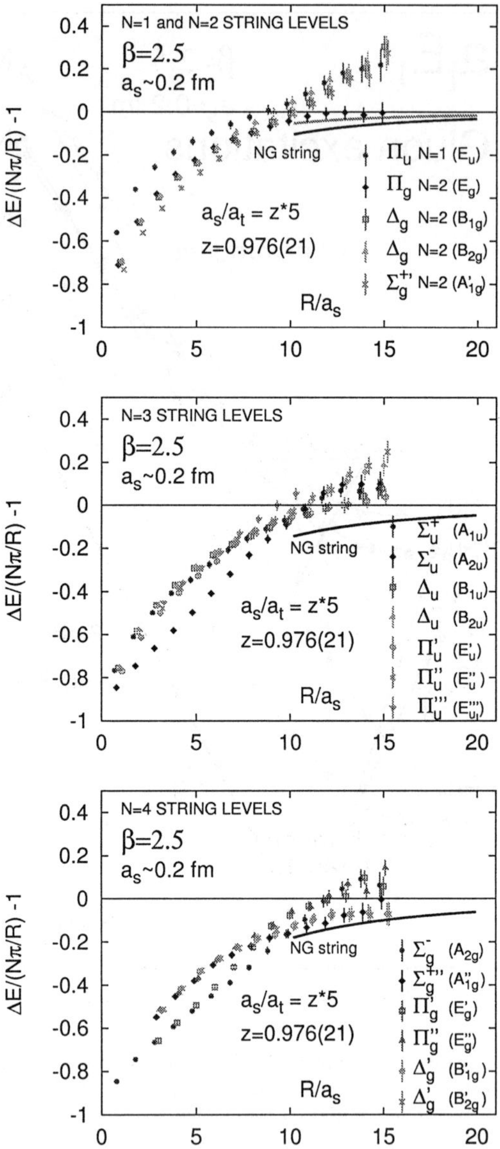

Figure 2. The energy gaps ΔE above the ground state Σ_g^+ of the stationary states of gluons in the presence of a static quark-antiquark pair in 4-dimensional $SU(3)$ gauge theory. The results at lattice spacing $a_s \sim 0.2$ fm are shown against the quark-antiquark separation R and are compared with the $N\pi/R$ splittings expected in an effective string theory at large R. The large-R results for a free Nambu-Goto (NG) string are also shown.

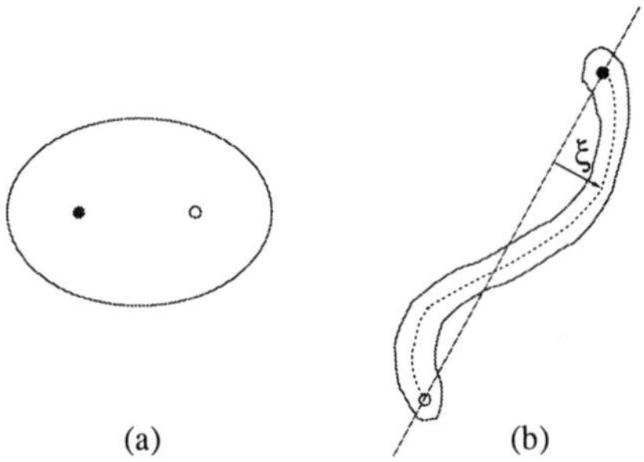

Figure 3. One possible interpretation of the spectrum in Fig. 1. (a) For small quark-antiquark separations, the strong chromoelectric field of the quark-antiquark pair repels the physical vacuum (dual Meissner effect) creating a bubble. Explanations of the low-lying stationary states must take into account both the gluonic modes inside the bubble and oscillations of the collective coordinates describing the bubble. (b) For large quark-antiquark separations, the bubble stretches into a thin tube of flux, and the low-lying states are explained by the collective motion of the tube since the internal gluonic excitations are much higher lying.

orderings and degeneracies are not consistent with the expectations from an effective string description. More importantly, the gaps differ appreciably from $N\pi/R$ with $N = 1, 2, 3, \ldots$, as clearly shown in Fig. 2. Such deviations, as large as 50% or more, cannot be considered mere corrections, making the applicability of an effective string description problematical. Between 0.5 to 2 fm, a dramatic level rearrangement occurs. For separations above 2 fm, the levels agree *without exception* with the ordering and degeneracies expected from an effective string theory. The gaps agree well with $N\pi/R$, but a fine structure remains. This first glimpse of such a fine structure offers the exciting possibility of deducing details of the effective QCD string action in future higher precision simulations.

It is reasonable to expect that the first few terms in the effective string action might predominantly arise from the geometric properties of the flux tube. The Nambu-Goto (NG) action is one of the simplest geometrical string models. The spectrum of the Nambu-Goto string with fixed ends in d dimensions has been calculated[6], with the result

$$E_N = \sigma R \left(1 - \frac{(d-2)\pi}{12\sigma R^2} + \frac{2\pi N}{\sigma R^2}\right)^{1/2}. \tag{4}$$

For small R, this model has a quantization problem[6] unless $d = 26$, but the problem disappears as R becomes large. The energy gaps expected for a Nambu-Goto string at large R are shown in Fig. 2. Deviations of the simulation results from the Nambu-Goto gaps suggests that physical properties, such as rigidity, may be relevant for the effective string action.

Fig. 3 illustrates one possible interpretation of the results shown in Fig. 1. At small quark-antiquark separations, the strong chromoelectric field of the quark-antiquark pair repels the physical vacuum in a dual Meissner effect, creating a bubble surrounding the pair. Descriptions of the low-lying stationary states must take into account both the gluonic modes inside the bubble and the motion of the collective coordinates describing the bubble. For large quark-antiquark separations, the bubble stretches into a thin tube of flux, and the low-lying states could then be explained by the collective motion of the tube since the internal gluonic excitations, being typically of order 1 GeV, are now much higher lying. We caution the reader that the above interpretation is simply speculation based on observations to date. Although the simulation results rule out the usefulness at small R of an effective string action constructed as a $1/R$ expansion, they do not actually rule out the unlikely possibility of a string description based on some other expansion parameter.

5. $SU(2)$ and compact $U(1)$ gauge theories in 3 dimensions

The spectra of three-dimensional $SU(2)$ and compact $U(1)$ gauge theories were also studied to address questions about the dependence of the results on the gauge group and the dimensionality of space-time. Due to the reduced dimensionality, higher statistical precision was possible in these calculations. These simulations also served to check various systematic errors.

The excitation gaps ΔE above the S_g ground state of ten levels were computed and are compared with $N\pi/R$ in Fig. 4. Again, the large-R results are consistent with the expectations from an effective string description without exception. A fine structure is also observed, but it is less pronounced than that in four-dimensional $SU(3)$. Unlike in four-dimensional $SU(3)$, no dramatic level rearrangements occur between small and large separation, but deviations from $N\pi/R$ are significant for small R. There is remarkable agreement between the $SU(2)$ and compact $U(1)$ results. A detailed examination of these results is still work in progress.

We have also pursued the spectrum in three-dimensional $Z(2)$ lattice gauge theory. These results are reported elsewhere[3]. Extremely precise determinations are possible in $Z(2)$ by exploiting a duality transformation into an Ising model. In the critical region, the resulting Ising model admits

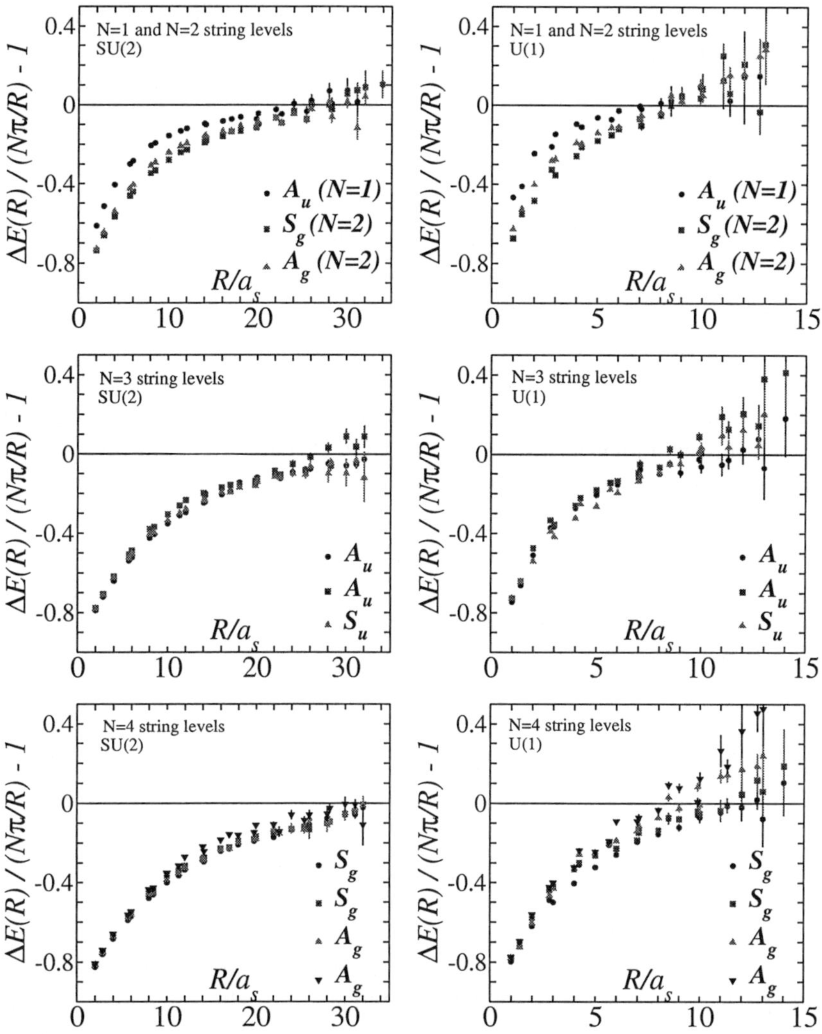

Figure 4. The energy gaps ΔE above the ground state S_g of the stationary states of the gauge field in the presence of a static source pair in 3-dimensional $SU(2)$ and compact $U(1)$ gauge theories. These gaps are shown against the separation R of the static sources and are compared with the $N\pi/R$ splittings expected in an effective string theory at large R. The results in $SU(2)$ were obtained on an 84^3 lattice using an anistropic improved lattice action with coupling $\beta = 5.6$ and bare aspect ratio $\xi = 2$, so that $a_s \sim 0.1$ fm. The compact $U(1)$ results were obtained on a $28^2 \times 224$ lattice using an anisotropic improved lattice action for $\beta = 0.5$ and $\xi = 8$.

a description in terms of a ϕ^4 real scalar field theory, allowing the possibility of understanding the underlying microscropic origins of confinement in a rigorous field-theoretical setting. The details of this work in progress are presented elsewhere[3].

6. Conclusion

In this talk, Monte Carlo computations of the energies of sixteen stationary states of the gluon field in the presence of a static quark-antiquark pair separated by a distance R were presented for a wide range of R from 0.1 to 3 fm. Striking confirmation of string-like flux formation of the gluon field surrounding a quark-antiquark pair separated by distances larger than 2 fm was presented. A tantalizing fine structure was revealed, suggesting the possibility of identifying the effective QCD string action in future higher precision simulations. A dramatic level rearrangement between small and large quark-antiquark separations was observed in a crossover region around 2 fm. The observed pattern of energy levels at small R strongly challenges an effective string description.

Eleven levels in three-dimensional $SU(2)$ and compact $U(1)$ lattice gauge theory were also studied. String formation was once again confirmed at large separations, with a fine structure less pronounced than in four-dimensional $SU(3)$. No dramatic level rearrangement was found between large and small separations. These studies are ongoing, and we are also vigorously pursuing the spectrum and other observables in three-dimensional $Z(2)$ gauge theory with the goal of determining the effective string action. Future work also includes calculating the three-dimensional $SU(3)$ spectrum, torelon (flux loops winding around the lattice) spectra, and studying the spatial structures of these gluonic excitations. This work was supported by the U.S. National Science Foundation under award PHY-0099450, the U.S. DOE, Grant No. DE-FG03-97ER40546, and the European Community's Human Potential Programme under contract HPRN-CT-2000-00145, Hadrons/Lattice QCD.

References

1. G.S. Bali, K. Schilling, C. Schlichter, Phys. Rev. D **51**, 5165 (1995).
2. K.J. Juge, J. Kuti, and C. Morningstar, Phys. Rev. Lett. **90**, 161601 (2003).
3. K.J. Juge, J. Kuti, and C. Morningstar, these proceedings.
4. M. Lüscher, K. Symanzik, P. Weisz, Nucl. Phys. **B173**, 365 (1980).
5. M. Lüscher, Nucl. Phys. **B180**, 317 (1981).
6. J. F. Arvis, Phys. Lett. **127B**, 106 (1983).

QCD STRING FORMATION AND THE CASIMIR ENERGY*

K. JIMMY JUGE

Institute for Theoretical Physics, Univ. of Bern, CH-3012 Bern, Switzerland

J. KUTI

Dept. of Physics, Univ. of California at San Diego, La Jolla, CA 92093-0319

C. MORNINGSTAR

Department of Physics, Carnegie Mellon University, Pittsburgh, PA 15213, USA

Three distinct scales are identified in the excitation spectrum of the gluon field around a static quark-antiquark pair as the color source separation R is varied. The spectrum, with string-like excitations on the largest length scales of 2–3 fm, provides clues in its rich fine structure for developing an effective bosonic string description. New results are reported from the three–dimensional Z(2) and SU(2) gauge models, providing further insight into the mechanism of bosonic string formation. The precocious onset of string–like behavior in the Casimir energy of the static quark-antiquark ground state is observed below R=1 fm where most of the string eigenmodes do not exist and the few stable excitations above the ground state are displaced. We find no firm theoretical foundation for the widely held view of discovering string formation from high precision ground state properties below the 1 fm scale.

1. QCD String Spectrum and the Casimir Energy

Last year, we presented a new analysis of the fine structure in the QCD string spectrum at the Lattice 2002 conference. Shortly afterwards, two papers were submitted using complementary methods for finding definitive signals of bosonic string formation from the rich excitation spectrum[1] and the ground state Casimir energy.[2]

QCD String Spectrum

Three exact quantum numbers which are based on the symmetries of the problem determine the classification scheme of the gluon excitation spectrum in the presence of a static $q\bar{q}$ pair.[1] We adopt the standard no-

*Talk presented by J.Kuti. This work was supported by the DOE, Grant No.DE-FG03-97 ER40546, the NSF under Award PHY-0099450, and the European Community's Human Potential Programme under contract HPRN-CT-2000-00145, Hadrons/Lattice QCD.

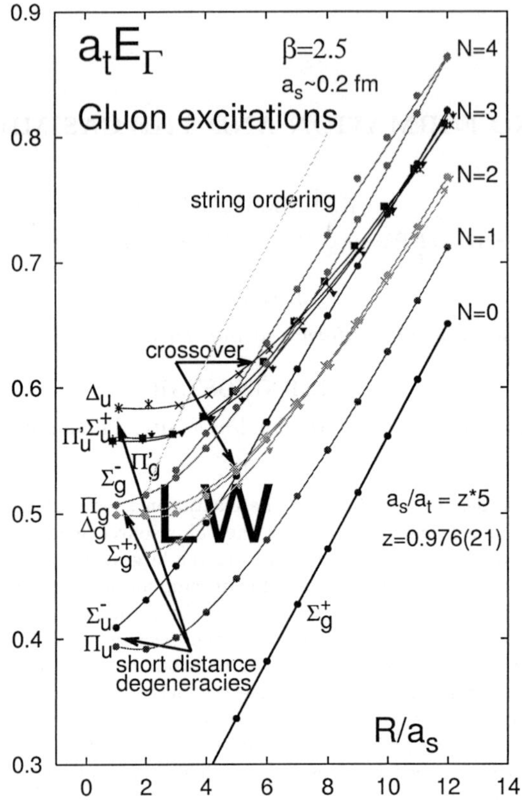

Figure 1. Short distance degeneracies which are not string-like and their crossover towards the QCD string spectrum are shown from Ref. 1 where further details are explained. The solid curves with simulation points, which identify energy levels degenerate in the asymptotic string limit, are only shown for visualization and do not represent fits to the data. The thin line without data points marks a lower bound for the onset of mixing effects with glueball states which requires careful interpretation. The symbol LW indicates the R range of high precision Casimir energy calculations from Ref. 2.

tation from the physics of diatomic molecules and use Λ to denote the magnitude of the eigenvalue of the projection $\mathbf{J}_g \cdot \hat{\mathbf{R}}$ of the total angular momentum $\mathbf{J}_g$ of the gluon field onto the molecular axis with unit vector $\hat{\mathbf{R}}$. The capital Greek letters $\Sigma, \Pi, \Delta, \Phi, \ldots$ are used to indicate states with $\Lambda = 0, 1, 2, 3, \ldots$, respectively. The combined operations of charge conjugation and spatial inversion about the midpoint between the quark and the antiquark is also a symmetry and its eigenvalue is denoted by η_{CP}. States with $\eta_{CP} = 1(-1)$ are denoted by the subscripts $g(u)$. There is an additional label for the Σ states; Σ states which are even (odd) under a reflection in a plane containing the molecular axis are denoted by a superscript + (−). Hence, the low-lying levels are labeled Σ_g^+, Σ_g^-, Σ_u^+, Σ_u^-, Π_g, Π_u, Δ_g, Δ_u, and so on, Σ_g^+ designating the ground state. For convenience, we use Γ to denote these labels in general. For better resolution of the fine structure in the spectrum, the gluon excitation energies $E_\Gamma(R)$ were extracted from Monte Carlo estimates of generalized large Wilson loops on lattices with small a_t/a_s aspect ratios and improved action. Restricted to the R=0.2–3 fm range of a selected simulation, the energy spectrum is shown in Fig. 1 for 10 excited states.

On the shortest length scale, the excitations are consistent with short distance physics without string–like level ordering in the spectrum. A crossover region below 2 fm is identified with a dramatic rearrangement of the short distance level ordering. On the largest length scale of 2–3 fm, the spectrum exhibits string-like excitations with asymptotic π/R string gaps which are split and slightly distorted by a fine structure. It is remarkable that the torelon spectrum of a closed string, with one unit of winding number around a compactified direction, exhibits a similar fine structure on the 2–3 fm scale, as reported for the first time at Lattice 2003.[3] This finding eliminates the boundary effects of fixed color charges as the main source of the fine structure in the distorted spectrum.

Casimir Energy

In a complementary study,[2] a string–like Casimir energy and the related effective conformal charge, $C_{\rm eff}(R) = -12R^3F'(R)/(\pi(D-2))$, were isolated where F(R) is the force between the static color sources and D is the space-time dimension of the gauge theory with bosonic string formation. With unparalleled accuracy, $C_{\rm eff}(R)$ was determined for the gauge group SU(3) in D=3,4 dimensions, in the range $0.2\ \text{fm} < R < 1.0\ \text{fm}$, below the crossover region of the string spectrum. A sudden change with increasing R, well below 1 fm, was observed in $C_{\rm eff}(R)$, breaking away from the the short distance running Coulomb law towards the string-like $C_{\rm eff}(R) \approx 1$ behavior. This was interpreted as a signal for early bosonic string formation. The results are surprising because the scale R is not large compared with the expected width of the confining flux, and more quantitatively, the string-like Casimir energy behavior is observed in the R range where the spectrum exhibits complex non-string level ordering, as shown in Fig. 1. We will try to develop now a better understanding of the seemingly paradoxical situation.

2. Bosonic String Formation in the Z(2) Gauge Model

The three-dimensional Z(2) lattice gauge model represents considerable simplification in comparison with four-dimensional lattice QCD. The SU(3) group elements on links are replaced by Z(2) variables, and the reduction to three dimensions implies a nontrivial continuum limit with a finite fixed point gauge coupling.

Dual ϕ^4 Field Theory Representation

The main features are easily seen from the dual transformation of the Z(2) gauge model to Ising variables which can be replaced by the real scalar field of ϕ^4 field theory in the critical region.[a] The continuum model exhibits confinement and bosonic string formation in the broken phase of the Ising representation. In addition, a nontrivial glueball spectrum is observed[5] with finite masses when measured in units of the string tension. The string tension of the confining flux in the Z(2) gauge model becomes the interface energy in the dual Ising-ϕ^4 representation, and the lowest mass 0^+ glueball state of the gauge theory with mass m maps into the elementary scalar of the dual lattice, with inverse correlation length m in the critical region. Higher glueball states are Bethe-Salpeter bound states of the elementary scalar.[5] The dual ϕ^4 Lagrangian of the Z(2) gauge model, in rigorous theoretical setting, is in analogy with the dual Landau-Ginzburg superconductivity model which attempts to describe the unknown microscopic quark confinement mechanism of QCD. [b]

The Ising-ϕ^4 field theory model is particularly intriguing from the microscopic string theory viewpoint, if we recall Polyakov's work on the connection with the theory of random surfaces. Using loop equations of closed Wilson loops near the continuum limit, he conjectured the equivalence of the three-dimensional Z(2) lattice gauge theory to a fermionic string theory.[7]

Renormalization Scheme

The euclidean 3d Z(2) model is described in the critical region (continuum limit) by a real order parameter field ϕ with the Lagrangian

$$\mathcal{L} = -\frac{1}{2}(\partial_\mu \phi_0)^2 - \frac{g_0}{4!}(\phi_0^2 - \frac{3m_0^2}{g_0})^2 \,. \qquad (1)$$

---

[a]References to earlier work on the three-dimensional Z(2) gauge model can be found in a recent paper on the finite temperature properties of the Z(2) string.[4]

[b]A recent review of quark confinement and dual superconductivity is given in Ref. 6 with discussion of earlier work and references.

The most frequently used renormalization scheme requires in the broken phase that the tadpole diagrams completely cancel without coupling constant renormalization ($g = g_0$) and with wave function renormalization $\phi_0 = \sqrt{Z}\phi$. In the following, with lattice cutoff, we define a scheme with finite coupling constant renormalization keeping the renormalized mass of the elementary scalar exactly at the pole of its propagator. Since the wave function renormalization is finite to every order, for convenience we choose Z=1 in 1-loop calculations. With $\mathrm{g_0 = g + \delta g}$, $\mathrm{v_0^2 \equiv 3m_0^2/g_0 = v^2 + \delta v^2}$, the renormalized Lagrangian for elementary excitations η around the vacuum expectation value $\phi = v$ is the starting point of the renormalized loop expansion with two counterterms to one-loop order,

$$\delta \mathrm{v}^2 = \lim_{\mathrm{V}\to\infty} \frac{3}{2}\mathrm{g} \cdot \frac{1}{\mathrm{V}} \sum_{\vec{\mathrm{k}}} \frac{1}{E^0_{\vec{\mathrm{k}}}} , \qquad \delta \mathrm{g} = \frac{\mathrm{g}^2}{\mathrm{m}} \frac{\ln 3}{32\pi} . \tag{2}$$

The infinite spatial volume limit is taken in the sum over the spectrum of inverse lattice energies of free massive excitations $E^0_{\vec{\mathrm{k}}}$ with periodic boundary conditions. The coupling constant counterterm $\delta\mathrm{g}$ satisfies the renormalization condition on the physical propagator pole to one-loop order.

In the presence of a pair of static sources, represented by a Wilson loop in the Z(2) gauge model, the renormalization procedure is unchanged in the Ising-ϕ^4 field theory description. The only change is in the lattice Lagrangian where the sign of the nearest neighbor interaction term is flipped on links which puncture the surface of the Wilson loop on the dual Z(2) gauge lattice. This flip represents a disorder line, or seam, between the two static sources on the spatial lattice. The end points of the seam are fixed but otherwise it is deformable by a "gauge transformation" of variables without changing the partition function. This invariance is inherited from the gauge invariant representation of the Wilson surface in the Z(2) gauge model.

Numerical Implementation of the Loop Expansion

The dual transformation of the Z(2) model to Ising variables facilitates very efficient simulations with multispin coding. The loop expansion provides theoretical insight into Monte Carlo simulations of the excitation spectrum using high statistics multispin Ising codes complemented by ϕ^4 field theory codes. Since the fixed point value u^* of the dimensionless coupling constant $\mathrm{u = g/m}$ is not small, the simulations provide an important cross-check on the convergence of the loop expansion which itself has to be implemented in a numerical procedure. The renormalized loop expansion in the presence of static sources requires the following three steps.

(i) First, for a given physical mass m and renormalized coupling g, the time independent renormalized classical field equation,

$$-\frac{\partial^2\phi_s(x,y)}{\partial x^2}-\frac{\partial^2\phi_s(x,y)}{\partial y^2}-\frac{1}{2}m^2\phi_s(x,y)+\frac{g}{6}\phi_s^3(x,y)=0\ , \tag{3}$$

of the static soliton ϕ_s is solved on the lattice in the $\vec{r}=(x,y)$ plane with flipped nearest neighbor interaction links along the seam between static sources. In the Z(2) gauge model, the two sources can be interpreted as opposite sign charges with an electric flux connecting them. In the ϕ^4 representation we refer to the ϕ_s classical solution as a static soliton, rather than the earthy flux–tube term. In the numerical procedure, a generalized Newton type nonlinear iterative scheme was implemented to obtain ϕ_s to double precision accuracy.

(ii) Second, the fluctuation spectrum around the static soliton $\phi_s(x,y)$ is determined by splitting the field into the classical solution plus fluctuations, $\phi(x,y,t)=\phi_s(x,y)+\eta(x,y,t)$, with the eigenmodes of the fluctuation field $\eta(x,y,t)=\sum_n[a_n(t)\psi_n(x,y)+a_n^\dagger(t)\psi_n^*(x,y)]$ satisfying the eigenvalue equation

$$-\frac{\partial^2\psi_n}{\partial x^2}-\frac{\partial^2\psi_n}{\partial y^2}+U''(\phi_s)\psi_n=E_n^2\cdot\psi_n\ . \tag{4}$$

The time dependence of the fluctuation field η is given in interaction picture by $a_n(t)=a_n(0)e^{-iE_nt}$ where the Hamiltonian is split into a quadratic part and an interaction part of the ψ_n eigenmodes. The second derivative of the $U(\phi)=\frac{g}{4!}(\phi^2-\frac{3m^2}{g})^2$ renormalized field potential energy is taken with respect to ϕ in Eq. (4), with $\phi=\phi_s$ substituted subsequently. Two parity quantum numbers P_x, P_y split the eigenmodes into four separate symmetry classes. With the two sources located at $(x,y)=(R/2,0)$ and $(x,y)=(-R/2,0)$, the quantum number $P_x=\pm1$ corresponds to the reflection symmetry $x\to-x$ of $\psi_n(x,y)$ and $P_y=\pm1$ corresponds to the $y\to-y$ reflection symmetry. The full spectrum of eigenvalues and eigenfunctions of Eq. (4) are computed by an Arnoldi diagonalization procedure in the finite volume of the lattice. Using the parity symmetries of the theory, diagonalization of large lattices with sizes up to 200x200 in the (x,y) plane were performed.

(iii) Third, the systematic renormalized loop expansion with the ϕ_s static soliton background is developed by building the finite volume field propagator $D_F(\vec{r}, t; \vec{r}', t')$ in Minkowski time,

$$D_F(\vec{r}, t; \vec{r}', t')) = \sum_n i \int \frac{dp_0}{2\pi} \frac{\psi_n(\vec{r})\psi_n^*(\vec{r}')}{p_0^2 - E_n^2 + i\epsilon} e^{-ip_0(t-t')} , \tag{5}$$

from the static $\psi_n(\vec{r})$ eigenmodes. An euclidean rotation is performed on the propagator during the numerical evaluation of the loop diagrams. The counterterms δv^2 and δg are used to remove loop divergences in the continuum limit and to keep the exact pole location at the physical mass m. Using the propagator of Eq. (5), the fluctuation correction to the static soliton profile ϕ_s was calculated to one-loop order, together with similar calculations of the ground state energy and excitation energies. In this work, we only report numerical results on the fluctuation spectrum of Eq. (4) and its 1-loop contribution to the ground state energy.

String Excitations in the Loop Expansion

For sufficiently large R, the discrete $P_y = -1$ bound state spectrum of Eq. (4) is expected to evolve into the asymptotic $E_N = \pi N/R$ $(N = 1, 2, \ldots)$ Dirichlet string spectrum of massless string excitations which originate from the translational mode of the well-known one-dimensional ϕ^4 soliton by the following simple consideration.

Consider first the spatial lattice in the finite (x,y) plane with a seam of flipped links winding around the compact x–direction with periodic boundary condition. The classical solution $\phi_{tor}(y) = m\sqrt{3/g}\,\tanh(m|y|/2)$ of Eq. (3) defines the torelon which is independent of x and winds around the compact x–direction with a seam positioned at y=0. We use continuum notation for the torelon and its excitations, but finite cutoff and volume effects are included in the numerical work. For $x > 0$, the transverse profile of the torelon is identical to that of the well-known one-dimensional soliton, and for $x < 0$ a sign flip is involved because of the seam at y=0. The torelon eigenmodes of Eq.(4) with $P_y = -1$ have the simple form

$$\psi_N^{tor}(x, y) = \sqrt{\frac{g}{2m^3}}\, \phi'_{tor}(y) \cdot \sqrt{1/R}\, e^{ip_N \cdot x}, \tag{6}$$

with quantized momenta, $p_N = 2\pi N/R$, $N = \pm 1, \pm 2, \ldots$, running along x in the compact interval R with periodic boundary condition. The energy spectrum is given by $E_N = 2\pi N/R$, with positive N values.

The classical transverse profile $\phi_{tor}(y)$ of the torelon coincides, to a good approximation, with that of the static soliton ϕ_s, if the separation between the sources is large enough. The static soliton profile ϕ_s does not interpolate from -v at large negative y to +v for large positive y at fixed x because of the flipped links along the seam. Rather, ϕ_s approaches v everywhere, far away from the seam line. The eigenmodes of the fluctuation operator are restricted now between the two sources and they are close to the form

$$\psi_{\rm N}^{\rm static}({\rm x},{\rm y}) = \sqrt{\frac{\rm g}{2{\rm m}^3}}\,\phi'_{\rm tor}({\rm y})\cdot\sqrt{2/{\rm R}}\,\sin(\pi{\rm N}/{\rm x}),\quad -R/2 \leqq x \leqq R/2\ , \quad (7)$$

with N taking positive integer values. The spectrum of these standing waves is the same as that of a massless Dirichlet string oscillating in the (x,y) plane with fixed ends. The excited eigenmodes of the effective Schrödinger equation, like the one of Fig. 2b for N=8, are therefore in one-to-one correspondence with massless Dirichlet string oscillations. The spectrum and the wave functions are expected to be somewhat distorted at finite R because of the distortions of the effective Schrödinger potential around the sources.

Representative examples of the numerical work are shown in Fig. 2 where the static soliton solution $\phi_{\rm s}({\rm x},{\rm y})$ on a 160x80 spatial lattice in the (x,y) plane corresponds to source separation R=100 and physical mass m=0.319 in the critical region (all dimensional quantities are expressed in lattice spacing units). For later comparisons, the lattice correlation length $\rm m^{-1}$ and the renormalized coupling g were chosen in the critical region to match one of our Monte Carlo simulations with v=0.45 and string tension $\sigma = 0.0101$. The static soliton solution $\phi_s(\vec{\rm r})$ determines the attractive potential energy of the effective Schrödinger eigenvalue problem in Eq. (4) which has a discrete bound state spectrum and a nearly continuous dense spectrum above the glueball threshold m representing scattering states on the static soliton in the infinite lattice volume limit.

The one–dimensional soliton, with classical mass $\rm 2m^3/g$, has a massive intrinsic excitation, or breathing mode, whose excitation energy is $(\sqrt{3}/2)\rm m$. In the large R limit, the intrinsic excitations of ϕ_s with $\rm P_y = +1$ become massive breathing modes of the Dirichlet string. The asymptotic spectrum of a massive string, given by $\rm E_K = \sqrt{3m^2/4 + \pi^2 K^2/R^2}$, $\rm K = 1, 2, \ldots$, is associated with eigenmodes like the K=2 wave function of Fig. 2c. The corresponding standing wave solutions,

$$\psi_{\rm K}({\rm x},{\rm y}) = \sqrt{3{\rm m}/4}\,{\rm sech}({\rm my}/2)\,\tanh({\rm m}|{\rm y}|/2)\cdot\sqrt{2/{\rm R}}\,\sin(\pi{\rm K}/{\rm x})\ , \quad (8)$$

originate from the massive excitations of the torelon with restriction to standing waves in the $\rm -R/2 \leqq x \leqq R/2$ interval.

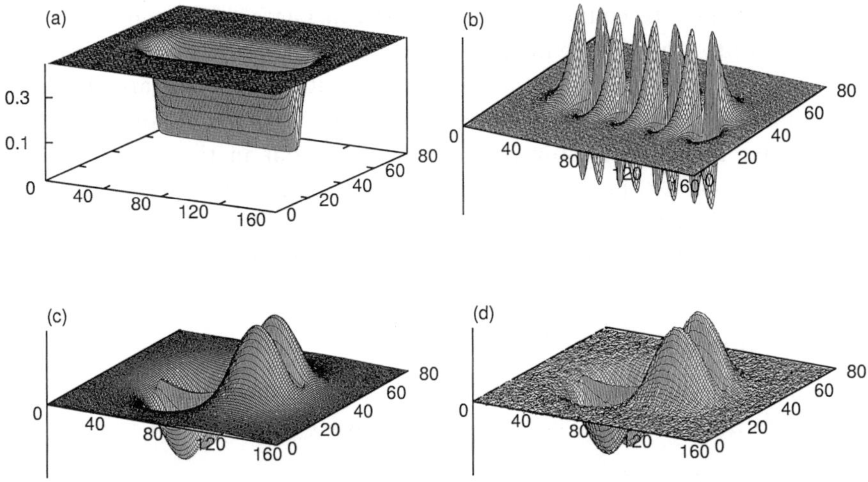

Figure 2. The static soliton solution ϕ_s of field equation (3) is shown in (a) with the choice R=100 and renormalized parameters given in the text; (b) shows the N=8 massless string excitation of the static soliton from the numerical diagonalization of the eigenvaule equation (4); the second massive string excitation (K=2 breathing mode) is shown in (c) from the numerical solution of the eigenvalue equation, and compared in (d) to exact Monte Carlo simulation of the same state with remarkable agreement.

3. Probing the String Theory Limit

We describe the $\eta(\vec{\mathrm{r}},\mathrm{t})$ fluctuations around the static soliton ϕ_s by the sum of three fields, $\eta(\vec{\mathrm{r}},\mathrm{t}) = \xi(\vec{r},t) + \chi(\vec{r},t) + \varphi(\vec{r},t)$, with

$$\xi(\mathrm{x},\mathrm{y},\mathrm{t}) = \sum_{\mathrm{N}} [\mathrm{a_N}(\mathrm{t})\psi_{\mathrm{N}}(\mathrm{x},\mathrm{y}) + \mathrm{a}^\dagger_{\mathrm{N}}(\mathrm{t})\psi^*_{\mathrm{N}}(\mathrm{x},\mathrm{y})] \,,$$

$$\chi(\mathrm{x},\mathrm{y},\mathrm{t}) = \sum_{\mathrm{K}} [\mathrm{a_K}(\mathrm{t})\psi_{\mathrm{K}}(\mathrm{x},\mathrm{y}) + \mathrm{a}^\dagger_{\mathrm{K}}(\mathrm{t})\psi^*_{\mathrm{K}}(\mathrm{x},\mathrm{y})] \,,$$

$$\varphi(\mathrm{x},\mathrm{y},\mathrm{t}) = \sum_{\mathrm{n}} [\mathrm{a_n}(\mathrm{t})\psi_{\mathrm{n}}(\mathrm{x},\mathrm{y}) + \mathrm{a}^\dagger_{\mathrm{n}}(\mathrm{t})\psi^*_{\mathrm{n}}(\mathrm{x},\mathrm{y})] \,,$$

where ξ is restricted to bound states with negative $\mathrm{P_y}$ parity which are expected to evolve into massless string excitations for large R. The field χ is restricted to $\mathrm{P_y} = +1$ parity bound states which evolve into massive string excitations, and φ is a sum over scattering states above the 0^+ glueball threshold m in the continuum.

These fields are coupled in the interaction Lagrangian, and when the massive fields χ and φ are integrated out, we get a nonlocal Lagrangian in the $\xi(\mathrm{x,y,t})$ field describing massless string excitations in the large R limit. As indicated by Eq. (7), the y–dependence in all the $\mathrm{P_y} = -1$ parity bound state wave functions is approximately factored out in the large R limit. Hence, the $\xi(\mathrm{x,y,t})$ field can be replaced on large length scales by the field $\mathrm{f(x,t)}$ which becomes the geometric string variable of low energy excitations measuring the displacements of the flux center–line in the y–direction as a function of x and t.

Effective String Action

The nonrenormalizable effective action of the $\mathrm{f(x,t)}$ field, with the massive fields integrated out, is given in a derivative expansion by

$$\begin{aligned} S_{\mathrm{eff}}(\mathrm{f}) = -\int \mathrm{dxdt}\Bigg[\frac{1}{2\pi\alpha'}\left(1-\frac{1}{2}(\partial\mathrm{f})^2-\frac{1}{8}((\partial\mathrm{f})^2)^2-\frac{1}{16}((\partial\mathrm{f})^2)^3+\cdots\right) \\ + \mathrm{const}\cdot(\partial\mathrm{f})^2\Box(\partial\mathrm{f})^2+\cdots\Bigg], \end{aligned} \tag{9}$$

where $\partial \sim m^{-1}$ is a long wavelength expansion. The string tension σ can be expressed as $\sigma = (2\pi\alpha')^{-1}$, and the notation $(\partial\mathrm{f})^2 = \partial_\mu \mathrm{f} \partial^\mu \mathrm{f}$ is used in Eq. (9). Since f is related to massless Goldstone excitations, originating from the restoration of translation invariance in torelon quantization, only derivatives of f appear in S_{eff}. The first three terms in the derivative expansion come from the kinetic terms in the original ϕ^4 field theory action. They are independent of the details of the field potential except for the overall factor of $(2\pi\alpha')^{-1}$. The first line in Eq. (9) agrees with the equivalent terms of the Nambu-Goto (NG) action, $S_{\mathrm{NG}} = (2\pi\alpha')^{-1}\sqrt{1-(\partial\mathrm{f})^2}$, when expanded in $(\partial\mathrm{f})^2$. The second line in Eq. (9) has contributions from the geometric curvature, but new contributions also appear whose geometric origin remains unclear. This is where the effective string action begins to show deviations from the NG string. To correct for end effects around the static sources, the effective action of Eq. (9) has to be augmented by boundary operators for the complete description of the Dirichlet string.

Exact Excitation Spectrum from Simulations

First, the physical mass, the vacuum expectation value v, and the renormalized coupling g were determined in high precision Monte Carlo simulations of the bulk euclidean lattice action in the Ising-ϕ^4 field theory representation without the seam of flipped links. The renormalized physical parameters were used as input to compare the loop expansion with simulations in the presence of static sources. The excitation spectrum around the static sources was determined from Monte Carlo estimates of correlation matrices which included an extended set of optimized operators. These operators were built from eigenfunctions of the fluctuation operator on two-dimensional time slices of the lattice.

Simulation results are shown in Fig. 3 for string formation as R is stretched. The shape of the static soliton profile $\phi_s(x, y)$ around the sources is depicted in the (x,y) plane of the two-dimensional spatial lattice for R=10 and R=100. With the scale set by the string tension, the two R values correspond to 0.3 fm and 3 fm, respectively. The renormalized bulk physical parameters of the simulations were given in the discussion of Fig. 2 as m=0.319, v=0.45, and $\sigma = 0.0101$. At the smaller R value, the static soliton with bag–like shape is not stretched, and there is only one bound state excitation below the glueball threshold. At R=3 fm, the stretched static soliton supports many string excitations, with the N=1,4,8 wave functions displayed. The Monte Carlo simulations are in good agreement with results from the eigenmodes of the fluctuation operator. This is indicated by the good match of Fig. 2a versus Fig. 3c and Fig. 2b versus Fig. 3f.

Matching the Excitation Spectrum to the String Action

A large number of excitation spectra were obtained from highly accurate simulations as R was varied in a wide range from 0.3 fm to approximately 10 fm. A good test for string formation is provided by the behavior of the spectrum as a function of R. The NG spectrum,

$$E_N = \sigma R \left(1 - \frac{D-2}{12\sigma R^2}\pi + \frac{2\pi N}{\sigma R^2}\right)^{\frac{1}{2}}, \tag{10}$$

with fixed end boundary conditions in D dimensions, was calculated in Ref. 8. N=0 corresponds to the string ground state and positive integer N values label the excitations of the Dirichlet string. Although there exists an inconsistency in the quantization of angular momentum rotations around the $q\bar{q}$-axis at finite R values unless $D = 26$, the problem asymptotically disappears in the $R \to \infty$ limit.[8] This is expected from the earlier discussion on string formation in the loop expansion. Indeed, derived in a consistent D=3 ϕ^4 field theory, the first few terms of the effective string action match the coefficients of their NG counterparts as seen in Eq. (9). If a string limit is reached for a large enough R range, the expansion of E_N into inverse powers of R from Eq. (9) should agree with the simulations at least to one

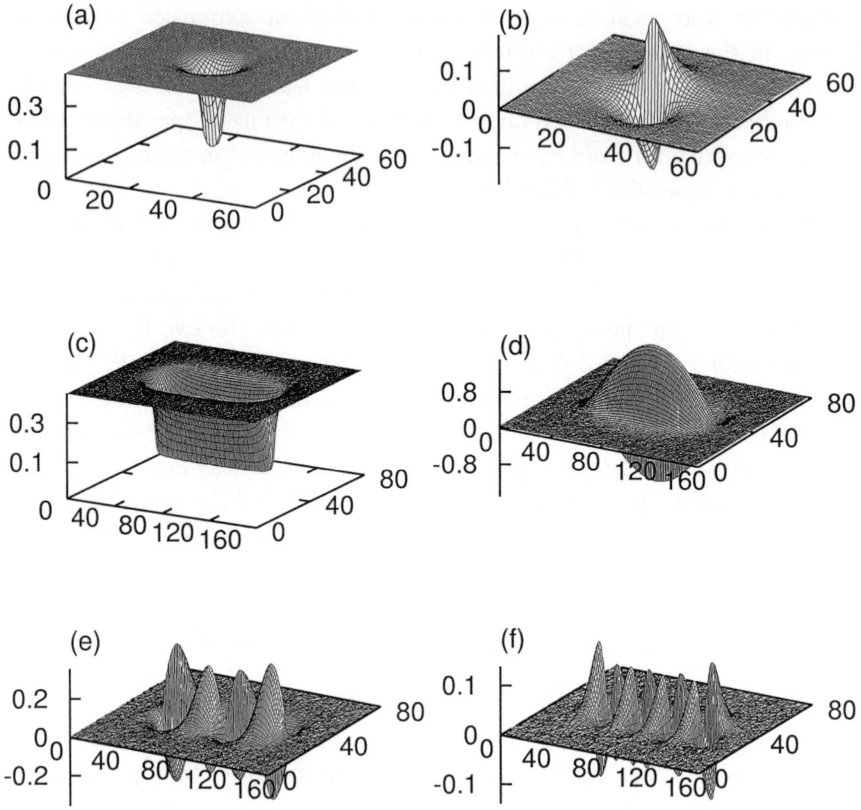

Figure 3. The soliton profile for R=10 is shown in (a), with the only bound state wave function below the glueball threshold depicted in (b). At R=100, the stretched soliton configuration shown in (c) exhibits several string like excitation with the N=1 wave function shown in (d), N=4 in (e), and N=8 in (f). The simulation results match the loop expansion of Fig. 2 with common renormalized parameters.

nontrivial order. At small R values the expansion will break down since the mathematical NG string will be the inconsistent description of the bag-like soliton and its excitations in three dimensions.

The comparison of simulations to string theory is illustrated in Fig. 4

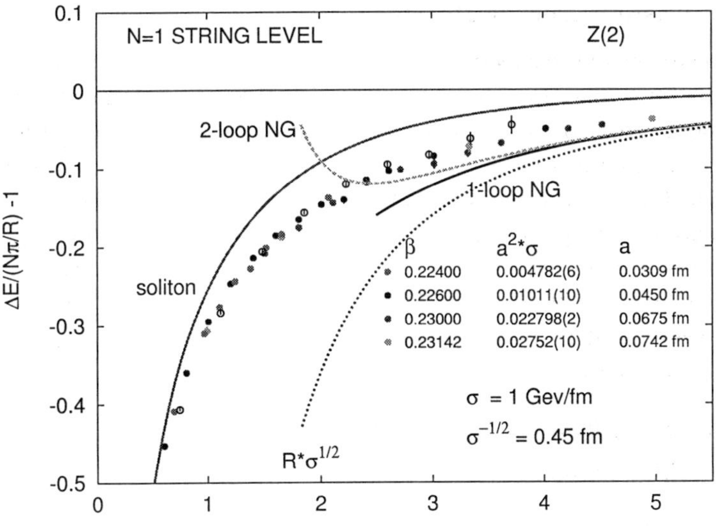

Figure 4. The energy gap ΔE above the ground state is plotted as $\Delta E/(N\pi/R) - 1$ to show percentage deviations from the asymptotic $N = 1$ string level. Several Z(2) simulations are combined with good scaling properties. The open circles represent D=3 SU(2) results after readjusting the ratio of the string tension σ to the glueball mass in Z(2). The null line coresponds to the tree level π/N NG string gaps. The dashed lines are 1–loop and two–loop NG approximations, repectively, and the black line between them is the full NG prediction.

where the exact N=1 excitation is plotted against the numerical spectrum of the fluctuation operator and the predictions of the Nambu-Goto string model. The numerical spectrum from Eq. (4) is the renormalized tree level starting point of the loop expansion in ϕ^4 field theory setting without assuming string formation. It is close in shape and details to the exact results for the entire R range, as shown by the solid line with the soliton tag in Fig. 4. It is expected that higher loop corrections will bring the agreement even closer. The simulations, however, deviate substantially from the predictions of the loop expansion in the NG string model, particularly at smaller R values below 1 fm where the loop expansion suddenly begins to diverge. The string formation for large R is clearly seen in the spectrum of Fig. 4 and all the other spectra we obtained, but further work is needed for quantitative matching of the coefficients in the effective string action.

Our results differ from the findings of Ref. 4 where simulations of the finite temperature Z(2) string were reported in good agreement with the expansion of the NG model to 1–loop order below the R=1 fm scale.

4. The Casimir Energy Puzzle

The breakdown of the effective string description below 1 fm and the related Casimir energy paradox are illustrated with the calculation of the ground state energy from the renormalized fluctuation operator around the static ϕ_s soliton. In the renormalized loop expansion, the early onset of the string-like $C_{\rm eff}(R) \approx 1$ behavior in the R range below 1 fm is not associated with massless string eigenmodes which are mostly missing for $R \lesssim 1$ fm. The results are consistent with the direct simulations of $C_{\rm eff}(R)$ and the spectrum in our Z(2) model. The simulations of Refs. 1, 2 present the same puzzle in QCD. If there is any physics associated with this puzzle, it remains unresolved.

Casimir Energy from the Fluctuation Operator

The soliton ground state energy E_s in Eq. (11),

$$E_s = E_s^{cl} + \frac{1}{2}\sum_\alpha E_\alpha - \frac{1}{2}\sum_{\vec{k}} E^0_{\vec{k}}$$
$$- \frac{g}{12}\delta v^2 \int \left[\phi_s - \frac{3m^2}{g}\right] dxdy + \frac{\delta g}{4}\int \left[\phi_s - \frac{3m^2}{g}\right]^2 dxdy\ , \qquad (11)$$

includes the renormalized classical energy E_s^{cl}, the sum of zero point energies summed over all eigenmodes α of the fluctuation operator of Eq. (4), and a sum over all momenta $\vec{k}$ in the zero point energy of the bulk vacuum without sources. The difference of the two eigenmode sums is still divergent in the continuum limit. The δv^2 counterterm removes this divergence and contributes a finite $\vec{r}$-dependent energy density to the ground state. The δg counterterm also contributes a finite and $\vec{r}$-dependent ground state energy density.

Exact Ground State Results

The simple definition $C_{\rm eff}(R) = -24R^2(E_s'(R) - \sigma)/(\pi(D-2))$ was used to isolate the effective Casimir energy term in the ground state of the soliton. The first derivative E_s' was directly simulated on the lattice by a special method we developed. The string tension σ was determined in high precision separate runs from the ground state of long torelons. The simulations in Fig. 5 are also compared with the predictions of the Nambu-Goto (NG) string model and the predictions of Eq. (11) from the numerical evaluation of the fluctuation operator spectrum.

The agreement of $C_{\rm eff}(R)$ with exact simulation results, as determined from the fluctuation operator spectrum, is quite good down to very small

R values. At smaller R values, full agreement can only be expected from higher loop corrections. It is interesting that the classical contribution of ϕ_s to $C_{\rm eff}(R)$ is significant below 1 fm. At large R, in the true string formation limit, the higher loop corrections should not contribute to the asymptotic 1/R Casimir term.

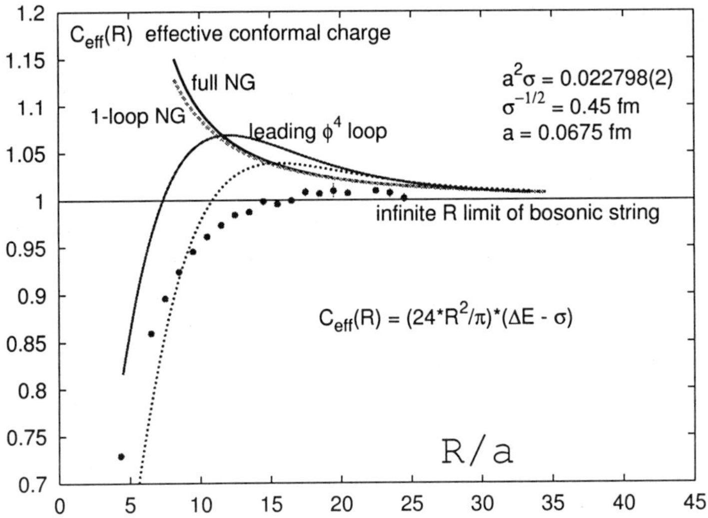

Figure 5. The black points are from high precision Ising-ϕ^4 simulations. The solid black curve with NG label is the full NG prediction, $C_{\rm eff}(R) = 1$ is the asymptotic string result which corresponds to the tree-level NG prediction. The dashed line close to the full NG curve shows the 1-loop approximation which includes the first correction to tree level from the R^{-1} expansion of Eq. (10). The other solid line with downturn for small R is calculated from the numerical evaluation of Eq. (11). The dashed line with downturn is obtained from the full line by subtracting the E_s^{cl} contribution.

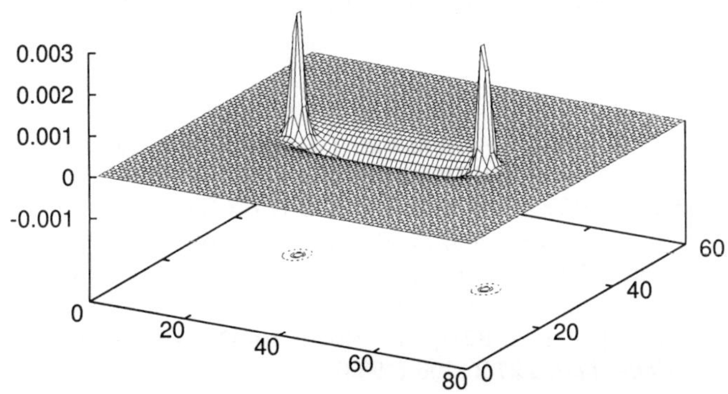

Figure 6. The ground state fluctuation energy density is plotted for R=40. The two peaks represent large fluctuation contributions around the static sources.

The dominant contribution to the ground state energy of the static soliton from the fluctuation operator is coming from the continuum spectrum but the ground state energy density remains concentrated around the static soliton. This is also seen in the Casimir energy density which is calculated from the first derivative of the energy density with respect to R, integrating to $C_{eff}(R)$ as we explicitly checked. For illustration, the Casimir energy density is shown in Fig. 6.

Although the 1/R expansion of $C_{eff}(R)$ from the NG prediction of Eq. (10) is not divergent below 1 fm, it breaks away from the data in a rather dramatic fashion.

5. Conclusions

We established bosonic string formation in a large class of gauge theory models from a direct study of the excitation spectrum at large separation of the static sources. The spectrum, with string-like excitations on the length scales of 2–3 fm and beyond, provides clues in its rich fine structure for developing an effective bosonic string description. The matching of the string-like spectrum to an effective string action remains a challenge.

Our results at small R differ from the findings of Ref. 4 where simulations of the finite temperature Z(2) string were reported in good agreement with the expansion of the NG model to 1–loop order below the R=1 fm scale. This agreement was interpreted as further support at finite temperature for the precocious onset of bosonic string formation in QCD below the 1 fm scale as reported in Ref. 2.

We find no firm theoretical foundation for discovering string formation from high precision ground state properties below the 1 fm scale. The explanation for string–like finite temperature free energy behavior below 1 fm also remains unclear. Further work is needed to understand the interpretation of the results from Refs. 2, 4 which present impressive high precision simulations.

References

1. K.J. Juge, J. Kuti, and C. Morningstar, *Phys. Rev. Lett.* **90**, 161601 (2003).
2. M. Lüscher and P. Weisz, *JHEP* **0207**, 049 (2002).
3. K.J. Juge *et al.*, *Lattice 2003*, *Nucl. Phys.* **B** (Proc.Suppl.), hep-lat/0309180.
4. M. Caselle, M. Hasenbusch, and M. Panero, *JHEP* **0301**, 057 (2003).
5. M.Caselle, M.Hasenbusch, P.Provero, K.Zarembo, *Nucl.Phys.***B623**, 474 ('02).
6. M. Baker, *Quark Confinement and the Hadron Spectrum V* (WSPC,'03) p.204.
7. A.M. Polyakov, *Phys. Lett.* **82B**, 247 (1979); **103B**, 211 (1981).
8. J. F. Arvis, *Phys. Lett.* **127B**, 106 (1983).

Y-TYPE FLUX-TUBE FORMATION AND GLUONIC EXCITATIONS IN BARYONS: FROM QCD TO QUARK MODEL

HIDEO SUGANUMA AND HIROKO ICHIE

Faculty of Science, Tokyo Institute of Technology, Ohokayama 2-12-1, Meguro, Tokyo 152-8551, Japan
suganuma@th.phys.titech.ac.jp

TORU T. TAKAHASHI

Yukawa Institute for Theoretical Physics, Kyoto University, Kitashirakawa-Oiwake, Sakyo, Kyoto 606-8502, Japan

Using SU(3) lattice QCD, we perform the first systematic study for the ground-state three-quark (3Q) potential $V_{3Q}^{\rm g.s.}$ and the 1st excited-state 3Q potential $V_{3Q}^{\rm e.s.}$, *i.e.*, the energies of the ground state and the 1st excited state of the gluon field in the presence of the static three quarks. From the accurate and thorough calculation for more than 300 different patterns of 3Q systems, the static ground-state 3Q potential $V_{3Q}^{\rm g.s.}$ is found to be well described by the Coulomb plus Y-type linear potential, *i.e.*, Y-Ansatz, within 1%-level deviation. As a clear evidence for Y-Ansatz, Y-type flux-tube formation is actually observed on the lattice in maximally-Abelian projected QCD. For more than 100 patterns of 3Q systems, we calculate the 1st excited-state 3Q potential $V_{3Q}^{\rm e.s.}$ in quenched lattice QCD, and find the gluonic excitation energy $\Delta E_{3Q} \equiv V_{3Q}^{\rm e.s.} - V_{3Q}^{\rm g.s.}$ to be about 1 GeV. This large gluonic-excitation energy is conjectured to ensure the success of the quark model for the low-lying hadrons even without gluonic excitations.

1. Introduction ~ Hadron Physics based on QCD

Quantum chromodynamics (QCD), the SU(3) gauge theory, was first proposed by Yoichiro Nambu[1] in 1966 as a candidate for the fundamental theory of the strong interaction, just after the introduction of the "new" quantum number, "color".[2] In spite of its simple form, QCD creates thousands of hadrons and makes the vacuum nontrivial, which leads to various interesting nonperturbative phenomena such as color confinement[3] and dynamical chiral-symmetry breaking.[4] This miracle of QCD is due to its strong-coupling nature in the infrared region, but the strong-coupling nature itself makes very difficult to deal with QCD.

In recent years, according to the remarkable progress of the computational power, the lattice QCD Monte Carlo calculation becomes a reliable and useful method for the analysis of nonperturbative QCD,[5] which indicates an important direction in the hadron physics. In this paper, using lattice QCD, we study the three-quark potential in detail.[6,7,8,9]

2. The Ground-State Three-Quark Potential in QCD

In general, the three-body force is regarded as a residual interaction in most fields in physics. In QCD, however, the three-body force among three quarks is a "primary" force reflecting the SU(3) gauge symmetry. In fact, the three-quark (3Q) potential is directly responsible for the structure and properties of baryons, similar to the relevant role of the Q-$\bar{\text{Q}}$ potential for meson properties, and both the Q-$\bar{\text{Q}}$ potential and the 3Q potential are equally important fundamental quantities in QCD. Furthermore, the 3Q potential is the key quantity to clarify the quark confinement in baryons. However, in contrast with a number of studies on the Q-$\bar{\text{Q}}$ potential using lattice QCD,[5,10] there was almost no lattice QCD study for the 3Q potential before our study in 1999,[9] in spite of its importance in the hadron physics.

2.1. *Theoretical Form for the 3Q Potential ~ Y-Ansatz*

From the detailed studies with lattice QCD, the Q-$\bar{\text{Q}}$ potential is known to be well described with the inter-quark distance r as [5,7]

$$V_{\mathrm{Q\bar{Q}}}(r) = -\frac{A_{\mathrm{Q\bar{Q}}}}{r} + \sigma_{\mathrm{Q\bar{Q}}} r + C_{\mathrm{Q\bar{Q}}}. \tag{1}$$

As for the 3Q potential form, there are two theoretical arguments for the limits of short and long distances.

(1) At the short distance, perturbative QCD is applicable, and therefore 3Q potential is expressed as the sum of the two-body Coulomb potential originating from the one-gluon-exchange process.
(2) At the long distance, the strong-coupling expansion of QCD is plausible, and it leads to the flux-tube picture.[11] For the 3Q system, there appears a junction which connects the three flux-tubes from the three quarks, and Y-type flux-tube formation is expected.[12,13,14]

Then, we theoretically conjecture the functional form of the 3Q potential as the Coulomb plus Y-type linear potential, *i.e.*, Y-Ansatz,

$$V_{\mathrm{3Q}}^{\mathrm{g.s.}} = -A_{\mathrm{3Q}} \sum_{i<j} \frac{1}{|r_i - r_j|} + \sigma_{\mathrm{3Q}} L_{\mathrm{min}} + C_{\mathrm{3Q}}, \tag{2}$$

where $L_{\rm min}$ is the minimal value of the total flux-tube length given by

$$L_{\rm min} = \overline{\rm AP} + \overline{\rm BP} + \overline{\rm CP} \tag{3}$$
$$= \left(\frac{1}{2}(a^2+b^2+c^2) + \frac{\sqrt{3}}{2}\sqrt{(-a+b+c)(a-b+c)(a+b-c)(a+b+c)}\right)^{1/2}$$

as shown in Fig.1.

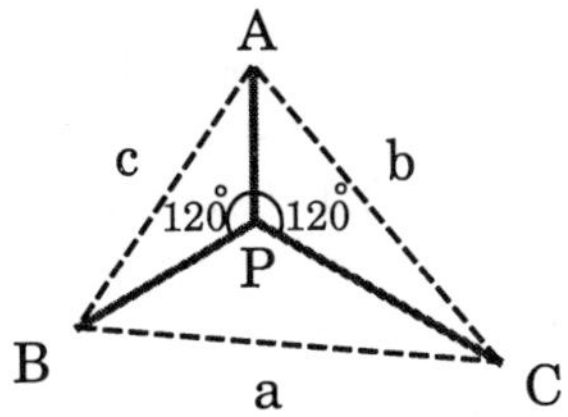

Figure 1. The flux-tube configuration of the 3Q system with the minimal value $L_{\rm min}$ of the total flux-tube length. There appears a physical junction linking the three flux-tubes at the Fermat point P, and one finds $L_{\rm min} = \overline{\rm AP} + \overline{\rm BP} + \overline{\rm CP}$.

Of course, it is nontrivial that these simple arguments on UV and IR limits of QCD hold for the intermediate region as $0.2{\rm fm} < r < 1{\rm fm}$. Then, we study the 3Q potential in lattice QCD. Note that the lattice QCD data itself is completely independent of any Ansatz for the potential form.

2.2. *The Three-Quark Wilson Loop*

Similar to the Q-$\bar{\rm Q}$ potential calculated with the Wilson loop, the 3Q potential can be calculated with the 3Q Wilson loop[6,7,8,13,14] defined on the contour of three large staples as

$$W_{\rm 3Q} \equiv \frac{1}{3!}\epsilon_{abc}\epsilon_{a'b'c'}U_1^{aa'}U_2^{bb'}U_3^{cc'} \tag{4}$$

with $U_k \equiv P\exp\{ig\int_{\Gamma_k} dx_\mu A^\mu(x)\}$ in Fig.2. The 3Q Wilson loop physically means that a color-singlet gauge-invariant 3Q state is created at $t=0$ and is annihilated at $t=T$ with the three quarks spatially fixed for $0<t<T$.

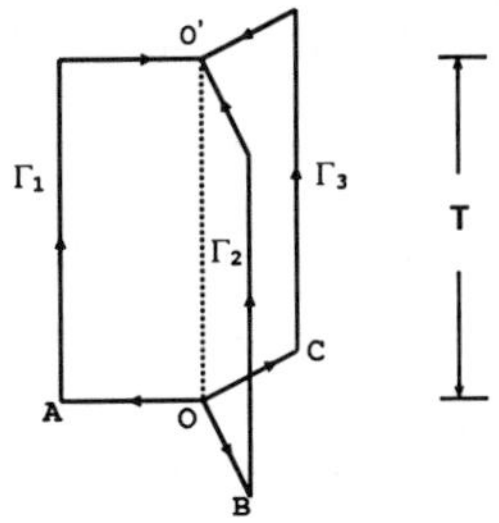

Figure 2. The 3Q Wilson loop. A color-singlet (gauge-invariant) 3Q state is created at $t=0$ and is annihilated at $t=T$. The three quarks are spatially fixed for $0<t<T$.

The vacuum expectation value of the 3Q Wilson loop is expressed as

$$\langle W_{3Q} \rangle = \sum_{n=0}^{\infty} C_n \exp(-V_n T), \tag{5}$$

where V_n denotes the n-th energy of the gauge-field configuration in the presence of the spatially-fixed three quarks.[6,7,8]

It is worth mentioning that, while V_n depends only on the 3Q location, C_n depends on the operator choice at $t = 0$ and T, *e.g.*, path linking between 3Q. In fact, the gauge-invariant 3Q state prepared at $t = 0$ generally includes excited-state contributions. In principle, by taking the large T limit, the ground-state potential V_0 can be extracted as $\langle W_{3Q} \rangle \sim C_0 \exp(-V_0 T)$. However, the large-$T$ limit calculation is difficult in the practical lattice-QCD calculation, since the signal decreases exponentially with T. Therefore, for the accurate calculation, it is desired to reduce the excited-state component in the 3Q state prepared at $t = 0$ and T.

2.3. *Smearing Method: Excited-State Component Reduction*

The smearing method is one of the most popular and useful techniques to extract the ground state in lattice QCD,[7] and is actually successful for the ground-state Q-$\bar{\text{Q}}$ potential.[7] The standard smearing for link-variables is expressed as the iterative replacement of the spatial link-variable $U_i(s)$ $(i = 1, 2, 3)$ by the obscured link-variable $\bar{U}_i(s) \in \text{SU}(3)$ which maximizes

$$\text{Re tr}\Big[\bar{U}_i^\dagger(s)\Big\{\alpha U_i(s) + \sum_{j \neq i, \pm} U_{\pm j}(s) U_i(s \pm \hat{j}) U_{\pm j}^\dagger(s + \hat{i})\Big\}\Big] \tag{6}$$

with the smearing parameter $\alpha \in \mathbf{R}$. Here, we define $U_{-\mu}(s) \equiv U_\mu^\dagger(s - \hat{\mu})$. This procedure is schematically illustrated in Fig.3. The n-th smeared link-variables $U_\mu^{(n)}(s)$ are iteratively defined as $U_i^{(n)}(s) \equiv \bar{U}_i^{(n-1)}(s) (i = 1, 2, 3)$ and $U_4^{(n)}(s) \equiv U_4(s)$ starting from $U_\mu^{(0)}(s) \equiv U_\mu(s)$.

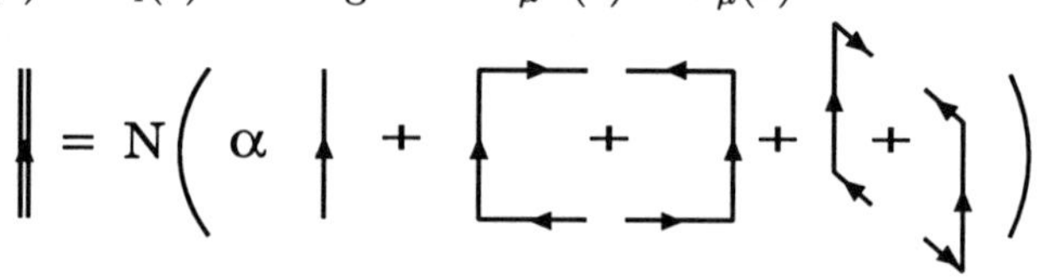

Figure 3. The schematic illustration for the smearing method for the link-variables.

Note that the smearing is just a method to construct the operator, and hence it never changes the physics itself such as the gauge configuration, unlike the cooling. As an important feature, the gauge-transformation property of $U_\mu^{(n)}(s)$ is just the same as that of $U_\mu(s)$, which ensures the gauge invariance of the n-th smeared 3Q Wilson loop $\langle W_{3Q}[U_\mu^{(n)}(s)] \rangle$.

The fat link-variable $U_\mu^{(n)}(s)$ includes a spatial extension in terms of the original link-variable $U_\mu(s)$, and the smeared "line" expressed with $U_\mu^{(n)}(s)$ physically corresponds to a Gaussian-distributed "flux-tube".[7] Therefore, the properly smeared line is expected to resemble the ground-state flux-tube. Here, the smearing parameter α and the iteration number n can be regarded as the variational parameters to enhance the ground-state overlap.

Thus, through the selection of the properly smeared 3Q Wilson loop $\langle W_{3Q}[U_\mu^{(n)}(s)]\rangle$, we can construct the ground-state-dominant 3Q operator for the accurate measurement of the ground-state 3Q potential.[6,7,8]

2.4. *Lattice QCD results for the Ground-State 3Q Potential*

For more than 300 different patterns of spatially-fixed 3Q systems, we perform the thorough calculation of the ground-state potential $V_{3Q}^{\rm g.s.}$ in SU(3) lattice QCD with the standard plaquette action with $12^3 \times 24$ at $\beta = 5.7$ and with $16^3 \times 32$ at β=5.8 and 6.0 at the quenched level. For the accurate measurement, we use the smearing method and construct the ground-state-dominant 3Q operator.[6,7,8]

To conclude, we find that the static ground-state 3Q potential $V_{3Q}^{\rm g.s.}$ is well described by the Coulomb plus Y-type linear potential, *i.e.*, Y-Ansatz, within 1%-level deviation,[6,7] as shown in Table 1.

Table 1. Examples of the 3Q potential $V_{3Q}^{\rm latt}$ for the 3Q system put on $(i,0,0)$, $(0,j,0)$, $(0,0,k)$ in $\mathbf{R}^3$ in lattice QCD at β=6.0. For each 3Q configuration, $V_{3Q}^{\rm latt}$ is measured from the single-exponential fit as $\langle W_{3Q}\rangle = \bar{C}e^{-V_{3Q}T}$. $\bar{C} \simeq 1$ physically means the ground-state dominance in the smeared 3Q Wilson loop. We add the difference from the best-fit Y-Ansatz, $\Delta V \equiv |V_{3Q}^{\rm latt} - V_{3Q}^{\rm Y}|$, which is only about 1% of the typical scale of V_{3Q}. The listed value is measured in the lattice unit.

(i,j,k)	$V_{3Q}^{\rm latt}$	$\bar{C}$	ΔV	(i,j,k)	$V_{3Q}^{\rm latt}$	$\bar{C}$	ΔV
(0,1,1)	0.6778(6)	0.9784(24)	0.0012	(0,2,3)	1.0259(24)	0.9607(91)	0.0045
(0,1,2)	0.8234(11)	0.9712(45)	0.0042	(0,2,4)	1.0946(32)	0.9657(120)	0.0003
(0,1,3)	0.9183(17)	0.9769(65)	0.0045	(0,2,5)	1.1454(41)	0.9282(149)	0.0064
(0,1,4)	0.9859(24)	0.9589(92)	0.0050	(0,2,6)	1.2075(28)	0.9464(76)	0.0018
(0,1,5)	1.0463(30)	0.9495(112)	0.0064	(0,2,7)	1.2563(33)	0.9262(90)	0.0012
(0,1,6)	1.1069(40)	0.9595(152)	0.0122	(0,3,3)	1.0999(23)	0.9566(62)	0.0031
(0,1,7)	1.1572(50)	0.9374(192)	0.0102	(0,3,4)	1.1595(25)	0.9454(67)	0.0044
(0,2,2)	0.9430(21)	0.9586(78)	0.0095	(0,3,5)	1.2170(25)	0.9426(65)	0.0026

We summarize in Table 2 the lattice QCD results for the string tension and the Coulomb coefficient, with comparing between 3Q and Q-$\bar{\rm Q}$ potentials. As remarkable features, we find the universality of the string tension between the 3Q and Q-$\bar{\rm Q}$ systems and the one-gluon-exchange result as [6,7]

$$\sigma_{3Q} \simeq \sigma_{Q\bar{Q}}, \qquad A_{3Q} \simeq \frac{1}{2}A_{Q\bar{Q}}. \tag{7}$$

Table 2. The best-fit parameter set (σ, A, C) in the function form of Y-Ansatz, $V_{3Q} = -A_{3Q}\sum_{i<j}\frac{1}{|\mathbf{r}_i-\mathbf{r}_j|} + \sigma_{3Q}L_{\min} + C_{3Q}$, where $L_{\min}$ denotes the minimal value of the Y-type flux-tube length. The similar fit on the Q-$\bar{\rm Q}$ potential (on-axis) is also listed. The physical unit is determined so as to reproduce $\sqrt{\sigma}$=427MeV for the Q-$\bar{\rm Q}$ potential. The universality of the string tension and the OGE result on the Coulomb coefficient are found as $\sigma_{3Q} \simeq \sigma_{Q\bar{Q}}$ and $A_{3Q} \simeq \frac{1}{2}A_{Q\bar{Q}}$, respectively.

		$\sigma\ [a^{-2}]$	$\sqrt{\sigma}$ [MeV]	A	$C\ [a^{-1}]$
$\beta = 5.7$	$3Q_Y$	0.1524(28)	413.0	0.1331(66)	0.9182(213)
$(a \simeq 0.186$fm)	$Q\bar{Q}$	0.1629(47)	427	0.2793(116)	0.6203(161)
$\beta = 5.8$	$3Q_Y$	0.1027(6)	416.6	0.1230(20)	0.9085(55)
$(a \simeq 0.152$fm)	$Q\bar{Q}$	0.1079(28)	427	0.2607(174)	0.6115(197)
$\beta = 6.0$	$3Q_Y$	0.0460(4)	407.1	0.1366(11)	0.9599(35)
$(a \simeq 0.104$fm)	$Q\bar{Q}$	0.0506(7)	427	0.2768(24)	0.6374(30)

2.5. *Further Investigations with various Ansätze*

Lattice Coulomb plus Y-type Linear Ansatz : Considering the lattice discretization error at a very short distance, we perform also more accurate fitting with the lattice Coulomb plus Y-type linear potential. We obtain almost the same result and confirm that Y-Ansatz is correct.[7]

Generalized Y-type Linear Ansatz : Considering a possible contamination of the Δ-type flux at a short distance, we study generalized Y-Ansatz,[7] which includes both Y and Δ Ansätze in some limits. The flux-tube core radius R is found to be rather small as $R \simeq 0.08$fm, which again supports Y-Ansatz.[7] (See Fig.4.)

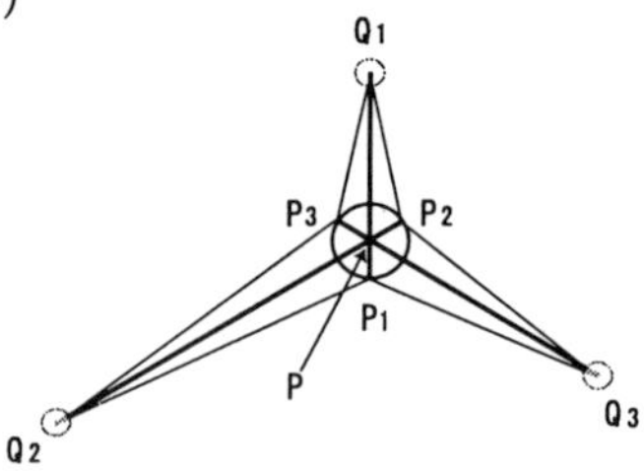

Figure 4. Generalized Y-Ansatz. The three quarks are spatially fixed at Q_1, Q_2, Q_3. The point P_1 is taken inside the circle around the physical junction (Fermat point) P so as to minimize $P_1Q_2 + P_1Q_3$. P_2 and P_3 are similarly defined. In generalized Y-Ansatz, $\overline{L_{\min}} \equiv \frac{1}{2}(\overline{P_1Q_2} + \overline{P_1Q_3} + \overline{P_2Q_1} + \overline{P_2Q_3} + \overline{P_3Q_1} + \overline{P_3Q_2})$ is used instead of $L_{\min}$

Yukawa plus Y-type Linear Ansatz : As other type of generalization, we investigate also the Yukawa plus Y-type linear potential,

$$V_{3Q} = -A_{3Q}^{\rm Yukawa}\sum_{i<j}\frac{e^{-m|r_i-r_j|}}{|r_i-r_j|} + \sigma_{3Q}^{\rm Yukawa}L_{\min} + C_{3Q}^{\rm Yukawa}, \tag{8}$$

as is indicated in the dual superconductor picture for color confinement.[3,15] However, we obtain $m \simeq 0$ for the screening mass m, which would support the type-II dual superconductor,[15] if this picture is correct. In any case, the Coulomb plus Y-type linear potential is confirmed once again.[7]

2.6. *Other Recent Studies on the 3Q Potential*

To clarify the current status of the 3Q potential, we introduce two recent studies on the 3Q potential.

de Forcrand's group : Recently, de Forcrand's group, who supported Δ-Ansatz in lattice QCD,[16] seems to change their opinion from Δ-Ansatz to Y-Ansatz[17] except for a very short distance, where the linear potential seems negligible compared with the Coulomb contribution. (As a problem of their argument, they relied on the continuum Coulomb potential even for the subtle argument at the very short distance, where the lattice Coulomb potential should be used.)

Cornwall : One of the theoretical basis of Δ-Ansatz was Cornwall's conjecture based on the vortex vacuum model.[18] Very recently, motivated by our studies, Cornwall re-examined his previous work and found an error in his model calculation. His corrected answer is Y-Ansatz instead of Δ-Ansatz.[19]

In this way, Y-Ansatz for the static 3Q potential seems almost settled both in lattice QCD and in analytic framework.

3. Y-type Flux-Tube Formation in Lattice QCD

Recently, as a clear evidence for Y-Ansatz, Y-type flux-tube formation is actually observed in the maximally-Abelian projected QCD from the direct measurement for the action density of the gluon field in the spatially-fixed 3Q system.[20,21,22] (See Fig.5)

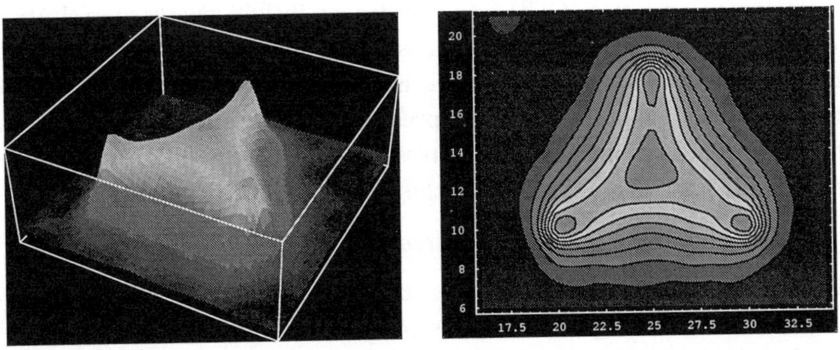

Figure 5. The lattice QCD result for Y-type flux-tube formation in the spatially-fixed 3Q system in maximally-Abelian projected QCD. The distance between the junction and each quark is about 0.5 fm.

4. The Excited-State Three-Quark Potential in QCD

In 1969, Y. Nambu first pointed out the string picture for hadrons[23,24] to explain the Veneziano amplitude[25] on hadron reactions and resonances. Since then, the string picture has been one of the most important pictures for hadrons and has provided many interesting ideas in the wide region of the elementary particle physics.

For instance, the hadronic string creates infinite number of hadron resonances as the vibrational modes, and these excitations lead to the Hagedorn "ultimate" temperature,[26] which gives an interesting theoretical picture for the QCD phase transition.

For the real hadrons, of course, the hadronic string is to have a spatial extension like the flux-tube, as the result of one-dimensional squeezing of the color-electric flux in accordance with color confinement.[27] Therefore, the vibrational modes of the hadronic flux-tube should be much more complicated, and the analysis of the excitation modes is important to clarify the underlying picture for real hadrons.

In the language of QCD, such non-quark-origin excitation is called as the "gluonic excitation", and is physically interpreted as the excitation of the gluon-field configuration in the presence of the quark-antiquark pair or the three quarks in a color-singlet state.

In the hadron physics, the gluonic excitation is one of the interesting phenomena beyond the quark model, and relates to the hybrid hadrons such as $q\bar{q}G$ and $qqqG$. In particular, the hybrid meson includes the exotic hadrons with $J^{PC} = 0^{--}, 0^{+-}, 1^{-+}, 2^{+-}, \cdots$, which cannot be constructed within the simple quark model.

In this section, we study the excited-state 3Q potential and the gluonic excitation using lattice QCD,[8] to get deeper insight on these excitations beyond the hypothetical models such as the string and the flux-tube models. Here, the excited-state 3Q potential is the energy of the excited state of the gluon-field configuration in the presence of the static three quarks, and the gluonic-excitation energy is expressed as the energy difference between the ground-state 3Q potential and the excited-state 3Q potential.

4.1. *Formalism to extract Excited-State 3Q Potentials*

We present the formalism to extract the excited-state potential.[8] For the simple notation, the ground state is regarded as the "0-th excited state". For the physical eigenstates of the QCD Hamiltonian $\hat{H}$ for the spatially-fixed 3Q system, we denote the n-th excited state by $|n\rangle$ $(n = 0, 1, 2, \cdots)$. Since the three quarks are spatially fixed in this case, the eigenvalue of $\hat{H}$ is expressed by a static potential as $\hat{H}|n\rangle = V_n|n\rangle$, where V_n denotes the n-th excited-state 3Q potential. Note that both V_n and $|n\rangle$ are universal physical quantities relating to the QCD Hamiltonian $\hat{H}$. In fact, V_n depends only on the 3Q location, and $|n\rangle$ satisfies the orthogonal condition as $\langle m|n\rangle = \delta_{mn}$.

Suppose that $|\Phi_k\rangle$ $(k = 0, 1, 2, \cdots)$ are arbitrary given independent spatially-fixed 3Q states. In general, each 3Q state $|\Phi_k\rangle$ can be expressed with a linear combination of the 3Q physical eigenstates $|n\rangle$ as

$$|\Phi_k\rangle = c_0^k|0\rangle + c_1^k|1\rangle + c_2^k|2\rangle + \cdots. \tag{9}$$

Here, the coefficients c_n^k depend on the selection of $|\Phi_k\rangle$, and hence they are not universal quantities.

The Euclidean-time evolution of the 3Q state $|\Phi_k(t)\rangle$ is expressed with the operator $e^{-\hat{H}t}$, which corresponds to the transfer matrix in lattice QCD. The overlap $\langle\Phi_j(T)|\Phi_k(0)\rangle$ is given by the 3Q Wilson loop with the initial state $|\Phi_k\rangle$ at $t = 0$ and the final state $|\Phi_j\rangle$ at $t = T$, and is expressed in the Euclidean Heisenberg picture as

$$\begin{aligned} W_T^{jk} &\equiv \langle\Phi_j|W_{3Q}(T)|\Phi_k\rangle = \langle\Phi_j(T)|\Phi_k(0)\rangle = \langle\Phi_j|e^{-\hat{H}T}|\Phi_k\rangle \\ &= \sum_{m=0}^{\infty}\sum_{n=0}^{\infty} \bar{c}_m^j c_n^k \langle m|e^{-\hat{H}T}|n\rangle = \sum_{n=0}^{\infty} \bar{c}_n^j c_n^k e^{-V_n T}. \end{aligned} \tag{10}$$

Using the matrix C satisfying $C^{nk} = c_n^k$ and the diagonal matrix Λ_T as $\Lambda_T^{mn} = e^{-V_n T}\delta^{mn}$, we rewrite the above relation as

$$W_T = C^\dagger \Lambda_T C. \tag{11}$$

Note here that C is not a unitary matrix, and hence this relation does not mean the simple diagonalization by the unitary transformation.

Since we are interested in the 3Q potential V_n in Λ_T rather than the non-universal matrix C, we single out V_n from the 3Q Wilson loop W_T as

$$W_T^{-1}W_{T+1} = \{C^\dagger\Lambda_T C\}^{-1}C^\dagger\Lambda_{T+1}C = C^{-1}\mathrm{diag}(e^{-V_0}, e^{-V_1}, \cdots)C, \tag{12}$$

which is a similarity transformation. Then, e^{-V_n} can be obtained as the eigenvalues of the matrix $W_T^{-1}W_{T+1}$, *i.e.*, solutions of the secular equation,

$$\det\{W_T^{-1}W_{T+1} - t\mathbf{1}\} = \prod_n (e^{-V_n} - t) = 0. \tag{13}$$

Thus, the 3Q potential V_n can be obtained from the matrix $W_T^{-1}W_{T+1}$.

In the practical calculation, we prepare N independent sample states $|\Phi_k\rangle$ $(k = 0, 1, \cdots, N-1)$. By choosing appropriate states $|\Phi_k\rangle$ so as not to include highly excited-state components, the physical states $|n\rangle$ can be truncated as $0 \leq n \leq N-1$. Then, W_T, C and Λ_T are reduced into $N \times N$ matrices, and the secular equation (13) becomes the N-th order equation.

4.2. *Lattice QCD results for the Excited-State 3Q Potential*

For more than 100 different patterns of spatially-fixed 3Q systems, we study the excited-state potential $V_{3Q}^{\rm e.s.}$ using lattice QCD with $16^3 \times 32$ at β=5.8 and 6.0 at the quenched level.[8] In Fig.6, we show the first lattice QCD results for the excited-state 3Q potential $V_{3Q}^{\rm e.s.}$ as well as the ground-state potential $V_{3Q}^{\rm g.s.}$. (In Fig.6, the minimal length $L_{\rm min}$ of the Y-type flux-tube is used as a label to distinguish the three-quark configuration.)

The energy gap between $V_{3Q}^{\rm g.s.}$ and $V_{3Q}^{\rm e.s.}$ physically means the excitation energy of the gluon-field configuration in the presence of the spatially-fixed three quarks, and the gluonic excitation energy $\Delta E_{3Q} \equiv V_{3Q}^{\rm e.s.} - V_{3Q}^{\rm g.s.}$ is found to be about 1GeV or more [8,21,22] in the typical hadronic scale as $L_{\rm min} \sim 1$ fm.

Note that the gluonic excitation energy of about 1GeV is rather large in comparison with the excitation energies of the quark origin, and such a gluonic excitation would contribute significantly in the highly-excited baryons with the excitation energy above 1 GeV. The present result predicts that the lowest hybrid baryon, which is described as $qqqG$ in the valence picture, has a large mass of about 2 GeV.[8]

Together with the recent lattice result[10] indicating the gluonic excitation energy $\Delta E_{Q\bar{Q}} \equiv V_{Q\bar{Q}}^{\rm e.s.} - V_{Q\bar{Q}}^{\rm g.s.}$ for the Q-$\bar{\rm Q}$ system to be in the order of 1GeV, the present result seems to suggest the constituent gluon mass of about 1GeV in terms of the constituent gluon picture.

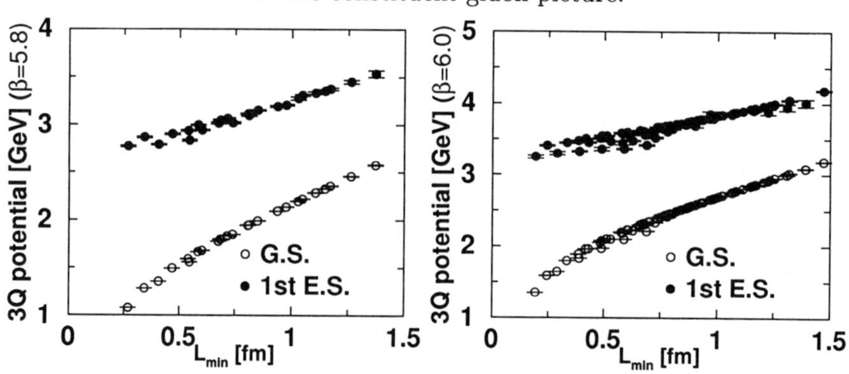

Figure 6. The lattice QCD results of the ground-state 3Q potential $V_{3Q}^{\rm g.s.}$ (open circles) and the 1st excited-state 3Q potential $V_{3Q}^{\rm e.s.}$ (filled circles) as the function of $L_{\rm min}$. These lattice results at $\beta = 5.8$ and $\beta = 6.0$ well coincide besides an irrelevant overall constant. The gluonic excitation energy $\Delta E_{3Q} \equiv V_{3Q}^{\rm e.s.} - V_{3Q}^{\rm g.s.}$ is found to be about 1GeV in the hadronic scale as $L_{\rm min} \sim 1$fm.

4.3. *Functional Form of the Excited-State 3Q Potential and Comparison with the Q-$\bar{Q}$ System*

From the lattice QCD data, we attempt to seek the functional form of the excited-state 3Q potential $V_{3Q}^{e.s.}$, but find no simple plausible form of $V_{3Q}^{e.s.}$, unlike the ground-state 3Q potential $V_{3Q}^{g.s.}$.

Next, we compare the 3Q gluonic excitation with the Q-$\bar{Q}$ gluonic excitation, considering the nature of the Y-type junction. If the Y-type junction behaves as a quasi-fixed edge of the three flux-tubes, these three flux-tubes would behave as independent three Q-$\bar{Q}$ systems, and therefore the 3Q gluonic excitation would be approximated as a simple incoherent sum of the three Q-$\bar{Q}$ gluonic excitations. If the Y-type junction behaves as a quasi-free edge, the 3Q gluonic-excitation energy ΔE_{3Q} would be smaller than each of Q-$\bar{Q}$ gluonic-excitation energies $\Delta E_{Q\bar{Q}}$ corresponding to the three flux-tubes, since the string with fixed edges has a larger vibrational energy.

Through the comparison of $V_{3Q}^{e.s.}$ or ΔE_{3Q} with several possible linear combinations of $V_{Q\bar{Q}}^{e.s.}$ or $\Delta E_{Q\bar{Q}}$, we find no simple relation between them. This fact is conjectured to reflect the complicated vibrational mode on the Y-type flux-tube, due to the interference among the vibrational modes on the three flux-tubes through the junction, which may indicate the quasi-free behavior of the Y-type junction.

5. Behind the Success of the Quark Model

Finally, we consider the connection between QCD and the quark model in terms of the gluonic excitation.[8,21,22] While QCD is described with quarks and gluons, the simple quark model successfully describes low-lying hadrons even without explicit gluonic modes. In fact, the gluonic excitation seems invisible in the low-lying hadron spectra, which is rather mysterious.

On this point, we find the gluonic-excitation energy to be about 1GeV or more, which is rather large compared with the excitation energies of the quark origin, and therefore the effect of gluonic excitations is negligible and quark degrees of freedom plays the dominant role in low-lying hadrons with the excitation energy below 1GeV.

Thus, the large gluonic-excitation energy of about 1GeV gives the physical reason for the invisible gluonic excitation in low-lying hadrons, which would play the key role for the success of the quark model without gluonic-excitation modes.[8,21,22]

In Fig.7, by way of the flux-tube picture, we present a possible scenario from QCD to the massive quark model in terms of color confinement and dynamical chiral-symmetry breaking.[21,22]

6. Summary and Concluding Remarks

Using SU(3) lattice QCD, we have studied the ground-state 3Q potential $V_{3Q}^{\text{g.s.}}$ and the 1st excited-state 3Q potential $V_{3Q}^{\text{e.s.}}$.

From the accurate and thorough calculation for more than 300 different patterns of 3Q systems, we have found that the static ground-state 3Q potential $V_{3Q}^{\text{g.s.}}$ is well described by the Coulomb plus Y-type linear potential, *i.e.*, Y-Ansatz, within 1%-level deviation. As a clear evidence for Y-Ansatz, Y-type flux-tube formation has been actually observed on the lattice in maximally-Abelian projected QCD.

For more than 100 different patterns of 3Q systems, we have performed the first study of the 1st excited-state 3Q potential $V_{3Q}^{\text{e.s.}}$ in quenched lattice

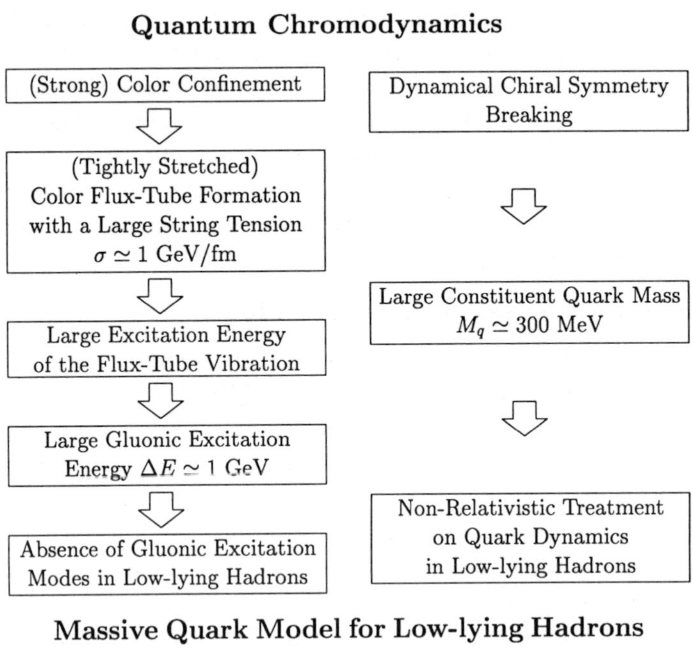

Figure 7. A possible scenario from QCD to the quark model in terms of color confinement and dynamical chiral-symmetry breaking (DCSB). DCSB provides a large constituent quark mass of about 300MeV, which enables the non-relativistic treatment for quark dynamics. Color confinement provides the color flux-tube formation among quarks with a large string tension of $\sigma \simeq 1$ GeV/fm. In the flux-tube picture, the gluonic excitation is described as the flux-tube vibration, and the flux-tube vibrational energy is expected to be large, reflecting the large string tension. The large gluonic-excitation energy of about 1GeV leads to the absence of the gluonic mode in low-lying hadrons, which would play the key role to the success of the quark model without gluonic excitation modes.

QCD, and have found the gluonic excitation energy $\Delta E_{3Q} \equiv V_{3Q}^{\rm e.s.} - V_{3Q}^{\rm g.s.}$ to be about 1 GeV. This indicates that the hybrid baryons ($qqqG$) are to be rather heavy and appear in the spectrum above 2GeV. We have conjectured that the large gluonic-excitation energy of about 1GeV leads to the success of the quark model for the low-lying hadrons even without gluonic excitations.

Acknowledgements H.S. thanks all the participants of Confinement2003.

References

1. Y. Nambu, in *Preludes in Theoretical Physics*, in honor of V.F. Weisskopf (North-Holland, Amsteldam,1966).
2. M.Y. Han and Y. Nambu, *Phys. Rev.* **139**, B1006 (1965).
3. For instance, articles in *Quantum Chromodynamics and Color Confinement*, edited by H. Suganuma, M. Fukushima and H. Toki (World Scientific, 2001).
4. Y. Nambu, G. Juna-Lasinio, *Phys. Rev.* **122**, 345 (1961); *ibid.* **124**, 246 (1961).
5. H.J. Rothe, *Lattice Gauge Theories*, 2nd edition (World Scientific, 1997) p.1.
6. T.T. Takahashi, H. Matsufuru, Y. Nemoto and H. Suganuma, *Phys. Rev. Lett.* **86**, 18 (2001).
7. T.T. Takahashi, H. Suganuma, Y. Nemoto and H. Matsufuru, *Phys. Rev.* **D65**, 114509 (2002) and references therein.
8. T.T. Takahashi and H. Suganuma, *Phys. Rev. Lett.* **90**, 182001 (2003).
9. T.T. Takahashi, H. Matsufuru, Y. Nemoto and H. Suganuma, Proc. of the TMU-Yale Symp. on *Dynamics of Gauge Fields*, Tokyo, Dec. 1999, edited by A. Chodos *et al.*, (Universal Academy Press, 2000) 179.
10. K.J. Juge, J. Kuti and C.J. Morningstar, *Phys. Rev. Lett.* **90**, 161601 (2003).
11. J. Kogut and L. Susskind, *Phys. Rev.* **D11**, 395 (1975).
12. S. Capstick and N. Isgur, *Phys. Rev.* **D34**, 2809, (1986).
13. M. Fable de la Ripelle and Yu. A. Simonov, *Ann. Phys.* **212**, 235 (1991).
14. N. Brambilla, G.M. Prosperi and A. Vairo, *Phys. Lett.* **B362**, 113 (1995).
15. H. Suganuma, S. Sasaki and H. Toki, *Nucl. Phys.* **B435**, 207 (1995).
16. C. Alexandrou, P. de Forcrand, A. Tsapalis, *Phys. Rev.* **D65**, 054503 (2002).
17. O. Jahn and P. de Forcrand, Proc. of *Lattice 2003*, hep-lat/0309115.
18. J.M. Cornwall, *Phys. Rev.* **D54**, 6527 (1996).
19. J.M. Cornwall, "On the Center-Vortex Baryonic Area Law", hep-th/0305101.
20. H. Ichie, V. Bornyakov, T. Streuer and G. Schierholz, *Nucl. Phys.* **A721**, 899 (2003); *Nucl. Phys.* **B** (Proc.Suppl.) **119**, 751 (2003).
21. T.T. Takahashi, H. Suganuma, H. Ichie, H. Matsufuru and Y. Nemoto, *Nucl. Phys.* **A721**, 926 (2003).
22. H. Suganuma, T.T. Takahashi and H. Ichie, *Nucl. Phys.* **A** in press.
23. Y. Nambu, in *Symmetries and Quark Models* (Wayne State University, 1969).
24. Y. Nambu, *Lecture Notes at the Copenhagen Symposium* (1970).
25. G. Veneziano, *Nuovo Cim.* **A57**, 190 (1968).
26. R. Hagedorn, *Nuovo Cim. Suppl.* **3**, 147 (1965).
27. Y. Nambu, *Phys. Rev.* **D10**, 4262 (1974).

LATTICE CALCULATION OF THE LOWEST ORDER HADRONIC CONTRIBUTION TO THE MUON ANOMALOUS MAGNETIC MOMENT: AN UPDATE WITH KOGUT-SUSSKIND FERMIONS *

T. BLUM

RIKEN BNL Research Center
Brookhaven National Laboratory
Upton NY 11973-5000, USA
E-mail: tblum@bnl.gov

I present a preliminary calculation of the hadronic vacuum polarization for 2+1 flavors of improved Kogut-Susskind quarks by utilizing a set of gauge configurations recently generated by the MILC collaboration. The polarization function $\Pi(q^2)$ is then used to calculate the lowest order (in α_{QED}) hadronic contribution to the muon anomalous magnetic moment.

1. Introduction

The anomalous magnetic moment of the muon, a_μ, is now known to fantastic precision, both experimentally[1] and theoretically[2]. This situation naturally provides an interesting test of the Standard Model. The largest uncertainty in the calculation of a_μ is associated with hadronic contributions; the lowest order in α contribution arises from the hadronic vacuum polarization of the photon. Using the analytic structure of the vacuum polarization and the optical theorem, this contribution is estimated from the experimentally measured total cross section of e^+e^- annihilation to hardrons[3,4]. Isospin symmetry relates the e^+e^- cross-section to the branching ratio of τ decay to hadrons which can also be used to calculate the hadronic contributions[3]. However, a purely theoretical, first principles treatment has been missing. Given the importance of the muon g-2 experiment, a completely independent theoretical calculation is desirable.

*I thank the US DOE, RIKEN, and NERSC for providing resources to complete this work.

Recently, the framework to calculate the hadronic piece of the anomalous moment, $a_\mu^{\rm had}$, in an entirely theoretical way from first principles using lattice QCD was given[5]. Encouraging quenched results were obtained using domain wall fermions[5] and improved Wilson fermions[6].

In a nut-shell, the first principles calculation is performed entirely in Euclidean space-time, so that the hadronic vacuum polarization computed on the lattice can be inserted directly into the one-loop vertex function for the muon. To obtain the physical result, at the end one continues back to Minkowski space-time so the external muon is on-shell.

The aim of the current study is to determine how accurate the lattice calculation can be in the near future. Toward that goal, I present preliminary results for the hadronic vacuum polarization calculated on a set of 2+1 flavor improved Kogut-Susskind fermion lattices generated by the MILC collaboration (see Table 1).

2. The vacuum polarization

The vacuum polarization tensor is defined as the Fourier transform of the two-point correlation function of the electromagnetic current, J^μ,

$$\Pi^{\mu\nu}(q^2) = \int d^4x\, e^{iq(x-y)} J^\mu(x)\, J^\nu(y), \tag{1}$$

$$= (q^\mu q^\nu - \delta^{\mu\nu})\Pi(q^2), \tag{2}$$

$$J^\mu = \frac{2}{3}\bar{u}\,\gamma_\mu\, u - \frac{1}{3}\bar{d}\,\gamma_\mu\, d - \frac{1}{3}\bar{s}\,\gamma_\mu\, s. \tag{3}$$

To satisfy the Ward-Takahashi identity,

$$q^\mu\, \Pi^{\mu\nu} = 0, \tag{4}$$

the exactly conserved lattice vector current is used which makes the extraction of the polarization function from Eq. (3) straightforward. In general, the lattice conserved current is point-split; it depends on the fields at a point x and its neighbors:

$$J^\mu(x) = \frac{c_0}{2}\left(\bar{\psi}(x+\hat{\mu})\, U_\mu^\dagger(x)\, \gamma_\mu \psi(x) + \bar{\psi}(x)\, U_\mu(x)\, \gamma_\mu \psi(x+\hat{\mu})\right) \tag{5}$$

(for a single flavor) and satisfies $\Delta^\mu J^\mu = 0$. The gauge field $U_\mu(x)$ makes the current gauge-invariant, and c_0 depends on the lattice action ($c_0 = 1$ for ordinary Kogut-Susskind fermions). With this form for the lattice current, and after subtracting the contact terms that arise because of its point-split form[7,5], the lattice polarization tensor is given by Eq. 2 with $q^\mu = 2\sin(\pi\, n_\mu / L_\mu)$, $n_\mu = 0, 1, 2, ..., L_\mu - 1$. Adding to the lattice action the three-hop Naik term[8] to improve the discretization of the derivative in the Dirac operator generates an additional divergence. In coordinate space

$$J^{3\mu}(x) = \frac{c_1}{2}\left(\bar{\psi}(x+3\hat{\mu})V_\mu^\dagger(x)\gamma_\mu\psi(x) + \bar{\psi}(x)V_\mu(x)\gamma_\mu\psi(x+3\hat{\mu})\right), \quad (6)$$

$$V_\mu(x) = U_\mu(x)U_\mu(x+\hat{\mu})U_\mu(x+2\hat{\mu}), \quad (7)$$

with $c_0 = 9/8$, $c_1 = -1/24$ for Naik fermions, and which satisfies $\Delta^\mu J^\mu + \Delta^{3\mu} J^{3\mu} = 0$. The backward difference operators are defined as

$$\Delta^\mu f(x) = f(x) - f(x-\hat{\mu}), \quad (8)$$

$$\Delta^{3\mu} f(x) = f(x) - f(x-3\hat{\mu}). \quad (9)$$

Thus the simple form of the Ward-Takahashi identity, and hence $\Pi^{\mu\nu}$, no longer holds. To avoid this complication, in the following the Naik term is dropped when computing valence quark propagators even though it appears in the a^2-tad action[9] used to generate the 2+1 flavor configurations. Note that the discretization errors are then $O(a^2)$, not $O(g^2 a^2)$ as for the full a^2-tad action. This modification should not significantly affect the small q^2, or long-distance, behavior of $\Pi(q^2)$.

Table 1. MILC 2+1 flavor lattices generated with the a^2-tad action[9]. a is the lattice spacing[10]. m_l and m_s denote light and strange quark masses, respectively. In all cases the strange quark mass corresponds roughly to its physical value. m_v and # configs are the valence quark mass and number of configurations used in the calculation of $\Pi(q^2)$.

a (fm)	size	m_l	m_s	m_{val}	# configs
0.121(3)	$20^3 \times 64$	0.01	0.05	0.05	57
0.121(3)	$20^3 \times 64$	0.01	0.05	0.01	439
0.120(3)	$24^3 \times 64$	0.005	0.05	0.005	143
0.086(2)	$28^3 \times 96$	0.0062	0.031	0.031	41
0.086(2)	$28^3 \times 96$	0.0062	0.031	0.0062	248
0.085(2)	$28^3 \times 96$	quenched		0.031	29
0.085(2)	$28^3 \times 96$	quenched		0.0062	31

3. RESULTS and DISCUSSION

3.1. *vacuum polarization*

In Figure 1, I show $\Pi(q^2)$ calculated on the improved Kogut-Susskind lattices[a]. All results at a given lattice spacing approach a common value in

[a] Some of these results were obtained after the meeting.

the large q^2 limit, as they should. $\Pi(q^2)$ is logarithmically divergent in the lattice spacing, only the running with q^2 being physical. Therefore, the results for different lattice spacings differ by a constant plus terms of order a^2. Such a shift is clearly visible between the two sets of data plotted in Figure 1. For the coarser lattice spacing ($a \approx 0.121$ fm), $\Pi(q^2)$ was calculated for three valence quark masses, $m_v = 0.005$, 0.01, and 0.05, or $m_v \approx 0.1\,m_s$, $0.2\,m_s$, and m_s respectively. For the finer lattice ($a \approx 0.086$ fm) we have $m_v = 0.0062$ and 0.031, or $m_v \approx 0.2\,m_s$ and m_s. There is a significant dependence on the valence quark mass. As $q^2 \to 0$, $\Pi(q^2)$ rises much more steeply as $m_v \to 0$. However, for the larger lattice spacing this behavior appears to weaken sharply between $m_v = 0.01$ and 0.005. For m_v= 0.005 $\Pi(q^2)$ actually decreases at the lowest value of q^2 which is probably indicative of low statistics.

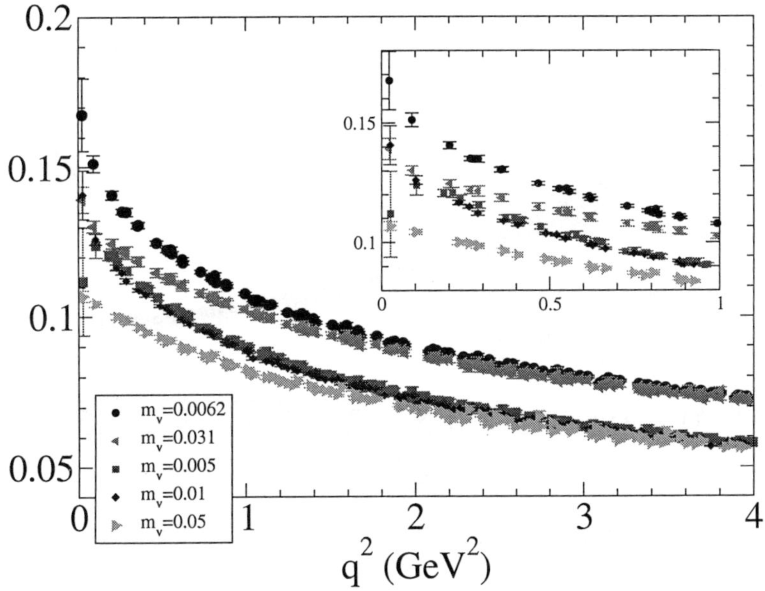

Figure 1. $\Pi(q^2)$ computed on 2+1 flavor lattices. The two upper curves correspond to the finer lattice spacing. Errors shown are statistical only.

In Figure 2, I compare $\Pi(q^2)$ computed on dynamical and quenched lattices at $a \approx 0.086$ fm ($a^{-1} \approx 2.239$ GeV). Discernible effects appear as $q^2 \to 0$ for $m_v = 0.0062$, the lightest quark mass for this lattice spacing.

More statistics on the quenched lattice are needed to quantify the effect. For $m_v = 0.031$ there is no apparent effect of unquenching. These results may indicate that the two-pion threshold is not lower than the mass of the vector particle. The two-pion state must have one unit of orbital angular momentum since the photon has $J = 1$, so the π's can not be at rest. The threshold for two non-interacting pions is $2\,E_\pi = 2\sqrt{m_\pi^2 + (2\pi/L)^2}$, or ≈ 0.54 and 0.61 for $m_v = 0.0062$ and 0.005, respectively. On the other hand, $m_\rho \approx 0.39$ and 0.53 in these cases[10]. Note that since $\Pi(q^2)$ is computed in Euclidean space, it should be a smooth function of q^2, even as such thresholds and resonances are crossed (in Minkowski space)[11].

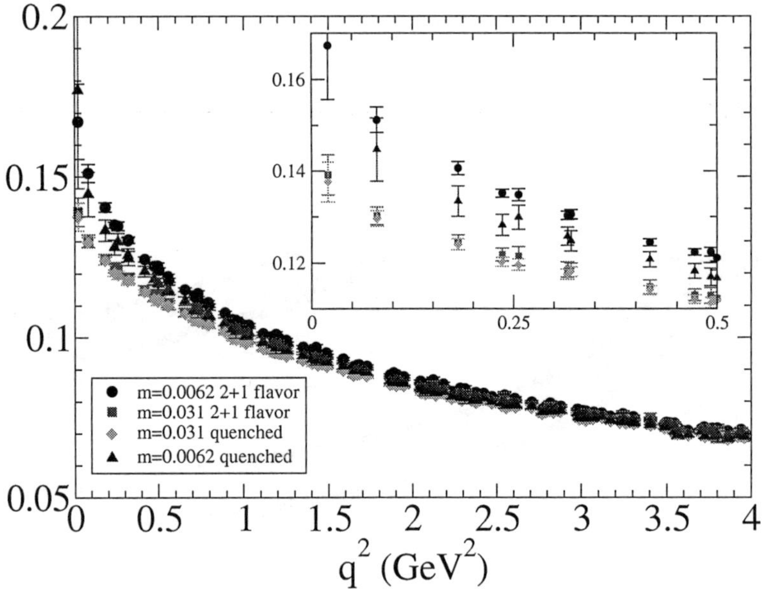

Figure 2. $\Pi(q^2)$. 2+1 flavor and quenched lattices at $a \approx 0.086$ fm.

Comparison of $\Pi(q^2)$ with continuum perturbation theory[12] as shown in Figure 3 indicates that the improved Kogut-Susskind results may suffer significant lattice artifacts (contrast with the quenched domain wall fermion results at roughly the same m_v but larger lattice spacing). Results are shifted by hand to account for the $\ln(a)$ term; there is no choice for this shift which yields good agreement over a large range of q^2 with continuum perturbation theory, unlike the case for domain wall fermions which compares quite well The agreement does improve as $m_v \to 0$ (Figure 4), in particular for the smaller lattice spacing. In Figure 4 I have chosen, arbitrarily, to match the lattice data and the continuum perturbation theory at $q^2 = 2$ GeV^2. This behavior may be indicative of an $a^2\, m\, q$ error. The perturbation theory results shown in Figures 3 and 4 are given in the $\overline{MS}$ scheme [b] , so m_v should also be given in this scheme. Except for the domain wall fermion results, this has not been done, but note that the quark mass dependence of the perturbation theory result is very mild for small mass.

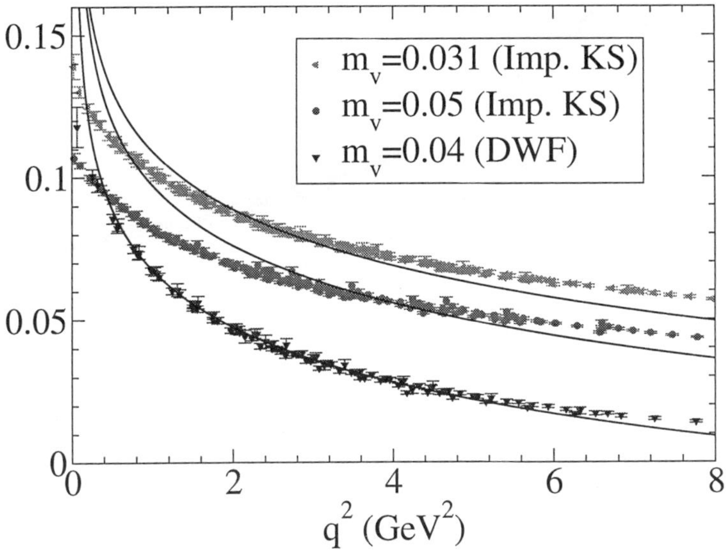

Figure 3. $\Pi(q^2)$. 2+1 flavor lattices. Comparison to continuum perturbation theory[12]. $m_v = 0.05$ ($a = 0.121$ fm), 0.031 ($a = 0.086$ fm), and 0.04 (quenched domain wall fermion results from [5], $a = 0.15$ fm).

The valence Kogut-Susskind fermions contribute like four flavors of continuum fermions to $\Pi(q^2)$ which therefore has to be scaled by 1/4. In the continuum limit, $a \to 0$, these four flavors are degenerate, so the scaling is exact. For $a \neq 0$ this scaling is not exact and leads to lattice spacing errors like the ones discussed above. It is just such flavor-symmetry breaking ar-

[b]Of course, the physical piece of $\Pi(q^2)$ is scheme and scale independent.

tifacts that are supposed to be suppressed by the a^2-tad action. Figure 5 shows that, indeed, the fat-link Kogut-Susskind fermions differ significantly from ordinary Kogut-Susskind fermions; the slope increases both at low and high values of q^2.

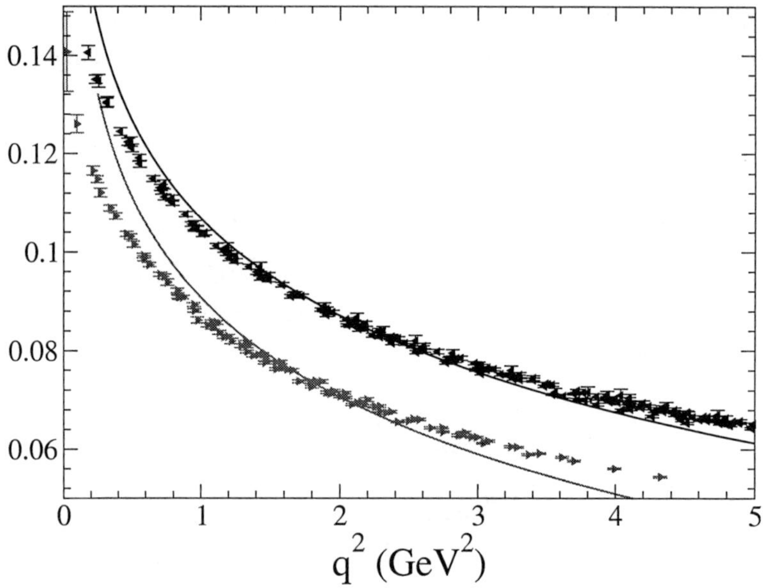

Figure 4. $\Pi(q^2)$. Same as Figure 3, but for $m_v = 0.01$ (lower) and 0.0062 (upper). The lattice (symbols) and perturbation theory (lines) results are arbitrarily matched at $q^2 = 2$ GeV^2 for comparison. Agreement with perturbation theory is better for the smaller lattice spacing, $a = 0.086$ fm (upper points).

As mentioned above, omission of the Naik term should not significantly alter the low q^2 behavior of $\Pi(q^2)$, even though the errors are now order a^2, not $g^2 a^2$. This is because the Naik term improves the derivative in the Dirac operator, but does not correct flavor symmetry. The reason for omitting the Naik term is that the simple ansatz, Eq. 2, no longer holds. Still, this ansatz should be a good estimate even for Naik fermions for small q. In Figure 6, $\Pi(q^2)$ ($m_v = 0.01$, $a = 0.121$ fm) is shown with and without the Naik term. For small q^2 the results appear similar, apart from a constant shift. There is also a hint that the Naik term makes $\Pi(q^2)$ steeper at larger values of q^2 which may improve the agreement with perturbation theory. However because only an approximate ansatz for $\Pi^{\mu\nu}$ is used in this case, this can only be taken as an indication.

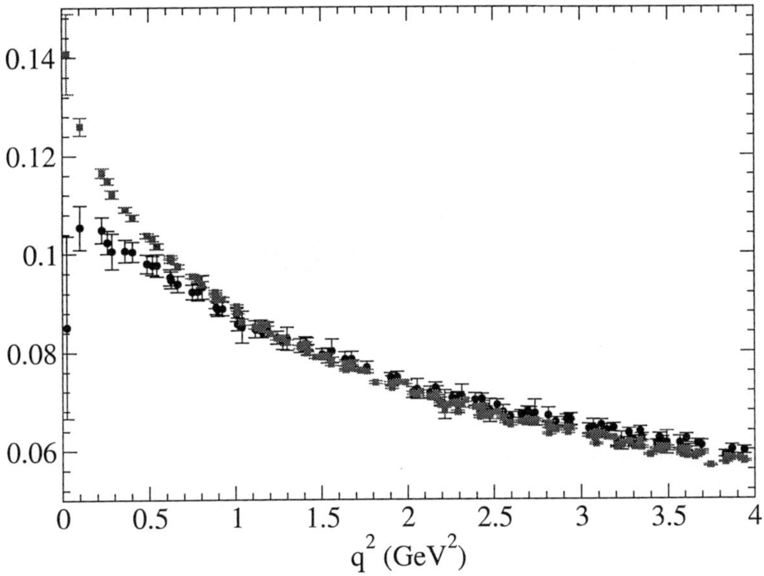

Figure 5. $\Pi(q^2)$. Comparison of fat-link improved (a^2-tad minus the Naik term) and ordinary Kogut-Susskind fermions. The smaller slope as $q^2 \to 0$ is indicative of lattice spacing artifacts that arise from flavor symmetry breaking.

3.2. *The anomalous magnetic moment*

The method for calculating $a_\mu^{\rm had}$ from the vacuum polarization is given in [5]. The central idea is to carry out the entire calculation in Euclidean space so that $\Pi(q^2)$ calculated on the lattice up to some momentum, $q^2_{\rm cut}$, can be directly inserted into the one-loop (QED) vertex of the muon which describes its interaction with an external magnetic field. A key feature is that the integral is dominated by the low q^2 region. If the lattice calculation is accurate enough, *i.e.* statistical errors are under good control, q^2 is low enough, *etc.*, then no additional theoretical input like a fit ansatz is necessary, and any faithful representation of the numerical data will suffice to calculate $a_\mu^{\rm had}$. I have chosen a simple polynomial to fit the lattice data,

$$\begin{aligned} a_\mu^{\rm had} &= a_0 + a_1\, x + a_2\, x^2 + a_3\, x^3 + a_4\, x^4, \qquad (10)\\ x &= q^2. \end{aligned}$$

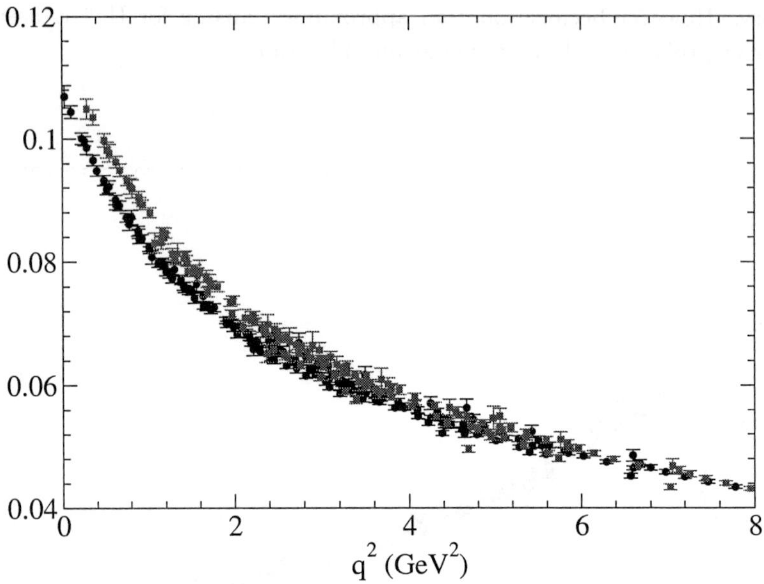

Figure 6. Effect of the Naik term (upper points) on $\Pi(q^2)$. $m_v = 0.05$, $m_l = 0.01$, and $m_s = 0.05$.

The results for $m_v = 0.0062$, $a = 0.086$ fm are shown in Figure 7. Values of $\Pi(q^2)$ at different q^2 are highly correlated, so I have fit only those results with $q^2 \lesssim 1$ GeV^2 in order to obtain a reasonable covariance matrix, and consequently, fits with acceptable χ^2. The fits shown in Figure 7 are representative of all the data; they tend to under-predict the data as $q^2 \to 0$. The discrepancy decreases as more terms are added to the fit function. As only its running with q^2 is physical, $\Pi(q^2)$ is renormalized by subtracting $\Pi(0)$, so the smaller the slope of $\Pi(q^2 \to 0)$, the smaller is $a_\mu^{\rm had}$.

Table 2 summarizes the preliminary values for $a_\mu^{\rm had}$ computed using improved Kogut-Susskind fermions on 2+1 flavor gauge configurations. The lattice results are used up to $q^2_{\rm cut} = 1$ GeV^2, and perturbation theory from there to ∞ to complete the integral. The matching with perturbation theory is not as good as for domain wall fermions, but since the perturbative contribution is already quite small at 1 GeV^2 this will not matter. For comparison, quenched results are also tabulated, including the domain wall fermion ones from [5]. The dependence on the polynomial degree of the fit function, seen in the fit results just described, shows up in $a_\mu^{\rm had}$ due to the sensitivity of $a_\mu^{\rm had}$ to the low q^2 region . Compared to the quenched domain wall fermion value, the improved Kogut-Susskind values are low, except the $m_v = 0.0062$ point. This is probably a consequence of the lattice spacing errors in the latter. For domain wall fermions, there is no indication of large scaling violations[5], though finite volume effects prevent ruling out this possibility. Calculating the quark mass dependence with domain wall fermions would also improve the comparison with the results presented here.

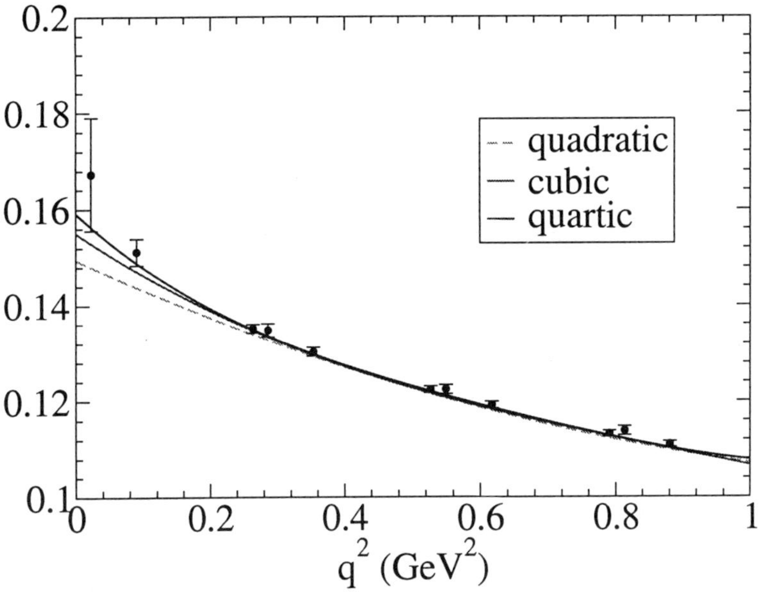

Figure 7. Covariant fits to $\Pi(q^2)$. 2+1 flavor lattice, $m_v = 0.0062$.

Using the quartic fit value, for 2+1 flavors, $m_{u,d} = 0.0062 \approx 1/5\, m_s$, $m_s = 0.031$, $a = 0.086$,

$$a_{\mu,2+1}^{\rm had} \times 10^{10} = \left(\left(\frac{2}{3}\right)^2 + \left(\frac{1}{3}\right)^2\right) \times 903(117) + \left(\frac{1}{3}\right)^2 \times 393(72)$$
$$= 545(65).$$

This should be taken only as a preliminary estimate. The fits to $\Pi(q^2)$ are not very stable; a better fit method should not under-predict the low q^2

region and may reduce the statistical errors. Concentrating on the light quark masses 0.0062 and 0.01 in Table 2, there is a significant increase in $a_\mu^{\rm had}$ at $a = 0.086$ fm over the $a = 0.121$ fm value; linear extrapolation to $a^2 = 0$ increases $a_\mu^{\rm had}$ by 100×10^{-10}. Finally, the disconnected part of the vacuum polarization has not been calculated. This piece is color and electric charge suppressed but could still contribute to $a_\mu^{\rm had}$. For comparison, the value computed from the e^+e^- cross-section is about 700×10^{-10} [13,14].

Table 2. $a_\mu^{\rm had} \times 10^{10}$ for a single quark flavor (entries correspond to Table 1). The eighth entry is for quenched domain wall fermions[5]. Values obtained from covariant polynomial fits to $\Pi(q^2)$. The instability in some cases stems from the under-estimation of the magnitude of the slope of $\Pi(q^2)$ as $q^2 \to 0$ (see Figure 7). In all fits $q^2 < 1$ GeV^2.

m_v	m_l/m_s	quadratic	cubic	quartic
0.05	0.01/0.05	252(5)*	310(11)*	252(56)*
0.01	0.01/0.05	433(10)	543(30)	610(84)
0.005	0.01/0.05	542(36)	671(158)	267(411)
0.031	0.01/0.05	249(8)*	383(24)	393(72)
0.0062	0.01/0.05	484(15)*	670(42)	903(117)
0.031	quenched	277(7)	315(18)	349(80)
0.0062	quenched	393(34)*	274(98)*	297(203)*
0.04	quenched	655(43)	772(52)	840(299)

* $\Pi(q^2)$ fit has poor χ^2

4. Summary

Using gluon configurations generated with the a^2-tad lattice action, the hadronic vacuum polarization was calculated for 2+1 flavor QCD. The contribution to the muon's anomalous magnetic moment was then obtained. The results are similar to quenched ones calculated previously, so more work, mostly to reduce the light quark masses, must be done. It was found that the improved Kogut-Susskind fermions still have significant lattice spacing errors. While domain wall fermions appear to exhibit better scaling, they are considerably more expensive than Kogut-Susskind fermions, especially when considering the small quark mass and large volume limits.

It appears that to compete with the accuracy of the usual dispersive method, which is quoted at about the one percent level, several improvements are still needed: (1) better statistics in the low q^2 regime, (2) a more accurate fit method, (3) smaller quark masses so the vector particles are truly unstable and have their physical widths, (4) a smaller lattice spacing if Kogut-Susskind fermions are used.

I am encouraged by on-going MILC Collaboration simulations at $a = 0.086$ fm, $m_l = 0.0031$, $m_s = 0.031$ on $40^3 \times 96$ lattices. This light quark mass is two times smaller than the quark mass used here and approximately 10 times lighter than the strange quark mass. Given the good scaling behavior of domain wall fermions, it would also be interesting to compute $\Pi(q^2)$ on 2 flavor lattices being generated by the RBC collaboration[15].

Acknowledgments

I warmly thank the MILC collaboration for use of their lattices, and in particular Doug Toussaint for his help. The MILC code was used as a basis for these computations.

References

1. Bennett, G. W., *et al.*, Phys. Rev. Lett. **89**, 101804 (2002).
2. For example, see talks by Kinoshita and de Rafael at The Lepton Moments International Symposium, Cape Cod, June 9-12, 2003, http://g2pc1.bu.edu/~leptonmom/.
3. M. Davier, S. Eidelman, A. Hocker and Z. Zhang, Eur. Phys. J. C **27**, 497 (2003)
4. K. Hagiwara, A.D. Martin, Daisuke Nomura, and T. Teubner, Phys. Lett. **B557**, 69 (2003).
5. T. Blum, Phys. Rev. Lett. **91**, 05209 (2003).
6. Paul Rakow, talk at Lattice 2003, Tsukuba, July 15-19, 2003, arXiv:hep-lat/0310027.
7. Gockeler, M., *et al.*, Nucl. Phys. B. Proc. Suppl. 94, 571 (2001).
8. S. Naik, Nucl. Phys. B **316**, 238 (1989).
9. K. Orginos and D. Toussaint [MILC collaboration], Phys. Rev. D **59**, 014501 (1999); K. Orginos, D. Toussaint and R. L. Sugar [MILC Collaboration], Phys. Rev. D **60**, 054503 (1999); G. P. Lepage, Phys. Rev. D **59**, 074502 (1999); For a recent review of a^2-tad results, see S. Gottlieb's talk at Lattice 2003, Tsukuba, July 15-19, 2003, arXiv:hep-lat/0310041.
10. Doug Toussaint, private communication.
11. See the talk by E. de Rafael, Lepton Moments International Symposium, Cape Cod, June 9-12, 2003, http://g2pc1.bu.edu/~leptonmom/.
12. K. G. Chetyrkin, J. H. Kuhn and M. Steinhauser, Nucl. Phys. B **482**, 213 (1996).
13. M. Davier, S. Eidelman, A. Hocker and Z. Zhang, arXiv:hep-ph/0308213.
14. F. Jegerlehner, arXiv:hep-ph/0310234.
15. Chris Dawson, talk at Lattice 2003, Tsukuba, July 15-19, 2003, arXiv:hep-lat/0310055.

Hot and Dense QCD

DECONFINEMENT IN QCD AND IN NUCLEAR COLLISIONS

HELMUT SATZ

Fakultät für Physik, Universität Bielefeld
Postfach 100 131, D-33501 Bielefeld, Germany

Centro de Física das Interacções Fundamentais (CFIF),
Instituto Superior Técnico, Av. Rovisco Pais, P-1049-001 Lisbon, Portugal

After summarizing the study of the states of matter in QCD, I address the efforts to create the predicted quark-gluon plasma with the help of high energy nuclear collisions and survey possible probes to investigate the produced medium. I then consider in some detail the partonic initial state provided by $A - A$ collisions and define the necessary conditions for the formation of a large-scale interacting system. Finally I discuss to what extent some present data reflect geometric critical behavior in the initial state or thermal critical behavior in the final state.

1. States of Matter in QCD

What happens to strongly interacting matter in the limit of high temperature and/or density? This question has fascinated physicists ever since short range strong interactions and the resulting multiple hadron production were discovered. Let us look at some of the features that have emerged.

- Hadrons have an intrinsic size, with a radius $r_h \simeq 1$ fm, and hence a hadron needs a space of volume $V_h \simeq (4\pi/3)r_h^3$ in order to exist. This leads quite naturally to a limiting density n_c of hadronic matter,[1] with $n_c = 1/V_h \simeq 1.5\ n_0$, where $n_0 \simeq 0.17\ \text{fm}^{-3}$ denotes the density of normal nuclear matter.
- Hadronic interactions provide abundant resonance production, and the resulting number $\rho(m)$ of hadron species increases exponentially as function of the resonance mass m, $\rho(m) \sim \exp(bm)$. Such a form for $\rho(m)$ appeared first in the statistical bootstrap model, based on self-similar resonance formation or decay.[2] It was then also obtained in the more dynamical dual resonance approach.[3] In hadron thermodynamics, the exponential increase of the reso-

nance degeneracy results in an upper limit for the temperature of hadronic matter, $T_c = 1/b \simeq 150 - 200$ MeV.[2]

- What happens beyond T_c? In QCD, hadrons are dimensionful color-neutral bound states of more basic pointlike colored quarks and gluons. Hadronic matter, consisting of colorless constituents of hadronic dimensions, can therefore turn into a quark-gluon plasma of pointlike colored quarks and gluons as constituents. This deconfinement transition is the QCD counterpart of the insulator-conductor transition in atomic matter.[4]
- A shift in the effective quark mass is a second transition phenomenon expected from the behavior of atomic matter. At $T = 0$, in vacuum, quarks dress themselves with gluons to form the constituent quarks that make up hadrons. As a result, the bare quark mass $m_q \sim 0$ is replaced by a constituent quark mass $M_q \sim 300$ MeV. In a hot medium, this dressing melts and $M_q \to 0$. Since the QCD Lagrangian for $m_q = 0$ is chirally symmetric, $M_q \neq 0$ implies spontaneous chiral symmetry breaking. The melting $M_q \to 0$ thus corresponds to chiral symmetry restoration.
- A third type of transition would set in if the attractive interaction between quarks leads in the deconfined phase to the formation of colored bosonic diquark pairs, the Cooper pairs of QCD. These diquarks can then condense at low temperature to form a color superconductor. Heating will dissociate the diquark pairs and turn the color superconductor into a normal color conductor.

Using the baryochemical potential μ as a measure for the baryon density of the system, we thus expect the phase diagram of QCD to have the schematic form shown in Fig. 1. Given QCD as the fundamental theory of strong interactions, we can use the QCD Lagrangian as dynamics input to derive the resulting thermodynamics of strongly interacting matter. For vanishing baryochemical potential, $\mu = 0$, this can be evaluated with the help of the lattice regularisation, leading to finite temperature lattice QCD. We summarize briefly the most important features thus obtained.

- The energy density of an ideal gas of massless pions is

$$\epsilon_h = 3\frac{\pi^2}{30}T^4 \simeq T^4, \tag{1}$$

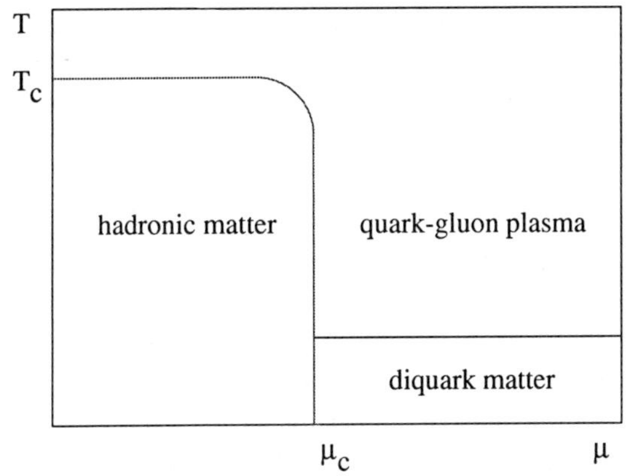

Figure 1. The phase diagram of QCD

while an ideal gas of massless quarks (for $N_f = 2$) and gluons gives

$$\epsilon_q = 37\frac{\pi^2}{30}T^4 \simeq 12\ T^4. \tag{2}$$

Deconfinement thus produces a sudden increase in energy density, corresponding to the latent heat of deconfinement. This behavior is illustrated in Fig. 2, using the results of Ref.[5]. For two light quark species, as well as for two light and one heavy species, the transition temperature becomes $T_c \simeq 175$ MeV, and the energy at this point is $\epsilon(T_c) \simeq 0.6$ GeV/fm^3.

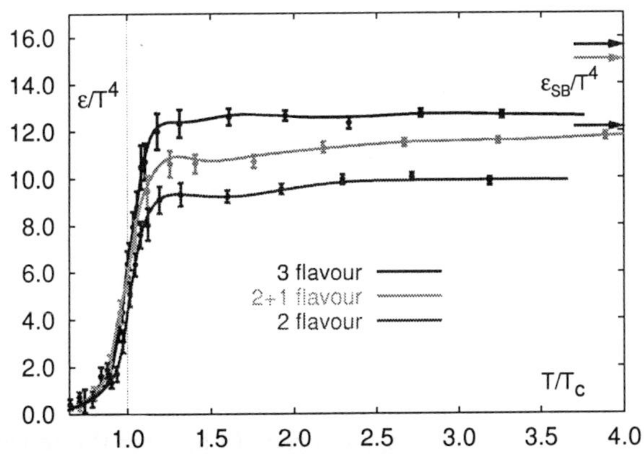

Figure 2. The energy density in QCD with dynamical quarks[5]

- To relate this abrupt change of behavior of the energy density more explicitly to deconfinement and/or chiral symmetry restoration, we turn to the corresponding order parameters.

  – For deconfinement, this is the Polyakov loop expectation value

$$L \sim \exp\{-V_{Q\bar{Q}}/T\} \tag{3}$$

where $V_{Q\bar{Q}}$ denotes the energy of a $Q\bar{Q}$ pair in the limit of infinite separation. In the confinement regime, $V_{Q\bar{Q}}$ diverges and hence $L = 0$; in a deconfined medium, color screening makes the energy finite and hence $L \neq 0$. Thus the change of behavior of L defines the deconfinement temperature T_L.

  – The order parameter for chiral symmetry restoration is given by the chiral condensate $\chi = \langle\psi\bar{\psi}\rangle \sim M_q$, which measures the dynamically generated constituent quark mass M_q. When $\chi \neq 0$, the chiral symmetry of the Lagrangian is spontaneously broken, and when $\chi = 0$, it is restored. Hence here the change of behavior defines the chiral symmetry restoration point T_χ.

  The relation between T_L and T_χ, as obtained in Ref.[6], is shown in Fig. 3. The two transitions clearly occur at the same temperature: at $\mu = 0$, chiral symmetry restoration and deconfinement coincide.

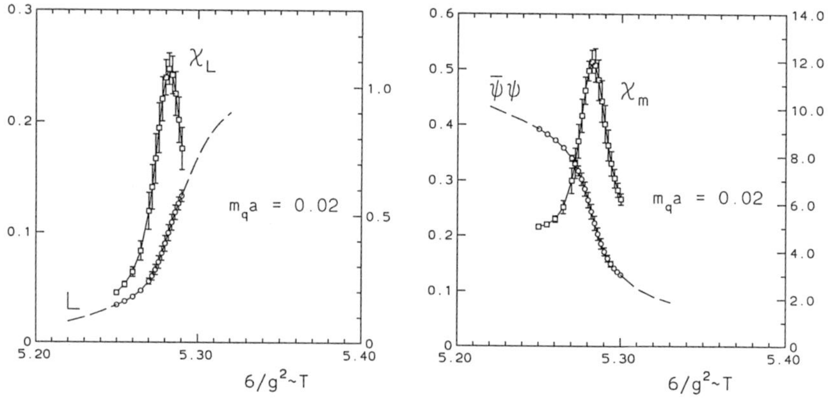

Figure 3. Polyakov loop and chiral condensate in QCD with dynamical quarks[6]

We can thus conclude that QCD predicts for $\mu = 0$ one thermal transition from hadronic matter to a quark-gluon plasma. For $N_f = 2$ or 2+1, it occurs at $T_c \simeq 175$ MeV; at this temperature, chiral symmetry is restored, deconfinement sets in, and the energy density increases quite suddenly by the latent heat of deconfinement. In the next section, we want to see if these predictions can be tested in the laboratory.

2. High Energy Nuclear Collisions

At high energies, nucleus-nucleus collisions produce many nucleon-nucleon interactions in the same space time region. Based on this, it is generally assumed

- that multiple parton collisions lead to the formation of a thermal medium,
- that the initial energy density of this medium is high enough to form a quark-gluon plasma, and
- that there exist observable signatures to test the two previous assumptions.

If thermalization occurs, the Bjorken relation

$$\epsilon_0 \simeq \frac{p_0}{\pi R_A^2 \tau_0} \left(\frac{dN_h^{AA}}{dy} \right) \tag{4}$$

provides an estimate of the initial energy density; here p_0 denotes the average energy of the secondaries, dN/dy their number per unit rapidity, and $\tau_0 \simeq 1$ fm the formation time. Expressing the nuclear collision multiplicity $(dN_h^{AA}/dy) \simeq A^\alpha (dN_h^{pp}/dy)$ in terms of the nucleon-nucleon multiplicity, with $\alpha \simeq 1.1$ from hadron-nucleus collisions, we get for the energy densities at the present and planned facilities the behavior shown in Fig. 4. The range of values shown in each case corresponds to AA collisions from $A = 30$ to $A = 200$. We conclude that the CERN-SPS and all higher energy accelerators should produce sufficient energy densities to form a quark-gluon plasma. What then are the possible probes of the produced medium?

2a. Hard Probes

We are here considering short distance phenomena, whose inherent spatial scale is much smaller than the typical hadronic scale of about 1 fm. These phenomena test the early, hot medium of high density.

- Quarkonium production: J/ψ suppression

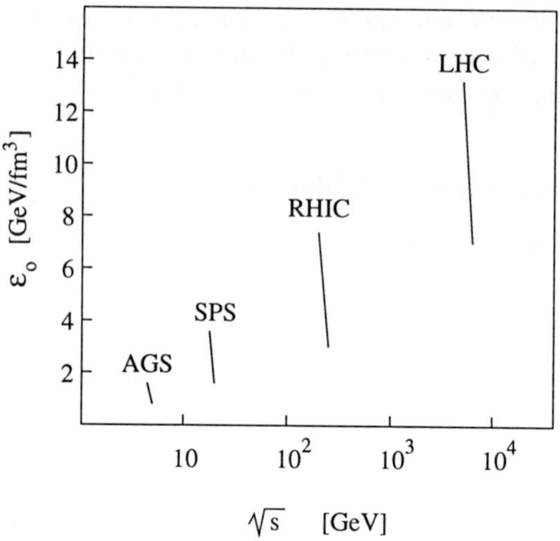

Figure 4. Energy densities for different accelerators

The J/ψ has different survival chances in confined and in deconfined media. In a confined environment, it is basically unaffected, since

$$r_{J/\psi} \ll r_h \sim \Lambda_{\rm QCD}^{-1}. \tag{5}$$

In a hot deconfined medium, color screening dissolves all $c\bar{c}$ binding, and hence hidden charm states eventually become open charm mesons. As a consequence, quark-gluon plasma formation should result in J/ψ suppression.[7] Such behavior is today confirmed by direct lattice calculations.[8]

We note here that different charmonium states (J/ψ, χ, ψ') have different inherent scales, so that they will give rise to different dissociation points. Since J/ψ production in hadronic interactions is in considerable part due to feed-down from the decay of higher excited states, this implies a stepwise J/ψ suppression as function of temperature.

- Hard jets: jet quenching

Hard partons lose energy when passing through a medium; this jet quenching depends on the density of the medium in question. Since deconfinement results in a rather sudden increase in constituent density, it is expected to lead to a sudden increase in jet quenching.[9,10,11]

To study this effect, we have to define a reference transverse momentum distribution, which is generally taken to be given by $p-p$ data scaled to nuclear interactions by assuming binary scaling. It is therefore necessary to specify how large a transverse momentum a hadron has to have to be considered as coming from jet production, *i.e.*, for what p_T the production process is governed by binary scaling.

- Normal effects: $p - A$ data

  Since J/ψ production is also reduced in cold nuclear matter, all studies of J/ψ suppression in nucleus-nucleus collisions must take this normal suppression into account before looking for anomalous effects due to color deconfinement. Similarly, there will be normal jet quenching in cold nuclear matter before deconfinement effects can appear. For both probes, we thus need data from $p-A$ studies, in order to be able to see if there is an onset of additional effects at higher constituent density in sufficiently central $A - A$ collisions.

2b. Soft Probes

These are formed by light hadrons, consisting of u, d, s quarks, in phenomena at low transverse momenta. The relevant scale is thus of hadronic nature, $\sim \Lambda_{\rm QCD}$, and hence such probes test the produced medium in its hadronic stage, for $T \leq T_c$.

- Hadron abundances

  At $T = T_c$, the quark-gluon plasma turns into an interacting hadron gas. If the interactions are resonance dominated, such a system of interacting hadrons can be replaced by an ideal gas of all possible resonance states.

  As a consequence, the relative abundances of all resonance species are specified in terms of only one parameter, the deconfinement temperature. In nuclear collisions at lower energies, the baryon density of the system will play a role, introducing the baryochemical potential μ as a second parameter.

  Furthermore, since $m_s \gg m_u, m_d$, strange baryon abundances may be systematically reduced.

- Transverse momentum spectra

  If $A - A$ collisions produce a hot thermal medium, subsequent collective expansion (flow) will lead to a broadening of the resulting p_T distributions. Higher initial energy densities are expected to produce more flow and hence more p_T broadening. In addition, hydrodynamic studies show that heavier particles exhibit more flow, and for non-central collisions, p_T spectra should show azimuthal anisotropies (elliptic flow).

- Source size

  From correlations between identical hadrons, one can through interferometry (HBT) determine the size of the emitting source. Here again a higher initial density should lead to more expansion before hadronization and hence to a larger source size.

2c. Electromagnetic Probes

Thermal dileptons and photons, produced by contituent interactions in a thermal medium, can escape from this medium without being effected by its subsequent evolution. They thus reflect the stage of the medium at the time of their emission. On the other hand, they can be emitted at any stage of the evolution, by hadronic as well as partonic processes, and are thus not easily connected to a specific stage. If and when this is possible, they provide interesting signals.

- Thermalization

  An unambiguous observation of thermal dileptons and/or photons is probably the only clear test for the production of a thermalized medium.
- Thermometer

  Since production rates are temperature-determined, such as

$$dN/dM_{e^+e^-} \sim \exp\{-M_{e^+e^-}/T\}, \tag{6}$$

  for the production of dileptons of mass $M_{e^+e^-}$, they can in principle determine the temperature of the medium.
- In-medium effects

  The ρ meson decay

$$\rho \to e^+e^- \tag{7}$$

  can occur inside the interacting hadronic medium and thus provide information of in-medium hadron modifications, leading, *e.g.*, to changes of masses or widths.

We thus see that there are indeed a variety of probes to study the different stages of nuclear collisions. Before turning to the status of the present experimental study, we want to consider in more detail the role of the initial state for the formation of the produced medium.

3. Initial State Conditions

The initial state of two colliding nuclei is clearly a non-equilibrium configuration. The canonical view of its evolution is schematically illustrated in Fig. 5. After the collision, there is a short pre-equilibrium stage, in which the primary partons of the colliding nuclei interact, multiply and then thermalize to form a quark-gluon plasma. This QGP then expands, cools and hadronizes. A prerequisite for the equilibration process is evidently that the partons originating from different nucleons form a large-scale interconnected system. If there is no "cross talk" between partons from different nucleons, thermalization is not possible.

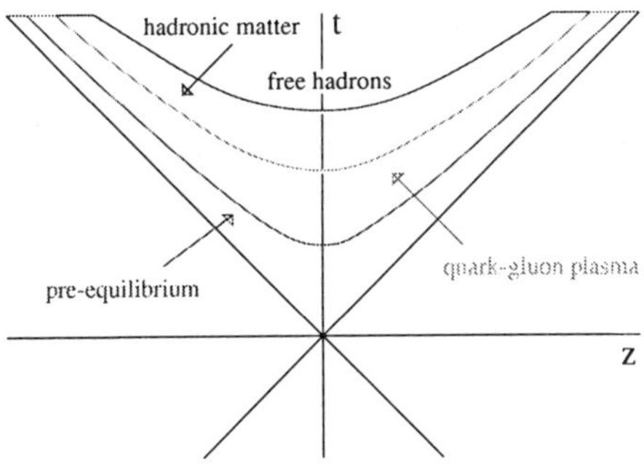

Figure 5. The expected evolution of a nuclear collision

The problem of color connection has been studied in hadron production through W^+W^- decay at LEP. The W's are produced essentially at rest in the annihilation of an energetic e^+e^- pair (see Fig. 6), and it is possible to compare the reaction in which both W's produce hadronic jets to that in which one decays leptonically. If there is cross talk between the decay quarks of one W with those from the other, the multiplicity of the four-jet decay is predicted to be less than twice that in the two-jet decay.[12] The data show no such reduction, suggesting that the decay quarks from different W's don't communicate.[13]

It is therefore necessary to determine under what conditions the initial state parton configurations can lead to color connection, and more specifically, if variations of the initial state can lead to a transition from discon-

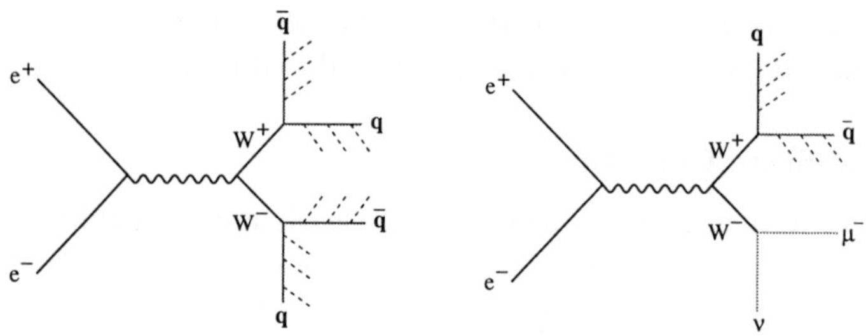

Figure 6. Four-jet and two-jet decays of W^+W^- pairs in e^+e^- annihilation at LEP

nected to connected partonic clusters. The results of such a study of the pre-equilibrium state in nuclear collisions do not depend on the subsequent evolution and thus in particular not require any kind of thermalization.

The structural problem underlying the transition from disconnected to connected systems of many components is a very general one, ranging from clustering in spin systems to the formation of galaxies. The formalism is provided by percolation theory, which describes geometric critical behavior.[14] We shall return to the basic idea very shortly.

Consider now the collision of two heavy nuclei at high energy, as seen in the overall center of mass. The Lorentz-contraction in the longitudinal direction makes it a collision of two thin discs, so that in the transverse plane, the parton density increases with A. The partons from different nucleon begin to overlap and form clusters: see Fig. 7. How does the cluster size grow with parton density, and when does it reach the dimension of the total transverse collision area? These are precisely the questions addressed by percolation theory, so that here we make a small interlude.

Consider placing N small circular discs ('partons') of radius r onto a large circular manifold ('the transverse nuclear plane') of radius $R \gg r$; the small discs may overlap. With increasing parton density $n \equiv N/\pi R^2$, this overlap will lead to more and larger connected partonic clusters. The striking feature of this phenomenon is that the average cluster size $S_{\rm cl}$ does not grow as some power of n; instead, it increases very suddenly from very small to very large values (see Fig. 8).

This suggests some kind of geometric critical behavior, and in fact in the 'thermodynamic limit' $N \to \infty$, $R \to \infty$, the cluster size diverges at a

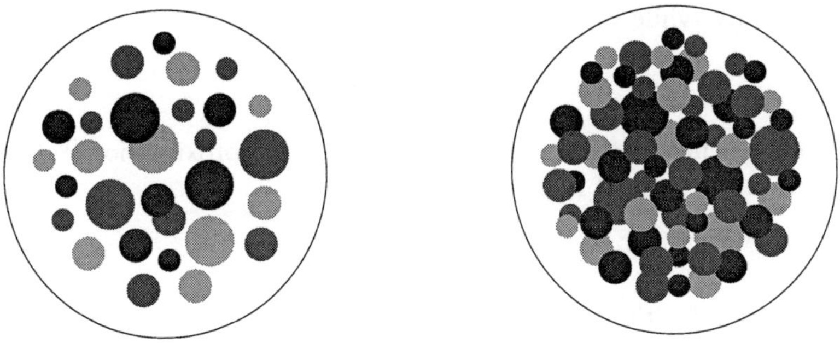

Figure 7. Partonic clusters in the transverse collision plane at low and high density

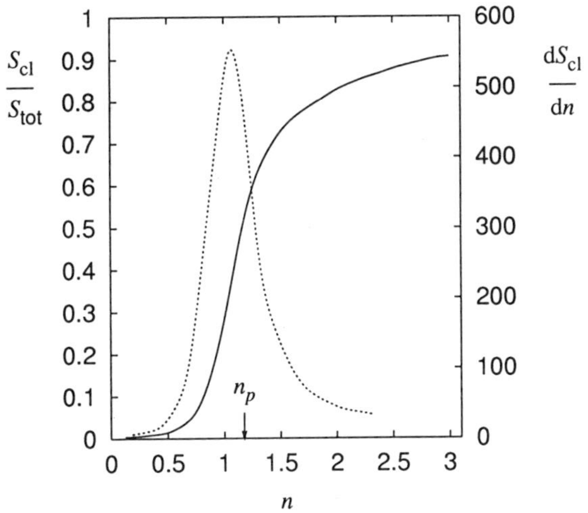

Figure 8. Fractional cluster size and its derivative as function of the parton density n

critical threshold value n_c of the density n,

$$S_{\text{cl}} \sim (n_c - n)^{-\gamma}, \tag{8}$$

as $n \to n_c$ from below. This appearence of infinite clusters at $n = n_c$ is defined as percolation: the size of the cluster reaches the size of the system. The divergence is governed by the critical exponent $\gamma = 43/18$, determined analytically, while the threshold

$$n_c = \frac{1.13}{\pi r^2} \tag{9}$$

is obtained numerically or through analytical approximation.[15] Hence we obtain in the limit of large systems

$$\frac{N}{\pi R^2} = \frac{1.13}{\pi r^2} \tag{10}$$

as condition for the onset of percolation.

After this brief interlude, we now return to nuclear collisions. The results just discussed tell us that in such an environment there is indeed, as function of parton density, a sudden onset of large-scale color connection. There is a critical density at which the partons form one large cluster, losing their independent existence and their relation to the parent nucleons. Parton percolation[16,17] is thus the onset of color deconfinement and although it is a necessary prerequisite for any subsequent QGP formation, it does not require or imply any kind of parton thermalization.

To obtain quantitative predictions, we have to specify the relevant numbers and scales. The parton content of a nucleon is determined in deep inelastic scattering experiments as function of the fractional parton momentum $x = k/p$, with k and p for parton and nucleon momenta, respectively, and of the resolution scale Q. The number of partons in a nucleon at rapidity y, as seen by a photon of scale Q, is thus given by

$$\frac{dN}{dy} = x\ \{g(x,Q) + \sum_i [q_i(x,Q) + \bar{q}_i(x,Q)]\}, \tag{11}$$

where g counts s gluons, $q/\bar{q}$ quarks and antiquarks. In a minimum bias nucleon-nucleon collision, the transverse parton size itself determines the resolution: it sets the scale at which partons 'probe each other' in the colliding nucleons. The parton size is through the uncertainty relation determined by its average transverse momentum,

$$\pi r^2 \simeq \pi / < k_T^2 >, \tag{12}$$

so that here the highest relevant k_T fixes Q. Since at $y = 0$, the fractional momentum is $x = k_T/\sqrt{s}$, Eq. (11) provides at given $\sqrt{s}$ the total number of partons of transverse momenta up to Q.

Given the parton density in a nucleon, we now have to specify the density in a nucleus-nucleus collision. Assuming the number of initial partons to be determined by the number of participant nucleons, we have in a central $A - A$ collision

$$\left(\frac{dN}{dy}\right)^{AA}_{y=0} \simeq 2\,A\left(\frac{dN}{dy}\right)_{y=0}. \tag{13}$$

These partons are distributed over the transverse nuclear size πR_A^2, with the actual distribution determined by the nuclear profile.

Combining these results, we obtain

$$\frac{2A}{\pi A^{2/3}}\left(\frac{dN}{dy}\right)_{y=0} = \frac{1.13}{\pi Q^{-2}} \tag{14}$$

for the parton percolation condition in central $A-A$ collisions, in terms of A, the resolution scale $Q_s \simeq k_T$ and the nucleonic parton density obtained in deep inelastic scattering. Fixing the incident c.m.s. energy to SPS ($\sqrt{s} \simeq$ 20 GeV) and RHIC (($\sqrt{s} \simeq$ 200 GeV) values, we find[17,18] the following thresholds

SPS ($\sqrt{s} \simeq 20$ GeV) $\quad A \geq 60,\ Q_s \leq 0.7$ GeV,

RHIC ($\sqrt{s} \simeq 200$ GeV) $\quad A \geq 40,\ Q_s \leq 1.0$ GeV.

In the case of non-central $A - A$ collisions, the transverse nuclear size πR_A^2 has to be replaced by the actual transverse overlap area at the given impact parameter. This overlap area can be determined in a Glauber study, using Woods-Saxon nuclear profiles,[19] and the resulting counterpart of eq. (14) then leads for $A - A$ collisions with $A \simeq 200$ to the thresholds

SPS ($\sqrt{s} \simeq 20$ GeV) $\quad N_{part} \geq 125,\ Q_s \leq 0.7$ GeV,

RHIC ($\sqrt{s} \simeq 200$ GeV) $\quad N_{part} \geq 100,\ Q_s \leq 1.0$ GeV,

with the centrality measured in terms of the number of participating nucleons.

Beyond the percolation point, we have a parton condensate, containing interacting and hence color-connected partons of all scales $k_T \leq Q_s$. The percolation point thus specifies the onset of color deconfinement; it says nothing about any subsequent thermalization. If there is eventual thermalization, the partonic momentum k_T will be related to the temperature T; hence the resolution scale Q_s, which determines the range of k_T, is in some sense a precursor of T. Since Q_s is found to increase with $\sqrt{s}$, A and centrality, increasing any of these leads in this pre-thermal sense to a 'hotter' parton condensate.

Summarizing this section, we note that the parton structure of nucleons leads to geometric critical behavior in nuclear collisions, as determined by parton percolation, and independent of any subsequent thermalization. The resulting parton condensate is closely related to the color glass condensate studied in similar initial state considerations.[20]

4. Questions and Answers

We thus find a possible two stage structure, with two distinct transitions, for the evolution of nuclear collisions.

- The initial state parton structure leads to geometric critical behavior, resulting in the formation of a parton condensate.
- This parton condensate can then thermalize, producing a quark-gluon plasma, which in a thermal transition will then eventually hadronize.

As a result, we have to address a number of crucial questions.

- How can we probe and distinguish geometric and thermal critical behavior?
- What observable features follow already from parton percolation and the presence of a parton condensate, and which need quark-gluon plasma formation?
- Do present data provide any evidence for or against thermalization?

We shall give two illustrations; clearly, much further study will be needed.

4.1. *J/ψ Suppression*

- We expect a J/ψ to be dissolved in a sufficiently hot quark-gluon plasma, while it will survive in hadronic matter.
- Since J/ψ, χ_c and ψ' have different dissociation thresholds, feed-down J/ψ production will result in a step-wise suppression pattern.
- Such a pattern is observed by the NA50 collaboration at CERN,[21] as shown in Fig. 9; normal nuclear absorption has been taken into account. What can we conclude from this pattern?

From thermal studies, one expects that in a quark-gluon plasma the higher excited states are dissociated essentially at T_c, directly produced J/ψ's at about 1.2 T_c or higher. This would lead to suppression onsets at

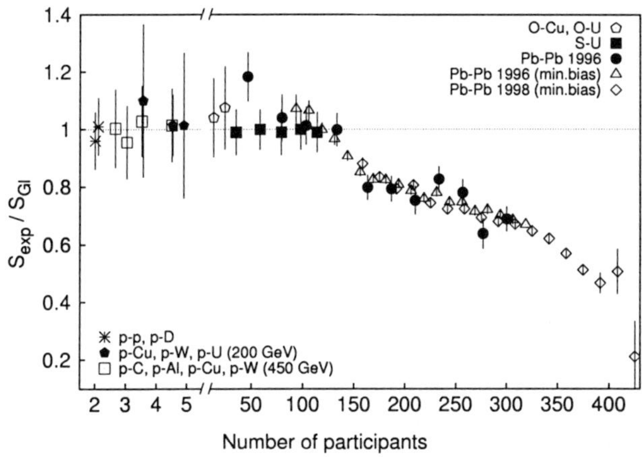

Figure 9. Centrality dependence of J/ψ survival in SPS data[21]

about 0.6 GeV/fm^3 and at 4.1 GeV/fm^3. The experimental onsets found by NA50 in $Pb - Pb$ collisions lead to 2.3 and 3.2 GeV/fm^3, *i.e.*, to rather different points.

In percolation, the higher excited states are expected to disappear at the first onset of percolation; direct J/ψ's, with a smaller scale, need a finer resolution and hence a more dense parton condensate. The resulting steps here are found to be[18] at about $N_{part} \simeq 125$ and $N_{part} \simeq 300$, which is seen to agree quite well with the pattern of Fig. 9.

The observed pattern of J/ψ suppression may thus well be due to initial state parton percolation. Further data for different A at the SPS (NA60) and for higher energy at RHIC should clarify this more.

4.2. *Expansion Features*

In the usual picture of the collision evolution, higher initial energy $\sqrt{s}$ produce a hotter initial medium and hence should results in more expansion. More expansion in turn will lead to

– larger source radii,

– more transverse flow,

– more elliptic flow.

Present data show:

- The source radii appear to be independent of $\sqrt{s}$ and are roughly of size R_A, as seen in Fig. 10 for central $A - A$ collisions with

$A \simeq 200$. It seems difficult to reconcile this with the formation of an expanding hot thermal medium.

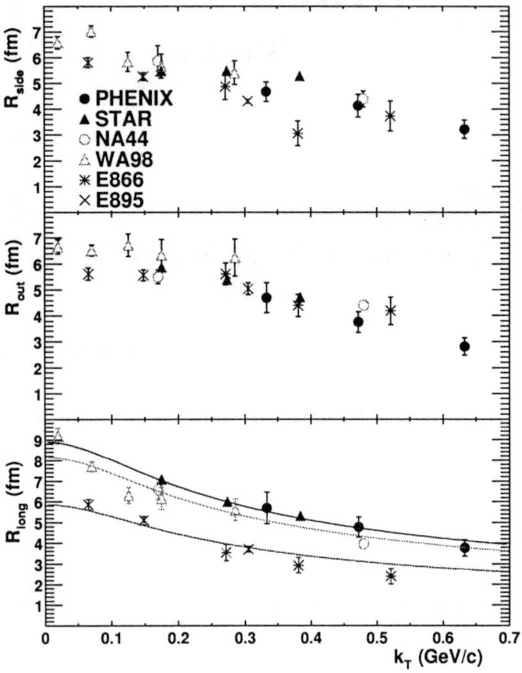

Figure 10. Source radii from HBT studies at different energies[22]

- The transverse flow does increase with energy below the SPS range, but this increase becomes much weaker with increasing energy. A saturation as function of energy is not excluded and would be difficult to accommodate for an expanding thermal medium. More studies of the behavior at very high energies will thus be very important. The same holds for elliptic flow.

4.3. *Conclusions*

At present we can thus only recall the main features here discussed.

- Strongly interacting matter in equilibrium, as described by statistical QCD, should show thermal deconfinement with quark-gluon plasma formation and a subsequent thermal hadronization transition.

- The parton structure of the initial state of nucleus-nucleus collsions leads to geometric deconfinement, described by percolation theory, and the formation of a parton condensate.
- The parton condensate formed in nuclear collisions can subsequently thermalize to produce a deconfined thermal medium. The future will tell if it really does.

References

1. I. Ya. Pomeranchuk, *Doklady Akad. Nauk. SSSR* **78**, 889 (1951).
2. R. Hagedorn, *Nuovo Cim. Suppl.* **3**, 147 (1965);
   Nuovo Cim. **56A**, 1027 (1968).
3. K. Bardakci and S. Mandelstam, *Phys. Rev.* **184**, 1640 (1969);
   S. Fubini and G. Veneziano, *Nuovo Cim.* **64A**, 811 (1969).
4. H. Satz, *Fortsch. Physik* **33**, 259 (1985).
5. F. Karsch, E. Laermann and A. Peikert, *Phys. Lett.* **B478**, 447 (2000).
6. F. Karsch and E. Laermann, *Phys. Rev.* **D50**, 6954 (1994).
7. T. Matsui and H. Satz, *Phys. Lett.* **178B**, 416 (1986).
8. M. Asakawa and T. Hatsuda, hep-lat/0308034;
   P. Petreczky *et al.*, hep-lat/0309012.
9. J. D. Bjorken, Fermilab-Pub-82/59-THY (1982) and Erratum, unpublished.
10. M. Gyulassy and X.-N. Wang, *Nucl. Phys.* **B420**, 583 (1994);
    M. Gyulassy, M. Plümer and X.-N. Wang, *Phys. Rev.* **D51**, 3436 (1995).
11. R. Baier *et al.*, *Phys. Lett.* **B345**, 277 (1995); *Nucl. Phys.* **B483**, 291 (1997); *Nucl. Phys.* **B484**, 265 (1997); *Nucl. Phys.* **B531**, 403 (1998).
12. J. R. Ellis and K. Geiger, *Phys. Lett.* **B404**, 230 (1997).
13. For a recent survey, see P. Abreu, hep-ph/0111395.
14. D. Stauffer and A. Aharony, *Introduction to Percolation Theory*, (Taylor and Francis, London, 1994).
15. See *e.g.*, M. S. Isichenko, *Rev. Mod. Phys.* **64**, 961 (1992).
16. M. Nardi and H. Satz, *Phys. Lett.* **B442**, 14 (1998);
    H. Satz, *Nucl. Phys.* **A661**, 104c (1999).
17. S. Digal *et al.*, *Phys. Lett.* **B549**, 101 (2002).
18. S. Digal *et al.*, hep-ph/0310354, *Europ. Phys. J.* **C**, in press.
19. D. Kharzeev *et al.*, *Z. Phys.* **C74**, 307 (1997).
20. L. McLerran and R. Venugopalan, *Phys. Rev.* **D49**, 2233 and 3352 (1994);
    L. McLerran, *Lect. Notes Phys.* **583**, 291 (2002).
21. M.C. Abreu *et al.* (NA50), *Phys. Lett.* **B410**, 337 (1997);
    M.C. Abreu *et al.* (NA50), *Phys. Lett.* **B450**, 456 (1999);
    M.C. Abreu *et al.* (NA50), *Phys. Lett.* **B477**, 28 (2000).
22. Compiled in K. Adcox *et al.* (PHENIX), *Phys. Rev. Lett.* **88**, 192302 (2002).

CONFRONTING LATTICE RESULTS ON THE TRANSITION FROM THE HADRON GAS TO THE QUARK-GLUON PLASMA WITH HADRONIC RESONANCE GAS MODELS*

FRITHJOF KARSCH

Fakultät für Physik,
Universität Bielefeld,
D-33615 Bielefeld,
Germany
E-mail: karsch@physik.uni-bielefeld.de

We discuss the role of resonances for the occurrence of the transition to the quark-gluon plasma in hot and dense matter. Properties of a hadronic resonance gas are compared to lattice results on the equation of state at zero as well as non-zero baryon chemical potential. Furthermore, it is shown that the quark mass dependence of the transition temperature can be understood in terms of lines of constant energy density in a resonance gas.

1. Introduction

Today properties of strongly interacting matter at high temperature and/or density are analyzed within the field theoretic framework given by the theory of strong interactions – Quantum Chromo Dynamics (QCD). Two aspects of this theory, which describes the interaction among their elementary constituents, quarks and gluons, are of central importance for our understanding of the different phases of QCD at non-zero temperature and density as well as for our understanding of the experimentally observed spectrum of hadronic bound states – confinement and chiral symmetry breaking. The $SU_R(n_f) \times SU_L(n_f)$ chiral symmetry, which is spontaneously broken in QCD with n_f massless quark flavors, as well as the $Z(N_c)$ center symmetry, realized in the gauge field sector of N_c-color QCD in the limit of infinitely heavy quarks, put strong constraints on qualitative aspects of the

*This work has partly been supported by the DFG under grants FOR 339/2-1, KA 1198/9-1 and the BMBF under grant no. 06BI102.

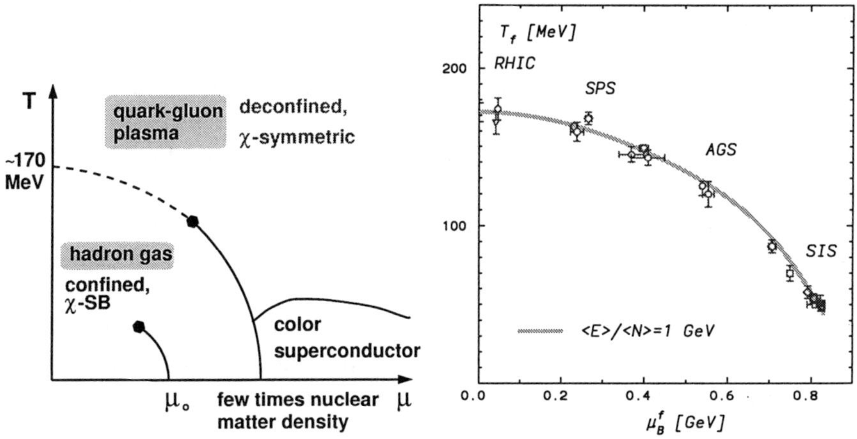

Figure 1. Sketch of the QCD phase diagram in the T-μ_B plane (left) and the freeze-out curve (right) determined from a comparison of experimentally observed particle abundances to the abundances of hadrons in a hadronic resonance gas at temperature T and a baryon chemical potential μ_B.

QCD phase diagram[1,2]. In fact, quite general considerations concerning universal properties of the QCD phase transition rely on these symmetries. Many of these predictions concerning the order of the QCD transition and concerning universal properties in the vicinity of second order phase transitions in QCD have been verified over the last 20 years in numerical studies of lattice regularized QCD[3].

In our attempts to reach a deeper understanding of the physics behind the occurrence of the QCD phase transition, mechanisms like the dual Higgs mechanism, monopole condensation or vortex percolation have been identified[4], which characterize the drastic modifications of the QCD vacuum that occur at the critical temperature. However, none of these considerations provides insight into the question which properties of QCD set the scale and control the quantitative aspects of the transition from hadronic matter at low temperature and density to the quark gluon plasma (QGP) at high temperature and/or density.

Long before lattice calculations provided first evidence for critical behavior in strongly interacting matter[5] the critical behavior in hadronic matter has been discussed in the framework of resonance gas models[6]. It has been noticed that ordinary hadronic matter cannot persist at arbitrary high temperatures and densities[6]; the copious production of resonances will lead to a natural end of the temperature and density regime in which hadrons can exist. It also has been suggested that the properties of dense hadronic matter and the role of resonances could be studied experimentally in heavy ion collisions[7]. Indeed, in such experiments it has been found that the abundances of various particle species are well described by a hadronic resonance gas model[8] which is characterized by two equilibrium parameters, temperature, T, and baryon chemical potential, μ_B. Both depend on the center of mass energy in these collisions; with increasing energy the temperature of the resonance gas increases and the relevant baryon chemical potential decreases. At RHIC the baryon chemical potential is quite small ($\mu_B \simeq 29$ MeV) while the temperature of the resonance gas reaches 177 MeV[9]. This is in good agreement with the transition temperature to the quark-gluon plasma phase, $T_c = (173 \pm 8 \pm sys)$ MeV, found in lattice calculations[10] of 2-flavor QCD at vanishing baryon chemical potential μ_B. Recent exploratory lattice calculations also show that the transition temperature drops slowly with increasing baryon chemical potential[11,12,13] and that the smooth transition at small μ_B turns into a second order transition at a critcal point[11] $(T_c, \mu_B^c) \simeq (160\text{ MeV}, 725\text{ MeV})$. The anticipated QCD phase diagram and the freeze-out curve determined from the particle abundances observed in various heavy ion experiments are shown in Fiq. 1.

In this paper we want to discuss evidence provided by lattice calculations for the contribution of hadron resonances to the thermodynamics of QCD. We will analyze the equation of state and the quark mass dependence of the QCD transition temperature. Furthermore, we will address the question to what extend the freeze out temperature observed in heavy ion experiments is related to the phase boundary for the transition to the QGP calculated in lattice QCD.

2. Lattice results on T_c and the equation of state

2.1. $\mu_B = 0$

Studies of the quark mass dependence of the transition to the high temperature phase of QCD show that the transition temperature, T_c, decreases gradually with decreasing quark mass[a]. Over a wide range of pion mass values, 300 MeV $\lesssim m_{PS} \lesssim 2$ GeV, the transition temperature depends

[a]Quark masses used in lattice calculations are bare parameters of the QCD Lagrangian which need to be renormalized. In order to avoid any discussion of the renormalization of quark masses it is more appropriate to discuss the quark mass dependence of ther-

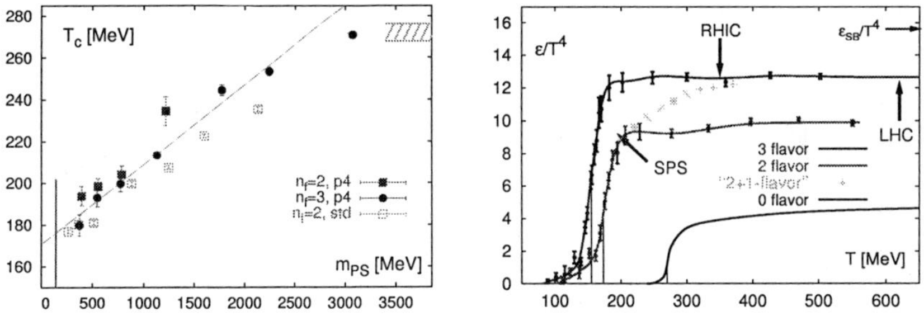

Figure 2. The transition temperature (left) as a function of the lightest pseudo-scalar meson mass (m_{PS}) in 2 and 3 flavor QCD and the energy density (right) in the SU(3) gauge theory ($n_f = 0$), as well as 2 and 3 flavor QCD for quark masses corresponding to $m_{PS} \simeq 770$ MeV.

linearly on m_{PS} and the slope seems to be more or less independent of the number of flavors (Fig. 2(left)). In Ref. 10 the mass dependence of T_c has been parametrized as,

$$T_c(m_{PS}) = T_c(0) + 0.04(1)m_{PS} \quad . \tag{1}$$

For smaller pion masses one may expect that chiral symmetry leads to modifications of this linear relation[b]; for $m_{PS} \gtrsim 2$ GeV the transition temperature will approach a constant as all meson and baryon masses become larger than the (almost) quark mass independent glueball-masses and thus will decouple from the thermodynamics.

The change in transition temperature between the light quark mass and infinite quark mass regime goes along with a change in ϵ_c/T_c^4, the critical energy density expressed in units of the transition temperature, by more than an order of magnitude (Fig. 2(right)), *i.e.* $\epsilon_c/T_c^4 \simeq (6 \pm 2)$ for $m_{PS} \simeq 770$ MeV and $\epsilon_c/T_c^4 \simeq (0.5 - 1)$ for $m_{PS} \equiv \infty$. This large change reflects the large difference in the number of degrees of freedom which control the high temperature ideal gas limit. However, it does not at all suggest that the critical energy density itself changes significantly. In fact, when taking into account the shift in T_c for both cases it seems that

modynamic observables in terms of a well controlled physical parameter. E.g. we will use here the lightest pseudo-scalar meson (pion) mass, m_{PS}. To express this in physical units (MeV) we use zero temperature lattice calculations of the string tension and set the scale by using $\sqrt{\sigma} = 420$ MeV.

[b]In the case of a second order transition in 2-flavor QCD one expects to find for small values of the pseudo-scalar mass, $T_c \sim m_{PS}^{2/\beta\delta}$, where $2/\beta\delta \simeq 1.1$ is a combination of critical exponents of the 3-d, $O(4)$-model.

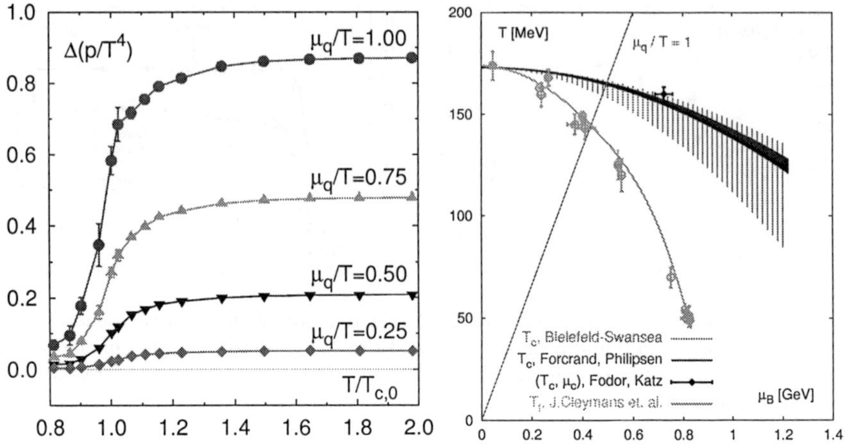

Figure 3. The change in pressure due to a non-zero quark chemical potential, $\mu_q = \mu_B/3$ calculated in 2-flavor QCD using a Taylor expansion to order $(\mu_q/T)^4$. As in the $\mu_B = 0$ case the simulations have been performed with quark masses corresponding to $m_{PS} \simeq 770$ MeV.

the critical energy density itself does not change much. Although statistical errors are still large both cases are consistent with a critical energy density in the range $\epsilon_c \simeq (0.5 - 1.0)$ GeV/fm^3.

2.2. $\mu_B > 0$

Lattice simulations at non-zero baryon chemical potential[c] generally suffer from the problem that the fermion contribution to the QCD partition function is no longer represented by a positive definite quantity; the fermion determinant becomes complex. This excludes the application of standard Monte Carlo simulation techniques. This problem has, however, been avoided in recent studies of the dependence of the transition temperature on the chemical potential[11,12,13] as well as calculations of the equation of state[14,15] at non-zero baryon chemical potential by using extrapolation techniques applied to numerical results obtained at $\mu_B = 0$.

In Fig. 3(left) we show results on the change in pressure,

$$\frac{\Delta p(T,\mu_q)}{T^4} = \frac{p(T,\mu_q) - p(T,0)}{T^4} \quad , \tag{2}$$

[c]We use here the notion of baryon chemical potential μ_B although lattice calculations are performed in terms of the quark chemical potential $\mu_q \equiv \mu_B/3$.

obtained from a Taylor expansion[15] of the pressure to order $(\mu_q/T)^4$ around $\mu_q = 0$. Similar results have been obtained using a reweighting technique[14]. A comparison with the pressure calculated at $\mu_q = 0$ shows that at $\mu_q/T = 1$ and for $T \gtrsim T_c$ the enforced presence of a non-zero baryon number adds about 30% to the overall pressure in the system.

The reweighting approach[11,12] as well as the analytic continuation of results obtained in simulations with an imaginary chemical potential[13] have also been used to determine the μ_B-dependence of the transition temperature. These results are shown in Fig. 3(right). We note that for small values of the chemical potential the lines shown in this figure do not correspond to a phase transition but rather characterize a rapid but smooth crossover from the hadronic to the plasma phase. Also shown in the figure is an estimate of the chiral critical point[11], *i.e.* a second order phase transition point, $(T_c, \mu_B^c) \simeq (160\ \text{MeV}, 725\ \text{MeV})$, in the QCD phase diagram. For $\mu_B > \mu_B^c$ the transition is expected to become a first order phase transition.

Although the different methods used to determine $T_c(\mu_B)$ do seem to give compatible results, a detailed quantitative comparison is difficult as all calculations performed so far have used different discretization schemes and/or quark mass values. Moreover, the estimates have partly been obtained on rather small lattices with unimproved gauge and fermion actions and/or too large quark masses. The current quantitative results thus need to be improved and confirmed in future calculations. A first analysis of the quark mass dependence of the transition line for 3-flavor QCD[17] seems to indicate that the μ_B-dependence becomes stronger with decreasing quark mass and that the chiral critical point shifts to smaller values of the baryon chemical potential.

3. Thermodynamics of the hadronic resonance gas

3.1. $\mu_B = 0$

In Hagedorn's approach to the thermodynamics of strongly interacting matter[6] critical behavior arises because any increase in energy of an ensemble of strongly interacting hadrons is predominantly used to generate new resonances rather than transforming it into kinetic energy of the constituents and thus using it for heating up the system. The exponentially rising spectrum of resonances, $\rho(m) \sim \exp(bm)$, leads to the occurrence of a critical temperature, $T_{c,\text{res}} = 1/b$.

A similar mechanism is known to lead to critical behavior in purely gluonic systems, *i.e.* in the $SU(N_c)$ gauge theories. Here the fluctuations

Table 1. Lattice results on the deconfinement temperature of $SU(N_c)$ gauge theories in d space-time dimensions.

d	N_c	$T_c/\sqrt{\sigma}$	d	N_c	$T_c/\sqrt{\sigma}$
3	2	1.08(1)	4	2	0.69(2)
	3	0.97(1)		3	0.632(2)
string	model:	0.977	string	model:	0.691

of color flux tubes (string) lead to an exponentially rising excitation spectrum which again leads to critical behavior. Calculations of the critical temperature, $T_{c,\text{string}}$, within the Nambu-Goto model yield[16],

$$\frac{T_{c,\text{string}}}{\sqrt{\sigma}} = \sqrt{\frac{3}{(d-2)\pi}} \quad , \tag{3}$$

which only depends on the space-time dimension, d, and, in particular, is independent of the color degrees of freedom. Lattice calculations of the phase transition temperature of $SU(N_c)$ gauge theories in 3 and 4 space-time dimensions, indeed, yield critical temperatures, which are in good agreement with the string model predictions. Some results from lattice calculations[18] are summarized in Table 1 and compared to the string model prediction. This clearly suggests that resonances play an essential role in triggering the occurrence of the deconfinement transition, although the order of the transition and, in those cases where the transition is second order, also universal properties of the transition are controlled by the global $Z(N_c)$ center symmetry of the $SU(N_c)$ gauge theories. It thus is interesting to explore what role the (exponentially) rising hadronic resonance spectrum plays for the occurrence of the transition to the plasma phase in the physical, light quark mass regime.

A first hint at the importance of resonances for the occurrence of the QCD transition may be obtained from the energy density at T_c. In the limit of n_f massless quark flavors the sector of massless Goldstone bosons would contribute to the energy density with $\epsilon/T_c^4 = (n_f^2-1)\pi^2/30$, if this contribution can be approximated by an ideal gas of non-interacting bosons. The calculations of the energy density shown in Fig. 2(right) have been performed with quark masses which correspond to a pion mass of about 770 MeV. Their contribution to the energy density thus is exponentially suppressed. A free gas of massive relativistic particles would contribute

$$\frac{\epsilon^1(m)}{T^4} = \frac{g}{2\pi^2}\sum_{k=1}^{\infty}(-\eta)^{k+1}\,\frac{1}{k}\left(\frac{m}{T}\right)^3\left[\frac{3T\,K_2(km/T)}{km} + K_1(km/T)\right] \tag{4}$$

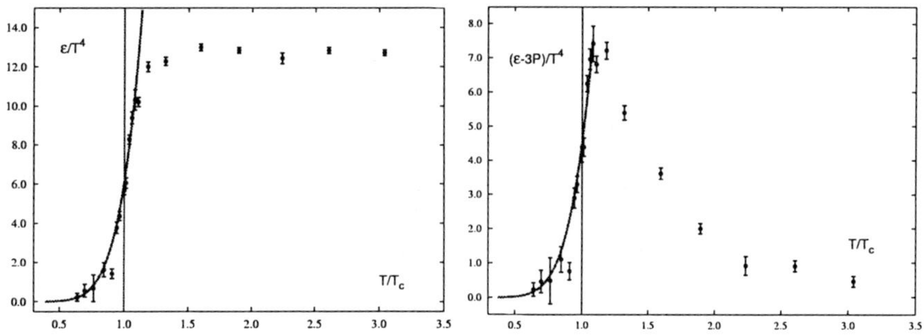

Figure 4. The energy density in (2+1)-flavor QCD (left) and the deviation from ideal gas behavior, $(\epsilon - 3p)/T^4$, (right) as a function of T/T_c. The solid line gives the result obtained from a resonance gas with hadron masses adjusted to compare to the lattice calculations which have been performed with too heavy up and down quarks.

to the energy density. Here $\eta = -1$ for bosons, $+1$ for fermions and g is the degeneracy factor of the particle state. For $m/T_c \simeq 4$ this yields $\epsilon/T_c^4 \simeq 0.083(n_f^2 - 1)$ which shows that quite a few hadronic degrees of freedom are needed to saturate the value of the energy density found in lattice calculations at T_c.

In a gas of non-interacting resonances energy density and pressure are given as sum over the single particle contributions, e.g. $\epsilon = \sum_i \epsilon^1(m_i)$, where i labels the masses, m_i, of experimentally known mesons and baryons. In order to compare the resonance gas model to lattice results we, however, have to take into account that the latter are not obtained from calculations performed with the physically realized light quark mass spectrum. A corresponding analysis has been performed in Refs. 19. The resulting comparison between a modified resonance gas model and lattice data for (2+1)-flavor QCD[10] is shown in Fig. 4. Similar agreement has been obtained for 2-flavor QCD by suppressing the contribution of "strange" hadrons. This shows that resonance can account for the rapid rise of the energy density observed in lattice calculations for $T \lesssim T_c$.

3.2. $\mu_B > 0$

The resonance gas model does make quite stringent predictions for the thermodynamics at non-vanishing chemical potential. As all baryons in QCD are heavy compared to the temperature regime of interest, *i.e.* $m_{\text{baryon}}/T \gtrsim 5$ for $T \leq T_c$, the contribution of baryons to the thermodynamics can be handled in the Boltzmann approximation. This leads to a factorization of the temperature and fugacity, $\exp(\mu_B/T)$, dependence of thermodynamic observables. The pressure of a gas of baryons and their resonances thus can be written as

$$\frac{p_B(T,\mu_B)}{T^4} = F_B(T)\cosh(\mu_B/T) \quad , \tag{5}$$

where $F_B(T)$ is given by a sum over all baryons and their resonances,

$$F_B(T) \equiv \sum_i \frac{g_i}{\pi^2}\left(\frac{m_i}{T}\right)^2 K_2(m_i/T) \quad . \tag{6}$$

As the mesonic part of a gas of non-interacting resonances does not depend on the baryon chemical potential the change in pressure due to a non-vanishing chemical potential is entirely determined by the baryonic sector, $\Delta p = p(T,\mu_B) - p(T,0) \equiv p_B(T,\mu_B) - p_B(T,0)$. This, of course, also holds for derivatives with respect to μ_B at fixed temperature. The factorization of the T and (μ_B/T)-dependent terms then leads to simple relations among various thermodynamic observables. For instance, one finds for the baryonic susceptibility,

$$\begin{aligned}\frac{\chi_B}{T^2} &= \left(\frac{\partial^2}{\partial(\mu_B/T)^2}\frac{p(T,\mu_B)}{T^4}\right)_{T\ fixed} \\ &= \frac{\Delta p}{T^4}\left(1-\cosh^{-1}(\mu_B/T)\right)^{-1} \quad .\end{aligned} \tag{7}$$

This relation is particularly interesting as it suggests that the ratio $\Delta p/T^2\chi_q \equiv \Delta p/9T^2\chi_B$ does not depend on details of the hadron mass spectrum and thus can directly be compared to lattice calculations performed with unphysically large quark masses. Such a comparison is shown in Fig. 5(left). The agreement between the lattice results and the resonance gas model relation is quite striking, although it has to be noted that the current lattice analysis has been performed using a Taylor series expansion to order μ_B^4 only. This truncation also eliminates large possibly fluctuations in the susceptibility in the vicinity of a phase transition, which can not be reproduced in the resonance gas model approach.

The resonance gas model also yields a reasonably good description of the temperature dependence of thermodynamic quantities at fixed fugacity. This, of course, involves information on the explicit form of the resonance spectrum. Taking again into account that lattice simulations so far have been performed with unphysical quark mass values one can directly compare the quark number susceptibilities calculated in the resonance gas model and on the lattice. This is shown in Fig. 5(right).

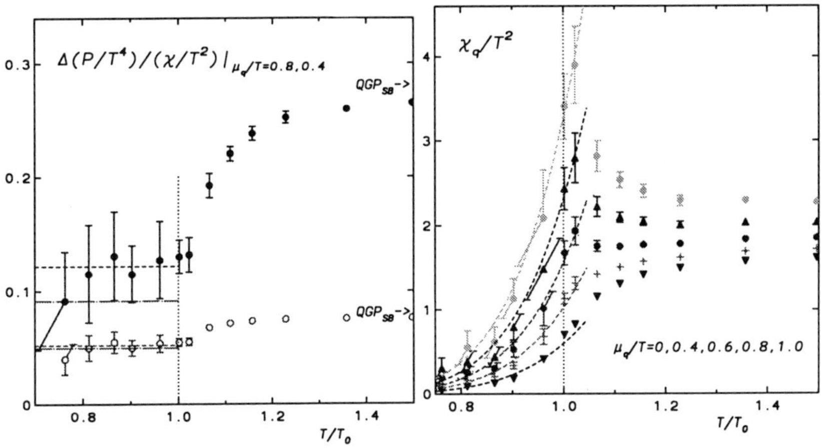

Figure 5. The ratio of pressure and quark number susceptibility $\chi_q \equiv 9\chi_B$ versus temperature for fixed values of the quark chemical potential $\mu_q/T = \mu_B/3T$ (left). The horizontal lines are the results of hadron resonance gas model calculations. The points are the lattice data from Ref. 15. While the dashed-dotted curves represent the complete expression given in Eq. 7 the dashed curves give the result of a Taylor expansion performed to the same order as that used in the lattice calculations. The right hand part of the figure shows the quark number susceptibility versus T/T_c for different values of the quark chemical potential. The lines give results from the resonance model calculation expanded in a Taylor series and truncated at the same order as used in the lattice calculation.

4. The critical temperature

The good agreement between lattice results on the QCD equation of state in the low temperature phase and the resonance gas model raises the question whether we can also understand the dependence of the QCD transition temperature on the quark mass in terms of a resonance gas. The similarity of the "critical" energy densities found in the infinite quark mass limit and for (moderately) light quark masses suggests that the transition to the quark gluon plasma phase occurs at approximately constant energy density, irrespective of the quark mass values or the physical masses of the hadronic resonances. This assumption is the basis for the comparison of transition temperatures in 2 and 3 flavor QCD with lines of constant energy density calculated in a resonance gas model[19]. Results are shown in Fig. 6. Up to pseudo-scalar masses $m_{PS} \simeq (3-4)\sqrt{\sigma} \simeq (1.3-1.7)$ GeV the agreement with the hadronic resonance gas is quite reasonable. For larger

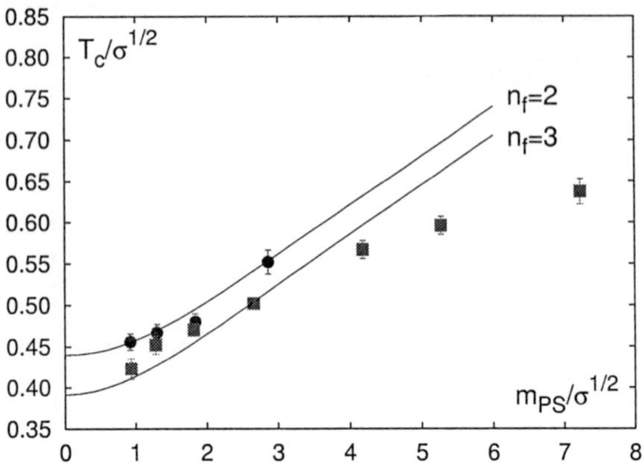

Figure 6. Transition temperatures in 2 and 3 flavor QCD versus the lightest pseudo-scalar meson mass. Both quantities are expressed in units of the square root of the string tension. The solid lines correspond to temperatures at which the energy density of a 2 and 3 flavor resonance gas attains the value $\epsilon = 0.8/\text{GeV}/\text{fm}^3$. The former consist of hadrons without strangeness content while the latter also includes strange mesons and baryons.

pseudo-scalar masses the glueball sector does start to play an increasingly important role as the heavy hadrons decouple from the thermodynamics and the "lighter" glueballs yield the largest contribution to the energy density. Aside from the inclusion of glueball states in the resonance gas[19] one eventually also has to take into account that thermal effects may strongly influence the glueball spectrum close to the transition temperature[21]. Doing so the qualitative features of the quark mass dependence of T_c can be modeled in the entire mass regime.

Finally, let us consider the lines of constant energy density in the T-μ_B plane. Separating the meson and baryon contribution to the energy density,

$$\epsilon(T, \mu_B) = \epsilon_M(T) + \epsilon_B(T) \cosh(\mu_B/T) \quad , \tag{8}$$

where $\epsilon_B(T)$ is the baryonic contribution to the energy density at $\mu_B = 0$, one now can follow the strategy applied also in the lattice calculations and determine the lines of constant energy density from a leading order Taylor expansion. Expanding around the transition point at vanishing chemical potential, $(T_c, \mu_B = 0)$, yields,

$$\frac{T_c(\mu_B)}{T_c} = 1 - \frac{1}{2} \frac{\epsilon_B(T_c)}{T_c \frac{\partial}{\partial T} \left(\epsilon_M(T) + \epsilon_B(T)\right)_{T=T_c}} \left(\frac{\mu_B}{T}\right)^2 \quad . \tag{9}$$

We note that the denominator on the right hand side is just the specific heat in a resonance gas at $(T_c, \mu_B = 0)$. This shows that one does expect a quite weak dependence of the transition temperature on the chemical potential, if the transition at μ_B is second order or close to a second order transition. Although this is in qualitative agreement with the weak μ_B-dependence of the transition line observed in lattice calculations it seems that the resonance gas still leads to a stronger variation of μ_B. A more direct comparison in the limit of physical quark masses would be desirable here.

5. Conclusions

We have shown that the hadronic resonance gas model is able to describe quite a few quantitative and qualitative results obtained in lattice calculations on the thermodynamics of the low temperature hadronic phase of QCD as well as basic properties of the transition line to the high temperature phase. This suggests that the copious production of hadronic resonance indeed plays an important role in triggering the transition to the quark gluon plasma phase of QCD. Unfortunately, a direct comparison of lattice calculations with an non-interacting gas of resonances given directly in terms of the experimentally known hadron spectrum is at present not yet possible. If one wants to compare the current lattice calculations, which still are being performed with too heavy quark masses, with the hadronic resonance gas some phenomenological input on the quark mass dependence of hadron resonances is required. However, it will soon be possible to further reduce or even eliminate the resulting ambiguities. With the availability of a new generation of Teraflops computers for lattice QCD it soon will become possible to perform studies of QCD thermodynamics with an "almost" realistic spectrum of quark masses.

References

1. B. Svetitsky and L.G. Yaffe, *Phys. Rev.* **D26**, 963 (1982).
2. R.D Pisarski and F. Wilczek, *Phys. Rev.* **D29**, 338 (1984).
3. F. Karsch, *Lect. Notes Phys.* **583**, 209 (2002).
4. for a discussion and further references see e.g., Proceedings of the International Workshop on *Understanding deconfinement in QCD* (World Scientific 2000, Edts: D. Blaschke, F. Karsch and C. D. Roberts).
5. L.D. McLerran and B. Svetitsky, *Phys. Lett.* **B98**, 195 (1981); J. Kuti, J. Polonyi and K. Szlachanyi, *Phys. Lett.* **B98**, 199 (1981); J. Engels, F. Karsch, I. Montvay and H. Satz, *Phys. Lett.* **B101**, 89 (1981).

6. R. Hagedorn, *Nuovo Cimento* **35**, 395 (1965).
7. G.F. Chapline, M.H. Johnson, E. Teller and M.S. Weiss, *Phys. Rev.* **D8**, 4302 (1973).
8. J. Cleymans and K. Redlich, *Phys. Rev. Lett.* **81**, 5284 (1998).
9. P. Braun-Munzinger et al., *Phys. Lett.* **B518**, 41 (2001).
10. F. Karsch, E. Laermann and A. Peikert, *Nucl. Phys.* **B605**, 579 (2001).
11. Z. Fodor and S. Katz, *JHEP* **0203**, 014 (2002).
12. C. R. Allton, S. Ejiri, S. J. Hands, O. Kaczmarek, F. Karsch, E. Laermann and C. Schmidt, *Phys. Rev.* **D66**, 074507 (2002).
13. Ph. deForcrand and O. Philipsen, *Nucl. Phys.* **B642**, 290 (2002).
14. Z. Fodor, S.D. Katz and K.K. Szabo, *Phys. Lett.* **B568**, 73 (2003).
15. C. R. Allton, S. Ejiri, S. J. Hands, O. Kaczmarek, F. Karsch, E. Laermann and C. Schmidt, *Phys. Rev.* **D68**, 014507 (2003).
16. R.D. Pisarski and O. Alvarez, *Phys. Rev.* **D26**, 3735 (1982).
17. F. Karsch, C. R. Allton, S. Ejiri, S. J. Hands, O. Kaczmarek, E. Laermann and C. Schmidt, hep-lat/0309116.
18. J. Fingberg, U. M. Heller and F. Karsch, *Nucl. Phys.* **B392**, 493 (1993) 493; B. Beinlich, F. Karsch, E. Laermann and A. Peikert, *Eur. Phys. J.* **C6**, 133 (1999); J. Engels, F. Karsch, E. Laermann, C. Legeland, M. Lütgemeier, B. Petersson and T. Scheideler, *Nucl. Phys. Proc. Suppl.* **53**, 420 (1997).
19. F. Karsch, K. Redlich and A. Tawfik, *Eur. Phys. J.* **C29**, 549 (2003).
20. F. Karsch, K. Redlich and A. Tawfik, *Phys. Lett.* **B571**, 67 (2003).
21. N. Ishii, H. Suganuma and H. Matsufuru, *Phys. Rev.* **D66**, 094506 (2002).

JET TOMOGRAPHY OF HOT AND DENSE MATTER*

XIN-NIAN WANG

Nuclear Science Division, MS 70R0319
Lawrence Berkeley National Laboratory,
Berkeley, CA 94720, USA
E-mail: xnwang@lbl.gov

Recent experimental measurements of high p_T hadron spectra and jet correlation at RHIC are analyzed within a parton model which incoporates initial jet production and final propagation in heavy-ion collisions. The suppresion of single hadron spectra, back-to-back correlation, their centrality dependence and azimuthal anisotropy point to a dense matter with an initial parton density about 30 times of that in a cold heavy nucleus.

1. Introduction

In high-energy heavy-ion collisions, a dense medium of quarks and gluons is expected to be produced and possibly a quark-gluon plasma is formed. One important step in establishing evidence of QGP formation is to charaterize the properties of the dense medium produced, for example, the parton and energy density and color confinement, among many other charateristics. Conventionally, one can study the properties of a medium via scattering experiments with particle beams. In deeply inelastic scattering (DIS) experiments, for example, leptons scatter off the nucleon medium via photon exchange with quarks. The response function or the correlation function of the electromagnetic currents,

$$W_{\mu\nu}(q) = \frac{1}{4\pi}\int d^4x e^{iq\cdot x}\langle A \mid j_\mu^{em}(0) j_\nu^{em}(x) \mid A\rangle \tag{1}$$

is a direct measurement of the quark distributions in a nucleon or nucleus. Such experiments have provided unprecedented information about partonic

*This work is supported by the director, office of energy research, office of high energy and nuclear physics, divisions of nuclear physics, of the u.s. department of energy under contract no. de-ac03-76sf00098.

structure of nucleons and nucleus and confirmed the prediction of QCD evolution[1].

For dynamic systems such as that produced in heavy-ion collisions, one can no longer use the technique of scattering with a beam of particles because of the transient nature of the matter. The lifetime of the system is very short, on the order of a few fm/c. The initial spatial size is only the size of the heavest nuclei, about 6 fm in diameter in the transverse dimension. The system expands very rapidly both in the longitudinal and transverse direction. These characteristics make it impossible to use external probes to study the properties of the produced dense matter in high-energy heavy-ion collisions. Fortunately, one can prove that the thermal average of the above correlation function gives the photon emission rate from the evolving system. The emission rate depends mainly on the local temperature or the parton density while the total yield also depends the whole evolution history of the system. Therefore, a strongly interacting system can reveal its properties and dynamics through photon and dilepton emission. One can further study the resonance properties of the emitted virtual photons and their medium modification. The screening of strong interaction in a color deconfined medium should lead to dissociation of the binding states and thus the quarkonia suppression [2]. Such color screening is a result of strong interaction between quarks and gluons at high density and temperature. The same interaction will also cause attenuation of fast and energetic partons propagating through the medium. Such an effect is the underlying physics of the jet quenching [3] phenomenon and jet tomography technique for studying properties of dense matter in high-energy heavy-ion collisions.

Jet quenching as a probe of the dense matter in heavy-ion collisions, takes advantage of the hard processes of jet production in high-energy heavy-ion collisions. Similar to the technology of computed tomography (CT), study of these energetic particles, their initial production and interaction with the dense medium, can yield critical information about the properties of the matter that is otherwise difficult to access through soft hadrons from the hadronization of the bulk medium. Though relatively rare with small cross sections, the jet production rate can be calculated perturbative in QCD and agrees well with experimental measurements in high-energy $pp(\bar{p})$ collisions. A critical component of the jet tomography is then to understand the jet attenuation through dense matter as it propagates through the medium.

2. Modified Fragmentation Function

A direct manifest of jet quenching is the modification of the fragmentation function of the produced parton, $D_{a\to h}(z,\mu^2)$ which can be measured directly. This modification can be directly translated into the energy loss of the leading parton.

To demonstrate medium modified fragmentation function and parton energy loss, one can study deeply inelastic scattering (DIS) eA [4,5,6]. Here, we consider the semi-inclusive processes, $e(L_1)+A(p) \longrightarrow e(L_2)+h(\ell_h)+X$, where L_1 and L_2 are the four-momenta of the incoming and the outgoing leptons, and ℓ_h is the observed hadron momentum. The differential cross section for the semi-inclusive process can be expressed as

$$E_{L_2}E_{\ell_h}\frac{d\sigma^h_{\rm DIS}}{d^3L_2 d^3\ell_h} = \frac{\alpha^2_{\rm EM}}{2\pi s}\frac{1}{Q^4}L_{\mu\nu}E_{\ell_h}\frac{dW^{\mu\nu}}{d^3\ell_h}\,, \tag{2}$$

where $p = [p^+, 0, \mathbf{0}_\perp]$ is the momentum per nucleon in the nucleus, $q = L_2 - L_1 = [-Q^2/2q^-, q^-, \mathbf{0}_\perp]$ the momentum transfer, $s = (p + L_1)^2$ and $\alpha_{\rm EM}$ is the electromagnetic (EM) coupling constant. $L_{\mu\nu}$ is the leptonic tensor while $W_{\mu\nu}$ is the semi-inclusive hadronic tensor.

In the parton model with the collinear factorization approximation, the leading-twist contribution to the semi-inclusive cross section can be factorized into a product of parton distributions, parton fragmentation functions and the hard partonic cross section. Including all leading log radiative corrections, the lowest order contribution from a single hard $\gamma^* + q$ scattering can be written as

$$\frac{dW^S_{\mu\nu}}{dz_h} = \sum_q e_q^2 \int dx f_q^A(x,\mu_I^2) H^{(0)}_{\mu\nu}(x,p,q) D_{q\to h}(z_h,\mu^2)\,, \tag{3}$$

$$\tag{4}$$

where $H^{(0)}_{\mu\nu}(x,p,q)$ is the hard part of the process in leading order, the momentum fraction carried by the hadron is defined as $z_h = \ell_h^-/q^-$ and $x_B = Q^2/2p^+q^-$ is the Bjorken variable. μ_I^2 and μ^2 are the factorization scales for the initial quark distributions $f_q^A(x,\mu_I^2)$ in a nucleus and the fragmentation functions $D_{q\to h}(z_h,\mu^2)$, respectively.

In a nuclear medium, the propagating quark in DIS will experience additional scatterings with other partons from the nucleus. The rescatterings may induce additional gluon radiation and cause the leading quark to lose energy. Such induced gluon radiations will effectively give rise to additional terms in the evolution equation leading to the modification of the fragmentation functions in a medium. These are the so-called higher-twist corrections since they involve higher-twist parton matrix elements and are power-suppressed. We will consider those contributions that involve two-parton correlations from two different nucleons inside the nucleus.

One can apply the generalized factorization to these multiple scattering processes[7]. In this approximation, the double scattering contribution to radiative correction can be calculated and the effective modified fragmentation function is

$$\widetilde{D}_{q\to h}(z_h,\mu^2) \equiv D_{q\to h}(z_h,\mu^2) + \int_0^{\mu^2} \frac{d\ell_T^2}{\ell_T^2}\frac{\alpha_s}{2\pi}\int_{z_h}^1 \frac{dz}{z}\left[\Delta\gamma_{q\to qg}(z,x,x_L,\ell_T^2)\right.$$
$$\left.\times D_{q\to h}(z_h/z) + \Delta\gamma_{q\to gq}(z,x,x_L,\ell_T^2) D_{g\to h}(z_h/z)\right] , \quad (5)$$

where $D_{q\to h}(z_h,\mu^2)$ and $D_{g\to h}(z_h,\mu^2)$ are the leading-twist fragmentation functions. The modified splitting functions are given as

$$\Delta\gamma_{q\to qg}(z,x,x_L,\ell_T^2) = \left[\frac{1+z^2}{(1-z)_+} T_{qg}^A(x,x_L) + \delta(1-z)\Delta T_{qg}^A(x,\ell_T^2)\right]$$
$$\times \frac{2\pi\alpha_s C_A}{\ell_T^2 N_c f_q^A(x,\mu_I^2)} , \quad (6)$$
$$\Delta\gamma_{q\to gq}(z,x,x_L,\ell_T^2) = \Delta\gamma_{q\to qg}(1-z,x,x_L,\ell_T^2). \quad (7)$$

Here, the fractional momentum is defined as $x_L = \ell_T^2/2p^+q^-z(1-z)$ and $x = x_B = Q^2/2p^+q^-$ is the Bjorken variable. The twist-four parton matrix elements of the nucleus,

$$T_{qg}^A(x,x_L) = \int \frac{dy^-}{2\pi} dy_1^- dy_2^- (1-e^{-ix_Lp^+y_2^-})(1-e^{-ix_Lp^+(y^- - y_1^-)})$$
$$\times e^{i(x+x_L)p^+y^-}\theta(-y_2^-)\theta(y^- - y_1^-)$$
$$\times \frac{1}{2}\langle A|\bar{\psi}_q(0)\,\gamma^+\,F_\sigma^{\ +}(y_2^-)\,F^{+\sigma}(y_1^-)\,\psi_q(y^-)|A\rangle , \quad (8)$$

has a dipole-like structure which is a result of LPM interference in gluon bremsstrahlung. In the limit of collinear radiation ($x_L \to 0$) or when the formation time of the gluon radiation, $\tau_f \equiv 1/x_Lp^+$, is much larger than the nuclear size, the destructive interference leads to the LPM interference effect.

Using the factorization approximation[4,5,7,8], we can relate the twist-four parton matrix elements of the nucleus to the twist-two parton distributions of nucleons and the nucleus,

$$T_{qg}^A(x,x_L) \approx \frac{\widetilde{C}}{x_A}(1-e^{-x_L^2/x_A^2}) f_q^A(x), \quad (9)$$

where $\widetilde{C} \equiv 2Cx_T f_g^N(x_T)$ is considered a constant. One can identify $1/x_L p^+ = 2q^- z(1-z)/\ell_T^2$ as the formation time of the emitted gluons. When it becomes comparable or larger than the nuclear size, the above matrix element vanishes, demonstrating a typical LPM interference effect.

Since the LPM interference suppresses gluon radiation whose formation time ($\tau_f \sim Q^2/\ell_T^2 p^+$) is larger than the nuclear size MR_A/p^+ in our chosen frame, ℓ_T^2 should then have a minimum value of $\ell_T^2 \sim Q^2/MR_A \sim Q^2/A^{1/3}$. Here M is the nucleon mass. Therefore, the leading higher-twist contribution is proportional to $\alpha_s R_A/\ell_T^2 \sim \alpha_s R_A^2/Q^2$ due to double scattering and depends quadratically on the nuclear size R_A.

With the assumption of the factorized form of the twist-4 nuclear parton matrices, there is only one free parameter $\widetilde{C}(Q^2)$ which represents quark-gluon correlation strength inside nuclei. Once it is fixed, one can predict the z, energy and nuclear dependence of the medium modification of the fragmentation function. Shown in Figs. 1 are the calculated nuclear modification factor of the fragmentation functions for ^{14}N and ^{84}Kr targets as compared to the recent HERMES data[9]. The predicted shape of the z- and ν-dependence agrees well with the experimental data. A remarkable feature of the prediction is the quadratic $A^{2/3}$ nuclear size dependence, which is verified for the first time by an experiment. By fitting the overall suppression for one nuclear target, we obtain the only parameter in our calculation, $\widetilde{C}(Q^2) = 0.0060$ GeV^2 with $\alpha_s(Q^2) = 0.33$ at $Q^2 \approx 3$ GeV^2.

We can quantify the modification of the fragmentation by the quark energy loss which is defined as the momentum fraction carried by the radiated gluon,

$$\begin{aligned}\langle \Delta z_g \rangle (x_B, \mu^2) &= \int_0^{\mu^2} \frac{d\ell_T^2}{\ell_T^2} \int_0^1 dz \frac{\alpha_s}{2\pi} z\, \Delta\gamma_{q\to gq}(z, x_B, x_L, \ell_T^2) \\ &= \widetilde{C} \frac{C_A \alpha_s^2}{N_c} \frac{x_B}{x_A Q^2} \int_0^1 dz \frac{1+(1-z)^2}{z(1-z)} \int_0^{x_\mu} \frac{dx_L}{x_L^2} (1 - e^{-x_L^2/x_A^2}), \end{aligned} \tag{10}$$

where $x_\mu = \mu^2/2p^+q^- z(1-z) = x_B/z(1-z)$ if we choose the factorization scale as $\mu^2 = Q^2$. When $x_A \ll x_B \ll 1$ we can estimate the leading quark energy loss roughly as

$$\langle \Delta z_g \rangle (x_B, \mu^2) \approx \widetilde{C} \frac{C_A \alpha_s^2}{N_c} \frac{x_B}{Q^2 x_A^2} 6\sqrt{\pi} \ln \frac{1}{2x_B}\,. \tag{11}$$

Since $x_A = 1/MR_A$, the energy loss $\langle \Delta z_g \rangle$ thus depends quadratically on the nuclear size.

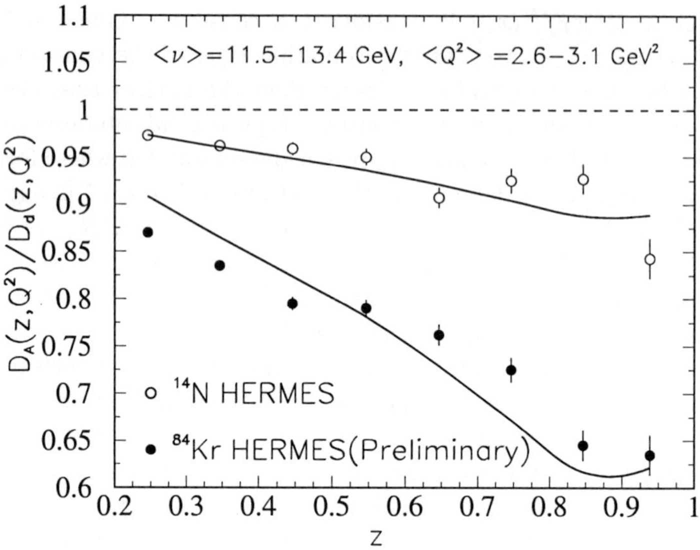

Figure 1. Predicted nuclear modification of jet fragmentation function is compared to the HERMES data [9] on ratios of hadron distributions between A and d targets in DIS.

In the rest frame of the nucleus, $p^+ = m_N$, $q^- = \nu$, and $x_B \equiv Q^2/2p^+q^- = Q^2/2m_N\nu$. One can get the averaged total energy loss as $\Delta E = \nu\langle\Delta z_g\rangle \approx \widetilde{C}(Q^2)\alpha_s^2(Q^2)m_N R_A^2(C_A/N_c)3\ln(1/2x_B)$. With the determined value of $\widetilde{C}$, $\langle x_B\rangle \approx 0.124$ in the HERMES experiment[9] and the average distance $\langle L_A\rangle = R_A\sqrt{2/\pi}$ for the assumed Gaussian nuclear distribution, one gets the quark energy loss $dE/dL \approx 0.5$ GeV/fm inside a Au nucleus.

2.1. *Jet Quenching in Hot Medium at RHIC*

To extend our study of modified fragmentation functions to jets in heavy-ion collisions, we can assume $\langle k_T^2\rangle \approx \mu^2$ (the Debye screening mass) and a gluon density profile $\rho(y) = (\tau_0/\tau)\theta(R_A - y)\rho_0$ for a 1-dimensional expanding system. Since the initial jet production rate is independent of the final gluon density which can be related to the parton-gluon scattering cross section[10] $[\alpha_s x_T G(x_T) \sim \mu^2\sigma_g]$, one has then

$$\frac{\alpha_s T_{qg}^A(x_B, x_L)}{f_q^A(x_B)} \sim \mu^2 \int dy\sigma_g\rho(y)[1 - \cos(y/\tau_f)], \qquad (12)$$

where $\tau_f = 2Ez(1-z)/\ell_T^2$ is the gluon formation time. One can recover the form of energy loss in a thin plasma obtained in the opacity expansion approach[11],

$$\langle \Delta z_g \rangle = \frac{C_A \alpha_s}{\pi} \int_0^1 dz \int_0^{\frac{Q^2}{\mu^2}} du \frac{1+(1-z)^2}{u(1+u)} \int_{\tau_0}^{R_A} d\tau \sigma_g \rho(\tau) \times \left[1 - \cos\left(\frac{(\tau - \tau_0)\, u\, \mu^2}{2Ez(1-z)} \right) \right]. \tag{13}$$

Keeping only the dominant contribution and assuming $\sigma_g \approx C_a 2\pi\alpha_s^2/\mu^2$ (C_a=1 for qg and 9/4 for gg scattering), one obtains the averaged energy loss,

$$\langle \frac{dE}{dL} \rangle \approx \frac{\pi C_a C_A \alpha_s^3}{R_A} \int_{\tau_0}^{R_A} d\tau \rho(\tau)(\tau - \tau_0) \ln \frac{2E}{\tau \mu^2}. \tag{14}$$

Neglecting the logarithmic dependence on τ, the averaged energy loss in a 1-dimensional expanding system can be expressed as $\langle \frac{dE}{dL} \rangle_{1d} \approx (dE_0/dL)(2\tau_0/R_A)$, where $dE_0/dL \propto \rho_0 R_A$ is the energy loss in a static medium with the same gluon density ρ_0 as in a 1-d expanding system at time τ_0. Because of the expansion, the averaged energy loss $\langle dE/dL \rangle_{1d}$ is suppressed as compared to the static case and does not depend linearly on the system size.

In order to calculate the effects of parton energy loss on the attenuation pattern of high p_T partons in nuclear collisions, we use a simpler effective modified fragmentation function[12,13],

$$D'_{h/c}(z_c, Q^2, \Delta E_c) = \left[\frac{z'_c}{z_c} D^0_{h/c}(z'_c, Q^2) + \langle \frac{\Delta L}{\lambda} \rangle \frac{z'_g}{z_c} D^0_{h/g}(z'_g, Q^2) \right] \times (1 - e^{-\langle \frac{\Delta L}{\lambda} \rangle}) + D^0_{h/c}(z_c, Q^2) e^{-\langle \frac{\Delta L}{\lambda} \rangle}, \tag{15}$$

where z'_c, z_g are the rescaled momentum fractions. This effective model is found to reproduce the pQCD result from Eq.(5) very well, but only when $\Delta z = \Delta E_c/E$ is set to be $\Delta z \approx 0.6\langle z_g \rangle$. Therefore the actual averaged parton energy loss should be $\Delta E/E = 1.6\Delta z$ with Δz extracted from the effective model. The factor 1.6 is mainly caused by the unitarity correction effect in the pQCD calculation.

Since gluons are bosons, there should also be stimulated gluon emission and absorption by the propagating parton because of the presence of thermal gluons in the hot medium. Such detailed balance is crucial for parton thermalization and should also be important for calculating the energy loss of an energetic parton in a hot medium[14]. Taking into account such detailed balance in gluon emission, one can then get the asymptotic behavior of the effective energy loss in the opacity expansion framework [14],

$$\frac{\Delta E}{E} \approx \frac{\alpha_s C_F \mu^2 L^2}{4\lambda_g E}\left[\ln\frac{2E}{\mu^2 L} - 0.048\right] - \frac{\pi\alpha_s C_F}{3}\frac{LT^2}{\lambda_g E^2}\left[\ln\frac{\mu^2 L}{T} - 1 + \gamma_{\rm E} - \frac{6\zeta'(2)}{\pi^2}\right], \tag{16}$$

where the first term is from the induced bremsstralung and the second term is due to gluon absorption in detailed balance which effectively reduce the total parton energy loss in the medium. Numerical calculations show that the effect of the gluon absorption is small and can be neglected for partons with very high energy. However, the thermal absorption reduces the effective parton energy loss by about 30-10% for intermediate values of parton energy. This will increase the energy dependence of the effective parton energy loss in the intermediate energy region. One can parameterize such energy dependence as,

$$\langle\frac{dE}{dL}\rangle_{1d} = \epsilon_0 (E/\mu - 1.6)^{1.2}/(7.5 + E/\mu), \tag{17}$$

The threshold is the consequence of gluon absorption that competes with radiation that effectively shuts off the energy loss. The parameter μ is set to be 1 GeV in the calculation.

To calculate the modified high p_T spectra in $A+A$ collisions, we use a LO pQCD model [15,16],

$$\frac{d\sigma^h_{AA}}{dyd^2p_T} = K\sum_{abcd}\int d^2b d^2r dx_a dx_b d^2k_{aT} d^2k_{bT} t_A(r) t_A(|\mathbf{b}-\mathbf{r}|) \times g_A(k_{aT}, r) g_A(k_{bT}, |\mathbf{b}-\mathbf{r}|) f_{a/A}(x_a, Q^2, r) f_{b/A}(x_b, Q^2, |\mathbf{b}-\mathbf{r}|) \times \frac{D'_{h/c}(z_c, Q^2, \Delta E_c)}{\pi z_c}\frac{d\sigma}{d\hat{t}}(ab \to cd), \tag{18}$$

with medium modified fragmentation funcitons $D'_{h/c}$ given by Eq. (15) and the fragmentation functions in free space $D^0_{h/c}(z_c, Q^2)$ are given by the BBK parameterization [17]. Here, $z_c = p_T/p_{Tc}$, $y = y_c$, $\sigma(ab \to cd)$ are elementary parton scattering cross sections and $t_A(b)$ is the nuclear thickness function normalized to $\int d^2b t_A(b) = A$. We will use a hard-sphere model of nuclear distribution in this paper. The $K \approx 1.5-2$ factor is used to account for higher order pQCD corrections. The parton distributions per nucleon $f_{a/A}(x_a, Q^2, r)$ inside the nucleus are assumed to be factorizable into the parton distributions in a free nucleon given by the MRS D$-'$ parameterization and the impact-parameter dependent nuclear modification factor which will given by the new HIJING parameterization [18]. The initial transverse momentum distribution $g_A(k_T, Q^2, b)$ is assumed to have a Gaussian form with a width that includes both an intrinsic part in a nucleon and nuclear broadening. This model has been fitted to the nuclear modification of the p_T spectra in $p + A$ collisions at up to the Fermilab energy $\sqrt{s} = 40$ GeV. Shown in Fig. 2 are the first prediction made in 1998 [15] of the Cronin effect at RHIC for $p + Au$ collisions at $\sqrt{s} = 200$ GeV as compared to the recent RHIC data. As one can see, the initial multiple scattering in nuclei can give some moderate Cronin enhancement of the high p_T spectra. Therefore, any suppression of the high p_T spectra in $Au + Au$ collisions has to be caused by jet quenching.

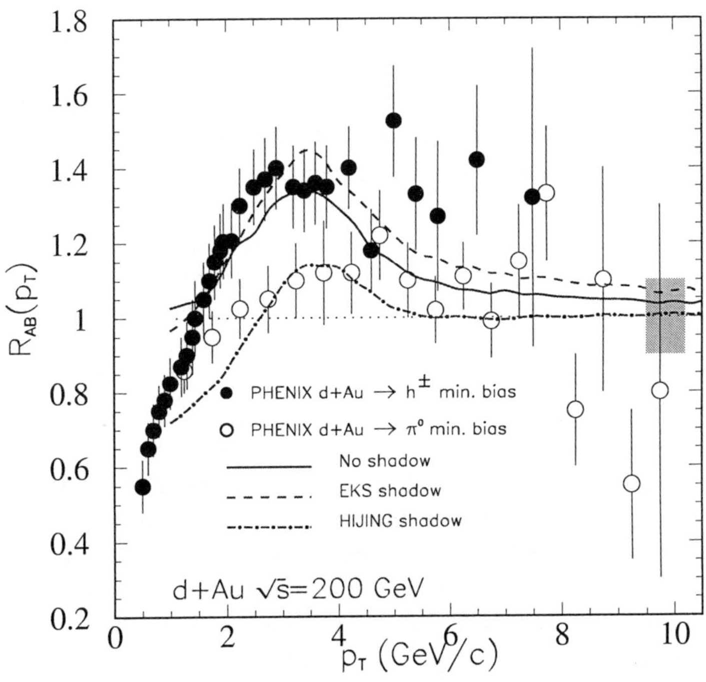

Figure 2. The first predictions [15] of the Cronin effect in $p + Au$ collisions at $\sqrt{s} = 200$ GeV are compared to the recent RHIC data from PHENIX [19] and STAR [20].

We assume a 1-dimensional expanding medium with a gluon density $\rho_g(\tau, r)$ that is proportional to the transverse profile of participant nucleons. According to Eq. 14, we will calculate impact-parameter dependence of the energy loss as

$$\Delta E(b,r,\phi) \approx \langle\frac{dE}{dL}\rangle_{1d} \int_{\tau_0}^{\Delta L} d\tau \frac{\tau-\tau_0}{\tau_0\rho_0}\rho_g(\tau,b,\vec{r}+\vec{n}\tau), \tag{19}$$

where $\Delta L(b,\vec{r},\phi)$ is the distance a jet, produced at $\vec{r}$, has to travel along $\vec{n}$ at an azimuthal angle ϕ relative to the reaction plane in a collision with impact-parameter b. Here, ρ_0 is the averaged initial gluon density at τ_0 in a central collision and $\langle dE/dL\rangle_{1d}$ is the average parton energy loss over a distance R_A in a 1-d expanding medium with an initial uniform gluon density ρ_0. The corresponding energy loss in a static medium with a uniform gluon density ρ_0 over a distance R_A is [21] $dE_0/dL = (R_A/2\tau_0)\langle dE/dL\rangle_{1d}$. We will use the parameterization in Eq. (17) for the effective energy dependence of the parton quark energy loss.

Shown in Fig. 3 are the calculated nuclear modification factors $R_{AB}(p_T) = d\sigma^h_{AB}/\langle N_{\rm binary}\rangle d\sigma^h_{pp}$ for hadron spectra ($|y|<0.5$) in $Au+Au$ collisions at $\sqrt{s}=200$ GeV, as compared to experimental data [23,22]. Here, $\langle N_{\rm binary}\rangle = \int d^2b d^2r t_A(r)t_A(|\vec{b}-\vec{r}|)$. To fit the observed π^0 suppression (solid lines) in the most central collisions, we have used $\mu = 1.5$ GeV, $\epsilon_0 = 1.07$ GeV/fm and $\lambda_0 = 1/(\sigma\rho_0) = 0.3$ fm. The hatched area (also in other figures in this paper) indicates a variation of $\epsilon_0 = \pm 0.3$ GeV/fm. The hatched boxes around $R_{AB}=1$ represent experimental errors in overall normalization. Nuclear k_T broadening and parton shadowing together give a slight enhancement of hadron spectra at intermediate $p_T = 2-4$ GeV/c without parton energy loss.

The flat p_T dependence of the π^0 suppression is a consequence of the strong energy dependence of the parton energy loss. The slight rise of R_{AB} at $p_T < 4$ GeV/c in the calculation is due to the detailed balance effect in the effective parton energy loss. In this region, one expects the fragmentation picture to gradually lose its validity and is taken over by other non-perturbative effects, especially for kaons and baryons. As a consequence, the $(K+p)/\pi$ ratio in central $Au+Au$ collisions is significant larger than in peripheral $Au+Au$ or $p+p$ collisions. To take into account this effect, we add a nuclear dependent (proportional to $\langle N_{\rm binary}\rangle$) soft component to kaon and baryon fragmentation functions so that $(K+p)/\pi \approx 2$ at $p_T \sim 3$ GeV/c in the most central $Au+Au$ collisions and approaches its $p+p$ value at $p_T > 5$ GeV/c. The resultant suppression for total charged hadrons (dot-dashed) and the centrality dependence agree well with the STAR data. One can directly relate $h^\pm$ and π^0 suppression via the $(K+p)/\pi$ ratio: $R^{h^\pm}_{AA} = R^{\pi^0}_{AA}[1+(K+p)/\pi]_{AA}/[1+(K+p)/\pi]_{pp}$. It is clear from the data that $(K+p)/\pi$ becomes the same for $Au+Au$ and $p+p$ collisions at $p_T > 5$ GeV/c. To demonstrate the sensitivity to the parameterized parton energy loss in the intermediate p_T region, we also show $R_{AA}^{h^\pm}$ in 0-5% centrality (dashed line) for $\mu = 2.0$ GeV and $\epsilon_0 = 2.04$ GeV/fm without the soft component.

In the same LO pQCD parton model, one can also calculate di-hadron spectra,

$$E_1 E_2 \frac{d\sigma_{AA}^{h_1 h_2}}{d^3p_1 d^3p_2} = \frac{K}{2} \sum_{abcd} \int d^2b d^2r dx_a dx_b d^2k_{aT} d^2k_{bT} t_A(r) t_A(|\mathbf{b}-\mathbf{r}|)$$
$$\times\, g_A(k_{aT}, r) g_A(k_{bT}, |\mathbf{b}-\mathbf{r}|) f_{a/A}(x_a, Q^2, r) f_{b/A}(x_b, Q^2, |\mathbf{b}-\mathbf{r}|)$$
$$\times\, D_{h/c}(z_c, Q^2, \Delta E_c) D_{h/d}(z_d, Q^2, \Delta E_d) \frac{\hat{s}}{2\pi z_c^2 z_d^2}$$
$$\times\, \frac{d\sigma}{d\hat{t}}(ab \to cd) \delta^4(p_a + p_b - p_c - p_d), \tag{20}$$

for two back-to-back hadrons from independent fragmentation of the back-to-back jets. Let us assume hadron h_1 is a triggered hadron with $p_{T1} = p_T^{\rm trig}$. One can define a hadron-triggered FF as the back-to-back correlation with respect to the triggered hadron:

$$D^{h_1 h_2}(z_T, \phi, p_T^{\rm trig}) = p_T^{\rm trig} \frac{d\sigma_{AA}^{h_1 h_2}/d^2p_T^{\rm trig} dp_T d\phi}{d\sigma_{AA}^{h_1}/d^2p_T^{\rm trig}}, \tag{21}$$

similarly to the direct-photon triggered FF [12,13] in γ-jet events. Here, $z_T = p_T/p_T^{\rm trig}$ and integration over $|y_{1,2}| < \Delta y$ is implied. In a simple parton model, the two jets should be exactly back-to-back. The initial parton transverse momentum distribution in our model will give rise to a Gaussian-like angular distribution. In addition, we also take into account transverse momentum smearing within a jet using a Gaussian distribution with a width of $\langle k_\perp \rangle = 0.6$ GeV/c. Hadrons from the soft component are assumed to be uncorrelated.

Shown in Fig. 4 are the calculated back-to-back correlations for charged hadrons in $Au + Au$ collisions as compared to the STAR data [24]. The same energy loss that is used to calculate single hadron suppression and azimuthal anisotropy can also describe well the observed away-side hadron suppression and its centrality dependence. In the data, a background $B(p_T)[1 + 2v_2^2(p_T)\cos(2\Delta\phi)]$ from uncorrelated hadrons and azimuthal anisotropy has been subtracted. The value of $v_2(p_T)$ is measured independently while $B(p_T)$ is determined by fitting the observed correlation in the region $0.75 < |\phi| < 2.24$ rad [24].

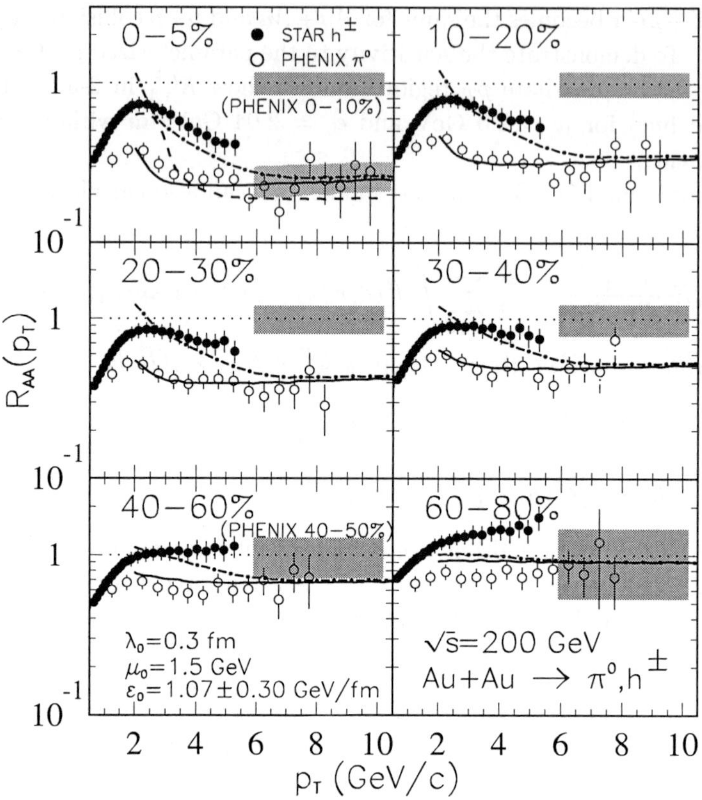

Figure 3. Hadron suppression factors in $Au + Au$ collisions as compared to data from STAR[23] and PHENIX [22]. See text for a detailed explanation.

With both the single spectra and dihadron spectra, the extracted average energy loss in this model calculation for a 10 GeV quark in the expanding medium is $\langle dE/dL \rangle_{1d} \approx 0.85 \pm 0.24$ GeV/fm, which is equivalent to $dE_0/dL \approx 13.8 \pm 3.9$ GeV/fm in a static and uniform medium over a distance $R_A = 6.5$ fm. This value is about a factor of 2 larger than a previous estimate [21] because of the variation of gluon density along the propagation path and the more precise RHIC data considered .

3. Summary

In summary, with the recent measurements of high p_T hadron spectra and jet correlations in $d + Au$ collisions at RHIC, the observed jet quenching in $Au + Au$ collisions has been established as a consequence of final state

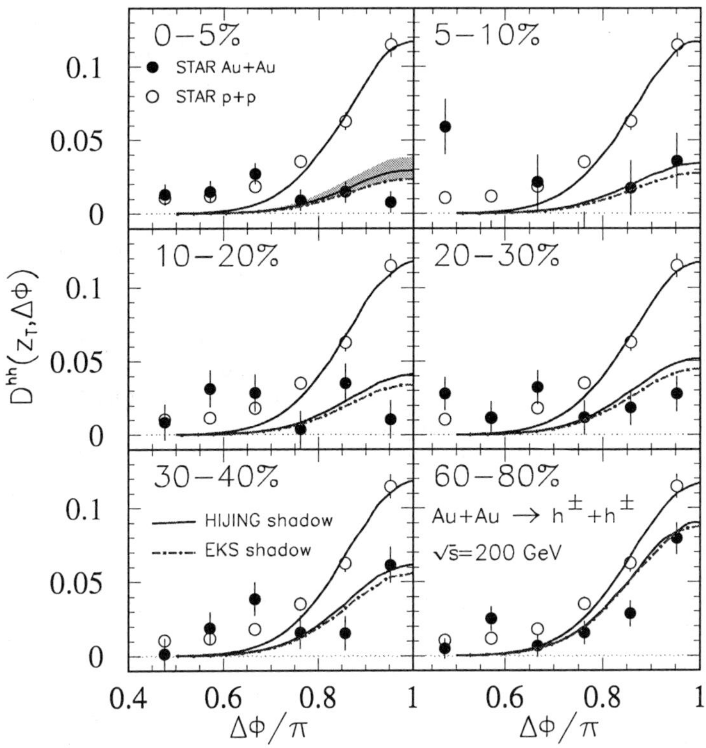

Figure 4. Back-to-back correlations for charged hadrons with $p_T^{\rm trig} > p_T > 2$ GeV/c, $p_T^{\rm trig} = 4 - 6$ GeV/c and $|y| < 0.7$ in $Au + Au$ (lower curves) and $p + p$ (upper curves) collisions as compared to the STAR[24] data.

interaction between jets and the produced dense medium. The collective body of data: suppression of high p_T spectra and back-to-back jet correlation, high p_T anisotrapy and centrality dependences of the observables, indicate that the cause of the jet quenching is due to parton energy loss rather than final state absorption of hadrons.

A simultaneous phenomenological study of the suppression of hadron spectra and back-to-back correlations, and high-p_T azimuthal anisotropy in high-energy heavy-ion collisions within a single LO pQCD parton model incorporating current theoretical understanding of parton energy loss can describe the experimental data of $Au + Au$ collisions very well. With HIJING (EKS) parton shadowing, the extracted average energy loss for a 10 GeV quark in the expanding medium is $\langle dE/dL\rangle_{1d} \approx 0.85(0.99) \pm 0.24$ GeV/fm, which is equivalent to $dE_0/dL \approx 13.8(16.1) \pm 3.9$ GeV/fm in a static and uniform medium over a distance $R_A = 6.5$ fm. This value is about a factor of 2 larger than a previous estimate [21] because of the variation of gluon density along the propagation path and the more precise RHIC data considered here .

References

1. Alan D. Martin, R. G. Roberts, W. J. Stirling and R. S. Thorne, Eur. Phys. J. C23 73, 2002; J. Pumplin *et al.*, JHEP 07 012,2002.
2. T. Matsui and H. Satz, Phys. Lett. B **178**, 416 (1986).
3. X. N. Wang and M. Gyulassy, Phys. Rev. Lett. **68**, 1480 (1992).
4. X. N. Wang and X. F. Guo, Nucl. Phys. A **696**, 788 (2001).
5. X. F. Guo and X. N. Wang, Phys. Rev. Lett. **85**, 3591 (2000).
6. B. W. Zhang and X.-N. Wang, arXiv:hep-ph/0301195.
7. M. Luo, J. Qiu and G. Sterman, Phys. Lett. **B279**, 377 (1992); M. Luo, J. Qiu and G. Sterman, Phys. Rev. D**50**, 1951 (1994); M. Luo, J. Qiu and G. Sterman, Phys. Rev. D**49**, 4493 (1994).
8. J. Osborne and X.-N. Wang, Nucl. Phys. A **710**, 281 (2002) [arXiv:hep-ph/0204046].
9. A. Airapetian *et al.* [HERMES Collaboration], Eur. Phys. J. C **20**, 479 (2001); V. Muccifora [HERMES Collaboration], arXiv:hep-ex/0106088.
10. R. Baier, Y. L. Dokshitzer, A. H. Mueller, S. Peigne and D. Schiff, Nucl. Phys. B **484**, 265 (1997)
11. M. Gyulassy, I. Vitev and X. N. Wang, Phys. Rev. Lett. **86**, 2537 (2001).
12. X. N. Wang, Z. Huang and I. Sarcevic, Phys. Rev. Lett. **77**, 231 (1996).
13. X. N. Wang and Z. Huang, Phys. Rev. C **55**, 3047 (1997).
14. E. Wang and X. N. Wang, Phys. Rev. Lett. **87**, 142301 (2001).
15. X. N. Wang, Phys. Rev. C **61**, 064910 (2000).
16. X. N. Wang, arXiv:nucl-th/0305010.
17. J. Binnewies, B. A. Kniehl and G. Kramer, Z. Phys. C **65**, 471 (1995).
18. S. Y. Li and X. N. Wang, Phys. Lett. B **527**, 85 (2002).
19. S. S. Adler *et al.* [PHENIX Collaboration], Phys. Rev. Lett. **91**, 072303 (2003) [arXiv:nucl-ex/0306021].
20. J. Adams *et al.* [STAR Collaboration], Phys. Rev. Lett. **91**, 072304 (2003) [arXiv:nucl-ex/0306024].
21. E. Wang and X. N. Wang, Phys. Rev. Lett. **89**, 162301 (2002).
22. K. Adcox *et al.*, Phys. Rev. Lett. **88**, 022301 (2002); S. S. Adler *et al.*, arXiv:nucl-ex/0304022.
23. C. Adler *et al.*, Phys. Rev. Lett. **89**, 202301 (2002); J. L. Klay, arXiv:nucl-ex/0210026.
24. C. Adler *et al.*, Phys. Rev. Lett. **90**, 082302 (2003).

A NEW METHOD FOR SIMULATING QCD AT FINITE DENSITY

JUN NISHIMURA

High Energy Accelerator Research Organization (KEK),
1-1 Oho, Tsukuba 305-0801, Japan
E-mail: jnishi@post.kek.jp

We propose a new method for simulating QCD at finite density, where interesting phases such as the color superconductivity phase is conjectured to appear. The method is based on a general factorization property of distribution functions of observables, and it is therefore applicable to any system with a complex action. The so-called overlap problem is completely eliminated by the use of constrained simulations. We test this method in a Random Matrix Theory for finite density QCD, where we are able to reproduce the exact results for the quark number density. The achieved system size is large enough to extract the thermodynamic limit. Our results provide a clear understanding of how the expected first order phase transition is induced by the imaginary part of the action. We also discuss the non-commutativity of the zero chemical potential limit ($\mu \to 0$) and the thermodynamic limit, which is relevant to recent Monte Carlo studies at small μ.

1. Introduction

Recently there are a lot of activities in QCD at finite density, where interesting phases such as a superconducting phase have been conjectured to appear [1]. At zero chemical potential Monte Carlo simulations of lattice QCD enable nonperturbative studies from first principles. It is clearly desirable to extend such an approach to finite density and explore the phase diagram of QCD in the T(temperature)-μ(chemical potential) plane. The main obstacle here is that the Euclidean action becomes complex once the chemical potential is switched on.

Nevertheless QCD at finite density has been studied by various approaches with exciting conjectures. First there are perturbative studies which are valid in the $\mu \to \infty$ limit [2,3]. Refs. [4] and [5] use effective theories with instanton-induced four-fermi interactions. As for Monte Carlo studies two directions have been pursued so far. One is to modify the model so that the action becomes real. This includes changing the gauge group from SU(3) to SU(2) [6], and introducing a chemical potential with opposite signs for up and down quarks [7]. The other direction is to explore the large T and small μ regime of lattice QCD, where the imaginary part of the action is not very large [8,9,10,11]. These studies already produced results relevant to heavy ion collision experiments, but more interesting physics will be uncovered if larger μ regime becomes accessible by simulations.

In Ref. [12] we have proposed a new method to simulate systems with a complex action, which utilizes a simple factorization property of distribution functions of observables. Since the property holds quite generally, the approach can be applied to any system with a complex action. The most important virtue of the method is that it eliminates the so-called overlap problem, which occurs in the standard re-weighting method. Ultimately we hope that this method will enable us, among other things, to explore the phase diagram of QCD at finite baryon density.

As a first step we have tested [13] the new approach in a Random Matrix Theory, which can be regarded as a schematic model for QCD at finite baryon density [14]. We also present some preliminary results [15], which reveal certain noncommutativity of the $\mu \to 0$ limit and the thermodynamic limit.

2. Brute-force approach —reweighting method—

Suppose we want to study the model defined by the partition function

$$Z = \int \mathrm{d}U \, \mathrm{e}^{-S_0 + i\,\Gamma} \,, \tag{2.1}$$

where S_0 and Γ are real. Since the weight $\mathrm{e}^{-S_0+i\,\Gamma}$ in (2.1) is not positive definite, we cannot regard it as a probability density. Hence it seems difficult to apply the idea of standard Monte Carlo simulations, which reduces the problem of obtaining vacuum expectation values (VEVs) to that of taking an average over an ensemble generated by the probability. One way to proceed is to apply the reweighting method and rewrite the VEV $\langle \mathcal{O} \rangle$ as

$$\langle \mathcal{O} \rangle = \frac{\langle \mathcal{O} \, \mathrm{e}^{i\Gamma} \rangle_0}{\langle \mathrm{e}^{i\Gamma} \rangle_0} \,, \tag{2.2}$$

where the symbol $\langle \, \cdot \, \rangle_0$ denotes a VEV with respect to the *phase-quenched* partition function

$$Z_0 = \int \mathrm{d}U \, \mathrm{e}^{-S_0} \,. \tag{2.3}$$

Since the system (2.3) has a positive definite weight, the VEV $\langle \, \cdot \, \rangle_0$ can be evaluated by standard Monte Carlo simulations. However, the fluctuations of the phase Γ in (2.2) grows linearly with the system size V. Due to huge cancellations, both the denominator and the numerator of the r.h.s. of (2.2) vanish as $\mathrm{e}^{-\mathrm{const}.V}$ as V increases, while the 'observables' $\mathrm{e}^{i\Gamma}$ and $\mathcal{O}\mathrm{e}^{i\Gamma}$ are of O(1) for each configuration. As a result, the number of configurations required to obtain the VEVs with some fixed accuracy grows as $\mathrm{e}^{\mathrm{const}.V}$. This is the notorious 'complex-action problem'. Moreover when one simulates the phase-quenched model (2.3), one cannot sample efficiently the configurations which are relevant to the calculation of the VEV $\langle \mathcal{O} \rangle$. This is the so-called 'overlap problem'.

3. New approach — factorization method —

In the factorization method proposed in Ref. [12], the fundamental objects are the distribution functions (we assume the observable $\mathcal{O}$ to be real)

$$\rho(x) \stackrel{\mathrm{def}}{=} \langle \delta(x - \mathcal{O}) \rangle \tag{3.1}$$

$$\rho^{(0)}(x) \stackrel{\mathrm{def}}{=} \langle \delta(x - \mathcal{O}) \rangle_0 \tag{3.2}$$

defined for the full model (2.1) and for the phase-quenched model (2.3), respectively. The important property of $\rho(x)$ is that it factorizes as

$$\rho(x) = \frac{1}{C}\, \rho^{(0)}(x)\, \varphi(x) \ , \tag{3.3}$$

where the constant C is given by $C \stackrel{\mathrm{def}}{=} \langle \mathrm{e}^{i\,\Gamma} \rangle_0$. The 'weight factor' $\varphi(x)$, which represents the effect of Γ, can be written as a VEV

$$\varphi(x) \stackrel{\mathrm{def}}{=} \langle \mathrm{e}^{i\,\Gamma} \rangle_x \tag{3.4}$$

with respect to yet another partition function

$$Z(x) = \int \mathrm{d}U\, \mathrm{e}^{-S_0}\, \delta(x - \mathcal{O}) \ . \tag{3.5}$$

The δ-function represents a constraint on the system. In actual simulation we replace the δ-function by a sharply peaked potential. The distribution $\rho^{(0)}(x)$ for the phase-quenched model can also be obtained from the same simulation. Then the VEV $\langle \mathcal{O} \rangle$ can be obtained by

$$\langle \mathcal{O} \rangle = \int \mathrm{d}x\, x\, \rho(x) = \frac{\int \mathrm{d}x\, x\, \rho^{(0)}(x)\, \varphi(x)}{\int \mathrm{d}x\, \rho^{(0)}(x)\, \varphi(x)} \ , \tag{3.6}$$

where the overlap problem is eliminated by forcing the simulation to sample the important configurations by the constraint. The knowledge of the weight factor $\varphi(x)$ is useful because it tells us precisely which values of $\mathcal{O}$

are favored or disfavored by the effects of the oscillating phase. Once a rough estimate of $\rho(x)$ is obtained, one may perform multi-canonical simulations with an appropriate weight (instead of simulating (3.5) for many x) to sample relevant configurations more efficiently. This has not yet been done, however.

4. Random Matrix Theory for finite density QCD

The Random Matrix Theory we study is defined by the partition function

$$Z = \int \mathrm{d}W \mathrm{e}^{-N\,\mathrm{tr}(W^\dagger W)} \det D \; , \tag{4.1}$$

where W is a $N \times N$ complex matrix, and D is a $2N \times 2N$ matrix given by

$$D = \begin{pmatrix} m & iW + \mu \\ iW^\dagger + \mu & m \end{pmatrix} \; . \tag{4.2}$$

The parameters m and μ correspond to the 'quark mass' and the 'chemical potential', respectively. The fermion determinant becomes complex for $\mu \neq 0$, so we write it as $\det D = \mathrm{e}^{i\Gamma} \, |\det D|$. The complex-action problem arises due to the phase Γ. In what follows we consider the massless case ($m = 0$) for simplicity and focus on the 'quark number density' defined by

$$\nu = \frac{1}{2N}\,\mathrm{tr}\,(\gamma_4 D^{-1}) \; , \qquad \gamma_4 = \begin{pmatrix} 0 & 1\!\!1 \\ 1\!\!1 & 0 \end{pmatrix} \; . \tag{4.3}$$

The model was first solved in the large-N limit [14], and turned out to be solvable later even for finite N [16]. The partition function can be expressed as

$$Z(\mu, N) = \pi \mathrm{e}^{\kappa} N^{-(N+1)}\, N! \left[1 + \frac{(-1)^{N+1}}{N!} \gamma(N+1, \kappa) \right] \; , \tag{4.4}$$

where $\kappa = -N\mu^2$ and $\gamma(n, x)$ is the incomplete γ-function defined by

$$\gamma(n, x) = \int_0^x \mathrm{e}^{-t}\, t^{n-1}\, \mathrm{d}t \; . \tag{4.5}$$

From this one obtains the VEV of the quark number density as

$$\langle \nu \rangle = \frac{1}{2N} \frac{\partial}{\partial \mu} \ln Z(\mu, N) \tag{4.6}$$

$$= -\mu \left[1 + \frac{\kappa^N \mathrm{e}^{-\kappa}}{(-1)^{N+1} N + \gamma(N+1, \kappa)} \right] \; . \tag{4.7}$$

Taking the large-N limit, one obtains

$$\lim_{N\to\infty}\langle\nu\rangle = \begin{cases} -\mu & \text{for } \mu < \mu_{\rm c} \\ 1/\mu & \text{for } \mu > \mu_{\rm c} \ , \end{cases} \tag{4.8}$$

where $\mu_{\rm c}$ is the solution to the equation $1+\mu^2+\ln(\mu^2) = 0$, and its numerical value is given by $\mu_{\rm c} = 0.527\cdots$. We find that the quark number density $\langle\nu\rangle$ has a discontinuity at $\mu = \mu_{\rm c}$. Thus the schematic model reproduces qualitatively the first order phase transition expected to occur in 'real' QCD at nonzero baryon density.

The phase-quenched model defined by the partition function

$$Z_0 = \int \mathrm{d}W \mathrm{e}^{-N\,\mathrm{tr}(W^\dagger W)}\,|\det D| \tag{4.9}$$

can be solved in the large N limit [14] and one obtains

$$\lim_{N\to\infty}\langle\nu\rangle_0 = \begin{cases} \mu & \text{for } \mu < 1 \\ 1/\mu & \text{for } \mu > 1 \ , \end{cases} \tag{4.10}$$

which is a continuous function of the chemical potential μ unlike (4.8). Thus the first order phase transition in the full model (4.1) occurs precisely due to the imaginary part Γ of the action. This model therefore provides a nice testing ground for simulation techniques for finite density QCD [14].

5. Testing the factorization method in the RMT

Since ν is complex for each configuration, we decompose it into the real and imaginary parts as $\nu = \nu_{\rm R} + i\nu_{\rm I}$ and calculate $\langle\nu_{\rm R}\rangle$ and $\langle\nu_{\rm I}\rangle$ by the factorization method ($\langle\nu_{\rm I}\rangle$ is purely imaginary). We introduce the distribution functions for $\nu_{\rm R}$ and $\nu_{\rm I}$ separately as

$$\rho_i(x) \stackrel{\rm def}{=} \langle\delta(x-\nu_i)\rangle \qquad i = \mathrm{R}, \mathrm{I}\ , \tag{5.1}$$

$$\rho_i^{(0)}(x) \stackrel{\rm def}{=} \langle\delta(x-\nu_i)\rangle_0 \qquad i = \mathrm{R}, \mathrm{I}\ . \tag{5.2}$$

The factorization holds for both $\rho_{\rm R}(x)$ and $\rho_{\rm I}(x)$ as

$$\rho_i(x) = \frac{1}{C}\,\rho_i^{(0)}(x)\,\varphi_i(x) \qquad i = \mathrm{R},\ \mathrm{I}\ , \tag{5.3}$$

where the constant C is given by $C \stackrel{\rm def}{=} \langle \mathrm{e}^{i\Gamma}\rangle_0$. The weight factors $\varphi_i(x)$ can be written as a VEV

$$\varphi_i(x) \stackrel{\rm def}{=} \langle \mathrm{e}^{i\Gamma}\rangle_{i,\,x} \qquad i = \mathrm{R},\ \mathrm{I} \tag{5.4}$$

with respect to the constrained phase-quenched model

$$Z_i(x) = \int \mathrm{d}W\, \mathrm{e}^{-N\,\mathrm{tr}(W^\dagger W)}\, |\det D|\, \delta(x - \nu_i) \qquad i = \mathrm{R}, \mathrm{I}\,. \tag{5.5}$$

Under the transformation $W \mapsto -W$, the Gaussian action is invariant, whereas the fermion determinant $\det D$ as well as the quark number density ν becomes complex conjugate. Due to this symmetry, we have

$$\varphi_\mathrm{R}(x)^* = \varphi_\mathrm{R}(x)\,, \tag{5.6}$$

$$\varphi_\mathrm{I}(x)^* = \varphi_\mathrm{I}(-x)\,, \tag{5.7}$$

$$\rho_\mathrm{I}^{(0)}(-x) = \rho_\mathrm{I}^{(0)}(x)\,. \tag{5.8}$$

Using these properties, we arrive at

$$\langle \nu_\mathrm{R} \rangle = \frac{1}{C} \int_{-\infty}^{\infty} \mathrm{d}x\, x\, \rho_\mathrm{R}^{(0)}(x)\, w_\mathrm{R}(x)\,, \tag{5.9}$$

$$\langle \nu_\mathrm{I} \rangle = \frac{2\,i}{C} \int_0^{\infty} \mathrm{d}x\, x\, \rho_\mathrm{I}^{(0)}(x)\, w_\mathrm{I}(x)\,, \tag{5.10}$$

$$C = \int_{-\infty}^{\infty} \mathrm{d}x\, \rho_\mathrm{R}^{(0)}(x)\, w_\mathrm{R}(x)\,, \tag{5.11}$$

where the weight factors $w_i(x)$ are defined by

$$w_\mathrm{R}(x) \stackrel{\mathrm{def}}{=} \langle \cos \Gamma \rangle_{\mathrm{R},x} \tag{5.12}$$

$$w_\mathrm{I}(x) \stackrel{\mathrm{def}}{=} \langle \sin \Gamma \rangle_{\mathrm{I},x} = -w_\mathrm{I}(-x)\,. \tag{5.13}$$

Table 1. Results of the analysis of $\langle \nu \rangle$ described in the text. Statistical errors computed by the jackknife method are shown. The last column represents the exact result for $\langle \nu \rangle$ at each μ and N. For $\mu = 0.2$ the exact result is $\langle \nu \rangle = -0.2$ with an accuracy better than 1 part in 10^{-9}.

μ	N	$\langle \nu_\mathrm{R} \rangle$	$i\,\langle \nu_\mathrm{I} \rangle$	$\langle \nu \rangle$	$\langle \nu \rangle$ (exact)
0.2	8	0.0056(6)	-0.1970(5)	-0.1915(7)	-0.20000...
0.2	16	0.0060(4)	-0.1905(13)	-0.1845(13)	-0.20000...
0.2	24	0.0076(9)	-0.1972(14)	-0.1896(17)	-0.20000...
0.2	32	0.0021(8)	-0.1947(19)	-0.1927(25)	-0.20000...
0.2	48	0.0086(37)	-0.2086(54)	-0.2000(88)	-0.20000...
1.0	8	0.8617(10)	0.1981(13)	1.0598(12)	1.066501...
1.0	16	0.8936(2)	0.1353(6)	1.0289(5)	1.032240...
1.0	32	0.9207(1)	0.0945(2)	1.0152(3)	1.015871...

In Table 1 we show our results for two values of μ, $\mu = 0.2$ and $\mu = 1.0$, which are on opposite sides of the first order phase transition point $\mu =$

$\mu_c = 0.527\cdots$. They are in good agreement with the exact results, and the achieved values of N are large enough to extract the large N limit.

In Fig. 1 we plot $w_R(x)$ for $N = 8$ at various μ. It is interesting that the $w_R(x)$ changes from positive to negative for $\mu < \mu_c$, but it changes from negative to positive for $\mu > \mu_c$. (Similarly $w_I(x)$ is positive at $x > 0$ for $\mu < \mu_c$, but it is negative at $x > 0$ for $\mu > \mu_c$.) Thus the behavior of $w_i(x)$ changes drastically as the chemical potential μ crosses its critical value μ_c. These results provide a clear understanding of how the first order phase transition occurs due to the effects of Γ. Fig. 2 shows the results for $\langle\nu\rangle$ obtained by the factorization method for $N = 8$ at various μ, which nicely reproduce the gap developing at the critical point.

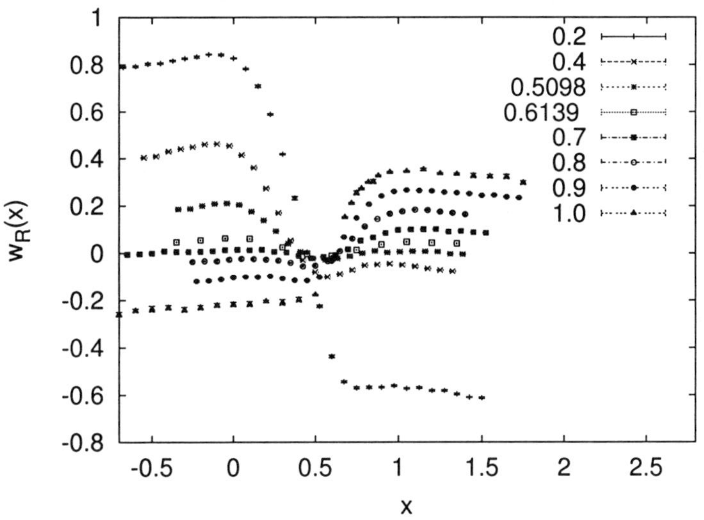

Figure 1. The weight factor $w_R(x)$ is plotted for $N = 8$ at various μ.

6. Noncommutativity of $\mu \to 0$ and $N \to \infty$

In this Section we discuss the noncommutativity of the two limits, $\mu \to 0$ and $N \to \infty$. The absence of such noncommutativity is implicitly assumed in most of the recent approaches used in simulating finite density QCD at small μ. This includes the multi-parameter reweighting approach [8], the Taylor expansion approach [9], and the imaginary μ approach [10,11]. In all

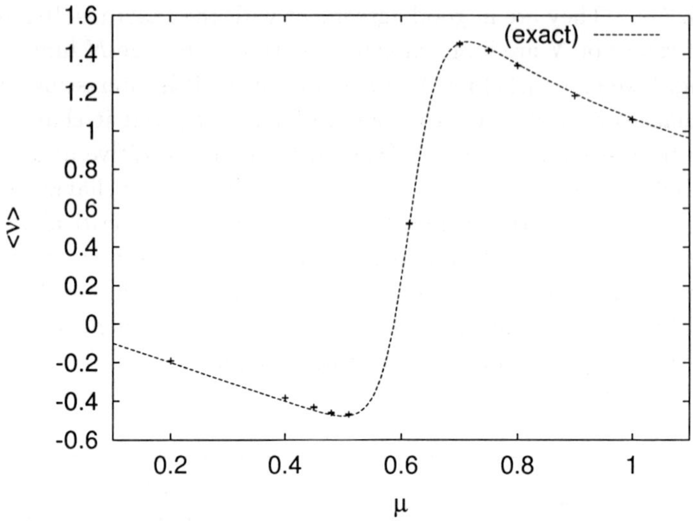

Figure 2. The VEV $\langle\nu\rangle$ obtained by the factorization method is plotted against μ for $N = 8$ including the critical regime. Statistical errors computed by the jackknife method are also shown. The dashed line represents the exact result (4.7) for $\langle\nu\rangle$ at $N = 8$.

these works one restricts oneself to the small μ regime where the fluctuation of the phase is still under control.

In fact the noncommutativity can be readily seen from the exact result (4.4) for the partition function (4.1). The phase of the determinant vanishes at $\mu = 0$ for finite N, and one obtains a nonzero result for the partition function $Z = 1$ in the large N limit. On the other hand, the oscillation of the phase becomes pronounced at sufficiently large N even for small but finite μ, and as a result one obtains $Z = 0$ in the large N limit as far as μ is kept finite. This implies in particular that the free energy

$$f(\mu) = -\lim_{N\to\infty} \frac{1}{N^2} \ln Z(\mu, N) \tag{6.1}$$

has a discontinuity at $\mu = 0$ as

$$\lim_{\mu\to 0} f(\mu) > f(0) = 0 \; . \tag{6.2}$$

We expect that, in general, the free energy of a system with a complex action has a discontinuity at a point in the parameter space where the imaginary part of the action vanishes identically.

With the factorization method we can take the two limits $\mu \to 0$, $N \to \infty$ in different orders and compare the results. As we will see, we do observe the noncommutativity in various ways. On the other hand, we know from the exact result (4.7) that the VEV $\langle \nu \rangle$ does not have the noncommutativity. In the factorization method $\langle \nu \rangle = \langle \nu_{\mathrm{R}} \rangle + i\langle \nu_{\mathrm{I}} \rangle$ is calculated by the formulae (5.9), (5.10) and (5.11). In fact the functions $w_{\mathrm{R}}(x)$, $w_{\mathrm{I}}(x)$ and $\rho_{\mathrm{R}}^{(0)}(x)$ have the noncommutativity, but these effects cancel each other in the end results for $\langle \nu_{\mathrm{R}} \rangle$ and $\langle \nu_{\mathrm{I}} \rangle$. In what follows we present preliminary results relevant to $\langle \nu_{\mathrm{R}} \rangle$, but similar results are obtained for $\langle \nu_{\mathrm{I}} \rangle$ as well [15].

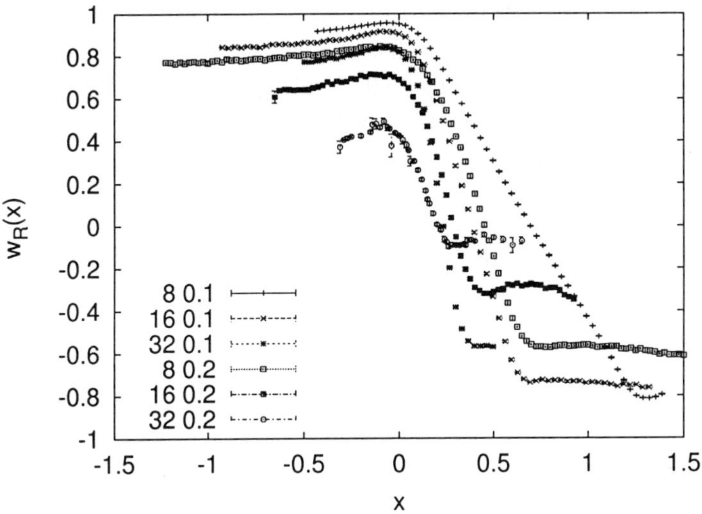

Figure 3. The weight factor $w_{\mathrm{R}}(x)$ is plotted for $\mu = 0.1$ and 0.2 at $N = 8, 16, 32$.

Let us first look at the weight factor $w_{\mathrm{R}}(x)$, which has the noncommutativity similar to the partition function. At $\mu = 0$ one obtains $w_{\mathrm{R}}(x) \equiv 1$ for any N, whereas in the large N limit one obtains $w_{\mathrm{R}}(x) \equiv 0$ for any μ. In Fig. 3 we plot $w_{\mathrm{R}}(x)$ for $\mu = 0.1$ and $\mu = 0.2$ at $N = 8, 16, 32$. It shows clearly that the behavior of $w_{\mathrm{R}}(x)$ depends much on the order of the two limits $\mu \to 0$, $N \to \infty$.

Let us next turn to $\rho_{\mathrm{R}}^{(0)}(x)$. In Fig. 4 we plot it for various N at $\mu = 0.2$. At small N the distribution is peaked near the origin and the dependence on N is small. At sufficiently large N the peak moves to $x \sim \mu$ and starts to grow, which is consistent with the large N result $\langle \nu_{\mathrm{R}} \rangle_0 = \mu$. Empirically we find that the transition occurs at

$$N_c = \frac{0.25 \sim 0.3}{\mu^2} . \tag{6.3}$$

Thus the distribution $\rho_R^{(0)}(x)$ for the phase-quenched model also depends much on the order of the two limits $\mu \to 0$, $N \to \infty$. This can be seen more clearly in Fig. 5, where we plot the VEV $\langle \nu_R \rangle_0$ against μ for various N. In particular the derivative $\frac{\partial}{\partial \mu} \langle \nu_R \rangle_0$ becomes 0 if one takes the $\mu \to 0$ limit first, but it becomes 1 if one takes the $N \to \infty$ limit first.

The product $\rho_R^{(0)}(x)\, w_R(x)$ gives the unnormalized distribution for ν_R in the full model, which we plot in Fig. 6. The distribution itself, even after appropriate normalization, has the noncommutativity, but the VEV $\langle \nu_R \rangle$ calculated by the formula (5.9) is always closed to zero (see Table 1). The reason depends on the order of the two limits. If we take the $N \to \infty$ limit first, the positive and negative regions of $\rho_R(x)$ cancel each other in the calculation of the first moment. If we consider small μ first, the distribution is peaked around the origin, which makes the first moment close to zero. Thus the noncommutativity cancels in the end result for the VEV $\langle \nu_R \rangle$.

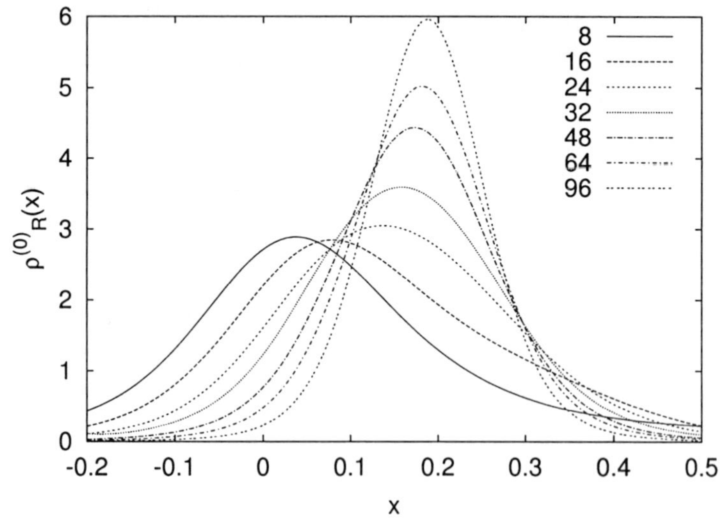

Figure 4. The function $\rho_R^{(0)}(x)$ is plotted for $\mu = 0.2$ at various N.

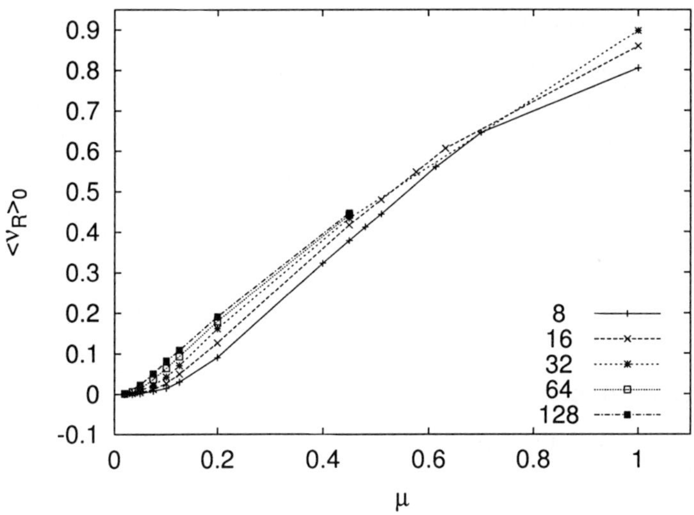

Figure 5. The VEV $\langle \nu_{\rm R} \rangle_0$ is plotted against μ for various N.

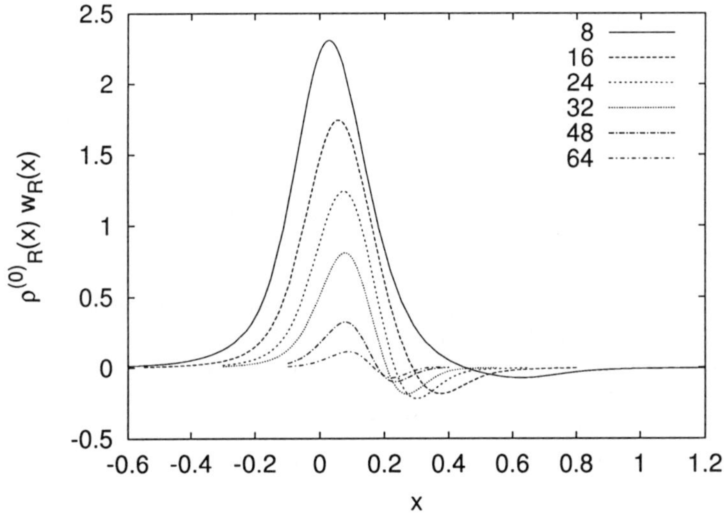

Figure 6. The product $\rho_{\rm R}^{(0)}(x) w_{\rm R}(x)$, which gives the unnormalized distribution for $\nu_{\rm R}$ in the full model, is plotted for $\mu = 0.2$ at various N.

7. Concluding remarks

The factorization method has been applied also to other systems with complex actions. In the original paper [12], it was used to study the dynamical generation of space time in superstring theory based on its matrix model formulation [17]. In this case the weight factors turned out to be positive definite, which enabled us to use their scaling property to make an extrapolation to larger system size. The method [18] proposed for simulating θ-vacuum like systems can be regarded as a special case of the factorization method. Promising results are obtained in the 2d CP^3 model etc.

Acknowledgments

This work is partially supported by Grant-in-Aid for Scientific Research (No. 14740163) from the Ministry of Education, Culture, Sports, Science and Technology.

References

1. D. Bailin and A. Love, *Phys. Rept.* **107**, 325 (1984).
2. D. T. Son, *Phys. Rev.* **D59**, 094019 (1999).
3. T. Schafer and F. Wilczek, *Phys. Rev.* **D60**, 114033 (1999).
4. M. G. Alford, K. Rajagopal and F. Wilczek, *Phys. Lett.* **B422**, 247 (1998).
5. R. Rapp, T. Schafer, E. V. Shuryak and M. Velkovsky, *Phys. Rev. Lett.* **81**, 53 (1998).
6. J. B. Kogut, D. Toublan and D. K. Sinclair, *Nucl. Phys.* **B642**, 181 (2002).
7. J. B. Kogut and D. K. Sinclair, *Phys. Rev.* **D66**, 034505 (2002).
8. Z. Fodor and S. D. Katz, *J. High Energy Phys.* **03**, 014 (2002).
9. C. R. Allton *et al.*, *Phys. Rev.* **D66**, 074507 (2002).
10. P. de Forcrand and O. Philipsen, *Nucl. Phys.* **B642**, 290 (2002).
11. M. D'Elia and M. P. Lombardo, *Phys. Rev.* **D67**, 014505 (2003).
12. K. N. Anagnostopoulos and J. Nishimura, *Phys. Rev.* **D66**, 106008 (2002).
13. J. Ambjørn, K. N. Anagnostopoulos, J. Nishimura and J. J. Verbaarschot, *J. High Energy Phys.* **10**, 062 (2002).
14. M. A. Stephanov, *Phys. Rev. Lett.* **76**, 4472 (1996).
15. J. Ambjørn, K. N. Anagnostopoulos, J. Nishimura and J. J. Verbaarschot, in preparation.
16. M. A. Halasz, A. D. Jackson and J. J. Verbaarschot, *Phys. Rev.* **D56**, 5140 (1997).
17. N. Ishibashi, H. Kawai, Y. Kitazawa and A. Tsuchiya, *Nucl. Phys.* **B498**, 467 (1997) .
18. V. Azcoiti, G. Di Carlo, A. Galante and V. Laliena, *Phys. Rev. Lett.* **89**, 141601 (2002).

DENSE QUARK MATTER IN COMPACT STARS

M. G. ALFORD

Physics Department
Washington University CB 1105
Saint Louis, MO 63130
USA

According to quantum chromodynamics (QCD), matter at ultra-high densities will take the form of a color-superconducting quark liquid, in which there is a condensate of of Cooper pairs of quarks near the Fermi surface. I present a review of the physics of color superconductivity, and discuss possible signatures by which it might be detected in compact stars. I give a detailed discussion of the effects of color superconductivity on the mass-radius relationship of compact stars, showing that one would have to fix the bag constant by other measurements in order to see the effects of color superconductivity. An additional parameter in the quark matter equation of state connected with perturbative corrections allows quark matter to imitate nuclear matter over the relevant density range, so that hybrid stars can show a mass-radius relationship very similar to that of nuclear matter, and their masses can reach 1.9 $M_{\odot}$.

1. Introduction

One of the most striking features of QCD is asymptotic freedom: the force between quarks becomes arbitrarily weak as the characteristic momentum scale of their interaction grows larger. This immediately suggests that at sufficiently high densities and low temperatures, matter will consist of a Fermi sea of essentially free quarks, whose behavior is dominated by the high-momentum quarks that live at the Fermi surface.

However, over the last few years it has become clear that the phase diagram of QCD is much richer than this. In addition to the hadronic phase with which we are familiar and the quark gluon plasma (QGP) that is predicted to lie at temperatures above 170 MeV, there is a whole family of "color superconducting" phases, that are expected to occur at high density and low temperature [1]. The essence of color superconductivity is quark pairing, driven by the BCS mechanism, which operates when there exists an attractive interaction between fermions at a Fermi surface. The QCD quark-quark interaction is strong, and is attractive in many channels, so we expect cold dense quark matter to *generically* exhibit color superconductivity. Moreover, quarks, unlike electrons, have color and flavor as well as spin degrees of freedom, so many different patterns of pairing are possible. This leads us to expect a rich phase structure in matter beyond nuclear density.

Color superconducting quark matter may occur naturally in the universe, in the cold dense cores of compact ("neutron") stars, where densities are above nuclear density, and temperatures are of the order of tens of keV. (It might conceivably be possible to create it in future low-energy heavy ion colliders, such as the Japan Proton Accelerator Research Complex (J-PARC) or the Compressed Baryonic Matter facility at GSI Darmstadt.) Up to now, most work on signatures has focussed on properties of color superconducting quark matter that would affect observable features of compact stars, and I will discuss some of these below.

2. Phase diagram of quark matter

In the real world there are two light quark flavors, the up (u) and down (d), with masses $\lesssim 10$ MeV, and a medium-weight flavor, the strange (s) quark, with mass ~ 150 MeV. The strange quark therefore plays a crucial role in the phases of QCD. In Fig. 1 I show two conjectured phase diagrams for QCD, one for a light strange quark and one for a heavy strange quark. In both cases, along the horizontal axis the temperature is zero, and the density rises from the onset of nuclear matter through the transition to quark matter. Compact stars are in this region of the phase diagram, although it is not known whether their cores are dense enough to reach the quark matter phase. Along the vertical axis the temperature rises, taking us through the crossover from a hadronic gas to the quark gluon plasma. This is the regime explored by high-energy heavy-ion colliders.

If the effective strange quark mass is low enough in quark matter at a few times nuclear density, then there is a direct transition from nuclear matter to color-flavor-locked (CFL) quark matter [2]. In the CFL phase the strange quark participates symmetrically with the up and down quarks in Cooper pairing—this is described in more detail below. If the strange quark is too heavy to pair symmetrically with the light quarks at these densities, then there will be an interval of some other phase. This may be crystalline color superconductivity [3] or some form of single-flavor pairing [4,5] (see below).

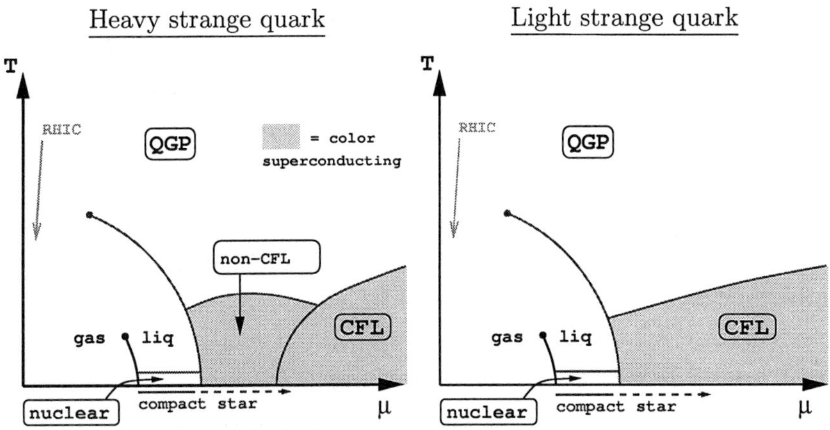

Figure 1. Conjectured phase diagrams for QCD in the real world. For a light strange quark, there is a direct transition from nuclar matter to color-flavor locked color superconducting quark matter. For a heavy strange quark, there is an intermediate phase where the strange quark pairs in some other way. Depending on the strength of instanton interactions, the CFL phase may include K^0 condensation.

3. Review of color superconductivity

The essential physics of color superconductivity is the same as that underlying conventional superconductivity in metals [6,5,7]. As mentioned above, asymptotic freedom of QCD means that at sufficiently high density and low temperature, there is a Fermi surface of almost free quarks. The interactions between quarks near the Fermi surface are certainly attractive in some channels (quarks bind together to form baryons) and it was shown by Bardeen, Cooper, and Schrieffer (BCS) [8] that if there is *any* channel in which the interaction is attractive, then there is a state of lower free energy than a simple Fermi surface. That state arises from a complicated coherent superposition of pairs of particles (and holes)—"Cooper pairs".

Attractive interactions play a crucial role in the BCS mechanism for the formation of Cooper pairs. This can easily be understood in an intuitive way. The Helmholtz free energy is $F = E - \mu N$, where E is the total energy of the system, μ is the chemical potential, and N is the number of fermions. The Fermi surface is defined by a Fermi energy $E_F = \mu$, at which the free energy is minimized, so adding or subtracting a single particle costs zero free energy. Now switch on a weak attractive interaction. It costs no free energy to add a pair of particles (or holes), and the attractive interaction between them then lowers the free energy of the system. Many such pairs will therefore be created in the modes near the Fermi surface, and these pairs, being bosonic, will form a condensate. The ground state will be a superposition of states with all numbers of pairs, breaking the fermion number symmetry.

Since pairs of quarks cannot be color singlets, the resulting condensate will break the local color symmetry $SU(3)_{\rm color}$. We call this "color superconductivity". Note that the quark pairs play the same role here as the Higgs particle does in the standard model: the color-superconducting phase can be thought of as the Higgs phase of QCD.

3.1. *Three flavors: Color-flavor locking (CFL)*

The favored pairing pattern at high densities, where the strange quark Fermi momentum is close to the up and down quark Fermi momenta, is "color-flavor locking" (CFL) [2]. This has been confirmed by both NJL [2,9,10] and gluon-mediated interaction calculations [11]. The CFL pairing pattern is

$$\begin{gathered}\langle q_i^\alpha q_j^\beta\rangle_{1PI} \propto C\gamma_5\Big((\kappa+1)\delta_i^\alpha\delta_j^\beta + (\kappa-1)\delta_j^\alpha\delta_i^\beta\Big) \\ [SU(3)_{\rm color}] \times \underbrace{SU(3)_L \times SU(3)_R}_{\supset [U(1)_Q]} \times U(1)_B \rightarrow \underbrace{SU(3)_{C+L+R}}_{\supset [U(1)_{\tilde Q}]} \times \mathbb{Z}_2\end{gathered} \tag{1}$$

Color indices α, β and flavor indices i, j run from 1 to 3, Dirac indices are suppressed, and C is the Dirac charge-conjugation matrix. The term multiplied by κ corresponds to pairing in the $(\mathbf{6}_S, \mathbf{6}_S)$, which although not energetically favored breaks no additional symmetries and so κ is in general small but not zero [2,11,12]. The Kronecker deltas connect color indices with flavor indices, so that the condensate is not invariant under color rotations, nor under flavor rotations, but only under simultaneous, equal and opposite, color and flavor rotations. Since color is only a vector symmetry, this condensate is only invariant under vector flavor+color rotations, and breaks chiral symmetry. The features of the CFL pattern of condensation are

- The color gauge group is completely broken. All eight gluons become massive. This ensures that there are no infrared divergences associated with gluon propagators.
- All the quark modes are gapped. The nine quasiquarks (three colors times three flavors) fall into an $\mathbf{8} \oplus \mathbf{1}$ of the unbroken global $SU(3)$, so there are two gap parameters. The singlet has a larger gap than the octet.

- A rotated electromagnetism ("$\tilde{Q}$") survives unbroken. It is a combination of the original photon and one of the gluons.
- Two global symmetries are broken, the chiral symmetry and baryon number, so there are two gauge-invariant order parameters that distinguish the CFL phase from the QGP, and corresponding Goldstone bosons which are long-wavelength disturbances of the order parameter. When the light quark mass is non-zero it explicitly breaks the chiral symmetry and gives a mass to the chiral Goldstone octet, but the CFL phase is still a superfluid, distinguished by its baryon number breaking.
- The symmetries of the 3-flavor CFL phase are the same as those one might expect for 3-flavor hypernuclear matter [9], so it is possible that there is no phase transition between them.

In a real compact star we must require electromagnetic and color neutrality [13,14] (possibly via mixing of oppositely-charged phases), allow for equilibration under the weak interaction, and include a realistic mass for the strange quark. These factors tend to pull apart the Fermi momenta of the different quark species, imposing an energy cost on cross-species pairing.

The requirement of neutrality penalizes the 2SC phase relative to the CFL phase. This can be shown by analyzing a generic expansion of the free energy in powers of m_s/μ [14] or by an NJL calculation [15] that handles $m_s \sim \mu$ and includes the coupling between the chiral condensate and quark condensate gap equations. The net result, assuming that mixed phases are excluded by the surface energy cost [16] (see Section. 4), is that there is no (or very little) density range in which 2SC is the phase with the lowest free energy: unpaired or CFL-paired quark matter are generally favored over 2SC.

Including a small strange quark mass may (depending on the size of instanton effects [17]) induce a flavor rotation of the condensate known as "kaon condensation" [18], which breaks isospin. A sufficiently large strange quark mass will prevent all pairing between the strange quark and the light quarks [9,14,19] yielding a two-flavor color superconductor (2SC) [7]. The true pairing pattern in the intermediate range of strange quark masses is a topic of ongoing inquiry. One way to achieve pairing between different flavors while accomodating the tendency for the Fermi momenta to separate is to only pair over part of the Fermi surface: This is crystalline ("LOFF") pairing [3,20] (see Section 4). Another possibility is that each flavor simply pairs with itself [4,5]. However, from now on we will assume that the strange quark is light enough so that quark matter always occurs in the CFL phase.

4. Compact star transport phenomenology

The high density and relatively low temperature required to produce color superconducting quark matter may be attained in compact stars. Typical compact stars have masses close to $1.4M_{\odot}$, and are believed to have radii of order 10 km.

Color superconductivity affects the equation of state at order $(\Delta/\mu)^2$. It also gives mass to excitations around the ground state: it opens up a gap at the quark Fermi surface, and makes the gluons massive. One would therefore expect it to have a profound effect on transport properties, such as mean free paths, conductivities and viscosities. Various observable consequences are under investigation.

- *r-mode spindown.* The r-mode is a bulk flow in a rotating star that, if the viscosity is low enough, radiates away energy and angular momentum in the form of gravitational waves. One can rule out certain models for compact stars on the grounds that they have such low damping that they could not support the high rotation rates observed in pulsars: r-mode spindown would have slowed them down. Madsen [21] has shown that for a compact star made *entirely* of quark matter in the CFL phase, even a gap as small as $\Delta = 1$ MeV is ruled out by observations of millisecond pulsars. It remains to extend this calculation to the more generic picture of a quark matter core surrounded by a nuclear mantle.
- *Interfaces and mixed phases.* These were studied in Ref. [16], and it was found that a mixed phase only occurs if the surface tension of the interface is less than about 40 MeV/fm^2 $= 0.2 \times (200\text{ MeV})^3$, a fairly small value compared to the relevant scales $\Lambda_{QCD} \approx$ 200 MeV, $\mu \sim 400$ MeV. A sharp nuclear-quark interface will have an energy-density discontinuity across it, which will affect gravitational waves emitted in mergers, and also the r-mode spectrum and the damping forces to which r-modes are subject.
- *Crystalline pairing (the "LOFF" phase).* This is expected to occur when two different types of quark have sufficiently different Fermi momenta that BCS pairing cannot occur [3]. This is a candidate for the intermediate phase of Fig. 1, where the strange quark mass, combined with requirements of weak equilibrium and charge neutrality, gives each quark flavor a different Fermi momentum. The

phenomenology of the crystalline phase has not yet been worked out, but recent calculations using Landau-Ginzburg effective theory indicate that the favored phase may be a face-centered cubic crystal [20], with a reasonably large binding energy. This raises the interesting possibility of glitches in quark matter stars.

– *Cooling by neutrino emission.* The cooling rate is determined by the heat capacity and emissivity, both of which are sensitive to the spectrum of low-energy excitations, and hence to color superconductivity. CFL quark matter, where all modes are gapped, has a much smaller neutrino emissivity and heat capacity than nuclear matter, and hence the cooling of a compact star is likely to be dominated by the nuclear mantle rather than the CFL core [22,23,24]. Other phases such as 2SC or LOFF give large gaps to only some of the quarks. Their cooling would proceed quickly, then slow down suddenly when the temperature fell below the smallest of the small weak-channel gaps. This behavior should be observable [25].

5. Mass-radius relationship for compact stars

Although the effects of color superconductivity on the quark matter equation of state are subdominant, they may have a large effect on the mass-radius relationship. The reason for this is that the pressure of quark matter relative to the hadronic vacuum contains a constant (the "bag constant" B) that represents the cost of dismantling the chirally broken and confining hadronic vacuum,

$$p = (1-c)\frac{3}{4\pi^2}\mu^4 - \frac{3}{4\pi^2}m_s^2\mu^2 + \frac{3}{\pi^2}\Delta^2\mu^2 - B\ . \tag{2}$$

If the bag constant is large enough so that nuclear matter is favored (or almost favored) over quark matter at $\mu \sim 320$ MeV, then the bag constant and μ^4 terms almost cancel, so if we can fix the bag constant by other means then the strange quark mass m_s and color superconducting gap Δ may have a large effect on the equation of state and hence on the mass-radius relationship of a compact star [26].

5.1. *$M(R)$ at fixed bag constant*

In Ref. [27] Sanjay Reddy and I explored the effect of quark pairing on the M-R relationship at fixed values of the bag constant that are consistent with nuclear phenomenology. Fig. 2 shows the mass-radius curve for the bag model of dense matter, in which there is competition between a nuclear matter phase and a quark matter phase. The nuclear matter was described either by the APR98 equation of state [28]. The quark matter equation of state was essentially that of equation (2), but we included the full (free-quark) correction due to the strange quark mass. The coefficient of the $\Delta^2\mu^2$ term is the one appropriate to CFL color superconductivity involving all nine colors/flavors of the quarks. We used physically reasonable values of the bag constant $B^{1/4} = 180$ MeV ($B = 137$ MeV/fm^3) and strange quark mass $m_s = 200$ MeV. We set the parameter c to zero: its effects will be discussed below.

Curves for unpaired ($\Delta = 0$) and color-superconducting ($\Delta = 100$ MeV) quark matter are shown. At these values the stars are typically "hybrid", containing both quark matter and nuclear matter. The solid lines in Fig. 2 correspond to stars that either have no QM at all, or a sharp transition between NM and QM: the core is made of quark matter, which is the favored phase at high pressure, and at some radius there is a transition to nuclear matter, which is favored at low pressure. The transition pressure is sensitive to Δ, for reasons discussed earlier. The dashed lines are for stars that contain a mixed NM-QM phase. In all cases we see that light, large stars consist entirely of nuclear matter. When the star becomes heavy enough, the central pressure rises to a level where QM occurs in the core. As can be seen from the figure the transition density is very sensitive to Δ. The line labeled "Cottam *et al.*" indicates the constraint obtained by recent measurements of the redshift on three spectral lines from EXO0748-676 [29].

5.2. *M(R) with non-free quarks: quark matter mimicking nuclear matter*

In Ref. [27] we kept the bag constant fixed, assuming that it could be fixed by other observations, and we treated the quark matter as free quarks with a pairing energy. It is interesting to see what happens when we relax these assumptions, since the bag constant is not easily measured, and even after taking pairing into account we expect remaining QCD interactions between the quarks in the Fermi sea.

To allow for effects of quark interactions beyond Cooper pairing, we follow the parameterization of Fraga et. al. [30], who find that the $O(\alpha_s^2)$ pressure for three unpaired flavors over the relevant range of μ is well-described by a bag-model-inspired form given by

$$P_{\alpha_s^2}(\mu) = \frac{3}{4\pi^2}\, a_{\text{eff}}\, \mu^4 - B_{\text{eff}}\,, \qquad a_{\text{eff}} \equiv 1 - c\,. \tag{3}$$

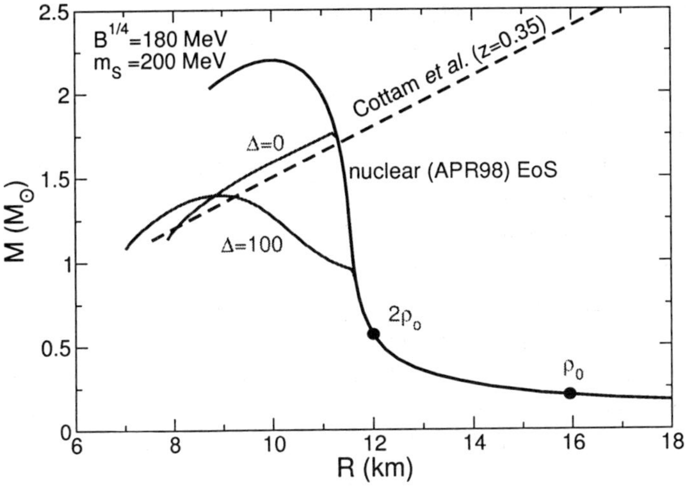

Figure 2. Mass-radius relationships for APR98 nuclear matter, competing with quark matter with fixed bag constant $B^{1/4} = 180$ MeV and $m_s = 200$ MeV, either unpaired ($\Delta = 0$) or CFL color-superconducting ($\Delta = 100$ MeV). The dots labeled ρ_0 and $2\rho_0$ on the nuclear matter mass-radius curve indicate that the central density at these locations correspond to nuclear and twice nuclear saturation density respectively.

They find $a_{\text{eff}} \approx 0.63$ ($c \approx 0.37$), but admit that at the density of interest for compact star physics the QCD coupling is strong, and there there is no reason to expect the leading order calculation to be accurate. We therefore take their result as indicating that it is reasonable to treat c as an additional parameter in the quark matter equation of state, as shown in Eq. (2), and we proceed to study its effects on the mass-radius relationship of compact stars.

To see how closely quark matter can mimic nuclear matter, we will not treat the bag constant as fixed, but tune it to keep the physics as constant as possible. Thus when we compare, say, $\Delta = 0$ (non-color-superconducting) quark matter with CFL ($\Delta = 50$) quark matter, we set the bag constant in each case so that the transition from nuclear to quark matter occurs at a given density of nuclear matter. This effectively "subtracts out" the part of any variation in Δ that simply corresponds to a renormalization of the bag constant, which is in any case very poorly known.

We explore the effect of a color superconducting gap Δ and perturbative correction c on the mass-radius relationship. We fix the bag constant by requiring that that the nuclear to quark matter phase transition occur at nuclear matter baryon density $\rho = 1.5n_{\text{sat}}$. The resultant $M(R)$ curves are shown in Fig. 3. The noticeable features of the plots are

(1) *Increasing c makes the stars smaller and lighter.*
In our parameterization the stars resulting from quark matter equations of state without perturbative correction ($c = 0$, blue lines) are smaller and lighter.

(2) *Color superconductivity acts like a change in the bag constant.*
In Ref. [27] we showed that at fixed bag constant, color superconductivity has a strong effect on the mass-radius relationship of compact stars. Here, by comparing the dashed lines with the dotted lines in Fig. 3, we see that it is difficult to distinguish the effect of color superconductivity from a change in the bag constant. In Fig. 3, as we vary parameters c and Δ of the quark matter equation of state, the bag constant is tuned to maintain a constant value of the nuclear density at the transition to quark matter, and in this situation color superconductivity only makes a small difference to the mass-radius relationship.

(3) *Quark matter with $c \approx 0.3$ looks just like APR98 nuclear matter.*
The stars with $c = 0.3$ have mass-radius relationships that are very similar to the pure nuclear APR98 matter. In fact, for the case where there are perturbative corrections but no color superconductivity the equations of state ($p(\mu)$) are so similar that our program found a series of phase transitions back and forth between CFL and APR98 up to $\mu = 546$ MeV (baryon density $\rho = 5.4n_{\text{sat}}$). This is why the $c = 0.3$ red dotted curve lies almost exactly on top of the solid black (APR98) curve, even though there was a phase transition from APR98 to CFL at $\rho = 1.5n_{\text{sat}}$ (which is first attained when the APR98 star reaches a mass of $0.315M_{\odot}$, $R = 13.3$ km).

5.3. $M(R)$ *measurements and quark matter*

We can now ask what significance mass and radius measurements will have for the presence of quark matter, and particularly color-superconducting quark matter, in compact stars.

- *What would rule out quark matter?*
From Fig. 3 we see that an observed mass $M \gtrsim 2\ M_{\odot}$ would be inconsistent with the star containing quark matter obeying the equation of state that we have studied here. However, we emphasize

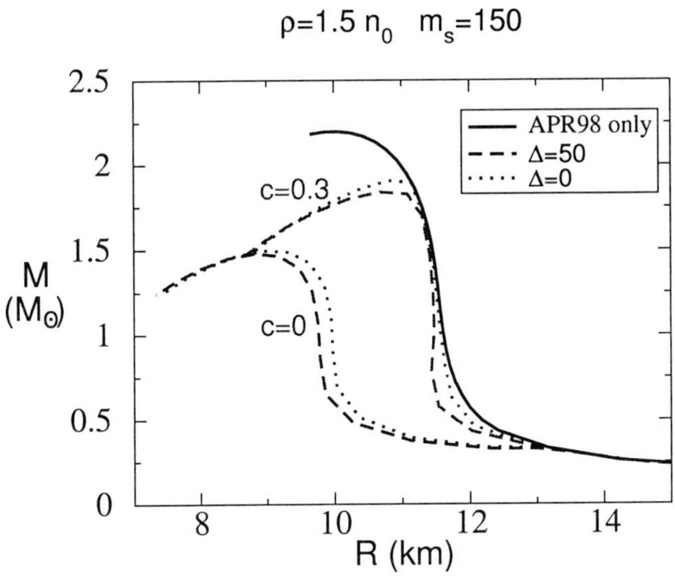

Figure 3. $M(R)$ relationship for APR98 nuclear matter with various quark matter equations of state. The strange quark is light, and the bag constant is tuned so that the nuclear matter to quark matter transition occurs at 1.5 times nuclear saturation density. Dotted lines are unpaired quark matter, dashed lines are CFL with gap of 50 MeV. Note how the curve for CFL quark matter with perturbative correction but no color superconductivity ($c = 0.3$, $\Delta = 0$; red dotted) closely follows the pure nuclear curve up to $M \approx 1.9 M_\odot$.

that by introducing the parameter c and setting it to a reasonable value $c \approx 0.3$ we have increased the mass range for hybrid stars, moving the upper limit from its old value around 1.6 $M_\odot$ up to about 1.9 $M_\odot$.

- *What would indicate the presence of quark matter?*
  This is difficult. Regions of M-R space that cannot be reached by any nuclear matter equation of state also cannot be reached by hybrid NM-QM equations of state. It is clear from Figs. 3 and 2 that hybrid stars are smaller than Walecka or APR98 nuclear matter stars, with radii of around 10 km at $M \approx 1.4$ $M_\odot$. But there are many other suggested nuclear equations of state, and the flattening of the $M(R)$ curve that appears in our plots to be characteristic of quark matter may easily be mimicked by kaon condensation in nuclear matter [31].

Obviously the region of pure quark matter objects which lie at very low mass and radius ("strangelets") is not attainable by nuclear matter, but the existence of such objects, unlike that of compact stars, remains a matter of speculation.

- *What would indicate the presence of color superconducting quark matter?*
  This is more difficult. Even if we found an $M(R)$ characteristic of quark matter, we would need an independent determination of the bag constant to claim that it was color-superconducting.

Acknowledgments

I thank the organizers of "Finite density QCD at Nara" and Confinement 2003. The work reported in Section 5 was performed in collaboration with Sanjay Reddy, and was supported by the UK PPARC and by the U.S. Department of Energy under grant number DE-FG02-91ER40628.

References

1. K. Rajagopal and F. Wilczek, hep-ph/0011333. M. G. Alford, Ann. Rev. Nucl. Part. Sci. **51** (2001) 131 [hep-ph/0102047]. T. Schaefer, hep-ph/0304281. D. H. Rischke, nucl-th/0305030. D. K. Hong, Acta Phys. Polon. B **32** (2001) 1253 [hep-ph/0101025].
2. M. Alford, K. Rajagopal and F. Wilczek, Nucl. Phys. **B537**, 443 (1999) [hep-ph/9804403].
3. M. Alford, J. Bowers and K. Rajagopal, Phys. Rev. D **63**, 074016 (2001) [hep-ph/0008208].
4. M. Alford, J. Bowers, J. Cheyne and G. Cowan, Phys. Rev. D **67**, 054018 (2003) [hep-ph/0210106]. M. Buballa, J. Hosek and M. Oertel, hep-ph/0204275. T. Schäfer, Phys. Rev. **D62**, 094007 (2000).
5. M. Iwasaki, T. Iwado, Phys. Lett. **B350**, 163 (1995); M. Iwasaki, Prog. Theor. Phys. Suppl. **120**, 187 (1995)
6. B. Barrois, Nucl. Phys. **B129** (1977) 390; "Nonperturbative effects in dense quark matter", Cal Tech PhD thesis, UMI 79-04847-mc (1979). S. Frautschi, Proceedings of workshop on hadronic matter at extreme density, Erice 1978, pp 19-27. D. Bailin and A. Love, Phys. Rept. **107** (1984) 325, and references therein.
7. M. Alford, K. Rajagopal and F. Wilczek, Phys. Lett. **B422**, 247 (1998) [hep-ph/9711395]. R. Rapp, T. Schäfer, E. V. Shuryak and M. Velkovsky, Phys. Rev. Lett. **81**, 53 (1998) [hep-ph/9711396].
8. J. Bardeen, L. Cooper, J. Schrieffer, Phys. Rev. **106**, 162 (1957); Phys. Rev. **108**, 1175 (1957)
9. T. Schäfer, F. Wilczek, Phys. Rev. Lett. **82**, 3956 (1999) [hep-ph/9903503].

10. N. Evans, J. Hormuzdiar, S. Hsu, M. Schwetz: Nucl. Phys. **B581**, 391 (2000) [hep-ph/9910313].
11. T. Schäfer, Nucl. Phys. **B575**, 269 (2000) [hep-ph/9909574]. I. Shovkovy, L. Wijewardhana, Phys. Lett. **B470**, 189 (1999).
12. R. D. Pisarski and D. H. Rischke, "Why color-flavor locking is just like chiral symmetry breaking". To be published in, *Proceedings of the Judah Eisenberg Memorial Symposium, "Nuclear Matter, Hot and Cold"*, Tel Aviv, April 14 - 16, 1999 [nucl-th/9907094].
13. K. Iida and G. Baym, Phys. Rev. D **63**, 074018 (2001) [hep-ph/0011229].
14. M. Alford and K. Rajagopal, JHEP **0206** (2002) 031 [hep-ph/0204001].
15. A. W. Steiner, S. Reddy and M. Prakash, Phys. Rev. D **66**, 094007 (2002) [hep-ph/0205201].
16. M. G. Alford, K. Rajagopal, S. Reddy and F. Wilczek, hep-ph/0105009.
17. T. Schäfer, Phys. Rev. D **65** (2002) 094033.
18. P. F. Bedaque and T. Schäfer, Nucl. Phys. A **697** (2002) 802 [hep-ph/0105150].
19. M. Alford, J. Berges and K. Rajagopal, Nucl. Phys. **B558**, 219 (1999) [hep-ph/9903502].
20. J. A. Bowers and K. Rajagopal, hep-ph/0204079.
21. J. Madsen, Phys. Rev. Lett. **85**, 10 (2000) [astro-ph/9912418].
22. D. Page, M. Prakash, J. Lattimer, A. Steiner, Phys. Rev. Lett. **85** (2000) 2048 [hep-ph/0005094].
23. I. Shovkovy, P. Ellis [astro-ph/0207346].
24. P. Jaikumar, M. Prakash and T. Schafer, Phys. Rev. D **66**, 063003 (2002) [astro-ph/0203088].
25. S. Reddy, M. Sadzikowski and M. Tachibana, nucl-th/0306015.
26. G. Lugones and J. E. Horvath, hep-ph/0211070.
27. M. Alford and S. Reddy, Phys. Rev. D **67**, 074024 (2003) [nucl-th/0211046].
28. A. Akmal, V.R. Pandharipande, D.G. Ravenhall, Phys.Rev. C58 1804 (1998) [nucl-th/9804027].
29. J. Cottam, F. Paerels, M. Mendez, Nature **420**, 51 (2002).
30. E. S. Fraga, R. D. Pisarski and J. Schaffner-Bielich, Phys. Rev. **D63**, 121702 (2001) [hep-ph/0101143].
31. J. M. Lattimer and M. Prakash, Astrophys. J. **550**, 426 (2001) [astro-ph/0002232].

THERMAL PHASE TRANSITION OF DENSE QCD*

T. MATSUURA and T. HATSUDA

Department of Physics, The University of Tokyo,
Tokyo 113-0033, Japan

K. IIDA

The Institute of Physical and Chemical Research (RIKEN),
Wako, Saitama 351-0198, Japan

G. BAYM

Department of Physics, University of Illinois at Urbana-Champaign,
Urbana, Illinois 61801-3080, USA

Using the Ginzburg-Landau (GL) free energy, we study the effects of thermal fluctuations of gluons and the diquark pairing field on the superconducting-to-normal state phase transition in a three-flavor color superconductor. At high baryon densities, the system becomes a Type I superconductor and the gluonic fluctuations dominate over diquark fluctuations. The thermal gluons induce a cubic term in the GL free energy, as well as large corrections to quadratic and quartic terms of the order parameter. The cubic term leads to a relatively strong first order transition in comparison to the very weak first order transitions in metallic Type I superconductors. The strength of the first order transition decreases with increasing baryon density. In addition, gluonic fluctuations lower the critical temperature of the first order transition. We derive explicit formulae for the critical temperature and the discontinuity of the order parameter at the critical point. The validity of the first order transition obtained in the one-loop approximation is also examined by estimating the size of the critical region.

1. Introduction

Degenerate quark matter at high baryon density is expected to undergo a phase transition to a color superconducting state.[1] The properties of color superconductors have been much studied in various approaches.[2,3,4,5,6] A major difference of color superconductors and metallic superconductors is

*talk presented by T. Hatsuda

that the former is a highly relativistic system in which the long-range magnetic interaction (dynamically screened only by Landau-damping[7]) is responsible for the formation of the non-standard form of the superconducting gap, Δ.[8] Despite this non-BCS feature of color superconductivity, the finite temperature transition in mean-field theory is of second order, with a BCS critical temperature $T_c \simeq 0.57\Delta$.[9]

In this paper, we address the question of thermal fluctuations of the gluons and of the diquark pairing. The similar thermal fluctuations were first studied in BCS superconductors in metals in Ref. 10, and in finite temperature field theory in Ref. 11. In metalic superconductors, Type I materials have a *weak* first order transition, characterized by a cubic term of the order parameter in the GL free energy induced by the thermal photons. A similar mechanism can be expected in color superconductors.[2] However, there are some crucial differences from metallic superconductors. Firstly, the fluctuations of the diquark field alone may lead to a first order phase transition.[13,14] Secondly, thermal gluon fluctuations may induce a relatively *strong* first order transition, in contrast to the metallic case partly because of the relativistic nature of the quarks and partly because of the large coupling constant α_s.

We study the effects of fluctuations of the diquark and gluon fields on the phase transition via their effects on the GL free energy, emphasizing the relative importance of the diquark and gluon fluctuations, and of a theoretical treatment of gluon fluctuations that consistently keeps all terms of the same order. We estimate, semi-quantitatively, the strength of first order transition as well as the modification of the transition temperature. Unlike the conclusion of Ref. 2, we find that the first order transition becomes weak with increasing baryon density, and that the transition temperature is lowered from its mean-field value. Full account of this work will be reported in Ref. 12.

2. Ginzburg-Landau Free Energy

Let us consider a system of degenerate massless u, d, s quarks with a common Fermi momentum. The pairng gap of the quark of color b and flavor j with that of color c and flavor k in the $J^P = 0^+$ channel is written as ϕ_{bcjk}. By further assuming that the pairing takes place in the color-flavor antisymmetric channel which is expected to be the most attractive in the weak coupling, the gap is parametrized as[4,5] $\phi_{bcjk} = \epsilon_{abc}\epsilon_{ijk}(\mathbf{d})^i_a$. Under $G = SU(3)_c \times SU(3)_{L+R} \times U(1)_B$, $\mathbf{d}^i_a$ transforms as a vector and belongs to the $(3^*, 3^*)$ representation of $SU(3)_c$ and $SU(3)_{L+R}$.

The GL free energy in three spatial dimensions, written in terms of $\mathbf{d}_a^i(\mathbf{x})$, with coupling to the $SU(3)_c$ gluon gauge fields, reads [5]

$$S = \bar{\alpha} \sum_a |\mathbf{d}_a|^2 + \beta_1 (\sum_a |\mathbf{d}_a|^2)^2 + \beta_2 (\sum_{ab} |\mathbf{d}_a^* \cdot \mathbf{d}_b|)^2 \qquad (1)$$

$$+2\kappa_T \sum_a |(D_l \mathbf{d})_a|^2 + \frac{1}{4} G_{lm}^\alpha G_{lm}^\alpha .$$

The parameters $\bar{\alpha}$, β_1, and β_2 characterize the homogeneous part of the free energy, while κ_T is the stiffness parameter. Since $\mathbf{d}_a^i$ is antisymmetric in color space, the color-covariant derivative reads $(D_l \mathbf{d})_a = \partial_l \mathbf{d}_a + \frac{i}{2} g A_l^\alpha (\lambda^{\alpha *} \mathbf{d})_a$, where the $\lambda^{\alpha *}$ are the complex conjugates of the Gell-Mann matrices. G_{lm} is the spatial part of the gluon field-strength tensor. The free energy density Eq.(1) may be interpreted as an $SU(3)_c \times SU(3)_f$ scalar field theory coupled to an $SU(3)_c$ gauge field in three-spatial dimensions. Eq. (1) is model independent and valid near the critical temperature of the second order transition.

Although the general analysis does not require specific values of the parameters in the GL free energy, it is useful to bear in mind their characteristic scales, as found in weak coupling:

$$\bar{\alpha} = 4N(\mu/3) \ln \left(\frac{T}{T_c} \right), \quad \beta_1 = \beta_2 = 3\kappa_T = \frac{7\zeta(3)}{8(\pi T_c)^2} N(\mu/3), \qquad (2)$$

where $N(\mu/3) = (\mu/3)^2/(2\pi^2)$, $\zeta(3) = 1.2020 \cdots$, μ the *baryon* chemical potential and T_c the critical temperature in the weak coupling.

The color-flavor locking (CFL) and isoscalar (IS) ordering in three-flavor matter are characterized as $\mathbf{d}_a^i = d\ \delta_{ai}$ (CFL), $d\ \delta_{a3}\delta_{i3}$ (IS), respectively. Since $\bar{\alpha}$ changes sign at the mean field T_c, it is useful to introduce the reduced tempetature as $\bar{\alpha} = \alpha_0 (T - T_c)/T_c \equiv \alpha_0 t$. Whether the paired state just below T_c is CFL or IS depends on the values of β_1 and β_2. In the weak coupling limit where $\beta_1 = \beta_2$ and $\alpha_0 \sim \mu^2$, the CFL ordering is favored with $d_{\rm CFL} \sim T_c \sqrt{|t|}$ for $T \sim T_c$.

3. Fluctuation about the mean-field

Let us consider the effect of thermal fluctuations of the spin-zero diquark (scalar) field and the spin-one gluon fields about their mean values, in the Gaussian approximation. The fluctuations of the gauge fields at the same spatial coordinate, are given by the thermal average, $\langle \mathbf{A}^\alpha \mathbf{A}^\beta \rangle$, of the product of the gauge fields:

$$\langle \mathbf{A}^\alpha \mathbf{A}^\beta \rangle = 2\delta_{\alpha\beta}\, T \int_{|\mathbf{k}|<\Lambda} \frac{d^3k}{(2\pi)^3} \frac{1}{\mathbf{k}^2 + (m_A)^2_{\alpha\alpha}}, \tag{3}$$

where we have taken the Coulomb gauge $\nabla \cdot \mathbf{A} = 0$. The momentum Λ is an ultraviolet cutoff, which corresponds to an upper bound on the wave numbers of the classical thermal fluctuations with zero Matsubara frequency. This cutoff is inversely proportional to the size of the quark pairs ($\Lambda \sim d \sim T_c$).[17,10] In the following we take $\Lambda = T_c$ for simplicity. In Eq. (3), $(m_A)_{\alpha\beta}$ is the Meissner mass matrix. Its components are listed in Table 1.

Table 1. The inverse squared correlation lengths of the scalar (d) and gauge (A) fluctuations, m_d^2 and m_A^2, together with the number of degenerate modes corresponding to each fluctuation.

	IS		CFL	
	$m^2_{d,A}$	degeneracy	$m^2_{d,A}$	degeneracy
d	$2(\beta_1+\beta_2)\, d^2/\kappa_T$	1	$2(3\beta_1+\beta_2)\, d^2/\kappa_T$	1
	$-\beta_2 d^2/\kappa_T$	8	$2\beta_2 d^2/\kappa_T$	8
	0	9	0	9
A	$\frac{4}{3}\kappa_T g^2 d^2$	1	$2\kappa_T g^2 d^2$	8
	$\kappa_T g^2 d^2$	4		
	0	3		

In weak coupling, we have $m_A \sim g\mu\sqrt{|t|}$ for $T \sim T_c$. Since the Meissner mass is vanishingly small compared to $\Lambda \sim T_c$ near the second order critical point, we may expand Eq. (3) in terms of $(m_A)_{\alpha\alpha}/T_c$ as

$$\langle \mathbf{A}^\alpha \mathbf{A}^\beta \rangle = \delta_{\alpha\beta} \frac{TT_c}{\pi^2} \left[1 - \frac{\pi}{2} \frac{(m_A)_{\alpha\alpha}}{T_c} + \left(\frac{(m_A)_{\alpha\alpha}}{T_c^2} \right)^2 + \cdots \right]. \tag{4}$$

One can similarly calculate the expectation value of the product of the fluctuations of the scalar diquark field.

The number of modes corresponding to a given correlation length is also indicated in Table 1. The ninefold massless scalar modes ($m_d = 0$) may be understood as follows. The IS state, characterized by $\mathbf{d}_a^i = \delta_{a3}\delta_{i3} d$, is invariant under $G' = SU(2)_c \times SU(2)_{L+R} \times U(1) \times U(1)$. Here the first $U(1)$ symmetry corresponds to a simultaneous rotation in baryon-color space, and the second to a simultaneous rotation in baryon-flavor space. Thus the number of Nambu-Goldstone bosons is dim[G]-dim[G'] $= 17 - 8 = 9$. The CFL state, characterized by $\mathbf{d}_a^i = d\delta_{ai}$, is invariant under $G' = SU(3)_{c+L+R}$. Thus one has $17 - 8 = 9$ Nambu-Goldstone bosons in this case too. Note that not all massless scalar modes with $m_d = 0$ in Table 1 are physical. Parts of them are absorbed in the longitudinal components of the gluon. As a result, only four scalar modes out of nine are physical in the IS state, while only one scalar mode is physical in the CFL state. However, as we show later, the physical massless modes do not modify the structure of the one-loop free energy. In weak coupling, the masses of the physical modes behave as $m_d \sim T_c\sqrt{|t|}$ for $T \sim T_c$.

As discussed in Refs. 10 and 17, the initial term in the expansion Eq. (4) proportional to TT_c simply shifts the critical temperature, T_c, of the second order transition. On the other hand, the terms proportional to TT_cm_d and TT_cm_A in Eq. (4) induce a cubic term of the order parameter in the GL potential, and thus generally drive the first order phase transition.[10] The terms $TT_cm_d^2$ and $TT_cm_A^2$ in Eq. (4) modify the coefficient of quadratic term of the order parameter in the GL potential, which turn out to be important in determining the strength of the first order transition.

4. Critical region

Let us now discuss the critical regions for scalar and gauge fluctuations. In the immediate vicinity of T_c, fluctuations of the soft modes become significant, leading to a breakdown of the Gaussian approximation.[17] The temperature span of this critical region can be determined from standard scaling arguments near the critical point. For our problem, the typical spatial scales of scalar and gauge field fluctuations are m_d^{-1} and m_A^{-1}, respectively. Using these scales, we can thus define the "effective" coupling strengths among the soft modes for the scalar and gauge fields as $\beta_i T_c/(32\pi^2\kappa_T^2 m_d)$ and $g^2T_c/(2\pi^2 m_A) = 2\alpha_s T_c/(\pi m_A)$, respectively. These coupling strengths should be small enough that the calculation of the free energy in a loop expansion is meaningful. Also the three dimensional effective theory for the soft modes is meaningful only when the masses of the soft mode are small enough compared to $\Lambda \sim T_c$. Combining the conditions discussed above, one finds necessary (but not sufficient) conditions for the Gaussian approximation to be valid,

$$\frac{|\beta_i|}{32\pi^2\kappa_T^2}T_c \ll m_d \ll T_c, \quad \frac{2\alpha_s}{\pi}T_c \ll m_A \ll T_c, \tag{5}$$

namely the temperature should be inside the appropriate region where the masses of the soft modes are not too small and not too large. Also the above equations imply that the coupling constants should be sufficiently small, $|\beta_i|/(32\pi^2\kappa_T^2) \ll 1$ and $2\alpha_s/\pi \ll 1$.

Let us introduce a ratio of the typical spatial scales of scalar and gauge field fluctuations

$$\kappa^2 \equiv \left(\frac{m_d}{m_A}\right)^2 \simeq \frac{|\beta_i|}{4\pi\alpha_s\kappa_T^2} \to \frac{10^3}{\alpha_s}\left(\frac{T_c}{\mu}\right)^2, \tag{6}$$

where the last equation is valid only in the weak coupling and a numerical coefficient of order unity is neglected. As we shall see later, the gauge (scalar) fluctuations are more important than the scalar (gauge) flucutations for $\kappa \ll 1(\gg 1)$. In the weak coupling, κ is considerably smaller than unity.

The parameter κ defined in Eq. (6) has also the meaning of Ginzburg-Landau parameter[19] that distinguishes the type of color superconductor under an external chromomagnetic field; $\kappa = \delta/\xi$ where δ is the penetration depth, and ξ the coherence length. It was recently shown, by the explicit calculation of the surface energy of a domain wall separating the normal and superconducting phases in the presence of an external magnetic field, that Type I color superconductor is realized for $\kappa_{\rm IS} < 1/\sqrt{2}$ in the IS state[5] and for $\kappa_{\rm CFL} < 0.589$ in the CFL state[20] with $\beta_1 = \beta_2$. As Eq. (6) indicates, $m_d \ll m_A$ is realized in the weak coupling. Therefore, the color superconductor is Type I at least at very high density. Whether the system is Type I or Type II at low densities is not known.

In the weak coupling limit, the Ginzburg criterion (the first of the inequalities in Eq. (5)) becomes

$$t \gg \epsilon_d \equiv 10^2\left(\frac{T_c}{\mu}\right)^4, \quad t \gg \epsilon_A \equiv 10\alpha_s\left(\frac{T_c}{\mu}\right)^2. \tag{7}$$

where we have neglected unimportant numerical coefficients of order unity. In the weak coupling, the ratio T_c/μ is exponentially suppressed as $\exp(-3\pi^2/\sqrt{2}g)$. Therefore one finds $1 \gg \epsilon_A \gg \epsilon_d$ at high baryon density. Note also the relation $\kappa^2 \sim 10^{-2}\epsilon_d/\epsilon_A$; the relative sizes of the critical regions are related to the types of color superconductor.

We briefly summarize the results obtained so far. For a Type I color superconductor, as realized in the high density region, gluon fluctuations give the dominant correction to the free energy. In the one-loop approximation, the gluon fluctuations change the order of the phase transition from second to first, and modify the critical temperature to T_c^*. If the relative shift of the critical temperature $(T_c^* - T_c)/T_c$ is well outside the critical region dictated by Eq. (5), the Gaussian approximation is consistent. This situation is quite analogous to the first order transition in Type I metallic superconductor.[10]

On the other hand, in a Type II color superconductor, which may be realized in the low density region for T_c comparable to μ, scalar fluctuations are not at all negligible. Furthermore, the Gaussian approximation becomes highly questionable. A renormalization group analysis with an ϵ expansion shows that, even without gauge fields, scalar fluctuations alone induce a first order transition[14] in an $SU(n) \times SU(n)$ model with $n \geq 3$. Our model falls in this category when the coupling of the gauge field with the diquarks is neglected. A further complication for Type II color superconductors is that the non-Abelian self-coupling of the gauge field may not be negligible.

Once we take into account fluctuations of the order-parameter and the gauge fields, we need to consider the phase structure in the four dimensional $(\beta_1, \beta_2, \kappa_T, \alpha_s)$ space. Fig. 1 shows its projection onto the two dimensional (β, α_s) space with a constraint $\beta \equiv \beta_1 = \beta_2 = 3\kappa_T$. The CFL phase in the weak coupling limit lies on this reduced space. The solid line in the figure is a boundary separating Type I and Type II color superconductors which is characterized by $\kappa = 1$ in Eq.(6). For the Gaussian approximation on the gauge field to be reliable, $2\alpha_s/\pi < 1$ should also be satisfied. Therefore, as long as we stay in the shaded region in the figure, the one-loop approximation taking into acount only the gauge field is reliable to study the effect of the thermal fluctuation. Also shown in the figure by the dashed line is a relation between α_s and β (both functions of μ) in the weak coupling. The weak coupling regime is found to be well inside the shaded area.

In Fig. 2, we show κ calculated in weak coupling. The dependences of μ on T_c and α_s are taken from the weak coupling results. The figure indicates that $\kappa \ll 1$ (Type I superconductor) is satisfied not only at high density but also at moderate densities, to the extent that one can rely on the extrapolation using the weak coupling formulae.

5. First order transition induced by gauge field

In this section, we assume a Type I color superconductor and evaluate the free energy of the CFL state up to the one-loop order, taking into account only the gauge field fluctuations. The free energy difference between the superfluid and normal phases, $F_{\rm eff}(d)$, in this approximation reads, in the

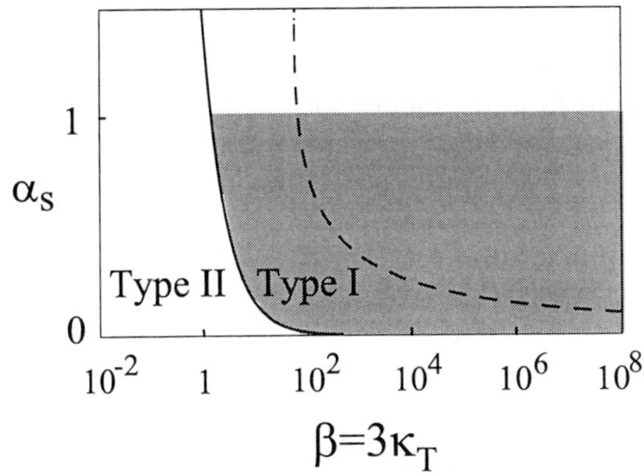

Figure 1. The phase structure in the two dimensional (β, α_s) space. The solid line characterized by $\kappa = 1$ is the boundary between Type I (right side) and Type II (left side) superconducting behavior. The dashed line shows α_s as a function of β, calculated in the weak coupling. The loop expansion for gauge field fluctuations is valid in the shaded area.

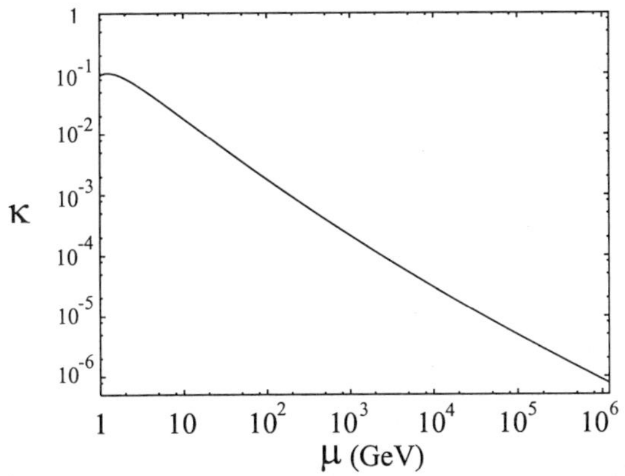

Figure 2. The parameter κ calculated as a function of μ in the weak coupling approximation.

Coulomb gauge,

$$F_{\text{eff}}(d) = S(d) + \frac{1}{2}\ \ln\ \det\left(\frac{\partial^2 S}{\partial \mathbf{A}^\alpha \partial \mathbf{A}^\beta}\right) - F_{\text{eff}}(d=0) \tag{8}$$

$$= S(d) + T\sum_\alpha \left[a_2(m_A)^2_{\alpha\alpha} - a_3(m_A)^3_{\alpha\alpha} + a_4(m_A)^4_{\alpha\alpha}\right], \tag{9}$$

$$\equiv 3\tilde{\alpha}d^2 - qd^3 + rd^4. \tag{10}$$

where $a_2 = T_c/(2\pi^2), a_3 = 1/(6\pi), a_4 = 1/(4\pi^2 T_c)$ and m_A is given in Table 1. At each order of the expansion of $F_{\rm eff}(d)$ in terms of d, dominant contribution comes from the modes with larger mass which is evident from Eq.(9).

The terms in Eq.(10) are explicitly written as

$$3\tilde{\alpha}d^2 = \left[3\bar{\alpha} + \frac{TT_c}{\pi^2}(32\pi\kappa_T\alpha_s)\right]d^2, \tag{11}$$

$$qd^3 = \left[\frac{T}{\pi}\frac{8\sqrt{2}}{3}(4\pi\kappa_T\alpha_s)^{\frac{3}{2}}\right]d^3, \tag{12}$$

$$rd^4 = \left[3(3\beta_1+\beta_2) + \left(\frac{T}{T_c}\right)128\kappa_T^2\alpha_s^2\right]d^4. \tag{13}$$

The effects of gauge fluctuations are three-fold: First, they increase the size of the quadratic term proportional to d^2. This increase implies that thermal fluctuations tend to make the superconducting phase less energetically favorable. Second, we find a cubic term with a negative coefficient, $-q$, which favors the superconducting phase. Furthermore, the second-order transition found in mean-field theory turns into a first order one due to this term, independent of the magnitude of μ. Finally, we find a positive correction to the quartic term from the fluctuations. Like the correction to the quadratic term, this term acts against the superconducting phase. The sum of these three corrections leads to a first order transition which is significantly stronger than the metallic superconductor, but is much weaker than that claimed in Ref. 2 where the quartic correction was neglected.

For later convenience, we define a "renormalized" critical temperature T_c' at which $\tilde{\alpha} = 0$. Then, one finds, $T_c'/T_c = [1 + (T_c^2/\alpha_0)(32\kappa_T\alpha_s/3\pi)]^{-1}$ and $\tilde{\alpha} = \alpha_0(T - T_c')/T_c' \equiv \alpha_0 t'$. The decrease of the renormalized critical temperature from the mean-field one ($T_c' < T_c$) is consistent with the fact that the contribution of the fluctuations to the quadratic term makes the superconducting state less favorable. On the other hand the true critical temperature of the first order transition, T_c^*, is the temperature where the free energy has two degenerate minima, $d = 0$ and $d = d^* \neq 0$. Then one finds

$$\frac{T_c^*}{T_c'} = 1 + \frac{q^2}{12\alpha_0 r}, \quad d^* = \frac{q}{2r}. \tag{14}$$

Due to the effect of the cubic term $-qd^3$ in the free energy which tends to stabilize the superconducting phase, the critical temperature of the first order transition is increased from T_c' as seen in Eq. (14). Thus the ratio

T_c^*/T_c' is one of the measures of the strength of first order transition induced by the thermal gluons. One may also define another measure such as the jump of the diquark condensate at $T = T_c^*$ relative to its $T = 0$ value, d^*/d_0, where $d_0 = d(T=0) = (\pi/e^\gamma)T_c$ with γ being the Euler constant. Using the formulae for β_i and κ_T in the weak coupling, we find the following estimates at high density;

$$\frac{T_c^*}{T_c} - 1 \sim -\frac{\alpha_s}{10^2}, \quad \frac{d^*}{d_0} \sim \frac{10}{\sqrt{\alpha_s}}\left(\frac{T_c}{\mu}\right) \sim 0.3\kappa, \tag{15}$$

The jump of the order parameter at the critical point, d^*/d_0, is at most a few percent level for $\mu > 1$ GeV, but is much larger than that expected in Type I metalic superconductors. Also, the first order transition becomes weaker logarithmically as μ increases, and approaches a second order transition at $\mu = \infty$. As we have mentioned, this is in contrast to the result of Ref. 2 in which the first order transition becomes strong as μ increases. Such behavior is unreasonable because the coupling of gluons and diquarks becomes weak at high density due to the asymptotic freedom. The discrepancy between our result and that in Ref. 2 originates from the fact that thermal corrections to the quartic term in the free energy were not taken into account in Ref. 2. As we approach to the baryon density close to the confinement-deconfinement transition, $\mu \sim 1$ GeV, the fluctuations of the scalar field as well as the non-Abelian interactions of the gluons neglected in our calculation become important. Therefore the results here may be modified qualitatively in such region. Study of the super-to-normal transition in the low density region remains an interesting open question.

Let us now discuss the reliability of the first order phase transition obtained in the weak coupling, from the point of view of the critical region examined in Sec. 4. The first inequalities in Eq. (5) can be interpreted as conditions for the size of $T - T_c$ or alternatively the conditions for the size of d: If $T_c^* - T_c$ or d^* is too small, critical fluctuations are not negligible and one cannot trust the result of the one-loop approximation. In the weak coupling, we have $t^* \equiv |T_c^* - T_c|/T_c \simeq 10^{-2}\alpha_s$ as well as $T_c/\mu \ll 1$, which implies that the conditions given in Eq.(7), $t^* \gg \epsilon_{A,d}$, are well satisfied. Namely the critical temperature of the first order transition is outside the critical region of the diquark and gauge fluctuations. Also, by substituting d^*/d_0 in Eq. (15) into the first inequalities in (5), one finds $\alpha_s/\beta \ll 1.6\,\pi^5$ and $\alpha_s \ll \pi^2/6$, which should be satisfied for the effective three-dimensional approach to be valid. They are in fact well satisfied, insofar as the couplings stay in the shaded region in Fig. 1.

6. Summary

We have studied the effect of thermal fluctuations of diquarks and the gluons on the superconducting-to-normal phase transition in a three-flavor color superconductor. For this purpose, we adopted the Ginzburg-Landau free energy in three-spatial dimensions. The relative importance of the two types of fluctuations is controlled by κ, the ratio of the masses of the scalar field and the gluon just below the critical temperature; κ is also identified as the Ginzburg-Landau parameter which differentiates Type I and Type II color superconductors. In the high density regime where the weak coupling approximation is valid, the system is Type I and the gauge fluctuations dominate over the scalar fluctuation.

The thermal phase transition of Type I color superconductor, which is second order in the mean field theory, becomes first order due to the cubic term in the GL free energy. The first order transition weakens (strengthens) as the baryon density increases (decreases). This is in sharp contrast to that found in Ref. 2. The difference stems from the thermal correction to the quartic term missing in Ref. 2.

Our general considerations in this paper for Type I superconductor are valid insofar as the parameters in the Ginzburg-Landau free energy stay in the shaded area in Fig. 1, which corresponds to high density region. On the other hand, in the low density, strong coupling region, not only scalar fluctuations but also non-Abelian interactions among thermal gluons are not negligible. Furthermore, the strange quark mass plays an important role in the unlocking transition from the CFL state to the IS state. The anti-quark pairing and the non-perturbative running of α_s at low momentum are also not negligible at low density.[21] These effects may change the nature of the phase transition at low density from what is realized at high densities. Lattice simulations of the $SU(3)_c \times SU(3)_{L+R} \times U(1)_B$ sigma-model + $SU(3)_c$ gauge field introduced in the present paper would be a good starting point to analyse the phase structure in the strong coupling region.

Acknowledgments

This work was supported in part by RIKEN Special Postdoctoral Researchers Grant No. 011-52040 by the Grants-in-Aid of the Japanese Ministry of Education, Culture, Sports, Science, and Technology (No. 15540254), and by the U.S. National Science Foundation Grant PHY00-98353.

References

1. K. Rajagopal and F. Wilczek, hep-ph/0011333.
   M.G. Alford, *Ann. Rev. Nucl. Part. Sci.* **51**, 131 (2001).
2. D. Bailin and A. Love, *Phys. Rep.* **107**, 325 (1984).
   M. Iwasaki and T. Iwado, *Phys. Lett.* **B350**, 163 (1995).
3. M. Alford, K.Rajagopal, and F. Wilczek, *Phys. Lett.* **B422**, 247 (1998).
   R. Rapp, T. Schäfer, E. V. Shuryak and M. Velkovsky, *Phys. Rev. Lett.* **81**, 53 (1998).
4. K. Iida and G. Baym, *Phys. Rev.* **D63**, 074018 (2001); *Erratum Phys. Rev.* **D66**, 059903 (2002).
5. K. Iida and G. Baym, *Phys. Rev.* **D66**, 014015 (2002).
6. K. Iida and G. Baym, *Phys. Rev.* **D65**, 014022 (2002).
7. G. Baym, H. Monien, C. J. Pethick and D. G. Ravenhall, *Phys. Rev. Lett.* **64**, 1867 (1990).
8. D. T. Son, *Phys. Rev.* **D59**, 094019 (1999).
9. R. D. Pisarski and D. H. Rischke, *Phys. Rev.* **D61**, 074017 (2000).
10. B.I. Halperin, T.C. Lubensky and S.-k. Ma, *Phys. Rev. Lett.* **32**, 292 (1974).
11. G. Baym and G. Grinstein, *Phys. Rev.* **Dl5**, 2897 (1977).
12. T. Matsuura, K. Iida, T. Hatsuda and G. Baym, paper in preparation.
13. R. D. Pisarski and D. H. Rischke, *Phys. Rev. Lett.* **83**, 37 (1999).
14. A.J. Paterson, *Nucl. Phys.* **B190**, 188 (1981).
15. M.G. Alford, J. Berges and K. Rajagopal, *Nucl. Phys.* **B558**, 219 (1999).
    T. Schäfer and F. Wilczek, *Phys. Rev.* **D60** 074014 (1999).
    H. Abuki, *Prog. Theor. Phys.* **110**, 937 (2003).
16. R.D. Pisarski, *Phys. Rev.* **C62**, 035202 (2000).
17. V.L. Ginzburg, *Fiz. Tverd. Tela* **2**, 2031 (1960) [*Sov. Phys. Solid State* **2**, 1824 (1961)].
18. W.E. Brown, J.T. Liu, and H.-c. Ren, *Phys. Rev.* **D62**, 054016 (2000).
19. V. L. Ginzburg and L. D. Landau, *Zh. Eksp. Theor. Fiz.* **20**, 1064 (1950).
    A.A. Abrikosov, *Zh. Eksp. Teor. Fiz.* **32**, 1442 (1957) [*Soviet Physics JETP* **5**, 1174 (1957)].
20. I. Giannakis and H-c. Ren, *Nucl. Phys.* **B669**, 462 (2003).
21. H. Abuki, T. Hatsuda and K. Itakura, *Phys. Rev.* **D65**, 074014 (2002).

Recent Experiments on QCD

(Quark Gluon Plasma & Penta-Quark System)

EXPERIMENTAL RESULTS FROM RHIC

Y. AKIBA

RIKEN, Institute of Chemical and Physical Research
Wako, Saitama351-0198, Japan
E-mail: akiba@bnl.gov

Recent experimental results of nucleus-nucleus collisions at RHIC is reviewed. The suppression of high p_T particles and away-side jets in central Au+Au collisions and their absence in d+Au collisions have established formation of a high density matter at RHIC. The strong elliptic flow is a strong evidence for rapid thermalization. The hadron yield ratio and p_T distributions provide further evidences for thermalization. The first measurement of open charm and charmonium have been achieved. They are important to further investigate the nature of the matter formed.

1. Introduction

Relativistic Heavy Ion collider collides heavy nuclei, up to gold on gold (Au+Au), at extremely high energy. The maximum energy of the Au+Au collision is 200 GeV per nucleon pair, or 40 TeV in total. This is more than 10 times higher than that of Pb+Pb collision at SPS. In such a high energy collision, it is expected that a state of matter made of deconfined quarks and gluons, the quark-gluon plasma (QGP), is formed. Recent lattice QCD calculation shows that the matter with temperature above 170 MeV and energy density above 1 GeV/fm^3 is the QGP. The goal of RHIC physics program is to look for the evidence of QGP formation, and then to investigate the property of the QGP.

The figure 1 shows a schematic picture of time evolution of the system formed in a high energy nuclear-nuclear collision. After the collision of two large nuclei, a high energy density matter with almost zero baryon density is formed near the central rapidity. The matter is presumably made of quarks and gluons. Due to frequent re-scattering among these partons, the system approaches to a local thermal equilibrium. If the thermalization is achieved and if the energy density of the matter is above the critical density, the QGP is formed. As the remnant of the collided nuclei receded at the speed of light, the system expand rapidly. Strong pressure inside of the high density matter also causes a radial expansion of the system. Due to the expansion, the temperature and the density of the system decreases. When the energy density becomes below the critical density, the system become a gas of interacting hadrons. When the density decreases to a point where the inelastic scattering among hadrons ceases to occur, the abundances of the hadrons are fixed (chemical freeze-out). The density further decreases and the system finally breaks up into non-interacting hadrons (thermal freeze-out).

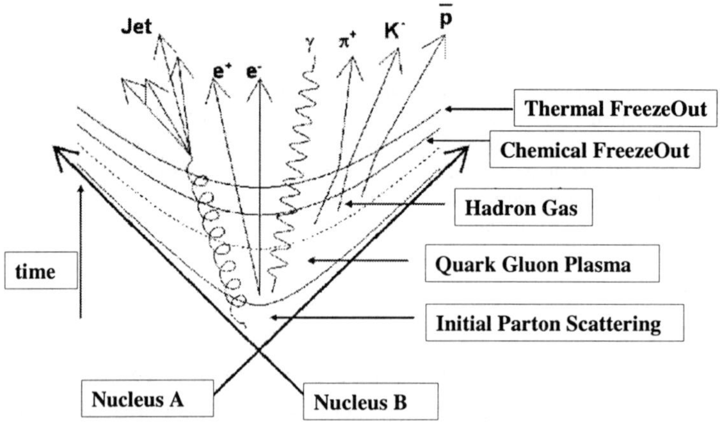

Figure 1. Time Evolution of a high energy nucleus-nucleus collision.

The system emits variety of signal during its time evolution. In the earliest stages of the reaction, signals like high p_T particles, J/ψ, charm, and direct photon are produced. In the QGP phase, electromagnetic radiation such as thermal direct photon and thermal di-lepton pairs are emitted. Collective flow of particle can also be produced in early stage of the collision. In the final phase of the reaction, hadrons like pions and kaons are produced. We detect those signals and analyze them to reconstruct the time evolution of the system. In particular, we look for evidences for a high density mater and its thermalization.

It is three years since we started physics at RHIC. A very large amount of data have been collected in Au+Au, d+Au, and p+p collisions. We now have strong evidences for formation of a very dense matter at RHIC as well as evidences for thermalization of the matter. In this talk, I will review the experimental data from RHIC and to examine how well the picture shown in the figure 1 is supported by the data.

2. High p_T physics at RHIC – Evidence for dense matter

High p_T particles are produced from direct scattering of high energy quarks and gluons at the early stage of heavy ion collisions. These high p_T partons traverse the hot dense matter created in the collision, and they fragment into hadrons outside of the medium, producing high p_T particles. Thus, we can use high p_T particles as a sensitive probe of the dense matter. It is predicted that fast partons lose energy in dense matter via medium induced gluon radiation. Therefore, if a very dense matter is formed, the production rate of high p_T particle should be suppressed.

To quantify the suppression factor, we define the nuclear modification factor, R_{AA}, as follows:

$$R_{AA} = \frac{dN^{AA}/dp_T}{T_{AA} d\sigma^{pp}/dp_T} \tag{1}$$

Here dN^{AA}/dp_T is the yield in AA collision, $d\sigma^{pp}/dp_T$ is the cross section in pp collision, and T_{AA} is the nuclear overlap function, which is the nucleon-nucleon luminosity per one nucleus-nucleus collision.

The R_{AA} of π^0 in central Au+Au collision at $\sqrt{s_{NN}} = 130$ GeV[1] and 200 GeV[2] measured by PHENIX experiment are shown in the lower part of Figure 2 as triangles. A very strong suppression of π^0, with value of R_{AA} from 1/3 to 1/5 in p_T from 1 GeV/c to 10 GeV/c, is observed.

It is possible that the high p_T suppression is caused by nuclear effects other than the formation of a high density matter. For example, reduction of the parton density in heavy nuclei (shadowing) can also cause the suppression. In order to test the possibility that the suppression is caused by such an initial state effect, we studied deuteron and gold (d+Au) collision at RHIC. If the suppression seen is due to initial state effect, it should be also observed in d+Au collisions.

Figure 2 compared R_{dAu} of π^0 with R_{AA} in the central Au+Au collision[3]. In d+Au collision, R_{dAu} is about 1.0 or higher in $p_T > 2$ GeV/c, and no strong suppression is observed. The slight increase of R_{dAu} to 1.1 can be attributed to the multiple scattering of incoming parton, which is known as Cronin effect in fixed target $p+A$ collision. This result has firmly established that the strong suppression observed in central Au+Au collision is not an initial state effect.

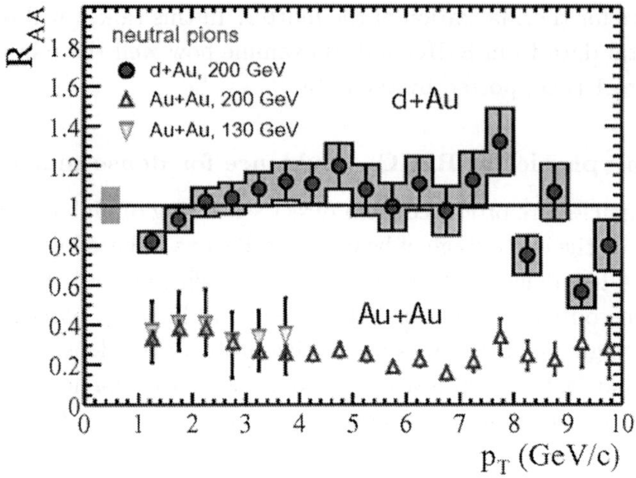

Figure 2. High p_T π^0 suppression in Au+Au (triangles) and d+Au (filled circles) measured by PHENIX.

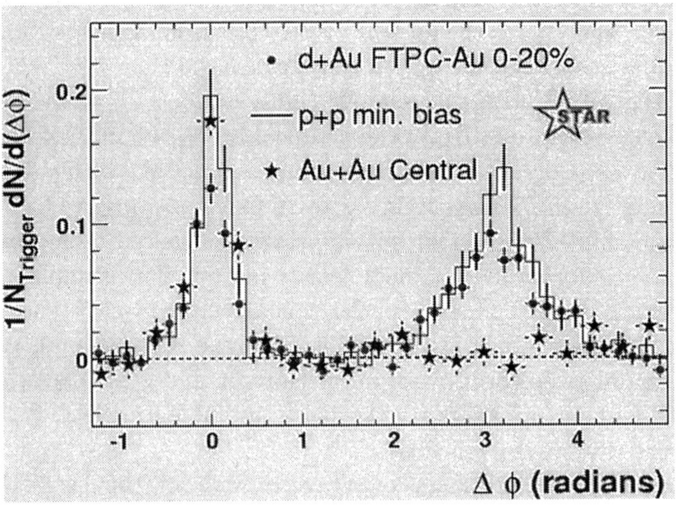

Figure 3. Two particle angular correlation measured by STAR.

Figure 3 shows the correlation between two high p_T particles in pp, dAu, and AuAu collisions measured by STAR experiment[4]. In pp collisions, there is a narrow peak near $\Delta\phi$=0 and a broader peak at $\Delta\phi = \pi$. They are caused by fragmentation of high p_T partons, or jets. In the central Au+Au collision, the peak at $\Delta\phi = \pi$ (the "away-side" jet peak) is strongly suppressed. The disappearance of the away-side jet is expected from the strong energy loss in the hot medium. The away-side jet is caused by the recoil parton, which must traverse the reaction zone and is subject to strong energy loss in the dense matter. On the other hand, in the d+Au collision, the correlation is very similar to that in pp, and little suppression is seen. Thus the disappearance of the "away-side" jet is unique in the Au+Au collision, and it should be a final state effect.

The strong suppression of high p_T particles and away-side jet in central Au+Au collisions and their absence in d+Au collision clearly have established that they are final state effect. They are very strong evidences for formation of a high density matter in Au+Au collisions at RHIC.

3. Evidence for Thermalization

3.1. *Elliptic Flow: evidence for early thermalization*

In non-central Au+Au collision at RHIC, a very strong collective motion, known as elliptic flow, is observed. The elliptic flow is important since the strength of the elliptic flow is thought to be sensitive to the degree of thermalization in early stages of the collision[5]. Experimentally, the elliptic flow is the second component of the Fourier expansion of the angular distribution of emitted particles with respect to the reaction plane.

$$\frac{dN}{d(\phi - \Phi_R)} \propto (1 + \sum_{n=1}^{\infty} 2v_n \cos[n(\phi - \Phi_R]). \tag{2}$$

Here ϕ is the azimuthal angle of an emitted particle and Φ_R is the azimuth of the reaction plane. The v_2 component is the elliptic flow.

PHENIX measurements of the elliptic flow for identified hardons (π, K, p) in Au+Au collision at $\sqrt{s_{NN}}$ = 200 GeV[6] are shown in figure 4. In the bottom left panel, the measured v_2 values as function of p_T are compared with the prediction of a hydrodynamic model. The model reproduces the data up to 2 GeV/c for pions and kaons, and 3 GeV/c for protons. The model assumes early thermalization ($\tau < 1$ fm/c) and the equation of state (EOS) of the QGP. If the start time of the hydrodynamic evolution is delayed or an EOS of hadronic gas is used, the model fails to reproduce the data. It should be noted a hydrodynamic model assumes local thermalization, and it gives the "maximum" elliptic flow since it is the limit of zero mean free path[5]. The success of hydrodynamic models in describing the elliptic flow is considered as a strong evidence for early thermalization at RHIC.

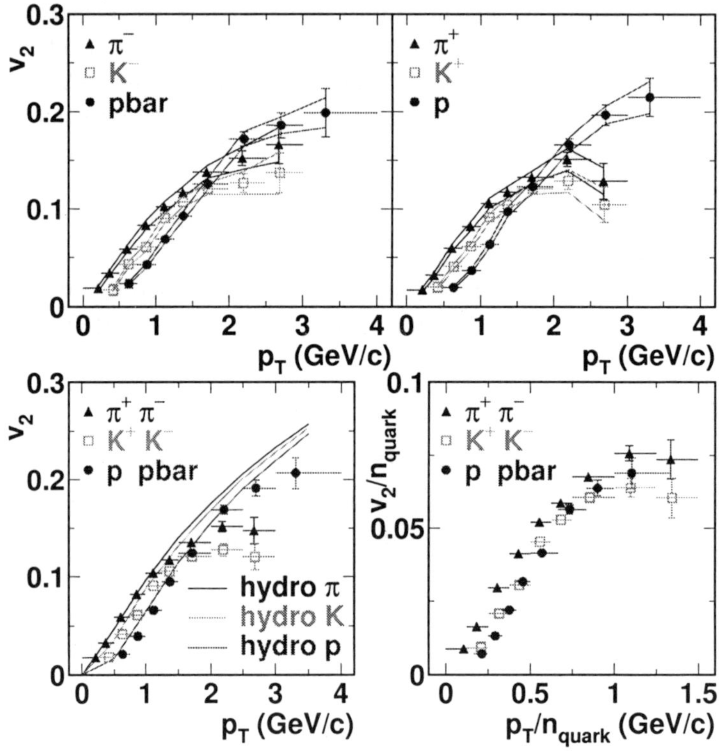

Figure 4. Elliptic flow (v_2) of π, K, p measured by PHENIX.

3.2. *Hadron yield and spectra: evidences for thermal equilibrium*

Hadron yield and spectra provide evidences for thermal equilibrium in the final stage of the reaction. Figure 5 compared the hadron yield ratios measured at RHIC with the prediction by a statistical hadron formation model[7]. The model describes the data very well with only two parameters, the temperature T and the baryon density μ_B. The obtained parameter T and μ_B

from a fit to the data are $T \sim 170$ MeV and $\mu_B \sim 30$ MeV, respectively. It is very interesting that the value for the temperature parameter is close to the critical temperature. This suggests that hadron freeze-out happens at the boundary between the QGP and hadron gas phase. Likewise, the transverse momentum distribution of hadrons are very well described by a model of thermal distribution plus radial flow[8], with a lower freeze-out temperature ($T \sim 110$MeV) and with a large average radial flow velocity $< \beta_T > \sim 0.5$. This is consistent with two step freeze-out scenario shown in figure 1.

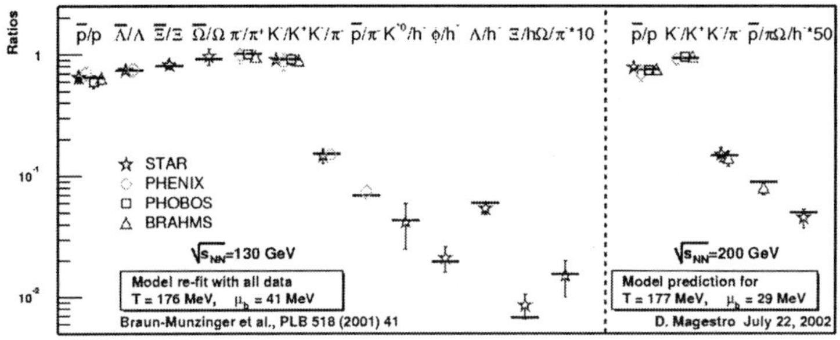

Figure 5. Statistical model prediction of the hadron yield ratios and the RHIC data.

3.3. *high p_T baryon: quark recombination?*

One of interesting and unexpected results from RHIC is observation of anomalously high p/π and $\bar{p}/\pi$ ratio at intermediately high p_T ($p_T = 2-4$ GeV/c). Figure 6 shows the proton to pion ratio and anti-proton to pion ratio in Au+Au collision as function of p_T in three centralities[9]. In the most peripheral collision, the ratio is consistent with pp data at the ISR ($\sqrt{s} = 53$ GeV). However, the ratio increases with centrality, and in the most central collision it is as large as $p/\pi \sim 1$. Such a large p/π ratio can not be explained by ordinary parton fragmentation picture. If hadrons in high p_T are produced by parton fragmentation, the particle composition should be determined by the fragmentation function alone, and p/π ratio should be similar in Au+Au and in pp. The observed large p/π ratio is a direct evidence for that a new mechanism other than parton fragmentation is responsible for baryon and anti-baryon production in intermediate p_T at

RHIC.

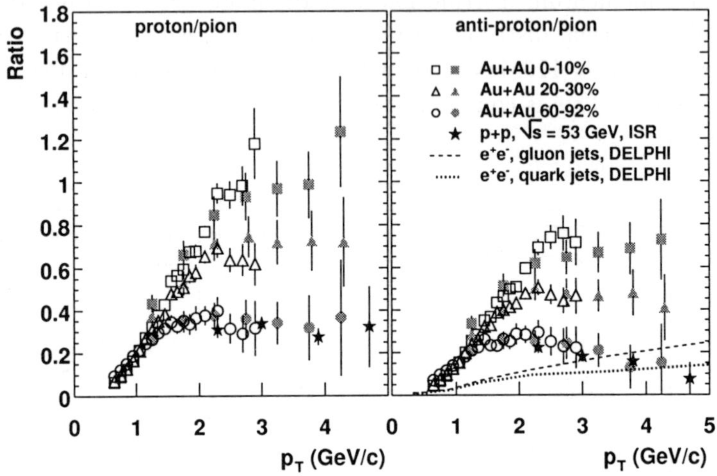

Figure 6. p/π ratio at high p_T measured by PHENIX.

The mechanism that is responsible for the large p/π ratio is unclear. One possibility is a very strong radial flow. Because of their large mass, protons and anti-protons obtain more p_T boosts from the radial flow than pions, and therefore p/π ratio increases. Recently, it is proposed that quark (anti-quark) re-combination is responsible for the large p/π ratio[10]. A pion are made of one quark and one anti-quark, and a proton is made of three quarks. Therefore, if hadrons are produced from thermalized medium of quarks and gluons, the momentum of a pion is the sum of a quark and anti-quark, while the momentum of a proton is the sum of three quarks. Thus, in this model, proton has harder momentum distribution and its yield become larger than pion. In both of the flow model and the re-combination model, the fragmentation of the jets wins at very high p_T, and it is predicted that p/π ratio should decrease above $p_T \sim 5$ GeV$/c$. Measurement of inclusive hadrons to pions (h/π) by PHENIX is consistent with this prediction. A measurement of p/π ratio in $p_T > 4$ GeV$/c$, which is forthcoming, will test this prediction.

3.4. *HBT puzzle*

Two particle interferometry of identical particle, know as HBT, is a powerful method to study the final state of the high energy nucleus-nucleus collisions. If the particles are emitted from a static source, the size of the source can be determined from the HBT measurement. Since the system produced in nucleus-nucleus collisions are very dynamic, the analysis is not as simple as the ideal case, but the system size, duration of the emission, and other properties of the final state can be extracted from the HBT analysis.

Figure 7 shows the interference radii of two pions in central Au+Au collision at $\sqrt{s_{NN}} = 200$ GeV as function of k_T, measured by PHENIX[11]. Here k_T is the average p_T of two pions. The hydrodynamic model prediction of the interference radii are also shown in curves. The model does not reproduce the data. This failure of hydrodynamic model description of the HBT data is know as HBT puzzle, and the cause of the discrepancy is not well understood. A proper treatment of the Coulomb correction to the data may reduce the discrepancy.

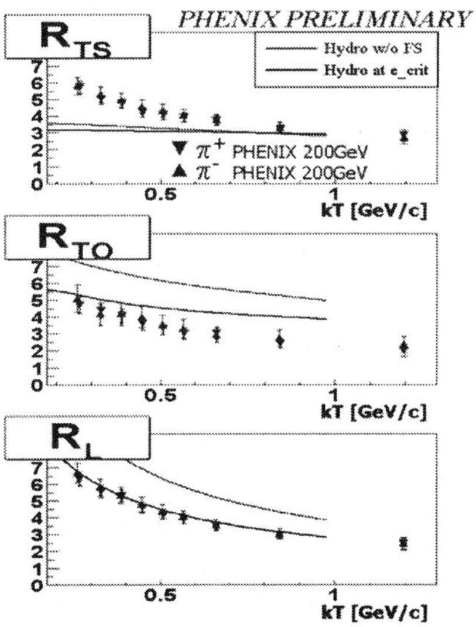

Figure 7. Two pion HBT radii measured by PHENIX.

4. Charm and charmonium at RHIC

4.1. *Open charm production at RHIC*

Open charm is an important probe of initial stage of nucleus-nucleus collisions. Most of charm is produced by collisions of high energy gluons at the initial or early stage of the collision. The charm quarks subsequently traverse and probe the dense matter formed in the collision. The production rate of charm is sensitive to the initial gluon density as well as possible pre-thermal contributions. The quark mass dependence of energy loss effect in the hot medium can be studied from the p_T distribution of charm.

PHENIX experiment measured the open charm production through their decay into single electrons[12]. Figure 8 shows the p_T distribution of non-photonic electrons in Au+Au collisions at $\sqrt{s_{NN}} = 200$ GeV in four centrality bins[13]. Here non-photonic electrons are electrons after subtractions of background from Dalitz decays of light mesons (π^0, η, etc) and photon conversions. The dominant contributor to the non-photonic component is charm decay, but in high p_T region ($p_T > 2.5$ GeV/c) the contribution from beauty decay becomes significant. In the figure, the measured electron distributions are compared with electron spectra from charm decay calculated by PYTHIA event generator. The PYTHIA calculation assumes that the total production cross section of $c\bar{c}$ pair is $\sigma_{c\bar{c}} = 650$ μb, and the calculation is scaled with the number of nucleon-nucleon collisions in each centrality bin. The PYTHIA model reasonably agree with the data in all of the four centralities. Contrary to the high p_T particles, the data shows that the charm production scales with number of binary collisions. This suggests that the energy loss of heavy quark is smaller than that of light quarks and gluons. A more accurate data, which is forthcoming, will answer the scaling property of the charm production.

4.2. *J/ψ production at RHIC*

There are a wide range of predictions on the J/ψ production at RHIC. Matsui and Satz predicted that J/ψ should be suppressed if a QGP is formed[14]. If the "anomalous" J/ψ suppression observed by NA50 is due to QGP formation, we expected almost complete suppression of initially produced J/ψ's. However, at RHIC energy, about 20 $c\bar{c}$ pairs are produced in a central Au+Au collision, and J/ψ can be formed from recombination of the $c\bar{c}$ pairs [15] or statistical hadronization [16,17].

Figure 9 [18] is the first result on J/ψ production in Au+Au collision at RHIC, measured by PHENIX. The figure shows J/ψ yield per number

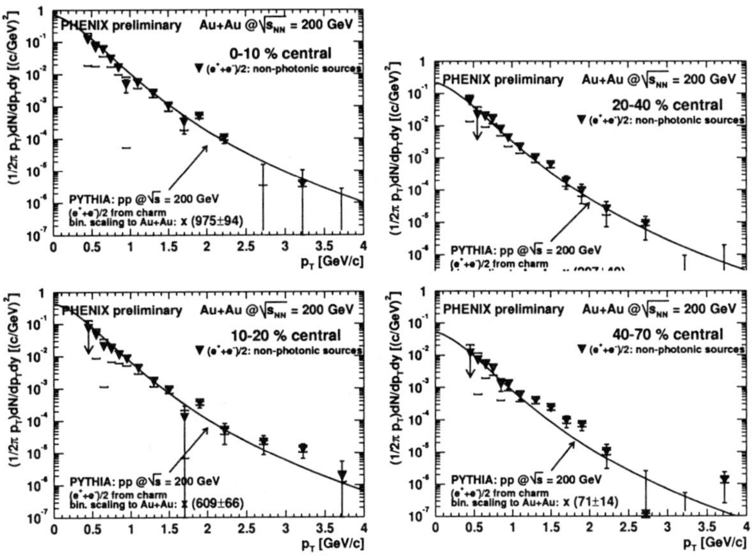

Figure 8. Invariant p_T spectra of electrons from non-photonic sources in Au+Au collisions at $\sqrt{s_{NN}} = 200$ GeV compared with the expected contributions from open charm decays for four different centralities.

of binary collisions in three centrality bins. In the most central bin (0-20%), no signal of J/ψ is observed and thus 90 % upper limit is given. Although the statistics of this data is limited, it already has some interesting constraint on theoretical models. In particular, the models that predict large enhancement of J/ψ, such as the model of $c\bar{c}$ coalescence in QGP [15] (shown as dotted line in the figure), are disfavored by the data. Future high statistics data from PHENIX will determine if the J/ψ is strongly suppressed, or it is formed by the quark recombination mechanism.

5. Summary

A very large amount of data have been collected from the first three years of RHIC. The high p_T data has established formation of a very dense matter in Au+Au collision at RHIC. The strong elliptic flow, hadron yield ratios and spectra are evidences for thermalization. The emerging data on charm and charmonium production can provide further insight in the nature of the matter formed in the collison.

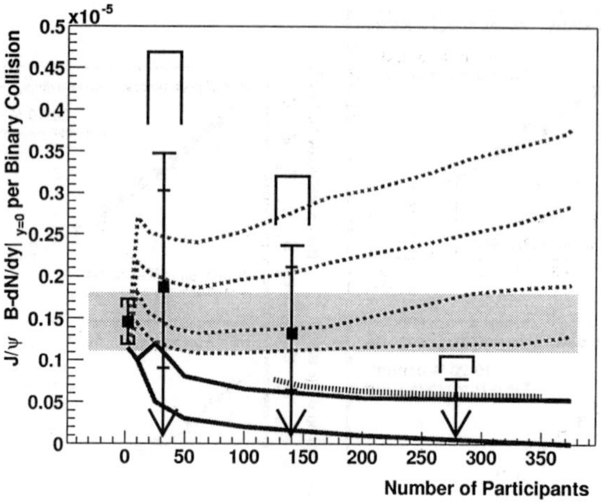

Figure 9. J/ψ yield in Au+Au collision at $\sqrt{s_{NN}}$ = 200 GeV measured by PHENIX.

References

1. K. Adcox *et al.*, *Phys. Rev. Lett.* **88**, 022301 (2002).
2. S. S. Adler *et al.*, *Phys. Rev. Lett.* **91**, 072301 (2003).
3. S. S. Adler *et al.*, *Phys. Rev. Lett.* **91**, 072303 (2003).
4. J. Adams *et al.*, *Phys. Rev. Lett.* **91**, 072304 (2003).
5. U. Heinz and P. Kolb, *Nucl. Phys.* **A702**, 269c (2002).
6. S. S. Adler *et al.*, *Phys. Rev. Lett.* **91**, 182301 (2003).
7. P.Braun-Munzinger *et al.*, *Phys. Lett.* **B518**, 41 (2001).
8. T. Chujo *et al.*, *Nucl. Phys.* **A715**, 151c (2003).
9. S. S. Adler *et al.*, *Phys. Rev. Lett.* **91**, 172301 (2003).
10. R.J. Fries *et al.*, *Phys. Rev. Lett.***90**, 202303 (2003).
11. A. Enokizono *et al.*, *Nucl. Phys.* **A715**, 595c (2003).
12. K. Adcox *et al.*, *Phys. Rev. Lett.***88**, 192303 (2002).
13. R. Averbeck *et al.*, *Nucl. Phys.* **A715**, 695c (2003).
14. T. Matsui and H. Satz, *Phys. Lett.* **B178**, 416 (1986).
15. R.L. Thews *et al.*, *Phys. Rev.* **C63**, 054905 (2001).
16. L. Grandchamp and R. Rapp, *Phys. Lett.* **B523**, 60 (2001).
17. A. Andronic *et al.*, *Phys. Lett.* **B518**, 36 (2003).
18. S. S. Adler *et al.*, *Phys. Rev.* **C** in press, nucl-ex/0305030.

EXPERIMENTAL EVIDENCE FOR PENTA-QUARK SYSTEM

T. NAKANO FOR THE LEPS COLLABORATION
RCNP, Osaka University
10-1 Mihogaoka, Ibaraki,
Osaka, 567-0047, Japan
E-mail: nakano@rcnp.osaka-u.ac.jp

In the K^- missing mass spectrum for the $\gamma + n \to K^+ + K^- + n$ reaction for which both the kaons are detected at forward angles, a peak is observed at 1540 ± 10 MeV, a width of lower or equal than 25 MeV and a statistical significance of 4.6σ. The strangeness quantum number (S) of the resonance is +1. It can be explained as an exotic 5-quark state, with constituents ($uudd\bar{s}$) that subsequently decays into a K^+ and a neutron.

1. Introduction

We searched for baryon resonances with strangeness quantum number S=+1 in the K^- missing mass spectrum for the $\gamma + n \to K^+ + K^- + n$ reaction [1]. The search was motivated in part by a recent paper by Diakonov, Petrov and Polyakov [2] where masses and widths of an anti-decuplet baryons were predicted from the chiral soliton model. The lightest member of the anti-decuplet is the Θ^+ which is an exotic 5-quark state with a quark configuration of $uudd\bar{s}$ that subsequently decays into a K^+ and a neutron. The model predicts the mass of the Θ^+ to be ~ 1530 MeV/c^2 with a narrow width of ≤ 15 MeV/c^2.

The search for baryon resonances with the strangeness quantum number S=+1, that cannot be formed by three quarks, has a long and interesting history. In fact, the summary of the S=+1 baryon resonance searches has been dropped from the Particle Data Group (PDG) listings although the possible exotic resonances were noted in the 1986 baryon listings [3]. Most of the previous searches were made using the partial wave analyses of kaon-nucleon (KN) scatterings [4]. These searches resulted in two possibilities, the isoscalar $Z_0(1780)$ and $Z_0(1865)$, for which the evidence of the existence was reviewed to be poor by PDG. The S=+1 baryon resonances in the mass region of ~ 1.5 GeV/c^2 have not been searched for in the KN scattering experiments in the past because momenta of kaons were too high as pointed out in Refs. [2,5].

2. Laser-Electron Photon beam facility at SPring-8 (LEPS)

The Spring-8 facility is the most powerful third-generation synchrotron radiation facility in the world. The energy of the electrons in the storage ring is 8 GeV and the beam current is 100 mA. The laser-electron photon (LEP) beam at the SPring-8 is generated by Backward-Compton scattering of laser photons with the 8-GeV electrons. The maximum energy of the beam is currently 2.4 GeV for a 351-nm (3.5 eV) argon laser, which is well above the threshold for $s\bar{s}$ productions. The polarization of the LEP beam is about 95 % at the the maximum energy when laser lights are linearly polarized. The polarization drops as the photon energy decreases and crosses zero at the half of the maximum energy. An energy of laser photons can be changed so that the polarization remains reasonably high in the energy region of interest.

The LEP energy is determined by measuring the energy of a recoil electron with a tagging counter which measures the deviation of the recoil electron from the 8-GeV electron beam orbit. Since the tagging counter does not cover the region near the beam orbit, a photon with an energy below 1.5 GeV cannot be tagged. The energy resolution of the tagged photon is 15 MeV (σ) mainly determined by the energy spread of the electron beam and an uncertainty of a photon-electron interaction point in the 7.8-m straight section.

The operation of the laser-electron photon beam at SPring-8 started in July, 1999. The intensity of the beam is about 2.5×10^6 photons/sec for a 5 W laser-output, and a typical tagger rate is 8×10^5/sec.

Figure 1 shows a schematic drawing of the LEPS detector. For tracking of charged particles, a silicon-strip vertex detector (SSD) and 3 drift chambers are used. The SSD consists of single-sided silicon-strip detectors (vertical and horizontal planes) with the strip pitch of 120 μm. The first drift chamber located before a 0.7-T magnet consists of 6 wire planes (3 vertical planes, 2 planes at $+45°$, and 1 plane at $-45°$), and the other two drift chambers after the magnet consist of 5 planes (2 vertical planes, 2 planes at $+30°$, and 1 plane at $-30°$). A time-of-flight (TOF) scintillator array is positioned 3 m behind the dipole magnet.

Electron-positron pairs produced before the target are rejected online

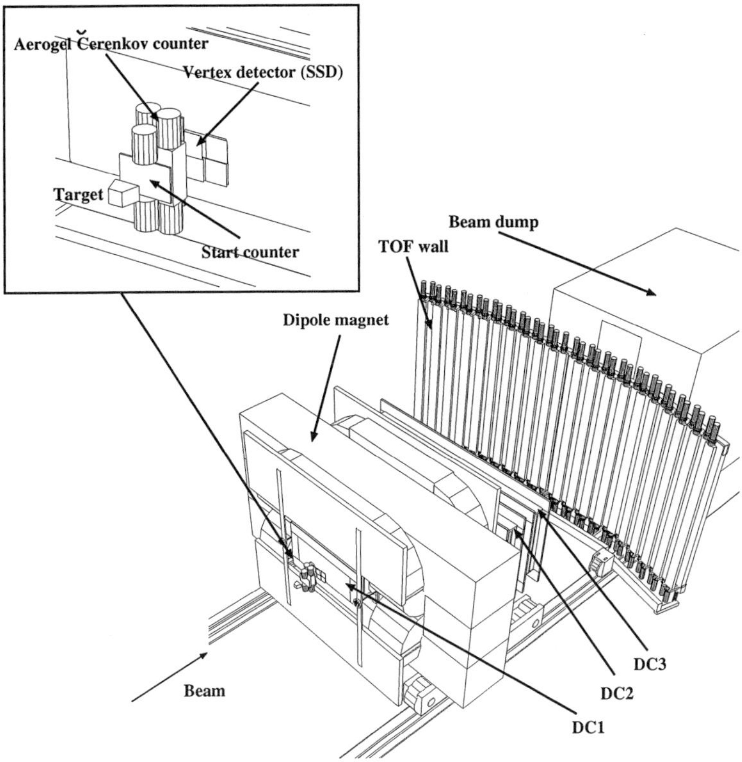

Figure 1. The LEPS detector setup.

by a plastic counter, and the pairs produced at the target are rejected by an aerogel Čerenkov counter (AC) with the index of 1.03. The AC also reject charged pions with momentum more than 0.6 GeV. The electron and positorons which escape the online trigger rejections are blocked by lead bars which were set horizontally along the median plane inside the magnet gap. Pions with a momentum higher than ~ 0.6 GeV/c are vetoed online by the aerogel Čerenkov counter (AC). A 0.5-cm thick plastic scintillator (SC) located 9.5 cm downstream from the 5-cm thick liquid-hydrogen (LH_2) target ensures at least one charged particle produced in the LH_2 target. The events from the SC is turned out to be very useful to study events generated from neutrons in carbon nuclei at the SC.

The angular coverage of the spectrometer is about ± 0.4 rad and ± 0.2 rad in the horizontal and vertical directions, respectively. The momentum resolution (σ) for 1-GeV/c particles is 6 MeV/c. The timing resolution (σ)

of the TOF is 150 psec for a typical flight length of 4 m from the target to the TOF. The momentum-dependent mass resolution is about 30 MeV/c^2 for a 1-GeV/c kaon.

The physics run with a 5-cm long liquid H_2 target wad carried out during December, 2000 to June, 2001. The trigger required a tagging counter hit, no charged particle before the target, charged particles after the target, no signal in the aerogel Čerenkov counter, at least one hit on the TOF wall. A typical trigger rate was about 20 counts per second.

For the present analysis, we selected K^+K^- pair events produced in the SC, which accounted for about half of the K^+K^--pair events (Fig. 2). The missing mass $MM_{\gamma K^+K^-}$ of the $\mathrm{N}(\gamma, K^+K^-)\mathrm{X}$ reaction was calculated by assuming that the target nucleon (proton or neutron) has the mean nucleon mass of 0.9389 GeV/c^2 (M_N) and zero momentum. Subsequently, events with $0.90 < MM_{\gamma K^+K^-} < 0.98$ GeV/c^2 were selected. The main physics background events due to the photo-production of the ϕ meson were eliminated by removing the events with the invariant K^+K^- mass from 1.00 GeV/c^2 to 1.04 GeV/c^2.

In order to eliminate photo-nuclear reactions of $\gamma p \to K^+K^-p$ on protons in ^{12}C and ^{1}H at the SC, the recoiled protons were detected by the SSD. The direction and momentum of the nucleon in the final state was calculated from the K^+ and K^- momenta. And we rejected such events in which the recoiled nucleon was out of the SSD acceptance or the recoiled proton hit was found in the SSD. A total of 109 events satisfied all the selection criteria ("signal sample").

In case of reactions on nucleons in nuclei, the Fermi motion has to be taken into account to obtain appropriate missing-mass spectra. The missing mass corrected for the Fermi motion, $MM^c_{\gamma K^\pm}$, is deduced as

$$MM^c_{\gamma K^\pm} = MM_{\gamma K^\pm} - MM_{\gamma K^+K^-} + M_N. \tag{1}$$

The validity of the correction was checked with the $\gamma n \to K^+\Sigma^- \to K^+\pi^-n$ sequential process, where the K^+ and π^- were detected as shown in Fig. 3. The mass spectrum for Σ^- with a measured width (σ) of 18 MeV/c^2 is well reproduced by a Monte Carlo simulation using the impulse approximation for the reaction from a neutron whose momentum distribution is generated according to a harmonic oscillator potential. The Monte Carlo simulation shows that the width is dominated by incomplete cancellation of the Fermi motion effect.

The corrected K^+ missing-mass distribution for the events that satisfy all the selection conditions is compared with that for the events for which

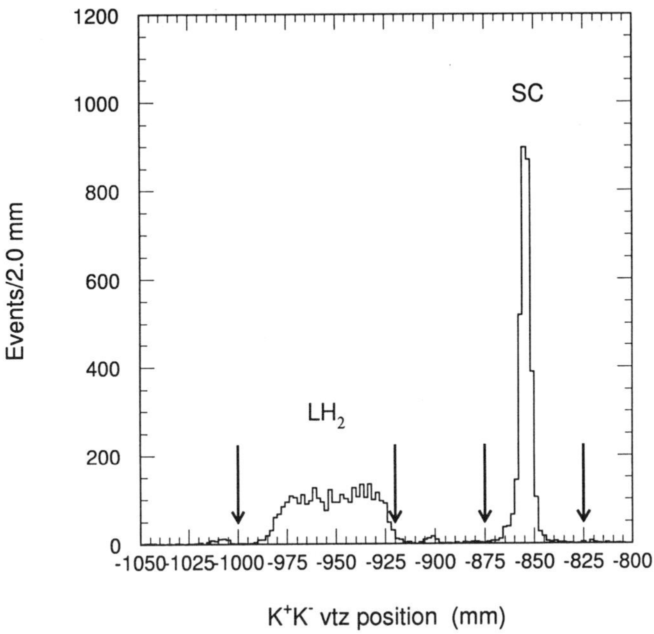

Figure 2. Vertex position (vtz) for K^+K^- events along the photon-beam direction. Cut points to select the SC events or the LH_2 events are indicated by the arrows.

a coincident proton hit was detected in the SSD. In the latter case, a clear peak due to the $\gamma + p \to K^+\Lambda(1520) \to K^+K^-p$ reaction is observed while the $\Lambda(1520)$ peak does not exist in the signal sample. This indicates that the signal sample is dominated by events produced by reactions on neutrons. Fig. 4 shows the corrected K^- missing mass distribution of the signal sample. A prominent peak at 1.54 GeV/c^2 is found. The broad background centered at $\sim$ 1.6 GeV/c^2 is most likely due to non-resonant K^+K^- production and the background shape in the region above 1.59 GeV/c^2 has been fitted by a distribution of events from the LH_2. The estimated number of the events above the background level is 19.0 $\pm$ 2.8, which corresponds to a Gaussian significance of 4.6 σ.

After subtracting the background from the signal sample, the spectrum in the region of $1.47 \leq MM^c_{\gamma K^-} < 1.61$ GeV/c^2 was compared with Monte Carlo simulations assuming a Breit-Wigner function for a resonance distri-

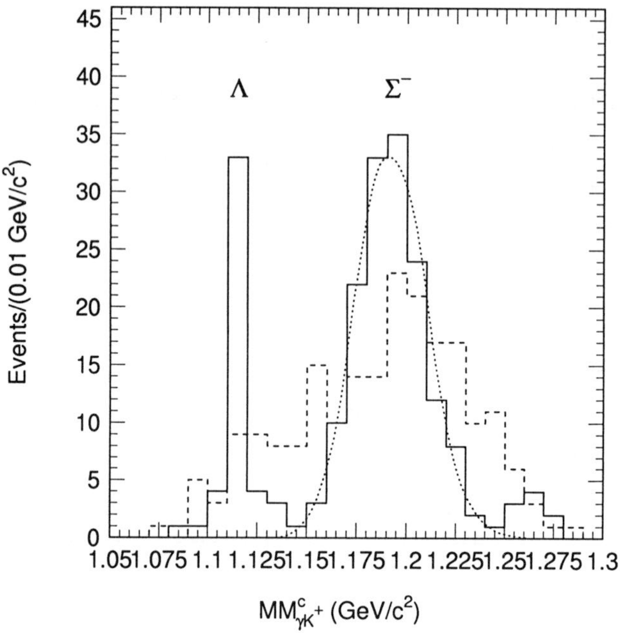

Figure 3. The missing mass, $MM^c_{\gamma K^+}$, spectra for the $K^+\pi^-$ events from the SC (solid histogram) and for Monte Carlo events for the $\gamma n \to K^+\Sigma^-$ channel (dotted curve) with the Fermi-motion correction. The dashed histogram shows the missing mass spectrum without the Fermi-motion correction.

bution. The best fit to the spectrum gives the mass of the resonance to be 1.54 ± 0.01 GeV/c^2. And the upper limit for the width was determined to be 25 MeV/c^2 with a 90 % C.L.. This narrow peak strongly indicates the existence of an $S = +1$ resonance which may be attributed to the exotic 5-quark baryon proposed as the Θ^+.

Soon after a preliminary result on the was announced by the LEPS collaboration at the international conference PANIC in October 2002, the CLAS collaboration at Jefferson Lab re-analyzed photo-reaction data which were collected in 1999 by using a liquid deuterium target. They found a 5.3 σ peak at 1542 MeV in the nK^+ invariant mass spectrum of the reaction $\gamma d \to K^+K^-pn$, where all charged particles in the final state were detected [6]. Since the momentum of the final neutron was fully determined

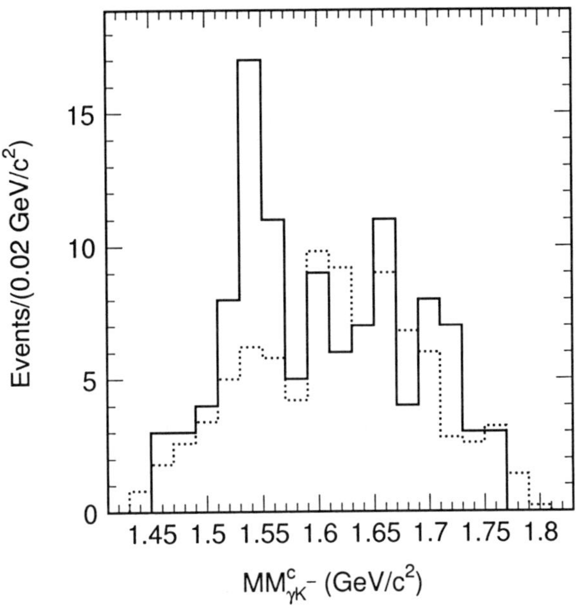

Figure 4. The $MM^c_{\gamma K^-}$ spectrum for the signal sample (solid histogram) and for events from the LH_2 (dotted histogram) normalized by a fit in the region above 1.59 GeV/c^2.

by using total momentum conservation law, their result was not affected by a Fermi motion of the initial neutron. The CLAS collaboration also announced an evidence for the Θ^+ by analyzing the reaction $\gamma p \to K^+K^-\pi^+n$ in a different data set [7].

Prior to the CLAS reports, the DIANA collaboration at the Institute of Theoretical and Experimental Physics (ITEP) in Russia reexamined low-energy K^+ Xe collision events in the Xennon bubble chamber, which were taken in 1986. They found a 4.4 σ peak at 1539 MeV with a very narrow width of < 9 MeV in the invariant pK^0 invariant mass spectrum of the charge-exchange reaction $K^+\mathrm{Xe}' \to \mathrm{K}^0\mathrm{pXe}'$ [8].

Most recently, the SAPHIR collaboration at ELSA in Germany reanalyzed old data and found a Θ^+ peak at 1540 MeV with a width less than 25 MeV in the reaction $\gamma p \to \bar{K}^0 K^+ n$ [9]. They found no evidence for a Θ^{++}

in the reaction and concluded that the Θ^+ should be isoscalar.

All experiments are very consistent with each other: The peak position of the Θ^+ is about 1540 MeV and the width is very narrow. However, further experimental studies are needed in order to establish the Θ^+. The determination of the spin and parity seems to be crucial to understand the nature of the state.

3. Acknowledgments

The authors gratefully acknowledge the dedicated efforts of the staff of the SPring-8 for providing a good quality beam. This work is supported by the Ministry of Education, Culture, Sports, Science and Technology of Japan, by the National Science Council of Republic of China (Taiwan), and by KOSEF of Republic of Korea.

References

1. T. Nakano *et al.* (LEPS Collaboration), *Phys. Rev. Lett.* **91**, 012002 (2003).
2. D. Diakonov, V. Petrov, and M. Polyakov, *Z. Phys.* **A359**, 305 (1997).
3. Particle Data Group, *Phys. Lett.* **170B**, 289 (1986).
4. R. Arndt *et al.*, *Phys. Rev.* D **46**, 961 (1992).
5. M. Polyakov *et al.*, *Eur. Phys.* J. A **9**, 115 (2000).
6. S. Stepanyan *et al.* (CLAS Collaboration), arXiv:hep-ex/0307018
7. V. Kubarovsky *et al.* (CLAS Collaboration), arXiv:hep-ex/0307088.
8. V.V. Barmin *et al.* (DIANA Collaboration), *Phys. At. Nucl.* **66**, 1715 (2003).
9. J. Barth *et al.* (SAPHIR Collaboration), arXiv:hep-ex/0307083.

Parallel Session Presentations

CALCULATION OF FINITE SIZE EFFECTS ON THE NUCLEON MASS IN UNQUENCHED QCD USING CHIRAL PERTURBATION THEORY*

A. ALI KHAN[A], T. BAKEYEV[B], M. GÖCKELER[C,D], T.R. HEMMERT[E], R. HORSLEY[F], A.C. IRVING[G], D. PLEITER[H], P.E.L. RAKOW[G], G. SCHIERHOLZ[H,I] AND H. STÜBEN[J]

[A] *Institut für Physik, Humboldt-Universität zu Berlin, 12489 Berlin, Germany,*
[B] *Joint Institute for Nuclear Research, 141980 Dubna, Russia,*
[C] *Institut für Theoretische Physik, Universität Leipzig, 04109 Leipzig, Germany,*
[D] *Institut für Theoretische Physik, Universität Regensburg, 93040 Regensburg, Germany,*
[E] *Physik-Department, Theoretische Physik T39, TU München, 85747 Garching, Germany,*
[F] *School of Physics, The University of Edinburgh, Edinburgh EH9 3JZ, UK,*
[G] *Theoretical Physics Division, Department of Mathematical Sciences, University of Liverpool, Liverpool L69 3BX, UK,*
[H] *John von Neumann-Institut für Computing NIC, 15738 Zeuthen, Germany,*
[I] *Deutsches Elektronen-Synchrotron DESY, 22603 Hamburg, Germany,*
[J] *Konrad-Zuse-Zentrum für Informationstechnik Berlin, 14195 Berlin, Germany*
(QCDSF and UKQCD Collaborations)

The finite size effects on nucleon masses are calculated in relativistic chiral perturbation theory. Results are compared with two-flavor lattice results.

1. Introduction

Finite size effects, in particular on dynamical configurations, can be a serious impediment to precision lattice calculations of hadron masses and matrix elements. On the lattice sizes used in production runs, the nucleons and pions are the relevant degrees of freedom for understanding the finite size effects on the nucleon mass. We calculate these effects here using two-flavor relativistic baryon chiral perturbation theory (χPT) at $O(q^3)$ (one loop).

*presented by A. Ali Khan

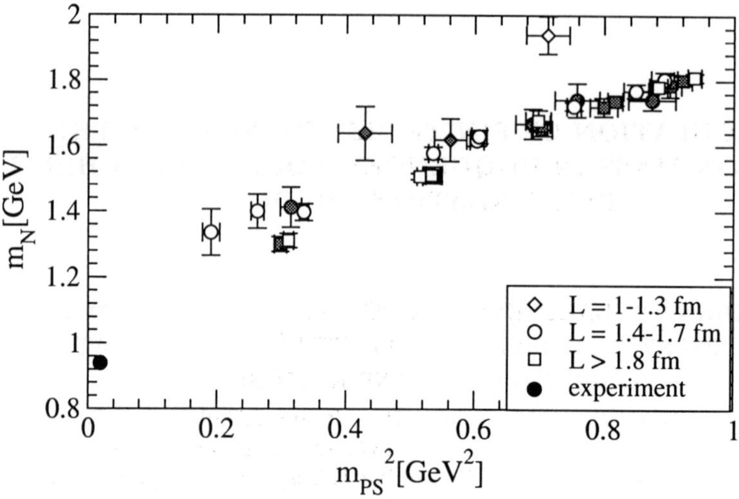

Figure 1. Nucleon masses from QCDSF and UKQCD (white symbols) and JLQCD (gray symbols).

This study of nucleon masses is based on QCDSF and UKQCD unquenched ($N_f = 2$) lattice data using a Wilson gauge action and two flavors of non-perturbatively $O(a)$-improved Wilson fermions (a denotes the lattice spacing). The corresponding pion masses are in the interval $0.45 - 1$ GeV. Valence and sea quark masses are equal. Lattice sizes are $1 - 2.2$ fm. The scale is set with r_0, using the physical value $r_0 = 0.5$ fm $\simeq 1/(395\text{MeV})$. JLQCD has undertaken a study with the same lattice actions and range of simulation parameters on varying lattice sizes [1]. In the comparison shown in Fig. 1 one can observe that masses from lattices smaller than 1.8 fm are consistently larger than the results from larger lattices.

The $a \to 0$ limit was not performed. Since discretization effects in unquenched simulations could be quite large [2] we have to keep an eye on the discretization errors in our results. Studying the variation of nucleon masses with a at a fixed lattice size we find that between two data points on approximately the same volume and at the same pion mass, but with a varying by $\sim 30\%$, the nucleon mass remains unchanged within the statistical errors. Moreover, we compare our results from large lattices ($L \geq 1.8$ fm) with two CP-PACS data sets [3] with renormalization group improved gauge fields and tree-level tadpole-improved clover quarks, $a^{-1} \approx 1.5$ and 2 GeV respectively and $L \geq 2.5$ fm, and find that they agree.

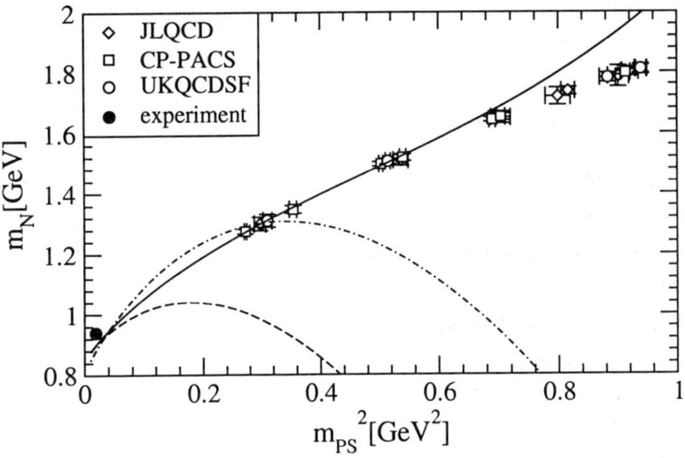

Figure 2. Nucleon masses on large lattices. The solid curve represents a fit with relativistic χPT, the dashed curve the NR limit using the same values of m_0 and c_1, and the dot-dashed curve the NR limit using $m_0 = 0.81$ GeV and $c_1 = -1.1$ GeV^{-1}.

2. χPT formalism

The one-loop contribution at $O(q^3)$ is generated by the $O(q^1)$ Lagrangian $\mathcal{L}_N^{(1)}$:

$$\mathcal{L}_N^{(1)} = \bar{\Psi}\left(i\gamma_\mu D^\mu - m_0\right)\Psi + \frac{1}{2} g_A \bar{\Psi}\gamma_\mu\gamma_5 u^\mu \Psi, \tag{1}$$

with $D_\mu = \partial_\mu + \frac{1}{2}[u^\dagger, \partial_\mu u]$, $u_\mu = iu^\dagger\partial_\mu U u^\dagger$ and $u^2 = U$. We use the infrared regularization scheme [4]. To compute the renormalized nucleon mass m_N, we include the tree-level $O(q^2)$ term $-4c_1 m_{PS}^2$ and an additional contribution of the form $e_1 m_{PS}^4$, which is needed for renormalization of the relativistic $O(q^3)$ result, although formally only entering at $O(q^4)$. The one-loop correction to m_0 is [5]:

$$\begin{aligned} m_N = m_0 - 4c_1 m_{PS}^2 &+ \left[e_1^r(\lambda) + \frac{3g_A^2}{64\pi^2 F_\pi^2 m_0}\left(1 - 2\ln\frac{m_{PS}}{\lambda}\right)\right] m_{PS}^4 \\ &- \frac{3g_A^2}{16\pi^2 F_\pi^2} m_{PS}^3 \sqrt{1 - \frac{m_{PS}^2}{4m_0^2}}\left[\frac{\pi}{2} + \arctan\frac{m_{PS}^2}{\sqrt{4m_{PS}^2 m_0^2 - m_{PS}^4}}\right]. \end{aligned} \tag{2}$$

$e_1^r(\lambda)$ denotes the renormalized e_1. The renormalization procedure is detailed in [5]. We employ $\lambda = 1$GeV, $g_A = 1.2$ [6], and $F_\pi = 92.4$ MeV.

The nucleon mass in the chiral limit, m_0, and the values of c_1 and e_1^r are determined by a fit to six lattice data points at the smallest masses. The fit is shown in Fig. 2. Relativistic χPT describes the data correctly for pion masses up to $\sim$ 700 MeV. For the values of the parameters, we find $m_0 = 0.85(14)\text{GeV}$, $c_1 = -0.80(18)\text{GeV}^{-1}$ and $e_1^r(1\,\text{GeV}) = 2.8(1.1)\text{GeV}^{-3}$. These results differ slightly from the ones quoted in [5] due to their fitting to a larger set of lattice points and using the value $g_A = 1.267$.

In the limit $m_{PS} \ll m_N$, keeping only terms up to $O(m_{PS}^3)$ of Eq. (2), one obtains non-relativistic (NR) χPT. In Fig. 2, we also plot NR results. We find that they approximate the data and the relativistic curve only for rather small pion masses. A method to extend the validity of the NR approximation to higher momentum scales within a cutoff scheme is described in [7]. The breakdown of the non-relativistic theory around $m_{PS} = 400$ MeV as shown in the plots can be avoided if momentum modes above a cutoff Λ_χ are absorbed in local counterterms (e.g. see the discussion in [7]). Then the nonrelativistic $O(q^3)$ calculation actually agrees quite well with the corresponding relativistic result up to $m_{PS} = 600$ MeV.

3. Finite Size Effects

We determine the finite size effect from the one-loop $O(q^3)$ contribution to the self-energy. Putting the external nucleon line on-shell, the self-energy is given by [4]

$$\Sigma(\not{q} = m_0) = \frac{3g_A^2 m_0 m_{PS}^2}{2F_\pi^2} \int_0^\infty dx \int \frac{d^4p}{(2\pi)^4} \left[p^2 + m_0^2 x^2 + m_{PS}^2(1-x)\right]^{-2} \quad (3)$$

in Euclidean space. The temporal extent of the lattice is taken to be infinite. We define $\delta = \frac{1}{m_0}\left(\Sigma(\not{q} = m_0, L) - \Sigma(\not{q} = m_0, \infty)\right)$, and using [8] express it as an integral over Bessel functions:

$$\delta = \frac{3g_A^2 m_{PS}^2}{16\pi^2 F_\pi^2} \int_0^\infty dx \sum_{\vec{n}\neq 0} K_0\left(L|\vec{n}|\sqrt{m_0^2 x^2 + m_{PS}^2(1-x)}\right). \quad (4)$$

Then we calculate the finite size behavior of the nucleon mass:

$$m_N(L) = (1+\delta) m_N(\infty). \quad (5)$$

We first extrapolate to $m_N(\infty)$ by using the lattice result from the largest available lattice as input to Eq.(5), and then determine $m_N(L)$ at finite lattice extent using the formula Eq.(5). In Fig. 3 we compare lattice nucleon masses from various volumes with both relativistic and NR χPT at $O(q^3)$. Relativistic χPT describes the lattice data very well up to pion masses $\sim$ 700 MeV. We find that for the finite size effect to be $< 1\%$, L should be ≥ 1.9 fm at the smallest pion mass shown in the plot.

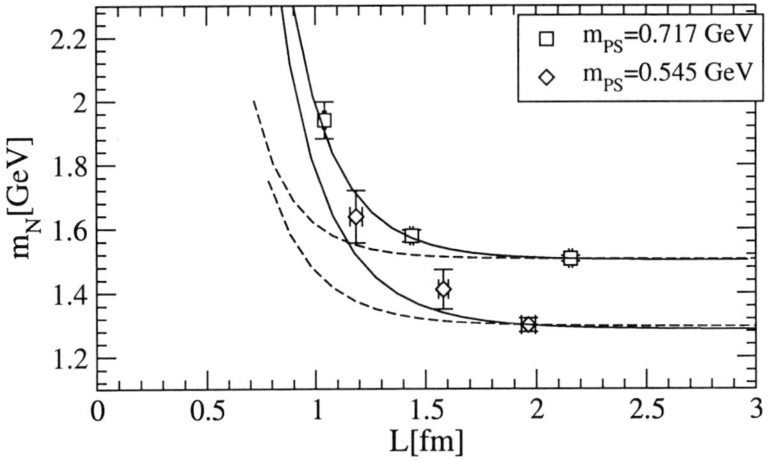

Figure 3. m_N as a function of size on the lattice and in χPT. Data sets correspond to a fixed (β, κ) value and are labeled by the pion mass on the largest volume. The dashed lines show the results from NR χPT.

The finite size effect calculated in the non-relativistic formalism (for discussion see also [9]) is considerably smaller compared to the relativistic result and the lattice data. On lattices of $1-2$ fm size it is $\sim 30-40\%$ of the relativistic effect.

Acknowledgements

This work is supported in part by the Deutsche Forschungsgemeinschaft. The simulations were done on the APEmille at NIC (Zeuthen), the Hitachi SR8000 at LRZ (Munich) and the Cray T3E at EPCC (Edinburgh) and NIC (Jülich). We thank all institutions for their support.

References

1. JLQCD Collaboration, S. Aoki *et al.*, *Phys. Rev.* **D68**, 054502 (2003).
2. R. Sommer *et al.*, proceedings Lattice 03, hep-lat/0309171.
3. CP-PACS Collaboration, A. Ali Khan *et al.*, *Phys. Rev.* **D65**, 054505 (2002); Erratum *ibid.* **D67**, 059901 (2003).
4. T. Becher and H. Leutwyler, *Eur. Phys. J.* **C9**, 643 (1999).
5. M. Procura, T.R. Hemmert and W. Weise, hep-lat/0309020.
6. T.R. Hemmert, M. Procura and W. Weise, hep-lat/0303002, *Phys. Rev.* **D** in press.
7. V. Bernard, T.R. Hemmert and U.G. Meißner, hep-ph/0307115.
8. P. Hasenfratz and H. Leutwyler, *Nucl. Phys.* **B343**, 241 (1990).
9. A. Ali Khan *et al.*, *Nucl. Phys.* **B**(Proc.Suppl.)**119**, 419 (2003).

THE FACTORIZATION METHOD FOR SIMULATING SYSTEMS WITH A COMPLEX ACTION

J. AMBJØRN

The Niels Bohr Institute, Blegdamsvej 17, DK-2100 Copenhagen Ø, Denmark

K. N. ANAGNOSTOPOULOS

Department of Physics, University of Crete, P.O. Box 2208, GR-71003 Heraklion, Greece

J. NISHIMURA

Theory Division, KEK, Tsukuba, Ibaraki 305–0801, Japan

J. J. M. VERBAARSCHOT

Department of Physics and Astronomy, SUNY, Stony Brook, NY 11794, USA

We propose a method for Monte Carlo simulations of systems with a complex action. The method has the advantages of being in principle applicable to any such system and provides a solution to the overlap problem. We apply it in random matrix theory of finite density QCD where we compare with analytic results. In this model we find non–commutativity of the limits $\mu \to 0$ and $N \to \infty$ which could be of relevance in QCD at finite density.

1. Introduction

There exist many interesting systems in high energy physics whose action contains an imaginary part, such as QCD at finite baryon density, Chern-Simons theories, systems with topological terms (like the θ-term in QCD) and systems with chiral fermions. This imposes a severe technical problem in the simulations, requiring an exponentially large amount of data for statistically significant measurements as the system size is increased or the critical point is approached. Furthermore, the overlap problem appears when standard reweighting techniques are applied in such systems and it becomes exponentially hard with system size to visit the relevant part of the configuration space. In Ref. [1] it was proposed to take advantage of a factorization property of the distribution functions of the observables one is interested to measure. This approach can in principle be applied to *any* system and it eliminates the overlap problem completely. In some cases it is possible to use finite size scaling to extrapolate successfully to large system sizes where it would have been impossible to measure oscillating factors directly. The method has been applied successfully in matrix models of non perturbative string theory (IKKT),[1] random matrix theory of finite density QCD (RMT) [2] as well as the 2d CP^3 model, the 1d antiferromagnetic model with imaginary B and the 2d compact U(1) with topological charge.[3] In this paper we present our results for RMT which we study in order to test the factorization method against known analytical results. We also discuss an observed non–commutativity in the limits $\mu \to 0$ and $N \to \infty$ which maybe relevant to Taylor expansion and imaginary μ approaches to the problem of finite density QCD.

The factorization method has the important property that it can be applied to any system featuring the complex action problem. Let a system be given by a partition function $Z = \int dA\, e^{-S_0}\, e^{i\Gamma}$ and the corresponding phase quenched model $Z_0 = \int dA\, e^{-S_0}$ where $S = S_0 - i\Gamma$ is the action of the system with its real and imaginary parts. A represents collectively the degrees of freedom of the model and in our case it corresponds to a set of $N \times N$ matrices. In case we are interested in measuring some observable $\mathcal{O}$, we consider the distribution functions $\rho_{\mathcal{O}}(x) = \langle \delta(x - \mathcal{O}) \rangle$ and $\rho^{(0)}_{\mathcal{O}}(x) = \langle \delta(x - \mathcal{O}) \rangle_0$, where $\langle \ldots \rangle_0$ refers to Z_0. Then we define the fiducial system $Z_{\mathcal{O},x} = \int dA\, e^{-S_0} \delta(x - \mathcal{O})$, the weight factor $w_{\mathcal{O}}(x) = \langle e^{i\Gamma} \rangle_{\mathcal{O},x}$ and the distribution $\rho_{\mathcal{O}}(x)$ factorizes $\rho_{\mathcal{O}}(x) = \frac{1}{C}\, \rho_{\mathcal{O}}(x)\, w_{\mathcal{O}}(x)$ where $C = \langle e^{i\Gamma} \rangle_0$. Then $\langle \mathcal{O} \rangle = \frac{1}{C} \int_{-\infty}^{\infty} dx\, x\, \rho^{(0)}_{\mathcal{O}}(x)\, w_{\mathcal{O}}(x)$. The δ–function constraint is implemented in our simulations by considering the system $Z_{\mathcal{O},V} = \int dA\, e^{-S_0}\, e^{V(\mathcal{O})}$ where $V(z) = \frac{1}{2}\gamma(z - \xi)^2$ and γ, ξ are parameters which control the constraining of the simulation. The results are insensitive to the choice of γ as long as it is large enough. Then we have that $w_{\mathcal{O}}(x = \langle \mathcal{O} \rangle_{i,V}) = \langle e^{i\Gamma} \rangle_{i,V}$. The distribution of $\mathcal{O}$ in $Z_{i,V}$ has a peak $\bar{x}$ and the quantity $V'(\bar{x})$ is the value of $f^{(0)}_{\mathcal{O}}(x) = \frac{d}{dx} \ln \rho^{(0)}_{\mathcal{O}}(x)$ at $x = \bar{x}$. The function $\rho^{(0)}_{\mathcal{O}}(x)$ can be obtained by integrating an analytic function to which we fit the $f^{(0)}_{\mathcal{O}}(x)$ data points.

By applying this method we force the system to sample configurations which give the essential contributions to $\langle \mathcal{O} \rangle$, something that would be exponentially difficult with system size in the phase quenched model, eliminating this way the overlap problem. This already allows us get close to the thermodynamic limit with modest computer resources. Furthermore we obtain direct knowledge of $w_{\mathcal{O}}(x)$ and $\rho_{\mathcal{O}}(x)$ which allows us to understand the effect of Γ. This is important for understanding the properties of the system when Γ plays a crucial role.

2. RMT of Finite Density QCD

We consider RMT with one quark flavour and zero quark mass.[4] The model is chosen in order to study the correctness and effectiveness of the factorization method, since one can compare results with known analytical solutions even for finite N. The observable we measure is the "quark number density" ν as a function of the chemical potential μ, and we consider the distribution functions $\rho_i(x)$, where $i = R, I$ corresponds to the real and imaginary parts of ν respectively. Notice that the effect of Γ is dramatic, causing a discontinuous transition in ν. Our results [2] nicely reproduce the exact results known for finite N and we are able to achieve large enough values of $N \leq 48$ to obtain the thermodynamic limit. In Figure 1 we show the plots of the distribution functions $\rho_{R,I}(x)$ for $\mu = 0.2$. Unfortunately, the function $w_R(x)$ is not positive definite and the important contributions come from the region where it changes sign. As expected, we find that finite size scaling does not work as well as in the case of the IKKT model[1] (although we obtain agreement up to order of magnitude for the values of $N \leq 96$ that we explored). We also find it very difficult to explore the critical region near the phase transition point $\mu_c = 0.527\ldots$ for $N > 8$ since $|w_i(x)|$ becomes very small. Since RMT is a schematic model of finite density QCD, we expect that the factorization method will be useful to explore the phase diagram of QCD.

In our simulations we find that for certain observables the limits $\mu \to 0$, $N \to \infty$ and $N \to \infty$, $\mu \to 0$ are not equivalent. In real QCD, the former is easy but not the latter. That this situation is possible can already be seen at the partition function level where $Z(\mu, N) = \mathrm{e}^{\kappa}[1 + \frac{(-1)^{N+1}}{N!}\gamma(N+1, \kappa)]$, $\kappa = -N\mu^2$ is equal to 1 and 0 respectively, but it turns out that the same is true also for observables like $\frac{\partial}{\partial\mu}\langle\nu_R\rangle_0$ and $w_R(x)$. $\langle\nu_R\rangle$, however, is well defined in this limit as expected. In the thermodynamic limit $\langle\nu_R\rangle_0 = \mu$ for $0 < \mu < 1$. In Figure 2 we see that this limit is approached like $\langle\nu_R\rangle_0 - \mu \sim \mathcal{O}(1/N)$ but only if $\mu > \mu_c(N)$. We find that the value of μ_c is consistent with $\mu_c^2 \sim 1/N$. A circle with radius $\mu_c = 1/\sqrt{N}$ contains only one eigenvalue of the matrix on average. For $N \ll 1/\mu_c^2$ we find that $\langle\nu_R\rangle_0 = 0$. This can be seen clearly from the second plot of Figure 2, where the distributions $\rho_R^{(0)}(x)$ for $N = 8$ peak around zero for $\mu < \mu_c$

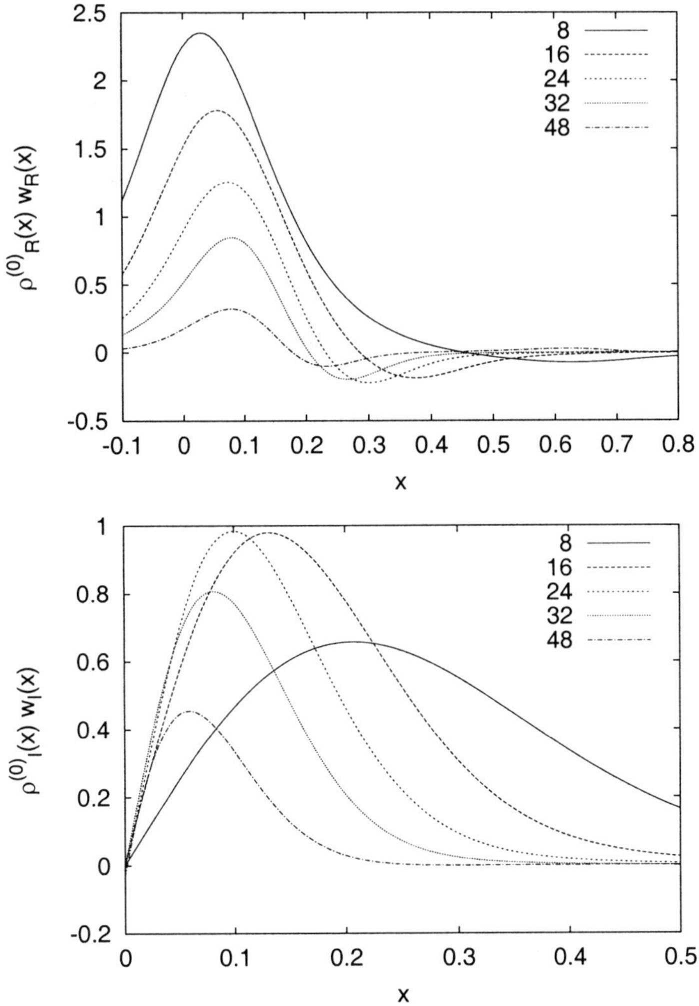

Figure 1. $\rho_R(x)$ and $\rho_I(x)$ for $\mu = 0.2$.

and their peaks get closer to μ as μ becomes larger than μ_c. Therefore $\lim_{N\to\infty}\lim_{\mu\to 0}\frac{\partial}{\partial\mu}\langle\nu_R\rangle_0 = 0 \neq \lim_{\mu\to 0}\lim_{N\to\infty}\frac{\partial}{\partial\mu}\langle\nu_R\rangle_0 = 1$. Similarly we find that $\lim_{N\to\infty}\lim_{\mu\to 0} w_R(x) = 0 \neq \lim_{\mu\to 0}\lim_{N\to\infty} w_R(x) = 1$. Details will be reported elsewhere.

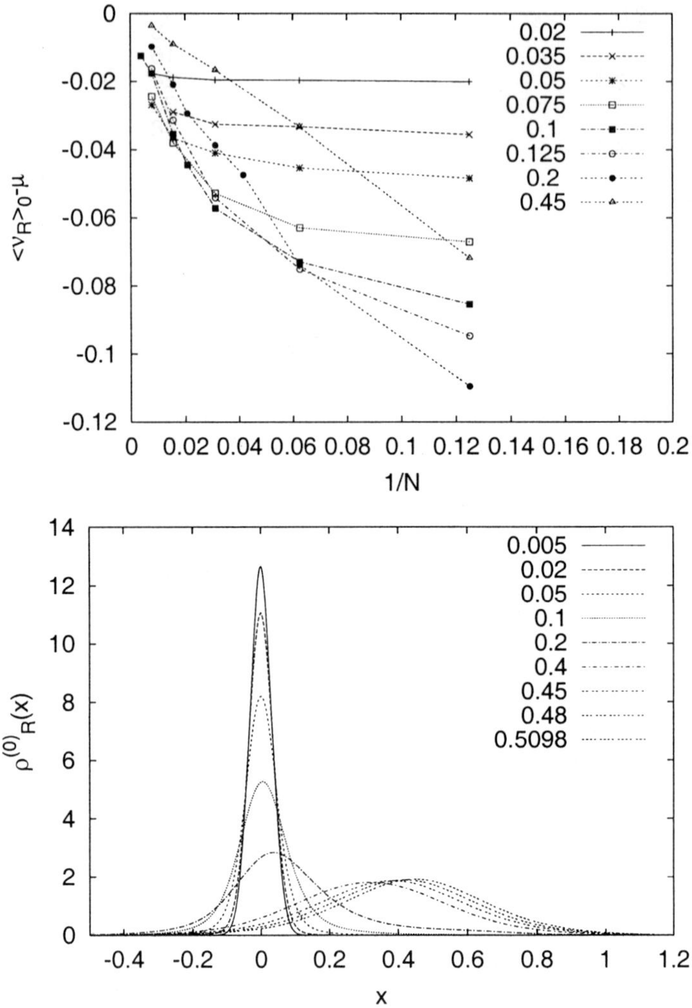

Figure 2. The difference of $\langle \nu_R \rangle_0$ from its $N = \infty$ value and $\rho_R^{(0)}(x)$ for $N = 8$ for various values of μ.

References

1. K. N. Anagnostopoulos and J. Nishimura, *Phys. Rev.* **D66**, 106008 (2002).
2. J. Ambjørn *et al.*, *JHEP* **0210**, 062 (2002).
3. V. Azcoiti *et al.*, *Phys. Rev. Lett.* **89**, 141601 (2002); hep-lat/0210004.
4. E. V. Shuryak and J. J. Verbaarschot, *Nucl. Phys.* **A560**, 306 (1993); J. J. Verbaarschot and T. Wettig, *Ann. Rev. Nucl. Part. Sci.* **50**, 343 (2000).

GLUODYNAMICS IN EXTERNAL FIELD: A TEST OF THE DUAL SUPERCONDUCTOR PICTURE*

M. CHERNODUB

ITP, Kanazawa University, Kanazawa, 920-1192, Japan
ITEP, B. Cheremushkinskaya 25, Moscow, 117259, Russia

We study gluodynamics in an external Abelian electromagnetic field within the dual superconductor approach. We show that the $SU(2)$ gluodynamics should possess a deconfining phase transition at zero temperature at certain value of the external field. A dual superconductor model for the $SU(3)$ gauge theory in external field predicts a rich phase structure containing confinement, asymmetric confinement and deconfinement phases. These results can be used to check the validity of the dual superconductor description of gluodynamics in external fields. We also discuss the gauge–independence of the obtained results.

1. We discuss the properties of the $SU(2)$ and $SU(3)$ gluodynamics in the external electromagnetic field using the dual superconductor approach[1] based on the Abelian monopole condensation. The condensate – observed[2] in various numerical simulations at low temperatures– forces a chromoelectric flux coming from (anti-)quarks to squeeze into a confining string. This picture of confinement has been confirmed in various lattice simulations[3].

A common feature of known superconductors is that at strong enough magnetic fields the superconductivity is destroyed and the superconductor goes in the normal (metal) state. In the dual superconductor model a similar effect [4] leads to the deconfinement phase transition. Below we study the phase diagram of gluodynamics within the dual superconductor approach. We work in the Bogomol'ny limit[5] supported by lattice simulations[6].

2. Let us first consider the $SU(2)$ gauge theory in the $4D$ Euclidean space. The infrared properties of the vacuum of this model can be described by the Abelian Higgs (or, Ginzburg–Landau) Lagrangian:

$$L_{GL}[B,\Phi] = \frac{1}{4}F^2_{\mu\nu} + \frac{1}{2}\left|D_\mu(B)\,\Phi\right|^2 + \lambda\Bigl(|\Phi|^2 - \eta^2\Bigr)^2\,, \tag{1}$$

*This work is supported by the JSPS Fellowship No. P01023.

where $F_{\mu\nu} = \partial_\mu B_\nu - \partial_\nu B_\mu$ is the field strength of the dual gauge field B_μ, Φ is the monopole field with the magnetic charge g_M, and $D_\mu = \partial_\mu + i g_M B_\mu$ is the covariant derivative. The gauge field B_μ is dual to the third component of the gluon field in an Abelian gauge. The model possesses the dual $U(1)$ gauge symmetry, $B_\mu \to B_\mu - \partial_\mu \alpha$, $\Phi \to e^{i g_M \alpha}\,\Phi$. The form of the potential implies the existence of the monopole condensate, $|\langle\Phi\rangle| = \eta > 0$.

Consider the four–dimensional sample of the (dual) superconductor occupying half–space, $x_2 \geqslant 0$. Let us apply the constant external EM field $F^{\mathrm{ext}}_{\mu\nu} = \varepsilon_{\mu\nu 34} H^{\mathrm{ext}}$ to the boundary of the superconductor. The external field is screened due to the induced superconducting current,

$$J_\mu = \Im m\big(\Phi^* D_\mu(B)\Phi\big) \equiv |\Phi|^2 \cdot v_\mu\,, \quad v_\mu = \partial_\mu \varphi + g_M B_\mu\,, \tag{2}$$

where we have set $\Phi = |\Phi| e^{i\varphi}$. The current is parallel to the boundary of the superconductor. The monopole kinetic term in Eq. (1) can be written as $|D_\mu \Phi|^2 = (\partial_\mu |\Phi|)^2 + |\Phi|^2 v_\mu^2$. Clearly, a non–zero current provides an additional positively–defined term in the Lagrangian ($\propto |\Phi|^2$). As a result, the external field destroys the monopole condensate.

We treat the model (1) classically. It is convenient to rewrite the action of the model (1) as an integral in a two-dimensional plane. One of the directions of the plane is the depth of the dual superconductor, x_2, while the second is given by the direction of the current (2). Choosing $J_\mu \propto \delta_{\mu,1}$ the first two terms of eq. (1) become, respectively, $F^2_{\mu\nu}/4 = H^2/2$ and

$$\big|D_\mu \Phi\big|^2 = \sum_{\alpha=1,2} \big|D_\alpha \Phi\big|^2 = \Big|\Big(D_1 \pm i D_2\Big)\Phi\Big|^2 \mp 2\varepsilon_{\alpha\beta}\partial_\alpha J_\beta \mp g_M H\,|\Phi|^2\,. \tag{3}$$

In the Bogomol'ny limit, $g_M^2/\lambda = 8$, the action of the model is:

$$S = \frac{L_3 L_4}{2}\int \mathrm{d}^2x \Big\{\big|(D_1 - iD_2)\Phi\big|^2 + \Big[H + \frac{g_M}{2}\big(|\Phi|^2 - \eta^2\big)\Big]^2\Big\} + S_f + S_J, \tag{4}$$

where L_i is the (infinitely large) length of the dual superconductor in i^{th} direction, $S_J = L_3 L_4 \int \mathrm{d}^2x\, \varepsilon_{\alpha\beta}\partial_\alpha J_\beta = L_1 L_3 L_4 J_1(x_2 = 0)$ is the action of the surface current and $S_f = g_M L_3 L_4 \eta^2 \int \mathrm{d}^2x\, H/2$ is the "flux" action.

The Bogomol'ny equations minimize the action (4),

$$(D_1 - D_2)\,\Phi = 0\,, \quad H + \frac{g_M}{2}\Big(|\Phi|^2 - \eta^2\Big) = 0\,.$$

The second equation gives the monopole condensate at the boundary, $|\Phi(x_2 = 0)|^2 = \eta^2 - 2H^{ext}/g_M$. The condensate disappears at the critical value of the external EM field, $H^{\mathrm{ext}} = H_{\mathrm{cr}} = g_M \eta^2/2$.

At zero temperature and in the absence of the external fields the tension of the string spanned on trajectories of the fundamental charges can be evaluated exactly in the Bogomol'ny limit[5,7], $\sigma = \pi\eta^2$. Using the Dirac relation between magnetic (g_M) and electric (g) charges, $g_M g = 2\pi$, we get the exact value of the critical EM field in terms of the string tension:

$$g H_{\rm cr}/\sigma = 1 \,, \qquad \text{for SU(2)}\,. \tag{5}$$

The dual superconductivity is destroyed in the *bulk* if $H \geqslant H_{\rm cr}$.

3. Now let us consider the Lagrangian of the $[U(1)]^2$ dual superconductor model corresponding to $SU(3)$ gluodynamics[8]:

$$L = \frac{1}{4} F^a_{\mu\nu} F^{a,\mu\nu} + \sum_{i=1}^{3} \Big[\frac{1}{2} \left| (\partial_\mu + i g_M \varepsilon^a_i B^a_\mu) \Phi_i \right|^2 + \lambda \Big(|\Phi_i|^2 - \eta^2 \Big)^2 \Big] \,, \tag{6}$$

where $F^a_{\mu\nu} = \partial_\mu B^a_\nu - \partial_\nu B^a_\mu$ is the field strength of the gauge field B^a_μ, $a = 3,8$. The phases of the monopole fields, Φ_i, $i = 1,2,3$, satisfy the relation $\sum_{i=1}^{3} \arg \Phi_i = 0$. The ϵ–symbols are the root vectors of the $SU(3)$ group: $\vec{\epsilon}_1 = (1,0)$, $\vec{\epsilon}_2 = (-1/2, -\sqrt{3}/2)$, $\vec{\epsilon}_3 = (-1/2, \sqrt{3}/2)$. In Eq. (6) no summation over the Latin index i is implied. The gauge fields $B^{3,8}_\mu$ are dual to the diagonal components $a = 3,8$ of the gluon field A^a_μ. Lagrangian (6) respects the dual $[U(1)]^2$ gauge invariance: $B^a_\mu \to B^a_\mu + \partial_\mu \alpha^a$, $\theta_i \to \theta_i + g_M(\varepsilon^3_i \alpha^3 + \varepsilon^8_i \alpha^8)$, $a = 3,8$, $i = 1,2,3$, where α^3 and α^8 are the parameters of the gauge transformation.

In the $[U(1)]^2$ Bogomol'ny limit, $g_M^2/\lambda = 16/3$, the equations of motion

$$\left(D^{(i)}_1 \pm i D^{(i)}_2 \right) \chi_i = 0 \,, \quad H^{(i)} \mp \frac{3 g_M}{4} \left(|\chi_i|^2 - \eta^2 \right) = 0 \,; \quad i = 1,2,3 \,, \tag{7}$$

give the same (due to the Weyl symmetry[9,10]) critical value for all components of the EM field, $H^{(i)} = \sum_{a=3,8} \epsilon^a_i H^a$, $i = 1,2,3$:

$$g H^{(i)}_{\rm cr} \equiv g \tilde{H}_{cr}/\sigma = 3/4 \,, \qquad \text{for SU(3)}\,. \tag{8}$$

Here we used the Dirac condition and the relation for the fundamental string tension, $\sigma = 2\pi\eta^2$, valid in the Bogomol'ny limit of the $[U(1)]^2$ model[11,9]. When the strength of the EM component $H^{(i)}$ reaches the $\tilde{H}_{cr}$ value then the monopole field, Φ_i, vanishes. Expressing the auxiliary fields $H^{(i)}$ in terms of the components the EM field, $H^{3,8}$, and using Eqs.(8) we get the phase diagram of the $[U(1)]^2$ dual superconductor depicted in Figure 1.

The phase diagram contains confinement (C), deconfinement (D) and the asymmetric confinement phases (A). The phase transition depends not

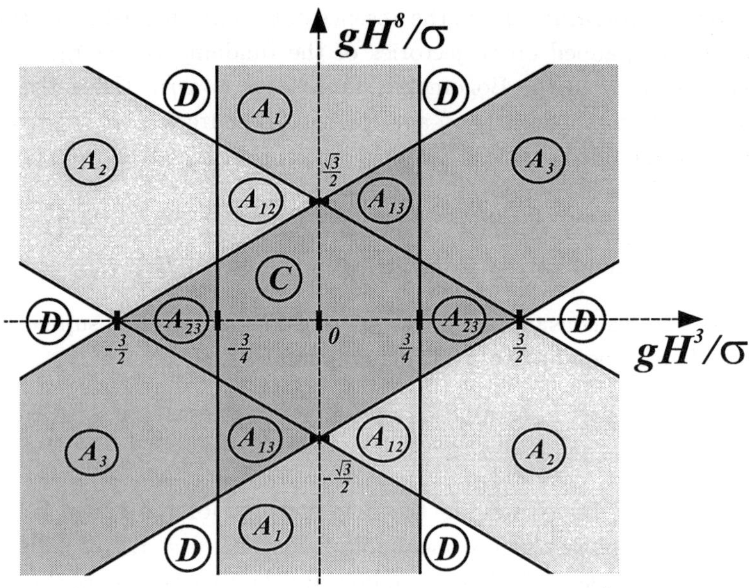

Figure 1. The zero temperature phase diagram of the dual $[U(1)]^2$ Higgs model in the Bogomol'ny limit in the presence of the external electromagnetic field.

only on the absolute value of the EM field but also on the ("color") orientation of this field in the Cartan subgroup. At low values of the field the model is always confining regardless of the color orientation. However, as the absolute value of the field is increased, the model enters – depending on color orientation – one of six (A_{12}, A_{13}, A_{23}, A_1, A_2 or A_3) asymmetric confinement phases. In the A_{ij} phase the ith and jth components of the monopole field are condensed while the expectation value of the third component is zero. In the phase A_i the ith component is condensed while the others two components are not. With further increase of the field the model either enters the deconfinement phase, D, or stays in one of the three asymmetric confinement phases, A_1, A_2 or A_3.

One can show[4] that three of six asymmetric confinement phases (A_{12}, A_{13} and A_{23}) contain one baryon and three meson states. These phases are confining since the quarks of all three colors are confined. The other three asymmetric confinement phases (A_1, A_2 and A_3) contain only two light meson states while the baryon bound state is absent.

4. Our results were obtained for the Abelian external fields which are applied to the dual superconductor corresponding to a *fixed* Abelian projection[a]. Although the fact of the monopole condensation seems to be projection–independent[2], important properties of the dual superconductor may depend on the on chosen Abelian projection[12,b].

We propose to investigate numerically the phase diagrams of the $SU(2)$ and $SU(3)$ gluodynamics in the Maximal Abelian projection and compare them with the predictions of Ref. [4]. This comparison should reveal whether the dual superconductor picture works at strong external fields or not.

References

1. Y. Nambu, *Phys. Rev. D10* (1974) 4262; G. 't Hooft, in *High Energy Physics*, ed. A. Zichichi, EPS International Conference, Palermo (1975); S. Mandelstam, Phys. Rept. **23**, 245 (1976).
2. M. N. Chernodub, M. I. Polikarpov, A. I. Veselov, *JETP Lett.* **63** (1996) 411; *Phys. Lett.* **B399** (1997) 267; A. Di Giacomo, G. Paffuti *Phys. Rev. D* **56** (1997) 6816; N. Nakamura, V. Bornyakov, S. Ejiri, S. Kitahara, Y. Matsubara and T. Suzuki, *Nucl. Phys. Proc. Suppl.* **53** (1997) 512.
3. For a review see, *e.g.*, M. N. Chernodub and M. I. Polikarpov, in *"Confinement, Duality and Non-perturbative Aspects of QCD"*, p.387, Plenum Press, 1998, hep-th/9710205; R. W. Haymaker, *Phys. Rept.* **315**, 153 (1999).
4. M. N. Chernodub, *Phys. Lett.* B**549**, 146 (2002).
5. E. B. Bogomolny, *Sov.J.Nucl.Phys.* **24** (1976) 449 [*Yad.Fiz.* **24** (1976) 861].
6. M. Baker, J.S. Ball, F. Zachariasen, *Phys. Rev.* **D31** (1985) 2575; *ibid.* **41** (1990) 2612; S. Maedan, Y. Matsubara, T. Suzuki, *Prog. Theor. Phys.* **84** (1990) 130; V. Singh, D.A. Browne, R.W. Haymaker, *Phys. Lett.* B **306**, 115 (1993); Y.Matsubara, S. Ejiri, T. Suzuki, *Nucl. Phys. Proc. Suppl.* **34** (1994) 176; G.S. Bali, C.Schlichter, K.Schilling, *Prog. Theor.Phys. Suppl.* **131** (1998) 645.
7. H. J. de Vega, F. A. Schaposnik, *Phys. Rev.* D**14** (1976) 1100.
8. S. Maedan and T. Suzuki, *Prog. Theor. Phys.* **81** (1989) 229 .
9. Y. Koma and H. Toki, *Phys. Rev.* D **62** (2000) 054027;
10. Y. Koma, E. M. Ilgenfritz, T. Suzuki, H. Toki, *Phys.Rev.* D **64** (2001) 014015.
11. M. N. Chernodub, *Phys. Lett.* B **474** (2000) 73.
12. M. N. Chernodub, M. I. Polikarpov and A. I. Veselov, *Phys. Lett.* B **342** (1995) 303; S. Fujimoto, S. Kato, T. Suzuki and T. Tsunemi, *Prog. Theor. Phys. Suppl.* **138** (2000) 36; J. M. Carmona, M. D'Elia, A. Di Giacomo, B. Lucini and G. Paffuti, *Phys. Rev.* D **64** (2001) 114507.
13. P. Cea, L. Cosmai, *JHEP* **0302** (2003) 031; *Phys.Rev.* D**60** (1999) 094506.

[a]Note that here we have implicitly assumed that we are working in the Maximal Abelian projection where the Bogomol'ny limit is realized[6].

[b]In particular, this implies that results of Refs.[13] for the phase diagram in the external Abelian fields can not be compared with our predictions.

THE STUDY OF CHIRAL RESTORATION USING THE QUARK-GLUON MIXED CONDENSATE $g\langle\bar{q}\sigma_{\mu\nu}G_{\mu\nu}q\rangle$ IN LATTICE QCD AT FINITE TEMPERATURE*

T. DOI, H. SUGANUMA, M. OKA

Department of Physics, Tokyo Institute of Technology, Ohokayama 2-12-1, Meguro, Tokyo 152-8551, Japan

N. ISHII

The Institute of Physical and Chemical Research (RIKEN), Hirosawa 2-1, Wako, Saitama, 351-0198, Japan

The quark-gluon mixed condensate $g\langle\bar{q}\sigma_{\mu\nu}G_{\mu\nu}q\rangle$ is studied using $SU(3)_c$ lattice QCD with the Kogut-Susskind fermion at the quenched level. Using the lattices as $\beta = 6.0$ with $16^3 \times N_t(N_t = 16, 12, 10, 8, 6, 4)$, $\beta = 6.1$ with $20^3 \times N_t(N_t = 20, 12, 10, 8, 6)$ and $\beta = 6.2$ with $24^3 \times N_t(N_t = 24, 16, 12, 10, 8)$ in high statistics of 100-1000 gauge configurations, we perform accurate measurement of the thermal effects on $g\langle\bar{q}\sigma_{\mu\nu}G_{\mu\nu}q\rangle$ as well as $\langle\bar{q}q\rangle$ in the chiral limit. We find that the thermal effects on both the condensates are very weak except for the vicinity of T_c, while both the condensates suddenly vanish around $T_c \simeq 280$MeV, which indicates strong chiral restoration near T_c. We also find that the ratio $m_0^2 \equiv g\langle\bar{q}\sigma_{\mu\nu}G_{\mu\nu}q\rangle/\langle\bar{q}q\rangle$ is almost independent of the temperature even in the very vicinity of T_c, which means that these two different condensates obey the same critical behavior. This nontrivial similarity between them would impose constraints on the chiral structure of the QCD vacuum near T_c.

1. Introduction

Fruitful phenomena in hadron physics, such as spontaneous chiral-symmetry breaking and color confinement, originate from the non-perturbative nature of QCD, or its nontrivial vacuum structure. The study of the vacuum structure of QCD is especially interesting at finite temperature/density, because it is expected that QCD has various phase structures in the phase diagram, and chiral restoration and color deconfinement can be observed. Considering that the on-going RHIC experiments are actually

*The lattice QCD Monte Carlo simulations have been performed on NEC SX-5 at Osaka University and IBM POWER4 at Tokyo Institute of Technology.

tackling this subject, we here present the theoretical study of finite temperature QCD. In this point, condensates are useful physical quantities because they characterize the nontrivial QCD vacuum. At finite temperature, we can probe the change of the QCD vacuum, by examining the thermal effects on the condensates.

Among various condensates, we study the quark-gluon mixed condensate $g\langle\bar{q}\sigma_{\mu\nu}G_{\mu\nu}q\rangle$ for the following reasons. First, we note that the mixed condensate is a chiral order parameter independent of $\langle\bar{q}q\rangle$, since it flips the chirality of the quark as

$$g\langle\bar{q}\sigma_{\mu\nu}G_{\mu\nu}q\rangle = g\langle\bar{q}_R(\sigma_{\mu\nu}G_{\mu\nu})q_L\rangle + g\langle\bar{q}_L(\sigma_{\mu\nu}G_{\mu\nu})q_R\rangle. \quad (1)$$

Therefore, $g\langle\bar{q}\sigma_{\mu\nu}G_{\mu\nu}q\rangle$, as well as $\langle\bar{q}q\rangle$, are important indicators on the chiral structure of the QCD vacuum. In particular, the thermal effects on both the condensates and the comparison of them will reveal mechanism of the chiral restoration of the QCD vacuum at finite temperature. Second, in contrast to $\langle\bar{q}q\rangle$, $g\langle\bar{q}\sigma_{\mu\nu}G_{\mu\nu}q\rangle$ represents a direct correlation between quarks and gluons in the QCD vacuum, and thus characterizes different aspect of the QCD vacuum. Third, the mixed condensate is relevant in various QCD sum rules, especially in baryons,[1] light-heavy mesons[2] and exotic mesons.[3] In this point, the quantitative estimate of the thermal effects on $g\langle\bar{q}\sigma_{\mu\nu}G_{\mu\nu}q\rangle$ has an impact on hadron phenomenology at finite temperature through the framework of the QCD sum rule. Therefore, it is desirable to estimate $g\langle\bar{q}\sigma_{\mu\nu}G_{\mu\nu}q\rangle$ at zero/finite temperature by a direct calculation from QCD, such as lattice QCD simulations.

So far, the mixed condensate $g\langle\bar{q}\sigma_{\mu\nu}G_{\mu\nu}q\rangle$ at zero temperature has been analyzed phenomenologically in the QCD sum rules.[4] In the lattice QCD, there has been no result except for a rather preliminary but pioneering work[5] done long time ago. Recently, new lattice calculations have been performed by us[6,7] using the Kogut-Susskind (KS) fermion, and by other group[8] using the Domain-Wall fermion. At finite temperature, however, there has been no result on $g\langle\bar{q}\sigma_{\mu\nu}G_{\mu\nu}q\rangle$ except for our early results.[7] Therefore, we here present the extensive results of the thermal effects on $g\langle\bar{q}\sigma_{\mu\nu}G_{\mu\nu}q\rangle$ as well as $\langle\bar{q}q\rangle$, including the analysis near the critical temperature.[9]

2. The Lattice Formalism

We calculate the condensates $g\langle\bar{q}\sigma_{\mu\nu}G_{\mu\nu}q\rangle$ as well as $\langle\bar{q}q\rangle$ using $SU(3)_c$ lattice QCD with the KS-fermion at the quenched level. We note that the KS-fermion preserves the explicit chiral symmetry for the quark mass

$m = 0$, which is a desirable feature to study both the condensates of chiral order parameters.

The Monte Carlo simulations are performed with the standard Wilson action for $\beta = 6.0, 6.1$ and 6.2. In order to calculate at various temperatures, we use the following lattices as

(1) $\beta = 6.0,\ 16^3 \times N_t\ \ (N_t = 16, 12, 10, 8, 6, 4)$,
(2) $\beta = 6.1,\ 20^3 \times N_t\ \ (N_t = 20, 12, 10, 8, 6)$,
(3) $\beta = 6.2,\ 24^3 \times N_t\ \ (N_t = 24, 16, 12, 10, 8)$.

We generate 100 gauge configurations for each lattice. However, in the vicinity of the chiral phase transition point, namely, $20^3 \times 8$ at $\beta = 6.1$ and $24^3 \times 10$ at $\beta = 6.2$, the fluctuations of the condensates get larger. We hence generate 1000 gauge configurations for the above two lattices to improve the estimate. The lattice units are obtained as $a \simeq 0.10, 0.09, 0.07$fm for $\beta = 6.0, 6.1, 6.2$, respectively, which reproduce the string tension $\sigma = 0.89$GeV/fm. We calculate the flavor-averaged condensates as

$$a^3\langle\bar{q}q\rangle = -\frac{1}{4}\sum_{f} \mathrm{Tr}\left[\langle q^f(x)\bar{q}^f(x)\rangle\right], \tag{2}$$

$$a^5 g\langle\bar{q}\sigma_{\mu\nu}G_{\mu\nu}q\rangle = -\frac{1}{4}\sum_{f,\ \mu,\nu} \mathrm{Tr}\left[\langle q^f(x)\bar{q}^f(x)\rangle\sigma_{\mu\nu}G_{\mu\nu}\right], \tag{3}$$

where $\mathrm{SU}(4)_f$ quark-spinor fields, q and $\bar{q}$, are converted into spinless Grassmann KS-fields χ and $\bar{\chi}$ and the gauge-link variable. We adopt the clover-type definition for the gluon field strength $G_{\mu\nu}$ on the lattice in order to eliminate $\mathcal{O}(a)$ discretization error. The more detailed formula for the condensates and gluon field strength are given in Ref. 6.

In the calculation of the condensates, we use the current-quark mass $m = 21, 36, 52$MeV. For the fields χ, $\bar{\chi}$, the anti-periodic condition is imposed at the boundary in the all directions. We measure the condensates on 16 different physical space-time points of x ($\beta = 6.0$) or 2 points of x ($\beta = 6.1, 6.2$) in each configuration. Therefore, at each m and temperature, we achieve high statistics in total as 1600 data ($\beta = 6.0$) or 200 data ($\beta = 6.1, 6.2$). Note that in the vicinity of the critical temperature, $20^3 \times 8$ ($\beta = 6.1$) and $24^3 \times 10$ ($\beta = 6.2$), we take up to 2000 data so as to achieve high statistics and to guarantee reliability of the results.

3. The Lattice Results and Discussions

We calculate the condensates $g\langle\bar{q}\sigma_{\mu\nu}G_{\mu\nu}q\rangle$ as well as $\langle\bar{q}q\rangle$ at each quark mass and temperature. We observe that both the condensates show a clear linear behavior against the quark mass m, which can be typically seen in Ref. 6. We therefore fit the data with a linear function and determine the condensates in the chiral limit. Estimated statistical errors are typically less than 5%, while the finite volume artifact is estimated to be about 1%, from the check on the dependence of the condensates on boundary conditions.[6]

For each condensate, we estimate the thermal effects by taking the ratio between the values at finite and zero temperatures. The renormalization constants are expected to be canceled in this ratio because there is no operator mixing for both the condensates in the chiral limit.[10] In figure 1, we plot the thermal effects on $g\langle\bar{q}\sigma_{\mu\nu}G_{\mu\nu}q\rangle$. We find a drastic change of $g\langle\bar{q}\sigma_{\mu\nu}G_{\mu\nu}q\rangle$ around the critical temperature $T_c \simeq 280$MeV. This is a first observation of chiral-symmetry restoration in terms of $g\langle\bar{q}\sigma_{\mu\nu}G_{\mu\nu}q\rangle$. We also find that the thermal effects on $g\langle\bar{q}\sigma_{\mu\nu}G_{\mu\nu}q\rangle$ are very weak below the critical temperature, namely, $T \lesssim 0.9T_c$. In figure 2, the same features can be also seen for $\langle\bar{q}q\rangle$.

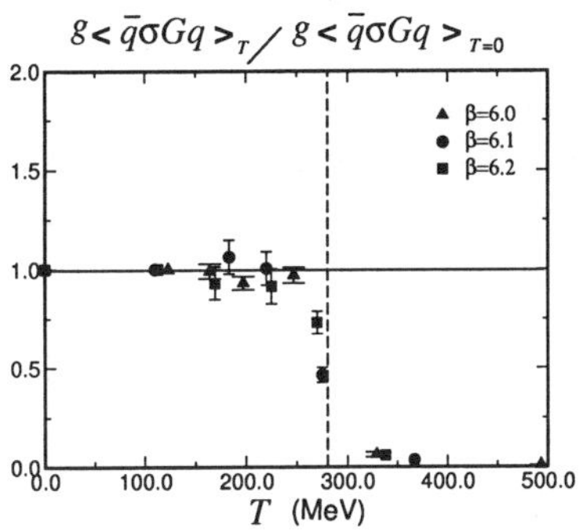

Figure 1. Thermal effects on the mixed condensate $g\langle\bar{q}\sigma_{\mu\nu}G_{\mu\nu}q\rangle$ v.s. the temperature T. The vertical dashed line denotes the critical temperature $T_c \simeq 280$MeV at the quenched level.

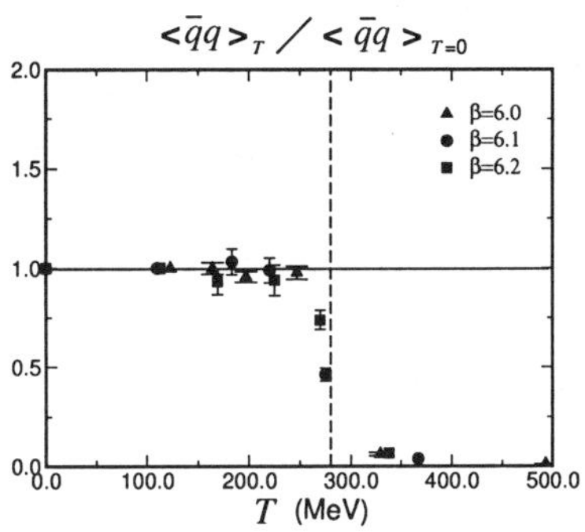

Figure 2. Thermal effects on the quark condensate $\langle\bar{q}q\rangle$ v.s. T.

We then compare the thermal effects on $g\langle\bar{q}\sigma_{\mu\nu}G_{\mu\nu}q\rangle$ and $\langle\bar{q}q\rangle$, because these two condensates characterize different aspects of the QCD vacuum. For this purpose, we plot the thermal effects on $m_0^2 \equiv g\langle\bar{q}\sigma_{\mu\nu}G_{\mu\nu}q\rangle/\langle\bar{q}q\rangle$ in figure 3. From figure 3, we observe that m_0^2 is almost independent of the temperature, even in the very vicinity of T_c. This very nontrivial result means that *both of* $g\langle\bar{q}\sigma_{\mu\nu}G_{\mu\nu}q\rangle$ *and* $\langle\bar{q}q\rangle$ *obey the same critical behavior.* This may indicate the existence of the universal behavior of chiral order parameters near T_c, and would impose constraints on the chiral structure of the QCD vacuum.

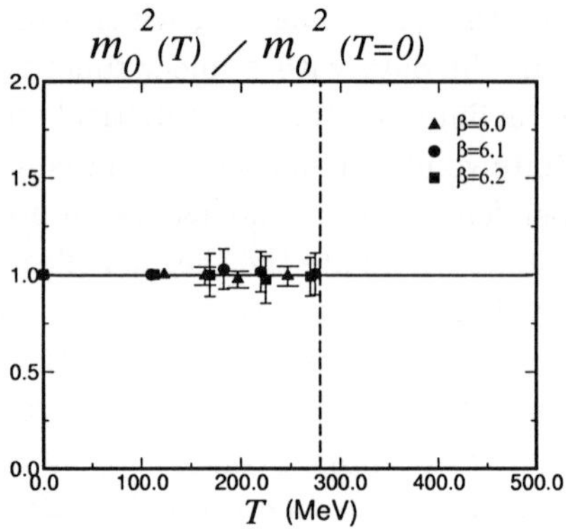

Figure 3. The thermal effects on $m_0^2 \equiv g\langle\bar{q}\sigma_{\mu\nu}G_{\mu\nu}q\rangle/\langle\bar{q}q\rangle$ v.s. the temperature T. This result indicates the same critical behavior between $\langle\bar{q}q\rangle$ and $g\langle\bar{q}\sigma_{\mu\nu}G_{\mu\nu}q\rangle$.

In summary, we have studied the thermal effects on $g\langle\bar{q}\sigma_{\mu\nu}G_{\mu\nu}q\rangle$ as well as $\langle\bar{q}q\rangle$ using SU(3)$_c$ lattice QCD with KS-fermion at the quenched level. While the thermal effects on both the condensates are very weak below T_c, we have observed a strong chiral phase transition near T_c as a sudden disappearance of both the condensates. We have also observed that $m_0^2 \equiv g\langle\bar{q}\sigma_{\mu\nu}G_{\mu\nu}q\rangle/\langle\bar{q}q\rangle$ is almost independent of the temperature even in the very vicinity of T_c, which means both the condensates obey the same critical behavior. This nontrivial similarity would impose constraints on the chiral structure of the QCD vacuum near T_c. For further studies, a full QCD lattice calculation is in progress in order to analyze the dynamical quark effects on the condensates.

References

1. H.G. Dosch, M. Jamin and S. Narison, *Phys. Lett.* **B220**, 251 (1989).
2. H.G. Dosch and S. Narison, *Phys. Lett.* **B417**, 173 (1998).
3. J.I. Latorre, P. Pascual and S. Narison, *Z. Phys.* **C34**, 347 (1987).
4. V.M. Belyaev and B.L. Ioffe, *Sov. Phys. JETP* **56**, 493 (1982).
5. M. Kremer and G. Schierholz, *Phys. Lett.* **B194**, 283 (1987).
6. T. Doi, N. Ishii, M. Oka and H. Suganuma, *Phys. Rev.* **D67**, 054504 (2003).
7. T. Doi, N. Ishii, M. Oka and H. Suganuma, *Nucl. Phys.* **A721**, 934 (2003); *Prog. Theor. Phys. Suppl.* **151**, 161 (2003); Proc. of *"Quark Confinement and the Hadron Spectrum V"* (World Scientific, 2003) p.381, hep-lat/0212025.
8. T.W. Chiu and T.H. Hsieh, hep-lat/0305016.
9. T. Doi, N. Ishii, M. Oka and H. Suganuma, *Nucl. Phys.* **B** (Proc. Suppl.) in press, hep-lat/0309124.
10. S. Narison and R. Tarrach, *Phys. Lett.* **B125**, 217 (1983).

THE 2PPI EXPANSION: DYNAMICAL MASS GENERATION AND VACUUM ENERGY*

D. DUDAL†AND H. VERSCHELDE

Ghent University
Department of Mathematical Physics and Astronomy
Krijgslaan 281-S9, B-9000 Gent, Belgium
E-mail: david.dudal@ugent.be,henri.verschelde@ugent.be

R. E. BROWNE AND J. A. GRACEY

Theoretical Physics Division, Department of Mathematical Sciences
University of Liverpool
P.O. Box 147, Liverpool, L69 3BX, United Kingdom
E-mail: rebrowne@amtp.liv.ac.uk, jag@amtp.liv.ac.uk

We discuss the $2PPI$ expansion, a summation of the bubble graphs up to all orders, by means of the $2D$ Gross-Neveu toy model, whose exact mass gap and vacuum energy are known. Then we use the expansion to give analytical evidence that a dimension two gluon condensate exists for pure Yang-Mills in the Landau gauge. This $\langle A_\mu^a A_\mu^a \rangle$ condensate consequently gives rise to a dynamical gluon mass.

1. Introduction

Lately, there was growing evidence for the existence of a condensate of mass dimension two in Yang-Mills (YM) theories in the Landau gauge. An obvious candidate for such a condensate is $\langle A_\mu^a A_\mu^a \rangle$. The phenomenological background of this type of condensate can be found in [1,2,3]. Also lattice simulations indicated a non-zero condensate $\langle A_\mu^a A_\mu^a \rangle$ [4]. See [5] for an overview of recent results.

*Talk given by D. Dudal at the International Conference on Color Confinement and Hadrons in Quantum Chromodynamics, Confinement 2003, TITech & Riken, Tokyo, Japan, July 21-24, 2003.
†Research Assistant of the Fund for Scientific Research-Flanders, Belgium.

Thinking of simpler models like massless $\lambda\phi^4$ or Gross-Neveu [6] and the role played by their quartic interaction in the formation of a (local) composite (in particular, containing two fields) condensate and the consequent dynamical mass generation for the originally massless fields [6,7], it is clear the possibility exists that the quartic gluon interaction gives rise to a two field composite operator condensate in YM (QCD) and mass generation for the gluons too.

In section 2, we give a short review of the $2PPI$ expansion by means of the $2D$ Gross-Neveu model. In section 3, we present our results for $\langle A^2 \rangle$ for $SU(N)$ Yang-Mills theory in the Landau gauge

2. The Gross-Neveu model

The $U(N)$ invariant Gross-Neveu Lagrangian in $2D$ Euclidean space time reads

$$\mathcal{L} = \overline{\psi}\partial\!\!\!/\psi - \frac{1}{2}g^2\left(\overline{\psi}\psi\right)^2. \tag{1}$$

This model possesses a discrete chiral symmetry $\psi \to \gamma_5\psi$, imposing $\langle\overline{\psi}\psi\rangle = 0$ perturbatively. We focus on the topology of vacuum diagrams. We can divide them into 2 disjoint classes: those diagrams falling apart in 2 separate pieces when 2 lines meeting at the same point are cut. We call those *2-point-particle-reducible* or $2PPR$. The second diagram of Figure 1 is a $2PPR$ diagram. The other type is the complement of the $2PPR$ class, we call these diagrams *2-point-particle-irreducible* ($2PPI$) diagrams. The first and third diagram of Figure 1 are $2PPI$ ones. We could now remove all $2PPR$ bubbles from the diagrammatic sum building up the vacuum energy by summing them in an effective mass m. Defining $\Delta = \langle\overline{\psi}\psi\rangle$, it can be shown[7] that

$$m = -g^2\left(1 - \frac{1}{2N}\right)\Delta. \tag{2}$$

Then the $2PPI$ vacuum energy E_{2PPI} is given by the sum of all $2PPI$ vacuum diagrams, now with a mass m running in the loops. It is important to notice that E_{2PPI} is *not* the vacuum energy due to a double counting ambiguity, already visible in the second diagram of Figure 1: each diagram can be seen as an insertion on the other one. This can be solved by considering $\frac{dE}{dg^2}$ instead of E. The g^2 derivative can hit $2PPR$ or $2PPI$ vertex. Hence

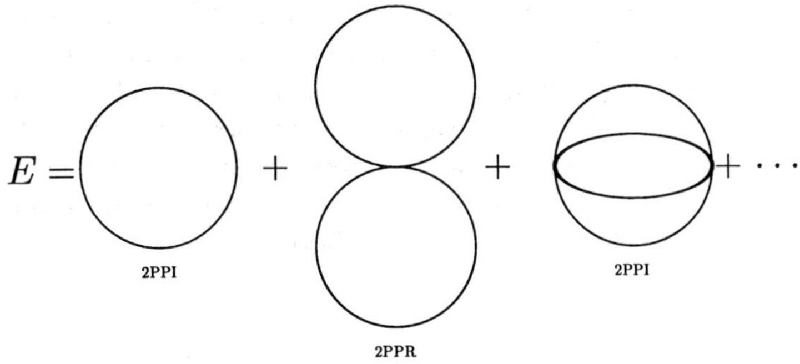

Figure 1. The vacuum energy as a sum of $2PPI$ and $2PPR$ diagrams.

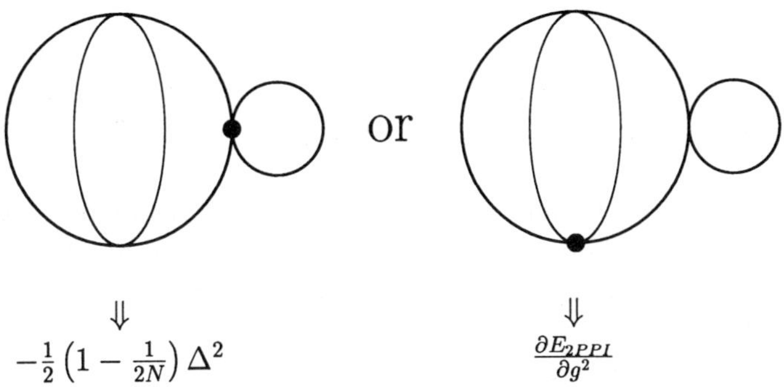

Figure 2. The $2PPI$ vacuum energy E_{2PPI}.

$$\frac{dE}{dg^2} = -\frac{1}{2}\left(1-\frac{1}{2N}\right)\Delta^2 + \frac{\partial E_{2PPI}(m)}{\partial g^2}. \tag{3}$$

This can be integrated using the ansatz

$$E = E_{2PPI} + cg^2\Delta^2. \tag{4}$$

It remains to determine the unknown constant c. Diagrammatically, it is easy to show that one has the following gap equation.

$$\frac{\partial E_{2PPI}}{\partial m} = \Delta. \tag{5}$$

Combination of the above formulae finally gives

$$E = E_{2PPI} + \frac{1}{2}g^2\left(1-\frac{1}{2N}\right)\Delta^2. \tag{6}$$

An important point is the renormalizability of the $2PPI$ expansion. Two possible problems could be mass renormalization and vacuum energy renormalization, since originally there was no external mass scale present. The proof is quite technical, but all formulae remain correct and are finite when the conventional counterterms of the Gross-Neveu model are included [7]. Essentially, the proof is based on coupling constant renormalization and the separation of $2PPI$ and $2PPR$ contributions.

It can be shown that

$$\frac{\partial E_{2PPI}}{\partial \Delta} = m \Leftrightarrow \frac{\partial E}{\partial m} = 0. \tag{7}$$

However, this does not mean that $E(\Delta)$ has the meaning of an effective potential, since $E(\Delta)$ is meaningless if the gap equation (7) is not fulfilled.

In Table 1, we list the numerical deviations in terms of percentage between our optimized 2-loop results[7] for the mass gap M and the square of minus the vacuum energy $\sqrt{-E}$ and the exact known values. We conclude that the $2PPI$ results are in relative good agreement with the exact values and converge to the exact $N \to \infty$ limit.

Table 1. Numerical results for M and $\sqrt{-E}$

N	deviation M (%)	deviation $\sqrt{-E}$ (%)
2	?	?
3	-4.5	47.7
4	-6.5	27.9
5	-6.1	19.9
10	-3.5	8.4
∞	0	0

3. $SU(N)$ Yang-Mills theory in the Landau gauge

Next, we consider the Euclidean Yang-Mills action in the Landau gauge where A^a_μ denotes the gauge field. Repeating the analysis of section 2 leads to[8]

$$\begin{aligned}\Delta &= \left\langle A^a_\mu A^a_\mu \right\rangle ,\\ m^2 &= g^2 \frac{N}{N^2-1}\frac{d-1}{d}\Delta,\\ E &= E_{2PPI} - \frac{g^2}{4}\frac{N}{N^2-1}\frac{d-1}{d}\Delta^2,\\ \frac{\partial E_{2PPI}}{\partial m^2} &= \frac{\Delta}{2} \Leftrightarrow \frac{\partial E}{\partial m^2} = 0. \end{aligned} \tag{8}$$

After some manipulation, the 2-loop results became

$$\frac{g^2 N}{16\pi^2} \approx 0.131 \qquad \sqrt{\Delta} \approx 536\text{MeV} \qquad E \approx -0.002\text{GeV}^4. \tag{9}$$

We notice that the relevant expansion parameter, $\frac{g^2N}{16\pi^2}$, is relatively small. As such, our results should be qualitatively trustworthy.

As a comparison, the value found by Boucaud et al from lattice simulations and an OPE treatment was $\left\langle A^2 \right\rangle_{\mu=10\text{GeV}} \approx 1.64\text{GeV}$. To find a value for the gluon mass m_g itself (as the pole of the gluon propagator) within the $2PPI$ framework, the diagrams relevant for mass renormalization of m should be calculated.

Acknowledgments

D. D. would like to thank the organizers of this conference. This work was supported in part by PPARC through a research studentship. (R.E.B.)

References

1. K. G. Chetyrkin, S. Narison and V. I. Zakharov, *Nucl. Phys.* **B550** (1999) 353.
2. F. V. Gubarev, L. Stodolsky and V. I. Zakharov, *Phys. Rev. Lett.* **86** (2001) 2220.
3. F. V. Gubarev and V. I. Zakharov, *Phys. Lett.* **B501** (2001) 28.
4. P. Boucaud, A. Le Yaouanc, J. P. Leroy, J. Micheli, O. Pène and J. Rodriguez-Quintero, *Phys. Rev.* **D63** (2001) 114003.
5. D. Dudal, H. Verschelde, V. E. R. Lemes, M. S. Sarandy, R. F. Sobreiro, S. P. Sorella, M. Picariello, A. Vicini and J. A. Gracey, to appear in these proceedings, hep-th/0308153.
6. D. J. Gross and A. Neveu, *Phys. Rev.* **D10** (1974) 3235.
7. D. Dudal and H. Verschelde, *Phys. Rev.* **D67** (2003) 025011.
8. D. Dudal, H. Verschelde, R. E. Browne and J. A. Gracey, *Phys. Lett.* **B562** (2003) 87.

NUMERICAL STUDY OF THE EQUATION OF STATE FOR TWO FLAVOR QCD AT FINITE DENSITY *

S. EJIRI,[a] C.R. ALLTON,[b] S.J. HANDS,[b] O. KACZMAREK,[a] F. KARSCH,[a] E. LAERMANN,[a] and C. SCHMIDT[a]

[a] *Fakultät für Physik, Universität Bielefeld, D-33615 Bielefeld, Germany*
[b] *Department of Physics, University of Wales Swansea, Singleton Park, Swansea, SA2 8PP, U.K.*

We discuss the equation of state for 2 flavor QCD at non-zero temperature and density. Derivatives of $\ln Z$ with respect to quark chemical potential μ_q up to fourth order are calculated, enabling estimates of the pressure, quark number density and associated susceptibilities as functions of μ_q via a Taylor series expansion. It is found that the fluctuations in the quark number density increase in the vicinity of the phase transition temperature and the susceptibilities start to develop a pronounced peak as μ_q is increased. This suggests the presence of a critical endpoint in the (T, μ_q) plane.

1. Introduction

Remarkable progress for QCD thermodynamics has been recently made by numerical studies of lattice QCD with small but non-zero baryon density[1]. The pseudocritical line has been investigated in the low density regime[2,3,4]. The phase transition for 2 flavor QCD is known to be crossover at $\mu_q = 0$ and expected to become a first order phase transition at a critical endpoint, and it might be possible to detect the endpoint experimentally via event-by-event fluctuations in heavy-ion collision experiments.

In this report, we discuss the equation of state at non-zero baryon number density. The study of the equation of state gives the most basic information for the experiments. Quantitative calculations of thermodynamic quantities such as pressure and energy density are indispensable. In particular, since the number density fluctuation should be large around the

*Presented by S. Ejiri. This work is supported by BMBF under grant No.06BI102, DFG under grant FOR 339/2-1 and PPARC grant PPA/a/s/1999/00026.

endpoint[5], the susceptibility of the quark number density, given by the second derivative of pressure with respect to μ_q, is an important quantity[6].

Several studies of the quark number susceptibility have been performed at $\mu_q = 0$ [7]. Moreover, measurements of pressure and energy density at $\mu_q \neq 0$ were done by Ref. 8 using the reweighting method, which allow the investigation of thermodynamic properties at non-zero baryon density. This approach, however, does not work for large μ_q and large lattice size due to the sign problem.

Here, we adopt the following strategy. We compute the derivatives of physical quantities with respect to μ_q at $\mu_q = 0$, and determine the Taylor expansion coefficients in μ_q[3,9], in which the sign problem does not arise. Because the pressure is an even function of μ_q, the μ_q^2-term is leading and μ_q^4-term is the next to leading, and in fact only these two terms are non-zero in the high temperature limit. We compute the Taylor expansion coefficients up to fourth order. Using the second derivative of pressure (energy density), we investigate the relation between the line of constant pressure (energy density) and the phase transition line[3]. The fourth order term enables us to evaluate the μ_q-dependence of the quark number susceptibility near $\mu_q = 0$. By estimating the change of the susceptibility, we discuss the possibility of the existence of the critical end point in the phase diagram of T and μ_q[10].

2. Taylor expansion in μ_q

Pressure is given in terms of the grand partition function $Z(T, V, \mu_q)$ by $p/T^4 = (1/VT^3) \ln Z$. However, the direct calculation of $\ln Z$ is difficult, hence most of the work done at $\mu_q = 0$ for the calculation of pressure is done by using the integral method[11,12], where the first derivative of pressure is computed by simulations, and the pressure is obtained by integration along a suitable integral path. For $\mu_q \neq 0$, direct Monte Carlo simulation is not applicable; in this case we proceed by computing higher order derivatives of pressure with respect to μ_q/T at $\mu_q = 0$, and then estimate $p(\mu_q)$ using a Taylor expansion,

$$\left.\frac{p}{T^4}\right|_{T,\mu_q} = \left.\frac{p}{T^4}\right|_{T,0} + \sum_{n=1}^{\infty} c_n(T) \left(\frac{\mu_q}{T}\right)^n, \qquad (1)$$

where $c_n = (1/n!) \partial^n (p/T^4)/\partial(\mu_q/T)^n|_{\mu_q=0}$. These derivatives can be computed by the random noise method, which saves CPU time, and also it can be proved that the odd terms are exactly zero. Furthermore, we do not

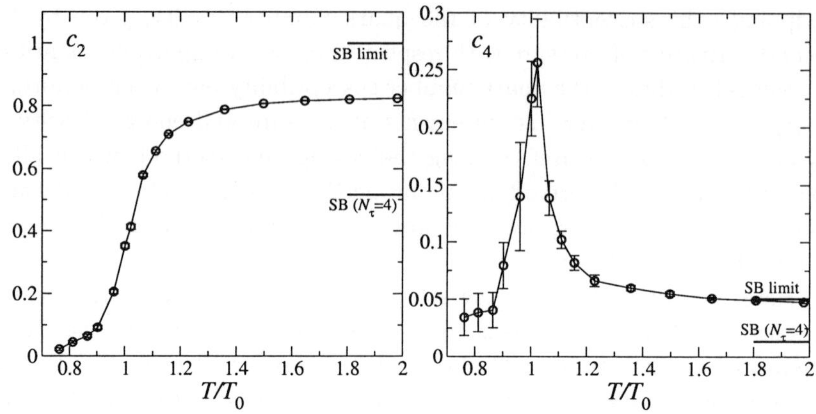

Figure 1. Coefficients of Taylor expansion, c_2 (left) and c_4 (right). T_0 is T_c at $\mu_q = 0$.

need simulations at $(T, \mu_q) = (0, 0)$ for the subtraction to normalize the value of p, since the derivatives of $p|_{T=0,\mu_q=0}$ with respect to μ_q are, of course, zero. This also reduces CPU time.

The quark number density n_q and quark number susceptibility χ_q are obtained by the derivatives of pressure; $n_q/T^3 = \partial(p/T^4)/\partial(\mu_q/T)$, $\chi_q/T^2 = \partial^2(p/T^4)/\partial(\mu_q/T)^2$. We compute the pressure up to $O(\mu_q^4)$ using 2 flavors of p4-improved staggered fermion[13] at $ma = 0.1$ on a $16^3 \times 4$ lattice. Then the quark number density and quark number susceptibility are obtained up to $O(\mu_q^3)$ and $O(\mu_q^2)$, respectively. The details of the simulations are given in Ref. 10. In Fig. 1, we plot the data for c_2 (left) and c_4 (right). Both of them are very small at low temperature and approach the Stefan-Boltzmann (SB) limit in the high temperature limit, as we expected. The remarkable point is a strong peak of c_4 around T_c.

We can understand this peak via two arguments. One is a prediction from the hadron resonance gas model[14], which is an effective model of the free hadron gas in the low temperature phase. The model study predicts $c_4/c_2 = 0.75$ and our results are consistent with this prediction for $T < T_c$; in fact, as T increases c_4/c_2 remains constant until $T \approx T_c$, whereupon it approaches the SB limit.

The other point is from a discussion of the convergence radius of the Taylor expansion. We expect that the crossover transition at $\mu_q = 0$ changes to first order transition at a point $\mu_q/T_c \sim O(1)$[2,15]. Then, the analysis by Taylor expansion must break down in that regime, i.e. the convergence radius should be smaller than the value of μ_q/T at the critical endpoint.

We define estimates for the convergence radius by $\rho_n \equiv \sqrt{|c_n/c_{n+2}|}$. We compute ρ_0 and ρ_2 from $c_0 \equiv p/T^4(0)$, c_2 and c_4. It is found that both ρ_0 and ρ_2 are quite large at high temperature as expected from the SB limit, $\rho_2^{SB} \simeq 2.01, \rho_4^{SB} \simeq 4.44$. On the other hand, around T_c, these are $O(1)$, since c_2 and c_4 are of the same order, so that our results around T_c suggest a singular point in the neighborhood of $\mu_q/T_c = 1$.

3. Equation of state at $\mu_q \neq 0$

Next, we calculate pressure and quark number susceptibility in a range of $0 \leq \mu_q/T \leq 1$ which is within the radius of convergence discussed above, using the data of c_2 and c_4; $\Delta(p/T^4) \equiv p(T,\mu_q)/T^4 - p(T,0)/T^4 = c_2(\mu_q/T)^2 + c_4(\mu_q/T)^4 + O(\mu_q^6)$, and $\chi_q/T^2 = 2c_2 + 12c_4(\mu_q/T)^2 + O(\mu_q^4)$.

We draw $\Delta(p/T^4)$ for each fixed μ_q/T in Fig. 2 (left) and find that the difference from $p|_{\mu_q=0}$ is very small in the interesting regime for heavy-ion collisions, $\mu_q/T \approx 0.1$ (RHIC) and $\mu_q/T \approx 0.5$ (SPS), in comparison with the value at $\mu_q = 0$, e.g. the SB value for 2 flavor QCD at $\mu_q = 0$: $p^{SB}/T^4 \simeq 4.06$. The effect of non-zero quark density on pressure at $\mu_q/T = 0.1$ is only 1%. Also, the result is qualitatively consistent with that of Ref. 8 obtained by the reweighting method.

Moreover, together with the data of derivative with respect to β, we can discuss the lines of constant pressure [energy density] at T_c. From $\mathrm{d}p(T,\mu^2) = 0$, the slope at $\mu_q = 0$ is

$$\frac{\mathrm{d}T}{\mathrm{d}(\mu_q^2)} = -\frac{\partial(p/T^4)}{\partial(\mu_q^2)} \Big/ \left(\frac{\partial(p/T^4)}{\partial T} + \frac{4p}{T^5}\right). \tag{2}$$

We obtain $T^2\partial^2(p/T^4)/\partial\mu_q^2 = 0.693(5)$, and use the data in Ref. 11, $p/T^4 = 0.27(5)$, $T\partial(p/T^4)/\partial T = -[a^{-1}(\partial a/\partial\beta)]^{-1}(\partial(p/T^4)/\partial\beta) = 2.2(6)$, at β_c for $ma = 0.1$. The same calculation is also performed for the energy density. We find that the slope of the constant pressure [energy density] line is $T(\mathrm{d}T/\mathrm{d}(\mu_q^2)) = -0.107(22)[-0.087(23)]$. Since the slope of T_c in terms of μ_q^2 is $T_c(\mathrm{d}T_c/\mathrm{d}(\mu_q^2)) = -0.07(3)$ [3], this result suggests that the line of constant pressure or energy density is parallel with the phase transition line, and the μ_q-dependence of the pressure or energy density at $T_c(\mu_q)$ is very small[3].

Figure 2 (right) is the result for χ_q/T^2 at fixed μ_q/T. We find a pronounced peak for $m_q/T > 0.5$, whereas χ_q does not have a peak for $\mu_q = 0$. This suggests the presence of a critical endpoint in the (T,μ_q) plane.

This discussion can be easily extended to the charge fluctuation χ_C. The spike of χ_C at T_c is weaker than that of χ_q, which means the increase

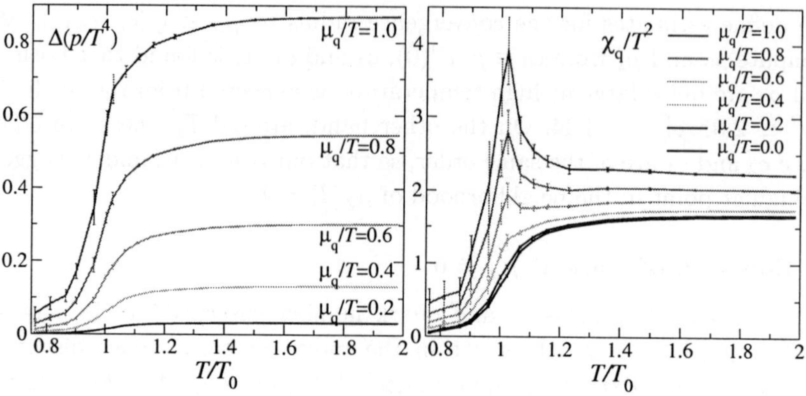

Figure 2. Difference of pressure from $\mu_q = 0$ (left) and Quark number susceptibility (right) as a function of T for each fixed μ_q/T. T_0 is T_c at $\mu_q = 0$.

of the charge fluctuation is smaller than that of the number fluctuation as μ_q increases[10].

References

1. S. Muroya, A. Nakamura, C. Nonaka and T. Takaishi, `hep-lat/0306031`; F. Karsch and E. Laermann, `hep-lat/0305025`.
2. Z. Fodor and S.D. Katz, *JHEP* **0203**, 014 (2002).
3. C.R. Allton *et al.*, *Phys. Rev.* **D66**, 074507 (2002).
4. P. de Forcrand and O. Philipsen, *Nucl. Phys.* **B642**, 290 (2002); `hep-lat/0307020`; M. D'Elia and M.-P. Lombardo, *Phys. Rev.* **D67**, 014505 (2003).
5. M. Stephanov, K. Rajagopal and E. Shuryak, *Phys. Rev. Lett.* **81**, 4816 (1998).
6. M. Asakawa, U. Heinz and B. Müller, *Phys. Rev. Lett.* **85**, 2072 (2000); S. Jeon and V. Koch, *Phys. Rev. Lett.* **85**, 2076 (2000); `hep-ph/0304012`; T. Kunihiro, `hep-ph/0007173`.
7. S. Gottlieb *et al.*, *Phys. Rev.* **D38**, 2888 (1988); C. Bernard *et al.* [MILC Collaboration], *Nucl. Phys. B (Proc. Suppl.)* **119**, 523 (2003).
8. Z. Fodor, S.D. Katz and K.K. Szabó, `hep-lat/0208078`.
9. S. Choe *et al.* [QCDTARO collaboration], *Phys. Rev.* **D65**, 054501 (2002).
10. C.R. Allton *et al.*, *Phys. Rev.* **D68**, 014507 (2003).
11. F. Karsch, E. Laermann and A. Peikert, *Phys. Lett.* **B478**, 447 (2000).
12. A. Ali Khan *et al.* [CP-PACS collaboration], *Phys. Rev.* **D64**, 074510 (2001).
13. U.M. Heller, F. Karsch, and B. Sturm, *Phys. Rev.* **D60**, 114502 (1999)
14. F. Karsch, K. Redlich and A. Tawfik, `hep-ph/0303108`; `hep-ph/0306208`.
15. C. Schmidt *et al.*, *Nucl. Phys. B (Proc. Suppl.)* **119**, 517 (2003); F. Karsch *et al.*, `hep-lat/0309116`.

TOPOLOGY AND CHIRALITY MIXING INTERACTION IN QCD

P. FACCIOLI*

E.C.T. Strada delle Tabarelle 286, Villazzano (Trento), I-38050, Italy.*
E-mail: faccioli@ect.it

T.A. DEGRAND

University of Colorado, Boulder, CO 80309, USA.
E-mail: degrand@aurinko.colorado.edu

We present the results of a recent investigation of the short-distance structure of non-perturbative interaction between light quarks. We analyze the connection between topology and the dynamics responsible for the spontaneous breaking of chiral symmetry. By measuring some appropriate correlation functions on the lattice, we provide compelling evidence for instanton-induced dynamics.

1. Introduction

Understanding the light-quark dynamics at all scales is a fundamental and open problem, in high-energy physics. In particular, the improving our knowledge of the non-perturbative interaction at intermediate and short distances has become essential, in order to explain the results of recent experiments on pion and proton electro-magnetic formfactors, performed at Jefferson Laboratory [1]. Indeed, these high-precision measurements have shown that the perturbative regime in QCD does not set-in until very large values of the momentum transfer ($Q^2 > 6$ GeV^2). The delay of the onset of the asymptotic regime implies that there are strong, short-scale non-perturbative interactions, inside ordinary hadrons.

It is commonly accepted that the physics of the light-quark sector of QCD is strongly influenced by the dynamics responsible for the chiral symmetry breaking (CSB). The characteristic energy scale associated to such a non-perturbative phenomenon is $4\pi f_\pi \sim 1.2$ GeV, sizably larger than

*speaker

the typical confinement scale, Λ_{QCD}. Such a separation of scales justifies attempting to understand the short-distance non-perturbative structure of light-hadrons, without having simultaneously to understand microscopic origin of color confinement. On the other hand, from the observation that $4\pi f_\pi \sim m_{\eta'}$ it follows that topological effects should be included in any effective description of the light-quark dynamics.

An effective interaction[a], which can simultaneously solve the U(1) problem and explain the CSB was derived by 't Hooft [2], by computing the Euclidean QCD generating functional in the semi-classical limit (i.e. accounting for small perturbations around the instanton solution): $L_{'t\,H} = G_{\bar{\rho},\bar{n}} \left([\,(\psi^\dagger \tau_a^- \psi)^2 - (\psi^\dagger i\gamma_5 \tau^- \psi)^2\,] + \frac{1}{2(2N_c-1)} (\psi^\dagger \sigma_{\mu\nu} \tau_a^- \psi) \right)$, where $\tau^- = (\vec{\tau}, i)$ ($\vec{\tau}$ are isospin Pauli matrices), and $G_{\bar{n}\,\bar{\rho}}$ is a coupling constant depending on the typical density of instantons in the vacuum $\bar{n}$, and their typical size $\bar{\rho}$. Notice that the finite size of the instanton field provides a natural cut-off scale for the interaction.

Unfortunately, the instanton density and size cannot be computed systematically in QCD, so the semi-classical 't Hooft interaction is not completely specified. The Instanton Liquid Model (ILM)[3] consists of assuming that the vacuum is saturated by instantons with typical size $\bar{\rho} \simeq 1/3$ fm and density $\bar{n} \simeq 1\,\text{fm}^{-4}$.

In a number of recent works it was shown that instantons can quantitatively explain the observed deviation of the pion and nucleon electromagnetic formfactor from the perturbative prediction, at high momentum transfer[4]. In the present work, we complement that analysis, and use lattice simulations to look directly for the specific signatures of the 't Hooft interaction.

2. Instantons and Chirality-Mixing Interactions

A characteristic property of the 't Hooft effective vertex is that it mixes only quarks of *different chirality*. Hence, if instantons are the dominant sources of the non-perturbative interaction, then one should be able to observe that quarks change their chirality after each non-perturbative interaction. On the contrary, if the non-perturbative vertex has the same structure of the perturbative gluon exchange (i.e. a purely vector or axial-vector coupling), then a single interaction should not flip the chirality of quarks. In this case, chirality mixing should arise only as a consequence of mass generation, i.e.

[a]For simplicity, in this work we shall restrict to the $N_f = 2$ case.

from a genuine many-body effect.

Based on this observations, one is led to consider the ratio:

$$R^{NS}(\tau) := \frac{A^{NS}_{flip}(\tau)}{A^{NS}_{non-flip}(\tau)} = \frac{\Pi_\pi(\tau) - \Pi_\delta(\tau)}{\Pi_\pi(\tau) + \Pi_\delta(\tau)}, \quad (1)$$

where $\Pi_\pi(\tau)$ and $\Pi_\delta(\tau)$ are pseudoscalar (PS) and scalar (S) iso-triplet two-point correlators, related to the currents $J_\pi(\tau) := \bar{u}(\tau)\, i\gamma_5\, d(\tau)$ and $J_\delta(\tau) := \bar{u}(\tau)\, d(\tau)$. If the propagation is chosen along the time direction, $A^{NS}_{flip(non-flip)}(\tau)$ represents the probability amplitude for a $|q\,\bar{q}\rangle$ pair with isospin 1 to be found after a "time" interval τ in a state in which the chirality of the quark and anti-quark is interchanged (not interchanged). Notice that the ratio $R^{NS}(\tau)$ must vanish as $\tau \to 0$ (no chirality flips), and must approach 1 as $\tau \to \infty$ (infinitely many chirality flips).

It can be shown that the correlator (1) receives no perturbative contribution. Moreover, a model-independent spectral analysis indicated that the interaction responsible for dynamical chirality-mixing is mediated by topological fields, and that the rate of chirality flips in a quark-antiquark system is proportional to the mass of the η' meson[5].

The chirality flip ratio $R^{NS}(\tau)$ was recently investigated in QCD, by means of lattice simulations performed with chiral fermions [6]. The results are the square points plotted in Fig. 1(A). As expected, the lattice data interpolate between 0 and 1. We observe that after few fractions of a fermi the quarks are more likely to be found in a configuration in which their chiralities is flipped, than to be found in their initial configuration. The presence of a maximum in $R^{NS}(\tau)$, and the subsequent fall-off toward 1 have a very simple explanation in the ILM: if quarks propagate in the vacuum for a time comparable with the typical distance between two neighbor instantons (i.e. two consecutive 't Hooft interactions), they have a large probability of crossing the field of the closest pseudo-particle. If so happens, they must necessarily flip their chirality. So, after some time, the quarks are most likely to be found in the configuration in which their chirality is flipped. On the other hand, if one waits for a time much longer than 1 fm, then the quarks will "bump" into many other such pseudo-particles, experiencing several more chirality flips. Eventually, either chirality configurations will become equally probable and $R^{NS}(\tau)$ will approach 1.

Let us now consider the result for $R^{NS}(\tau)$ obtained in the Random Instanton Liquid Model[b] (circles in Fig. 1(A)). The agreement with the

[b] In such a version of the ILM quarks are assumed to propagate in a vacuum populated by

lattice results is impressive, implying not only that instantons are the relevant sources of short-range chirality-mixing interaction, but also that the phenomenological estimates $\bar{\rho} \simeq 1/3$ fm, $\bar{n} \simeq 1\,\text{fm}^{-4}$ are realistic.

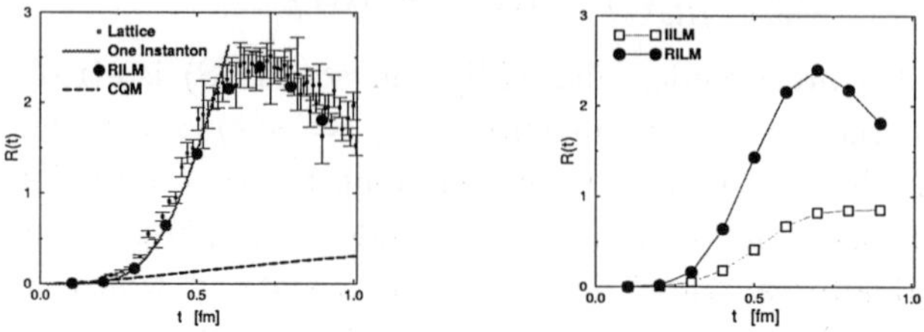

(A) (B)

Figure 1. (A): The chirality-flip correlator in lattice QCD (square points) and in the RILM (circles). The solid line represent the single-instanton contribution. The dashed curve was obtained from two free "constituent" quarks with a mass of 400 MeV. (B): Suppression of chirality flipping events, due to fermion-loop exchange in the ILM. Circles are RILM (quenched) results, squares are IILM (unquenched) results.

3. Instanton, Unitarity and Quark Loops

In this section we discuss an interesting link between topology and unitarity, which emerges when one analyzes the contribution of quark loops to chirality flips. We recall that lattice results reported in section 2 have been obtained in the quenched approximation. Unitarity requires that all correlation functions must be positive definite and this implies that $R^{NS}(\tau) \leq 1$, at all times. The existence of a maximum in which the chirality flip ratio is greater than one is a reflection of the fact that, in the quenched approximation, the unitarity of the theory is lost. We necessarily conclude that the fermionic determinant suppresses some chirality-flipping events, which are otherwise allowed in the quenched approximation. Clearly, *realistic descriptions of the effective quark-quark interaction must reproduce a dramatic enhancement of the chirality flipping amplitude, when quark loops are suppressed.*

Quite remarkably, this phenomenon is naturally explained in the instanton picture. If quark loops are allowed, then instantons and anti-instantons

an ensemble of randomly distributed instanton and anti-instantons. It can been shown that the RILM accounts for the 't Hooft interaction to all orders, but neglects quark loops (quenched approximation).

can interact through fermion exchange. Such an interaction generates an attraction between instantons and anti-instantons leading to a screening of the topological charge. As a result of such a screening, quarks crossing the field of an instanton are very likely to find, in the immediate vicinity, an anti-instanton which restores their initial chirality configuration. In Fig. 1(B) we compare the chirality flipping ratio $R^{NS}(\tau)$ obtained from a quenched (RILM) and unquenched (IILM[c]) version of the ILM. We observe that, with the inclusion of the fermionic determinant, the unitarity condition $R^{NS}(\tau) < 1$ is restored. We stress that, although such a restoration must necessarily take place in QCD, it represents a remarkable success of the ILM, which is not a unitary field theory.

4. Conclusions

We have presented a study of the quark dynamics, at intermediate and short distances, based on the results of lattice QCD simulations with chiral fermions. Our analysis provides strong evidence that instantons represent the main source of non-perturbative interaction between light quarks. We studied the contribution of quark loops, by comparing the results of quenched and full simulations. We observed dramatic quenching effects that can be naturally explained by instantons and suggests an interesting link between chiral dynamics, unitarity and topology.

References

1. The Jefferson Laboratory F_π Collaboration, J. Volmer *et al.*, *Phys. Rev. Lett.* **86**, 1713 (2001); O. Gayou *et al.*, *Phys. Rev. Lett.* **88**, 092301 (2002).
2. G.'t Hooft, *Phys. Rev.* **D14**, 3432 (1976).
3. T. Schaefer and E.V. Shuryak, *Rev. Mod. Phys.* **70**, 323 (1998), arXiv:hep-ph/9610451.
4. P. Faccioli, A. Schwenk and E.V. Shuryak, arXiv:hep-ph/0309129; P. Faccioli, A. Schwenk and E.V. Shuryak, *Phys. Rev.* **D67**, 113009 (2003), arXiv:hep-ph/0202027; P. Faccioli, A. Schwenk and E.V. Shuryak, *Phys. Lett.* **B528**, 34 (2002), arXiv:hep-ph/0205307.
5. P. Faccioli, arXiv:hep-ph/0211383.
6. P. Faccioli and T. A. DeGrand, to appear on *Phys. Rev. Lett.*, arXiv:hep-ph/0304219.
7. H. Martinez, Ph.D. Thesis, M.I.T. (unpublished)

[c]Interacting Instanton Liquid Model

THE STUDY OF THE CONTINUUM LIMIT OF THE SUPERSYMMETRIC WARD-TAKAHASHI IDENTITY FOR $N = 1$ SUPER YANG-MILLS THEORY

A. FEO

School of Mathematics, Trinity College, Dublin 2, Ireland and Dipartimento di Fisica, Università di Parma and INFN Gruppo Collegato di Parma, Parco Area delle Scienze, 7/A, 43100 Parma, Italy

The one-loop corrections to the supersymmetric Ward-Takahashi identity (WTi) are investigated in the off-shell regime in the Wilson formulation of the discretized $N = 1$ Super Yang-Mills (SYM) theory. The study of the continuum limit as well as the renormalization procedure for the supercurrent are presented.

1. Introduction

Recently, there have been a number of interesting results in lattice supersymmetry, as for example, in the two dimensional Wess-Zumino model[1], four dimensional N=1 SYM theory[2], with chiral fermions or not, as well as other supersymmetric theories[3]. Here, to formulate supersymmetry on the lattice we follow the ideas of Curci and Veneziano[4]. What they propose is to give up manifest supersymmetry on the lattice, and instead, to restore it in the continuum limit, by tunning the bare coupling g_0 and the gluino mass to the supersymmetric point which also coincides with the chiral point. In the Wilson formulation of Curci and Veneziano lattice supersymmetry can also be investigated by means of the WTi. On the lattice, the WTi contains explicit supersymmetric breaking terms and the supersymmetric limit is defined to be the point in the parameter space where these breaking terms vanish and the continuum supersymmetric WTi is recovered. Nice numerical results[5], are already in the literature, but still the study of the continuum limit of the supersymmetric WTi was missing. Here we want to shed some light on how to deal with this difficult issue, illustrating a recent result[6], that shows that, it is possible to write down the renormalized supersymmetric WTi and the general procedure to determine the renormalization coefficient for the supercurrent, Z_T.

2. Renormalized supersymmetric WTi on the lattice

The starting point of our calculation is the renormalized supersymmetric WTi on the lattice, which has been introduced in Ref. 6,

$$\begin{aligned}&Z_S\langle O\nabla_\mu S_\mu(x)\rangle + Z_T\langle O\nabla_\mu T_\mu(x)\rangle - 2(m_0-\tilde{m})Z_\chi^{-1}\langle O\chi^R(x)\rangle + \\ &Z_{CT}\langle\frac{\delta O}{\delta\bar{\xi}(x)}|_{\xi=0}\rangle - Z_{GF}\langle O\,\frac{\delta S_{GF}}{\delta\bar{\xi}(x)}|_{\xi=0}\rangle - Z_{FP}\langle O\,\frac{\delta S_{FP}}{\delta\bar{\xi}(x)}|_{\xi=0}\rangle + \\ &\sum_j Z_{B_j}\langle OB_j\rangle = 0\,,\end{aligned} \tag{1}$$

where $\xi(x)$ is a localized transformation parameter, $S_\mu(x) = -\frac{2i}{g_0}\,\mathrm{Tr}\left(G_{\rho\tau}(x)\sigma_{\rho\tau}\gamma_\mu\lambda(x)\right)$, is a local definition of the lattice supercurrent which mixes with $T_\mu(x) = -\frac{2}{g}\,\mathrm{Tr}\left(G_{\mu\nu}(x)\gamma_\nu\lambda(x)\right)$. $\chi(x) = \frac{i}{g_0}\,\mathrm{Tr}\left(G_{\rho\tau}(x)\sigma_{\rho\tau}\lambda(x)\right)$ is the gluino mass term, ∇_μ is the symmetric lattice derivative, $G_{\rho\tau}(x)$ is the clover plaquette operator and $\sigma_{\rho\tau} = \frac{i}{2}[\gamma_\rho,\gamma_\tau]$. Then, $\langle\frac{\delta O}{\delta\bar{\xi}(x)}|_{\xi=0}\rangle$, $\langle O\,\frac{\delta S_{GF}}{\delta\bar{\xi}(x)}|_{\xi=0}\rangle$ and $\langle O\,\frac{\delta S_{FP}}{\delta\bar{\xi}(x)}|_{\xi=0}\rangle$ are the contact terms, gauge fixing terms and Faddeev-Popov terms, respectively (we do not report them here, see Ref. 6). Notice that in Eq. (1) these terms are also renormalized. This is because their one-loop corrections are not just multiples of the corresponding tree-level values. Finally, $\sum_j Z_{B_j}\langle OB_j\rangle$ represent the mixing not only with non-gauge invariant operators (in the case the operator insertion O is non-gauge invariant), but also extra mixing with gauge invariant operators which do not vanish in the off-shell regime but vanish in the on-shell one. In principle, one require a complete list of them, or, as in our case, a sub-list of operators whose contributions are different from zero to the renormalization constant we are interested on. This point will become more clear in the following.

In the supersymmetric limit, the renormalized gluino mass is zero, so the third term in the first line of Eq. (1) vanish and we leave with a simple expression. From now on, when we refer to Eq. (1) we will assume the supersymmetric limit.

3. Renormalization constants for the supercurrent

We are now considering each matrix element in Eq. (1) with O a non-gauge invariant operator given by $O := A_\nu^b(y)\,\bar{\lambda}^a(z)$. Each matrix element in Eq. (1) is proportional to each element of the *Gamma*-matrix base $\Gamma = \{1,\gamma_5,\gamma_\alpha,\gamma_5\gamma_\alpha,\sigma_{\alpha\rho}\}$. To determine Z_T one needs the projections over γ_α and $\gamma_\alpha\gamma_5$ (the other ones are trivial).

In Fourier transformation (FT), we choose p as the outcoming momentum for the gluon field A_μ and q the incoming momentum for the fermion field λ. Thus, each matrix element can be written as $\langle A^b_\nu(y)\,\bar\lambda^a(z)\,C(x)\rangle \stackrel{FT}{\Longrightarrow} D_F(q)\cdot(C(p,q))_{amp}\cdot D_B(p)\cdot\delta_{ab}$, where $(C(p,q))_{amp}$ can be, i.e., $\nabla_\mu S_\mu$, $\nabla_\mu T_\mu$, etc., with the external propagators amputated, $D_F(q)$ and $D_B(p)$ are the full fermion and gluon propagators, respectively, while δ_{ab} is the color structure, similar to all diagrams. The non-trivial part of the calculation is the determination of $(C(p,q))_{amp}$ for each matrix element in Eq. (1). To calculate Z_T one should pick up from each matrix element of Eq. (1) those terms which contains the same Lorentz structure as S_μ and T_μ, to tree-level. Those operators which do not contain the same tree-level Lorentz structure as S_μ and T_μ do not enter in the determination of Z_T.

The renormalization constants as well as the operators, can be written as a power of ${g_0}^6$,

$$
\begin{aligned}
Z_{operator} &= Z^{(0)}_{operator} + g_0^2 Z^{(2)}_{operator} + \cdots\,, \\
\langle Operator\rangle &= \langle Operator\rangle^{(0)} + g_0^2\langle Operator\rangle^{(2)} + \cdots\,,
\end{aligned}
\tag{2}
$$

where $\langle Operator\rangle^{(2)}$, is the 1-loop correction while $\langle Operator\rangle^{(0)}$, is the tree-level value. It is easy to see that, for $p=q$, a condition which would greatly simplify the calculation because implies that the operator insertion is at zero momentum, $(\nabla_\mu S_\mu(x))^{(0)}_{amp} = (\nabla_\mu T_\mu(x))^{(0)}_{amp} = 0$. That means that the tree-level of $\nabla_\mu S_\mu$ and $\nabla_\mu T_\mu$ can not be distinguished at zero momentum transfer. To get Z_T we need to distinguish the tree-level values of S_μ and T_μ and for that reason we require general external momenta, p and q.

Substituting Eq. (2) into Eq. (1), up to order g_0^2, and using the projections over γ_α and $\gamma_\alpha\gamma_5$ we obtain, respectively (where the explicit expression for the tree-level operators in Eq. (1) are used[6]),

$$
\begin{aligned}
&\frac{1}{4}\mathrm{tr}\big(\gamma_\alpha(\nabla_\mu S_\mu)^{(2)}_{amp}\big) + Z^{(2)}_S 2i(p_\alpha p_\nu - p_\alpha q_\nu - p^2\delta_{\alpha\nu} + p\cdot q\delta_{\alpha\nu}) + \\
&Z^{(2)}_T i(p_\alpha p_\nu - p_\alpha q_\nu - p^2\delta_{\alpha\nu} + p\cdot q\delta_{\alpha\nu}) + \frac{1}{4}\mathrm{tr}\big(\gamma_\alpha\big(\frac{\delta O}{\delta\bar\xi(x)}|_{\xi=0}\big)^{(2)}_{amp}\big) + \\
&Z^{(2)}_{CT}2i(p_\alpha q_\nu - p\cdot q\delta_{\alpha\nu} + p^2\delta_{\alpha\nu}) - Z^{(2)}_{GF}2ip_\alpha p_\nu - \frac{1}{4}\mathrm{tr}\big(\gamma_\alpha\big(\frac{\delta S_{GF}}{\delta\bar\xi(x)}|_{\xi=0}\big)^{(2)}_{amp}\big) \\
&+\frac{1}{4}Z^{(2)}_{B_j}\mathrm{tr}\langle\gamma_\alpha OB_j\rangle^{(0)} = 0
\end{aligned}
\tag{3}
$$

and

$$\frac{1}{4}\mathrm{tr}\big(\gamma_\alpha\gamma_5\big(\nabla_\mu S_\mu\big)^{(2)}_{amp}\big) + Z^{(2)}_S 2ip_\rho q_\sigma \varepsilon_{\nu\alpha\rho\sigma} - Z^{(2)}_{CT} 2ip_\rho q_\sigma \varepsilon_{\nu\alpha\rho\sigma} +$$
$$\frac{1}{4}\mathrm{tr}\big(\gamma_\alpha\gamma_5\big(\frac{\delta O}{\delta\bar{\xi}(x)}|_{\xi=0}\big)^{(2)}_{amp}\big) - \frac{1}{4}\mathrm{tr}\big(\gamma_\alpha\gamma_5\big(\frac{\delta S_{GF}}{\delta\bar{\xi}(x)}|_{\xi=0}\big)^{(2)}_{amp}\big) +$$
$$\frac{1}{4}Z^{(2)}_{B_j}\mathrm{tr}\big\langle\gamma_\alpha\gamma_5 OB_j\big\rangle^{(0)} = 0\,. \tag{4}$$

More explicitly, the non-trivial part is the computation of the one-loop correction of the projections over $\Gamma_r = \{\gamma_\alpha, \gamma_\alpha\gamma_5\}$ (off-shell regime), for $\mathrm{tr}\big(\Gamma_r\big(\nabla_\mu S_\mu\big)^{(2)}_{amp}\big)$ (12 diagrams), $\mathrm{tr}\big(\Gamma_r\big(\frac{\delta S_{GF}}{\delta\bar{\xi}(x)}|_{\xi=0}\big)^{(2)}_{amp}\big)$ (12 diagrams) and $\mathrm{tr}\big(\Gamma_r\big(\frac{\delta O}{\delta\bar{\xi}(x)}|_{\xi=0}\big)^{(2)}_{amp}\big)$ (4 diagrams), for a total of 28 Feynman diagrams for the determination of $Z^{(2)}_T$ (see Ref. 6 for details).

The determination of $Z^{(2)}_T$ require the knowledge of which operators in $Z^{(2)}_{B_j}\mathrm{tr}\big\langle\Gamma_r OB_j\big\rangle^{(0)}$ should be included. Our claim[6] is that, to get $Z^{(2)}_T$ one can substitute without any ambiguity, $\frac{1}{4}Z^{(2)}_{B_i}\mathrm{tr}\big\langle\gamma_\alpha OB_i\big\rangle^{(0)} \to Z^{(2)}_{B_0} i(p_\alpha p_\nu - p^2\delta_{\alpha\nu}) + Z^{(2)}_{B_1} ip_\nu q_\alpha + Z^{(2)}_{B_2} iq_\nu q_\alpha + Z^{(2)}_{B_3} iq^2\delta_{\alpha\nu}$, and $\frac{1}{4}Z^{(2)}_{B_i}\mathrm{Tr}\big\langle\gamma_\alpha\gamma_5 OB_j\big\rangle^{(0)} \to 0$. From the explicit computation of the matrix elements of the WTi we get the following expressions,

$$\mathrm{tr}\big\langle\gamma_\alpha\Delta\big\rangle^{(2)} \stackrel{FT}{\Longrightarrow} A_1 q^2\hat{p}_\alpha\hat{p}_\nu + A_2 q^2\hat{p}_\alpha\hat{q}_\nu + (A_3 + M_3)q^2\delta_{\alpha\nu} + M_1 q^2\hat{p}_\nu\hat{q}_\alpha +$$
$$M_2 q^2\hat{q}_\alpha\hat{q}_\nu + P_1 q^2\hat{p}^2_\nu\delta_{\nu\alpha} + P_2 q^2\hat{q}^2_\nu\delta_{\nu\alpha} + \cdots \tag{5}$$

and

$$\mathrm{tr}\big\langle\gamma_\alpha\gamma_5\Delta\big\rangle^{(2)} \stackrel{FT}{\Longrightarrow} A_4 q^2\hat{p}_\rho\hat{q}_\sigma\varepsilon_{\nu\alpha\rho\sigma}\,, \tag{6}$$

where $\Delta \equiv O\nabla_\mu S_\mu(x) + \frac{\delta O}{\delta\bar{\xi}(x)}|_{\xi=0} - O\,\frac{\delta S_{GF}}{\delta\bar{\xi}(x)}|_{\xi=0}$ and the following sempli-fication, $p^2 = q^2$ and $p\cdot q = 0$, has been used. Notice that Eq. (5), even in the continuum limit, contains non-Lorentz covariant terms.

From the matching of Eqs. (3,4) with Eqs. (5,6), the following conditions can be derived,

$$A_1 = -2iZ^{(2)}_S - iZ^{(2)}_T + 2iZ^{(2)}_{GF} - iZ^{(2)}_{B_0}\,,$$
$$A_3 + M_3 = 2iZ^{(2)}_S + iZ^{(2)}_T - 2iZ^{(2)}_{CT} + iZ^{(2)}_{B_0} - iZ^{(2)}_{B_3}\,,$$
$$M_1 = -iZ^{(2)}_{B_1}\,,\;\; M_2 = -iZ^{(2)}_{B_2}\,,\;\; \cdots \tag{7}$$

and

$$A_2 = 2iZ^{(2)}_S + iZ^{(2)}_T - 2iZ^{(2)}_{CT}\,,\;\; A_4 = -2iZ^{(2)}_S + 2iZ^{(2)}_{CT}\,. \tag{8}$$

The last two conditions can be explicitly solved for $Z_T^{(2)}$,

$$Z_T^{(2)} = -iA_2 - iA_4 \,. \tag{9}$$

We see from Eqs. (8,9) that $Z_T^{(2)}$ does not need to know the Z_{B_j} which enter explicitly in Eq. (7). That is why we did not include the operators which produce non-Lorentz covariant terms in Eq. (7). Their contribution vanish to the determination of $Z_T^{(2)}$.

Our perturbative result (using a Vegas Monte Carlo routine) is $Z_T^{(2)} = Z_T^{(2)}|_{1-loop} = 0.664$. To compare it with the numerical one[5], $Z_T^{NUM} \equiv Z_T/Z_S = 0.185(7)$, one has to divide it by 2, $Z_T^{PT} = \frac{1}{2}Z_T^{(2)}|_{1-loop} = 0.332$.

4. Conclusions

A general procedure to calculate the renormalization constant Z_T has been presented. All the contribution to the supersymmetric WTi has been calculated to one-loop order in lattice perturbation theory. Comparison with numerical results[5] gives good agreement. An important point here is that, even in the continuum limit, Lorentz breaking terms appears in Eq. (5). The nice point is that, once the $Z_T^{(2)}$ has been determined, it is possible to impose the on-shell condition on the gluino mass. These Lorentz breaking terms cancel out from Eq. (5) and the continuum supersymmetric WTi is recovered. The determination of Z_S is under way.

Acknowledgments

A. F. would like to thank H. Suganuma for the opportunity to talk at this stimulating conference. It is also a pleasure to thank M. Bonini, R. De Pietri, S. Furui, K. Konishi, M. Shifman, I.-O. Stamatescu for useful discussions.

References

1. S. Catterall, hep-lat/0309040; S. Catterall and S. Karamov, *Phys. Rev.* **D68** 014503 (2003); M. Campostrini *et al.*, hep-lat/0309054, hep-lat/0209010.
2. A. Feo, hep-lat/0210015 (and references therein).
3. K. Itoh *et al.*, hep-lat/0210049; D. B. Kaplan *et al.*, hep-lat/0206019; D. B. Kaplan, hep-lat/0309099; J. Giedt, hep-lat/0307024, hep-ph/0309344; J. R. Hiller *et al.*, hep-ph/0304147; M. Harada and S. Pinsky, hep-lat/0303027.
4. G. Curci and G. Veneziano, *Nucl. Phys.* **B292** 555 (1987).
5. F. Farchioni *et al.* [DESY-Münster-Roma Coll.], hep-lat/0111008.
6. A. Feo, hep-lat/0305020.

WHEN IS THE DECONFINEMENT PHASE TRANSITION UNIVERSAL?

K. HOLLAND

Department of Physics, University of California at San Diego,
La Jolla CA 92093-0319, USA

M. PEPE AND U. -J. WIESE*

Institute for Theoretical Physics, University of Bern,
Sidlerstrasse 5, CH-3012 Bern, Switzerland

Pure Yang-Mills theory has a finite-temperature phase transition, separating the confined and deconfined bulk phases. Svetitsky and Yaffe conjectured that if this phase transition is of second order, it belongs to the universality class of transitions for particular scalar field theories in one lower dimension. We examine Yang-Mills theory with the symplectic gauge groups $Sp(N)$. We find new evidence supporting the Svetitsky-Yaffe conjecture and make our own conjecture as to which gauge theories have a universal second order deconfinement phase transition.

1. Introduction

Yang-Mills theories with gauge group G have a finite-temperature phase transition, separating the confined phase of colorless glueballs from the deconfined gluon plasma phase.[1] The transition is signalled by the spontaneous breaking of a global symmetry[2] related to H, the center of G. The Polyakov loop is the order parameter for the transition,[3] transforming as $\Phi(\vec{x})' = z\Phi(\vec{x}), z \in H$ under the global center transformation. Its expectation value is $\langle\Phi\rangle = \exp(-\beta F_q)$, where F_q is the free energy of a static quark in the gluon background, $\beta = 1/T$ is the time extent and T is the temperature. In the confined phase, there are no isolated quarks and $F_q \to \infty$ in the infinite volume limit, giving $\langle\Phi\rangle = 0$. In the deconfined phase, F_q is finite and $\langle\Phi\rangle \neq 0$, spontaneously breaking the global center symmetry.

Svetitsky and Yaffe conjectured that if Yang-Mills theory with gauge group G has a second order deconfinement transition, with the correlation

---

*on leave from MIT.

length $\xi \to \infty$, it should have universal properties belonging to the universality class of H-symmetric scalar field theories in one lower dimension.[4] When $\xi \gg \beta$, the system is no longer sensitive to the finite time extent and dimensional reduction occurs. The universality class is determined by quantities such as the critical exponents, e.g. how the correlation length diverges as the critical temperature is approached, $\xi \propto (T - T_c)^{-\nu}$. Note, however, that the conjecture does *not* state that the deconfinement transition *must* be second order.

Most studies have looked at 4-d and 3-d $SU(N)$ Yang-Mills theories. The center of $SU(N)$ is $\mathbb{Z}(N)$, the N roots of unity. We first consider 4-d theories. With $SU(2)$ as the gauge group, the theory has a second order deconfinement transition,[5] belonging to the universality class of 3-d $\mathbb{Z}(2)$-symmetric scalar field theory, i.e. the Ising universality class.[6] This fully supports the Svetitsky-Yaffe conjecture. For $SU(3)$, the transition is weakly first order with a large but finite correlation length ξ and no universal properties.[7] Studies have shown that for $SU(N)$ theories with $N = 4, 6, 8$, the deconfinement transitions are first order, with the strength increasing for larger N, again without any universal properties.[8] Interestingly, 3-d $\mathbb{Z}(N)$-symmetric scalar field theory for $N \geq 5$ is in the universality class of the 3-d $U(1)$-symmetric XY model.[9] The gauge theories simply don't make use of this universality class. There is a richer structure in 3 dimensions. For $SU(N)$ with $N = 2, 3, 4$, the deconfinement transitions are second order, belonging to the 2-d $\mathbb{Z}(N)$-symmetric universality classes with $N = 2, 3, 4$ respectively, again supporting the Svetitsky-Yaffe conjecture.[10]

2. $Sp(N)$ Gauge Theory

We look at Yang-Mills theories with the symplectic groups $Sp(N)$. These groups have the property that $Sp(N) \subset SU(2N)$ and $U \in Sp(N)$ satisfies the constraint

$$U^* = JUJ^\dagger, \quad J = \begin{pmatrix} 0 & \mathbb{1} \\ -\mathbb{1} & 0 \end{pmatrix} \quad \Rightarrow \quad U = \begin{pmatrix} W & X \\ -X^* & W^* \end{pmatrix}, \tag{1}$$

where W and X are complex $N \times N$ matrices. Since U and U^* are related by a unitary transformation $J \in Sp(N)$, the $2N$-dimensional fundamental representation of $Sp(N)$ is pseudo-real and charge conjugation is a global gauge transformation. These properties are familiar from $SU(2)$. Writing $U = \exp(iH)$, the Hermitean matrix H satisfies

$$H^* = -JHJ^\dagger = JHJ \quad \Rightarrow \quad H = \begin{pmatrix} A & B \\ B^* & -A^* \end{pmatrix}, \tag{2}$$

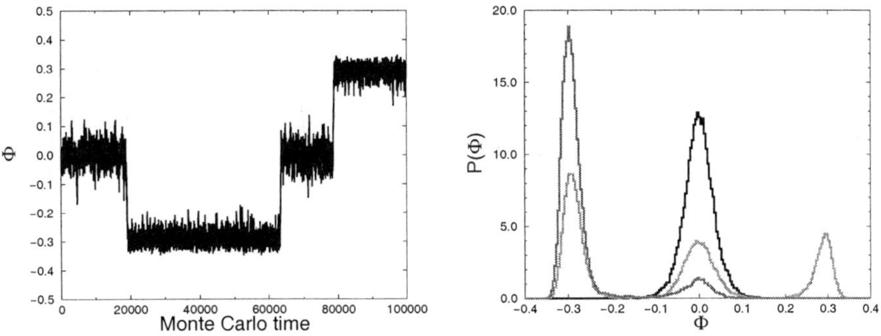

Figure 1. (a) Tunneling between coexisting confined and deconfined phases at T_c and (b) probability distributions of Φ for temperatures close to T_c.

where the $N \times N$ complex matrices A and B satisfy $A = A^\dagger$ and $B = B^T$. The N^2 and $N(N+1)$ degrees of freedom of A and B respectively mean that $Sp(N)$ has $N^2+N(N+1) = (2N+1)N$ generators. $Sp(N)$ has rank N. There are the special equivalent cases $Sp(1) = SU(2)$ and $Sp(2) \simeq SO(5)$. Most interestingly, unlike $SU(N)$, the center of $Sp(N)$ is $\mathbb{Z}(2)$ for all N. This allows us to disentangle the size of the group from the center and see what effect this has on the deconfinement transition. According to the Svetitsky-Yaffe conjecture, a second order transition should belong to the Ising universality class.

The lattice formulation of $Sp(N)$ Yang-Mills theory is straightforward. We use the Wilson action in our simulations. As $SU(2) \subset Sp(N)$, we can update the $Sp(N)$ gauge links using the standard heatbath[11] and overrelaxation[12] algorithms to update the various $SU(2)$ subgroups, in the same way as done for $SU(N)$ gauge theory. We find that there is no bulk phase transition between strong and weak coupling. Further details will be presented in a forthcoming paper.[13]

We first consider 4-d $Sp(2)$ gauge theory, where one might expect the deconfinement transition to be second order, just as for $SU(2)$. However, we find the transition to be first order, even with $\mathbb{Z}(2)$ as the center. In Fig. 1, we plot tunneling between coexisting confined ($\Phi = 0$) and deconfined ($\Phi \neq 0$) phases at the critical temperature, as well as the probability distributions of Φ close to T_c. Coexistence of the phases is a clear signal for a first order transition. Measurements of the Polyakov loop suscepti-

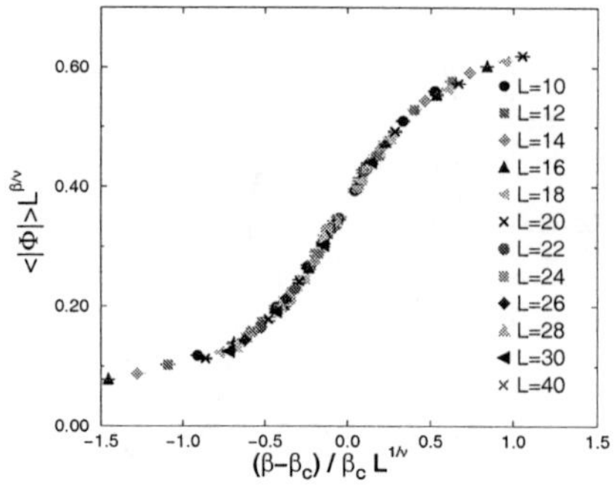

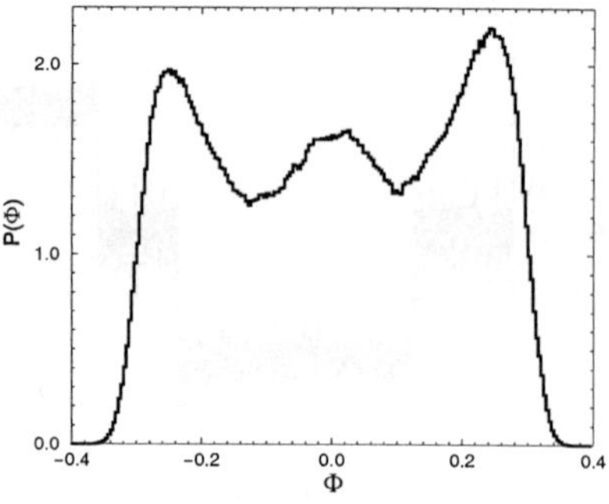

Figure 2. (a) The universal function $\langle|\Phi|\rangle L^{\beta/\nu}$ for 3-d $Sp(2)$ gauge theory and (b) coexisting confined and deconfined phases for 3-d $Sp(3)$ gauge theory.

bility, the specific and latent heats also show a clear first order transition. Similarly, we find that 4-d $Sp(3)$ Yang-Mills theory also has a first order deconfinement transition. Going from $Sp(1) = SU(2)$ to $Sp(2)$, the phase transition changes from second to first order, even though the center of the group is the same, indicating that the size of the group is more important.

In 3-d, we find a richer structure. Exactly like $Sp(1)$, we find that $Sp(2)$ Yang-Mills theory has a second order deconfinement transition. Measurements for various temperatures and volumes can be mapped onto one universal curve when rescaled using the critical exponents of the 2-d Ising universality class, as shown in Fig. 2(a). This is new evidence supporting the Svetitsky-Yaffe conjecture. However, for $Sp(3)$, the transition is weakly first order, as we see in Fig. 2(b), where one has to go to large volumes to distinguish the coexisting phases. Again, we find that the transition switches from second to first order as we increase the size of the group, even though the center is unchanged.

3. Summary

From our work and other studies, we conjecture that only for $Sp(1) = SU(2) \simeq SO(3)$ is there a universal second order deconfinement phase transition in 4 dimensions. In 3-d, we find that $Sp(2)$ gauge theory has a second order deconfinement transition belonging to the 2-d Ising universality class, which is new evidence supporting the Svetitsky-Yaffe conjecture. We expect that there are no other second order transitions with other gauge groups.[13,14] In both 4-d and 3-d $Sp(N)$ gauge theory, we find that the deconfinement transition changes from second to first order as we increase N, even though the center of the group is always $\mathbb{Z}(2)$. The order of the transition seems to be dictated by the size of the group, not the center. This is natural, as the number of glueballs in the confined phase is group-independent, whereas the number of deconfined gluons increases with the group size, leading to a larger mismatch in the number of degrees of freedom at the critical temperature as N increases.

Acknowledgments

This work has been supported under grant DOE-FG03-97ER40546, by the Schweizerischer Nationalfond and the European Community's Human Potential Program HPRN-CT-2000-00145 Hadrons/Lattice QCD.

References

1. K. Holland and U.-J. Wiese, in *At the Frontier of Particle Physics*, edited by M. Shifman, *Vol.* **3**, 1909, `arXiv:hep-ph/0011193`.
2. G. 't Hooft, *Nucl. Phys.* **B138**, 1 (1978); *Nucl. Phys.* **B153**, 141 (1979).
3. A. M. Polyakov, *Phys. Lett.* **B72**, 477 (1978); L. Susskind, *Phys. Rev.* **D20**, 2610 (1979).
4. B. Svetitsky and L. G. Yaffe, *Nucl. Phys.* **B210**, 423 (1982).
5. L. D. McLerran and B. Svetitsky, *Phys. Lett.* **B98**, 195 (1981); *Phys. Rev.* **D24**, 450 (1981); J. Kuti *et al.*, *Phys. Lett.* **B98**, 199 (1981); J. Engels *et al.*, *Phys. Lett.* **B101**, 89 (1981); R. V. Gavai, *Nucl. Phys.* **B215**, 458 (1983); R. V. Gavai *et al.*, *Nucl. Phys.* **B220**, 223 (1983).
6. J. Engels *et al.*, *Nucl. Phys.* **B332**, 737 (1990); *Nucl. Phys.* **B387**, 501 (1992).
7. T. Celik *et al.*, *Phys. Lett.* **B125**, 411 (1983); J. B. Kogut *et al.*, *Phys. Rev. Lett.* **50**, 393 (1983); S. A. Gottlieb *et al.*, *Phys. Rev. Lett.* **55**, 1958 (1985); F. R. Brown *et al.*, *Phys. Rev. Lett.* **61**, 2058 (1988); M. Fukugita *et al.*, *Phys. Rev. Lett.* **63**, 1768 (1989); N. A. Alves *et al.*, *Phys. Rev. Lett.* **64**, 3107 (1990).
8. J. F. Wheater and M. Gross, *Phys. Lett.* **B144**, 409 (1984); M. Wingate and S. Ohta, *Phys. Rev.* **D63**, 094502 (2001); B. Lucini, M. Teper and U. Wenger, *Phys. Lett.* **B545**, 197 (2002); `arXiv:hep-lat/0307017`.
9. J. Hove and A. Sudbo, `arXiv:cond-mat/0301499`.
10. J. Engels *et al.*, *Nucl. Phys. Proc. Suppl.* **53**, 420 (1997); P. de Forcrand and O. Jahn, to appear in *Lattice'03 proceedings*.
11. M. Creutz, *Phys. Rev.* **D21**, 2308 (1980).
12. N. Cabibbo and E. Marinari, *Phys. Lett.* **B119**, 387 (1982).
13. K. Holland, M. Pepe and U.-J. Wiese, in preparation.
14. K. Holland, P. Minkowski, M. Pepe and U.-J. Wiese, *Nucl. Phys.* **B668**, 207 (2003).

PROPERTIES OF THERMAL GLUEBALLS*

NORIYOSHI ISHII

Radiation Laboratory,
RIKEN (The Institute of Physical and Chemical Research),
2-1 Hirosawa, Wako, Saitama 351–0198, Japan
E-mail: ishii@rarfaxp.riken.go.jp

HIDEO SUGANUMA

Faculty of Science, Tokyo Institute of Technology,
2–12-1 Ohokayama, meguro, Tokyo 152–8550, Japan

We study the properties of the 0^{++} glueball at finite temperature using SU(3) quenched lattice QCD. We find a significant thermal effects near T_c. We perform the χ^2 fit analyses adopting two Ansätze for the spectral function, i.e., the conventional narrow-peak Ansatz and an advanced Breit-Wigner Ansatz. The latter is an extension of the former, taking account of the appearance of the thermal width at $T > 0$. We also perform the MEM analysis. These analyses indicate that the thermal effect on the glueball is a significant thermal-width broadening $\Gamma(T_c) \sim 300$ MeV together with a modest reduction in the peak center $\Delta\omega_0(T_c) \sim 100$ MeV.

Near the critical temperature T_c, the QCD vacuum begins to change its structure. For instance, the string tension reduces and the spontaneously broken chiral symmetry begins to restore partially. Since hadrons are composites of quarks and gluons, whose interaction depends on the properties of the QCD vacuum, it is natural to expect the structural changes of various hadrons. In this direction, there are a number of studies based on QCD motivated effective models, which predict the mass reductions of charmonium and light $q\bar{q}$ mesons near T_c.[1,2] In BNL-RHIC experiments, these changes are considered as important precritical phenomena of QCD phase transition. In addition, lattice QCD Monte Carlo calculations have been performed to examine the pole-mass shifts of hadrons at the quenched level.[3,4] The 0^{++} glueball is one of such hadrons, which are expected to ex-

*The numerical calculation has been done on NEC SX-5 at Osaka Univeristy.

hibit structural changes near T_c. Studies based on the effective models[5,6] suggest a large mass reduction of the glueball as a natural consequence of the sizable difference between the 0^{++} glueball mass $m_{\rm G} = 1500 - 1700$ MeV and the critical temperature $T_c = 260 - 280$ MeV. In this paper, we investigate the glueball properties at finite temperature by means of lattice QCD Monte Carlo calculation.[7,8,9]

We consider the temporal correlator $G(\tau) \equiv \langle \phi(\tau)\phi(0) \rangle$ of the glueball operator $\phi(\tau)$. The simplest construction of the glueball operator is the plaquette operator, which, however, is known to have only a negligible overlap to the lowest-lying 0^{++} glueball. To enhance the overlap, the smearing method provides us with a systematic procedure by extending the spatial size of the operator. In this paper, we adopt the APE smearing.[8,10] The continuum expression of the smeared glueball operator is obtained in the Coulomb gauge as

$$\phi \propto \int \frac{d^3y d^3z}{(2\pi)^{3/2}\rho^3} \exp\left(-\frac{(\vec{y}-\vec{z})^2}{2\rho^2} \right) G^a_{ij}(\vec{y}) G^a_{ij}(\vec{z}), \tag{1}$$

where G^a_{ij} is the field strength tensor, and $\rho \in \mathbb{R}$ controls the size of the spatial extension. (For detail, see Refs. 8.) By choosing ρ appropriately, we can maximize the contribution from the lowest-lying glueball in the glueball correlator $G(\tau)$.

We next consider the spectral representation of $G(\tau) \equiv \langle \phi(\tau)\phi(0) \rangle$ as

$$G(\tau) = \int_0^\infty d\omega K(\tau,\omega) A(\omega), \tag{2}$$

with $\beta \equiv 1/T$, $K(\tau,\omega) \equiv \frac{\cosh(\omega(\beta/2-\tau))}{\sinh(\beta\omega/2)}$, and $A(\omega)$ denotes the spectral function with its spatial momentum projected to zero. At low temperature, peak positions of $A(\omega)$ provide us with the pole masses of hadrons. Hence the simplest parameterization of $A(\omega)$ for the χ^2 fit analysis is given as[7]

$$A(\omega) = C\left\{ \delta(\omega - m) - \delta(\omega + m) \right\}, \tag{3}$$

where C and m are used as the fit parameters, corresponding to the strength and the pole mass, respectively. To adopt this Ansatz, the peak should be narrow enough. We will refer to Eq. (3) as the "*narrow-peak Ansatz*". At $T > 0$, however, through the interaction with the thermally excited particles, the thermal width is generated even for stable particles. As T gets larger, the effects of the thermal width become more important. To respect its existence, we adopt the following parameterization as[8]

$$A(\omega) = C\left\{ \delta_\Gamma(\omega - \omega_0) - \delta_\Gamma(\omega + \omega_0) \right\}, \tag{4}$$

where $\delta_\Gamma(\omega) \equiv \frac{1}{\pi}\mathrm{Im}\left(\frac{1}{\omega - i\Gamma}\right)$. C, ω_0, and Γ are used as the fit parameters corresponding to the strength, the peak center, and the thermal width, respectively. We will refer to Eq. (4) as the "*Breit-Wigner Ansatz*". Note that $\lim_{\Gamma\to 0} \delta_\Gamma(\omega) = \delta(\omega)$. Hence, the Breit-Wigner Ansatz is a natural extension of the narrow-peak Ansatz.

For these χ^2 fit analyses, proper Ansätze for $A(\omega)$ have to be provided. However, in the very vicinity of T_c, and above T_c, there may appear more complicated structures. In this sense, it is desirable to obtain $A(\omega)$ directly from $G(\tau)$, which, however, is known to be an ill-posed problem numerically. Recently developed Maximum Entropy Method (MEM)[11,12] can deal with this inverse problem adopting the following Shannon-Jaynes entropy as

$$S \equiv \int_0^\infty d\omega \left[A(\omega) - m(\omega) - A(\omega)\log\left(\frac{A(\omega)}{m(\omega)}\right)\right], \qquad (5)$$

where $m(\omega)$ is a real and positive function referred to as the default model function. $m(\omega)$ is required to reproduce the asymptotic behavior of $A(\omega)$ as $\omega \to \infty$.

We use the SU(3) anisotropic lattice plaquette action[13] as

$$S_{\mathrm{G}} = \frac{\beta_{\mathrm{lat}}}{N_c \gamma_{\mathrm{G}}} \sum_{s,i<j\le 3} \mathrm{ReTr}\left\{1 - P_{ij}(s)\right\} + \frac{\beta_{\mathrm{lat}}\gamma_{\mathrm{G}}}{N_c} \sum_{s,i\le 3} \mathrm{ReTr}\left\{1 - P_{i4}(s)\right\}, \qquad (6)$$

where $P_{\mu\nu}(s) \in \mathrm{SU}(3)$ denotes the plaquette operator in the μ-ν-plane. The lattice parameter and the bare anisotropic parameter are fixed as $\beta_{\mathrm{lat}} \equiv 2N_c/g^2 = 6.25$ and $\gamma_{\mathrm{G}} = 3.2552$, respectively, so as to reproduce the renormalized anisotropy as $a_s/a_t = 4$. The scale unit is introduced from the on-axis data of the static inter-quark potential with the string tension $\sqrt{\sigma} = 440$ MeV. The resulting lattice spacings are given as $a_t^{-1} = 9.365(66)$ GeV and $a_s^{-1} = 2.341(16)$ GeV. The critical temperature is estimated as $T_c \simeq 280$ MeV from the susceptibility of the Polyakov loop.[8] We generate 5,500–9,900 gauge configurations to construct the glueball correlators,[7,8] where statistical data are divided into bins of the size 100 to reduce possible auto-correlations near T_c. To enhance the low-energy spectrum, we adopt an appropriate smearing corresponding to $\rho \simeq 0.4$ fm in Eq. (1).

We present the numerical results. Whereas the narrow-peak analysis suggests that the thermal effect on the glueball is the pole-mass reduction[7] of about 300 MeV, the advanced Breit-Wigner analysis suggests that it is a thermal-width broadening[8] of about 300 MeV together with a modest reduction in the peak center of about 100 MeV. The numerical results of the χ^2 fit analyses with the Breit-Wigner Ansatz are shown in Fig. 1.

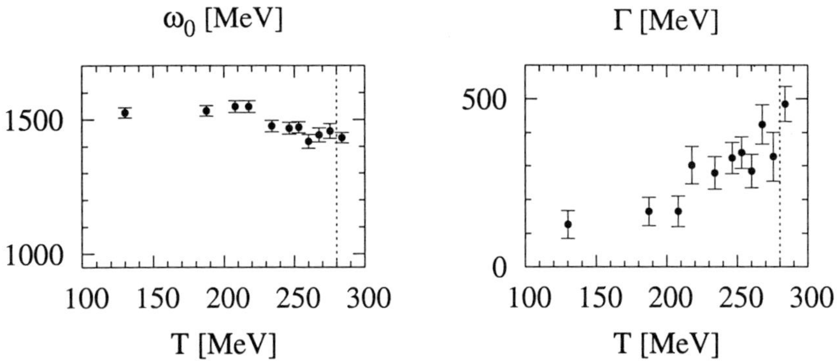

Figure 1. The peak center $\omega_0(T)$ and the thermal width $\Gamma(T)$ of the 0^{++} glueball against temperature T, obtained from the χ^2 fit analysis with the Breit-Wigner Ansatz. The vertical dotted lines indicate the critical temperature $T_c \simeq 280$ MeV.

We now turn to the result of MEM analysis. As the default model function, we adopted the perturbative expression of the smeared glueball correlator up to $O(\alpha_S^0)$ as[9]

$$m(\omega) = N\omega^4 \exp\left\{-(\omega\rho)^2/4\right\}, \tag{7}$$

where the normalization factor N is determined from $1 = \int_0^\infty d\omega K(\tau = 0, \omega) m(\omega)$ so as to mimic $G(\tau = 0) = 1$, which we adopt as the normalization of the correlator. Following the numerical procedure given in Ref. 12, we reconstruct $A(\omega)$. The results at various temperatures are shown in Fig. 2. We see that the peak tends to become broader with the increasing temperature,[9] which is consistent with the results of the Breit-Wigner fit analysis.

To summarize, we have studied the properties of thermal glueballs below T_c by using SU(3) anisotropic lattice QCD at the quenched level. We have performed χ^2 fit analyses adopting two Ansätze for the spectral function $A(\omega)$, i.e., the conventional narrow-peak Ansatz (δ-function type) and the advanced Breit-Wigner Ansatz so as to respect the appearance of the thermal width at $T > 0$. Whereas the former has suggested that a significant pole mass reduction of the 0^{++} glueball of about 300 MeV near T_c, the latter has suggested that the thermal effects on the glueball is rather a significant thermal-width broadening of about 300 MeV together with a modest reduction in peak center of about 100 MeV. We have also performed MEM analysis. We have seen that the peak of the 0^{++} glueball tends to

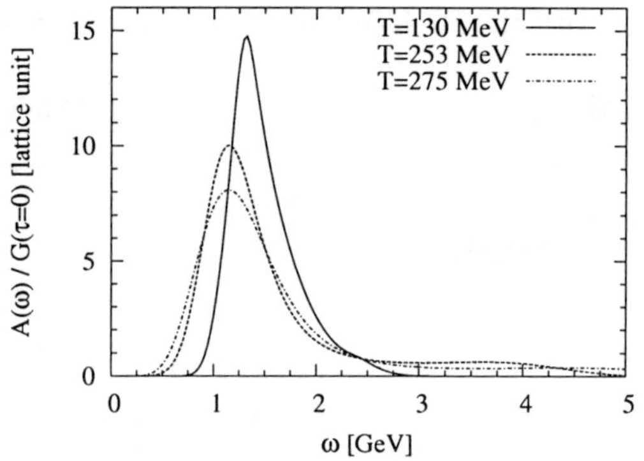

Figure 2. MEM results for the 0^{++} glueball spectrum at various temperatures below T_c.

become broader with the increasing temperature below T_c. This is consistent with the results of the advanced Breit-Wigner fit analysis, suggesting the importance of the thermal width. Experimentally, there are two glueball candidates,[14] i.e., $f_0(1500)$ and $f_0(1710)$, both of which have rather narrow decay width $\Gamma \simeq 100$ MeV. Hence, the experimental observation of thermal-width broadening of 300 MeV would be possible in the future experiments at RHIC. For more realistic comparison with experiments, it is desirable to estimate the effects of the dynamical quarks on the thermal properties of the glueball.

References

1. T. Hashimoto, K. Hirose, T. Kanki and O. Miyamura, *Phys. Rev. Lett.* **57**, 2123 (1986).
2. T. Hatsuda and T. Kunihiro, *Phys. Rev. Lett.* **55**, 158 (1985).
3. QCD-TARO Collaboration, *Phys. Rev.* **D63**, 054501 (2001).
4. T. Umeda, R. Katayama, O. Miyamura and H. Matsufuru, *Int. J. Mod. Phys.* **A16**, 2215 (2001).
5. H. Ichie, H. Suganuma and H. Toki, *Phys. Rev.* **D52**, 2944 (1995).
6. N. Ishii and H. Suganuma, *Eur. Phys. J.* **A17**, 77 (2003).
7. N. Ishii, H. Suganuma and H. Matsufuru, *Phys. Rev.* **D66**, 014507 (2002).
8. N. Ishii, H. Suganuma and H. Matsufuru, *Phys. Rev.* **D66**, 094506 (2002).
9. N. Ishii and H. Suganuma, *Nucl. Phys.* **B**(Proc. Suppl), hep-lat/0309102.
10. APE Collaboration, *Phys. Lett.* **B192**, 163 (1987).
11. M. Jarrell and J.E. Gubernatis, *Phys. Rept.* **269**, 133 (1996).
12. M. Asakawa, T. Hatsuda and Y. Nakahara, *Prog. Part. Nucl. Phys.* **46**, 459 (2001).
13. T. R. Klassen, *Nucl. Phys.* **B533**, 557 (1998).
14. K.K. Seth, *Nucl. Phys.* **A675**, 25c (2000).

FEATURES OF $SU(N)$ GAUGE THEORIES*†

B. LUCINI, M. TEPER AND U. WENGER

Theoretical Physics, University of Oxford,
1 Keble Road, Oxford OX1 3NP, United Kingdom
E-mail: lucini,teper,wenger@thphys.ox.ac.uk

We review recent lattice results for the large N limit of $SU(N)$ gauge theories. In particular, we focus on glueball masses, topology and its relation to chiral symmetry breaking (relevant for phenomenology), on the tension of strings connecting sources in higher representations of the gauge group (relevant for models of confinement and as a comparative ground for theories beyond the Standard Model) and on the finite temperature deconfinement phase transition (relevant for RHIC-like experiments). In the final part we present open challenges for the future.

1. Introduction

The possibility that observables in $SU(N)$ gauge theories are (at least for large enough N) smooth functions of N has been advocated a long time ago[1]. In particular, it has been proven diagrammatically that in the limit $N \to \infty$ the theory is simpler (only planar diagrams survive) and that in a neighbourhood of that limit the leading corrections due to a finite N go as $1/N^2$. This can have practical implications for our understanding of QCD if the gauge group of this theory, $SU(3)$, shares the bulk of the physics with $SU(\infty)$. To verify if this is the case, an investigation from first principles is mandatory. Another motivation for studying the physics of $SU(N)$ at large N within the conventional gauge theory comes from calculations performed in "beyond the Standard Model" frameworks: the bridge between the two approaches is often $SU(\infty)$ Yang-Mills[2].
Lattice calculations are the most reliable tool for investigations of gauge theories from first principles. In the following we give a quick overview of recent lattice results obtained by our group for $SU(\infty)$. For more details about the calculations we refer to the quoted literature.

---

*Summary of the talk presented by B. Lucini and of the poster presented by U. Wenger.
†Work partially supported by the Royal Society and the Lockey Committee.

2. The lowest-lying masses in the spectrum

We have studied $SU(N)$ gauge theories for $2 \leq N \leq 8$. At zero temperature, the theory is confined at all these N, which suggests that the large N limit is confining. We use the value of the string tension σ to set the scale for the studied observables.

A large N understanding of the mass spectrum can help in identifying glueball states in experiments. We have looked at the masses of the 0^{++}, 2^{++} and 0^{++*} glueballs as a function of N. We have measured those quantities for $2 \leq N \leq 5$. Our results for $m/\sqrt{\sigma}$ are plotted in fig. 1 as a function of $1/N^2$. The choice of the independent variable is dictated by the fact that the first expected correction to the large N limit value of an observable has this functional form. A fit to the data using only this correction generally works pretty well all the way down to $SU(2)$ (*i.e.* there is a somehow unexpected precocious onset of the large N behaviour) and allows us to extract the value of the spectrum at $N = \infty$. We find[3]

$$\begin{aligned} m/\sqrt{\sigma} &= 3.341(76) + 1.75/N^2 \\ m/\sqrt{\sigma} &= 4.93(13) + 2.58/N^2 \\ m/\sqrt{\sigma} &= 6.48(35) - 1.7/N^2 \end{aligned} \tag{1}$$

respectively for the 0^{++}, the 2^{++} and the 0^{++*} glueballs.

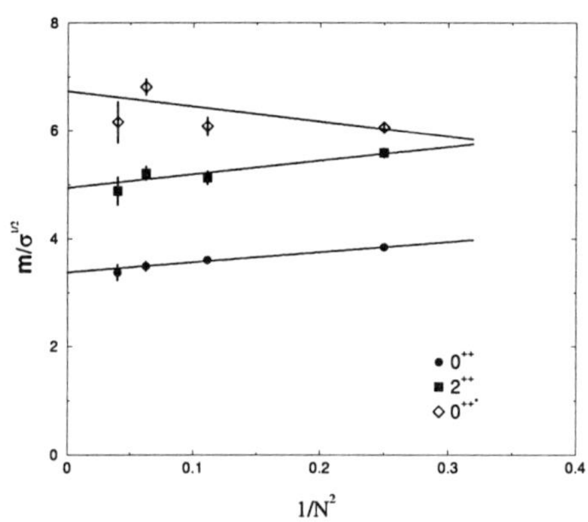

Figure 1. The mass of the lightest glueballs as a function of $1/N^2$. Solid lines are our best fits to the data.

3. The topological susceptibility

Another interesting quantity in the large N limit is the topological susceptibility χ_t, which is related in that limit to the η' mass by the Witten-Veneziano formula[4]. This formula is a good approximation even when the $SU(3)$ value of the topological susceptibility is plugged in. An explanation of this fact can be found by investigating how χ_t varies with N. We find[3]

$$\chi_t^{1/4}/\sqrt{\sigma} = 0.3739(59) + 0.439/N^2 \ . \tag{2}$$

The determination of χ_t at $N = \infty$ given in ref. [5] agrees with our result.

4. Topology and chiral symmetry breaking[a]

From the Banks-Casher relation we know that the chiral condensate is proportional to the density of small eigenmodes λ of the Dirac operator, $\langle\bar{\psi}\psi\rangle \sim \lim_{\lambda\to 0}\rho(\lambda)$. Hence a possible scenario for spontaneous chiral symmetry breaking is to assume that the non-vanishing density is due to exact zero modes of the Dirac operator which, through their interaction, are lifted away from zero yielding $\lim_{\lambda\to 0}\rho(\lambda) \neq 0$. As a consequence the near-zero modes would have a topological origin since they emerged from the topological zero modes. A comparison of the topological content of near-zero modes and zero modes of the $SU(N)$ Dirac operator can tell us qualitatively whether this scenario holds in the large N limit[6]. We find that the topo-

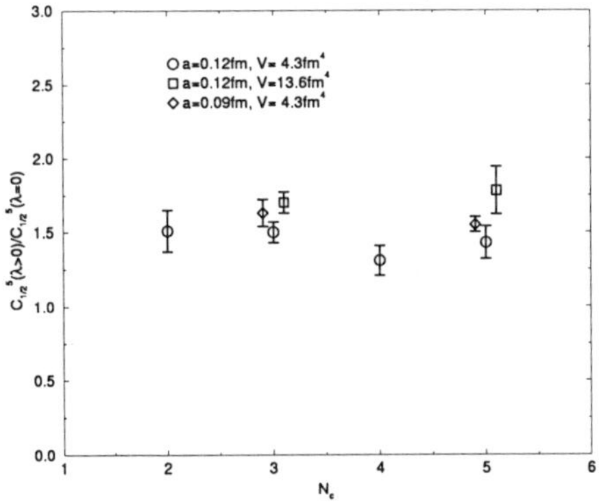

Figure 2. The normalised topological content of the near-zero modes responsible for chiral symmetry breaking for different volumes V, lattice spacings a and N.

[a]We thank N. Cundy who contributed to the results in this section.

logical and chiral contents of both the zero and near-zero modes become smaller as N increases, but at roughly the same rate. As a consequence the topological content of the near-zero modes normalised by the one of the zero modes (fig. 2) is constant for all N. We find that this remains qualitatively true as we vary the volume V and the lattice spacing a and we are therefore able to conclude that topology indeed drives chiral symmetry breaking for all N.

5. *k*-strings

For $N \geq 4$ strings connecting sources in higher representations of the gauge group can be stable. This is a consequence of the $\mathbb{Z}(N)$ symmetry of the confined phase. Taking into account also charge conjugation, it is easy to see that the number of stable stringy states for $SU(N)$ is the integer part of $N/2$. These states are referred to in the literature as k-strings and the rank of the representation of the sources is called N-ality. Within a class of strings with the same N-ality the string tension is unique. We indicate this tension by σ_k.

The *a priori* unknown value of the ratio σ_k/σ poses constraints to effective models of confinement. Moreover, it can shed light on the connection between QCD and some "beyond the Standard Model" theories[7]. Usually in these frameworks one obtains the so-called sine formula:

$$\sigma_k/\sigma = \sin(k\pi/N)/\sin(\pi/N) \; ; \tag{3}$$

however, this is not a universal feature[8]. The first lattice calculation of this ratio was performed in [9] for $SU(4)$. Although a useful continuum extrapolation could not be obtained, it was found that the $k = 2$ string is a genuine bound state. Recent calculations[10,11] have obtained a precise continuum determination of σ_k/σ. This is compatible with both the sine formula and the Casimir scaling ansatz, which predicts that σ_k/σ is equal to the ratio of the lowest quadratic Casimir operators in the class of the representations with the N-ality of the sources:

$$\sigma_k/\sigma = k(N-k)/(N-1) \; . \tag{4}$$

While the authors of [12] claim to be able to exclude Casimir scaling, in our opinion the question is far from being settled: the numerical values of (3) and (4) are close enough for systematic effects to become relevant. We are currently trying to deal with those issues.

Indirect information on σ_k/σ can be extracted from the size of the corresponding strings[11,13] or from the behaviour of the leading correction to its asymptotic value as N increases[13,14].

6. The deconfinement phase transition

The quark-gluon plasma phase of QCD is currently being investigated experimentally. Some of the related theoretical questions like the late onset of the Stephan-Boltzman law can be answered in the context of large N. As a first step in that direction, we have investigated the physics of the deconfinement phase transition as N varies and then we have extrapolated our results to the large N limit. For the deconfinement temperature T_c we find[15,16]

$$T_c/\sqrt{\sigma} = 0.596(4) + 0.453(30)/N^2 \ . \tag{5}$$

For $N \geq 3$ the transition is first order, with a monotonically increasing latent heat as a function of N, which suggests that also for $SU(\infty)$ the transition is first order. In fact, an extrapolation of the latent heat to $N = \infty$ using the leading expected $O(1/N^2)$ correction predicts a finite value for this quantity in the limiting case.

The interface tension between the confined and the deconfined phase grows

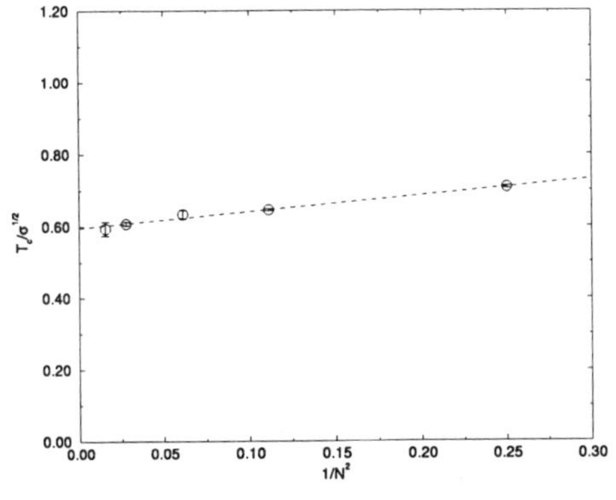

Figure 3. $T_c/\sqrt{\sigma}$ vs $1/N^2$. The solid line is our best fit to the data.

as N is increased and seems to diverge in the critical region at infinite N. Interpreted in terms of the Master Field[17], this gives rise to the speculation that in the large N limit there are several Master Fields separated by infinite energy barriers[16]. The rich physics of the limiting case is given by the interplay between those vacua at large but finite N.

7. Conclusions

$SU(N)$ gauge theories have a sensible large N limit that can be studied by lattice techniques. The results can have relevant implications for our understanding of QCD and of the physics beyond the Standard Model. Possible future directions of our investigations include the physics of confinement, the full mass spectrum of glueballs and the equation of state at finite temperature.

Acknowledgements

B.L. would like to thank P. Faccioli, J. Greensite, K. Konishi and M. Shifman for interesting discussions. B.L. is supported by a EU Marie Curie Fellowship and U.W. by a PPARC SPG Fellowship.

References

1. G. 't Hooft, *Nucl. Phys.* **B72**, 461 (1974); E. Witten, *Nucl. Phys.* **B160**, 57 (1979); S. Coleman, *1979 Erice Lectures*; A. Manohar, *1997 Les Houches Lectures*, hep-ph/9802419; S.R. Das, *Rev. Mod. Phys.* **59**, 235 (1987).
2. For a review, see O. Aharony, S.S. Gubser, J. Maldacena, H. Ooguri and Y. Oz, *Phys. Rept.* **323**, 183 (2000).
3. B. Lucini and M. Teper, *JHEP* **0106**, 050 (2001).
4. E. Witten, *Nucl. Phys.* **B156**, 269 (1979); G. Veneziano, *Nucl. Phys.* **B159**, 213 (1979).
5. L. Del Debbio, A. Panagopoulos and E. Vicari, *JHEP* **0208**, 044 (2002).
6. N. Cundy, M. Teper, U. Wenger, *Phys. Rev.* **D66** (2002), 094505; hep-lat/0209104; hep-lat/0309011.
7. A. Hanany, M. Strassler and A. Zaffaroni, *Nucl. Phys.* **B513**, 87 (1998); A. Armoni and M. Shifman, *Nucl. Phys.* **B664**, 233 (2003).
8. R. Auzzi and K. Konishi, *New J. Phys.* **4**, 59 (2002).
9. M. Wingate and S. Ohta, *Phys. Rev.* **D63**, 094502 (2001).
10. B. Lucini and M. Teper, *Phys. Lett.* **B501**, 128 (2001).
11. B. Lucini and M. Teper, *Phys. Rev.* **D64**, 105019 (2001).
12. L. Del Debbio, A. Panagopoulos, P. Rossi and E. Vicari, *Phys. Rev.* **D65**, 021501 (2002); *JHEP* **0201**, 009 (2002).
13. A. Armoni and M. Shifman, hep-th/0307020.
14. M. Teper, hep-ph/0203203.
15. B. Lucini, M. Teper and U. Wenger, *Phys. Lett.* **B545**, 197 (2002).
16. B. Lucini, M. Teper and U. Wenger, hep-lat/0307017.
17. E. Witten, in *Recent Developments in Gauge Theories*, ed. G. 't Hooft et al. (Plenum Press, 1980).

MONOPOLES IN REAL TIME FOR CLASSICAL U(1) GAUGE FIELD THEORY

TAMÁS S. BIRÓ

KFKI, Research Institute for Particle and Nuclear Physics, H-1525 Budapest, Hungary

HARALD MARKUM AND RAINER PULLIRSCH

Atominstitut, Technische Universität Wien, A-1040 Vienna, Austria

U(1) gauge fields are decomposed into a monopole and photon part across the phase transition from the confinement to the Coulomb phase. We analyze the leading Lyapunov exponents of such gauge field configurations on the lattice which are initialized by quantum Monte Carlo simulations. We observe that the monopole field carries the same Lyapunov exponent as the original U(1) field. As a long awaited result, we show that monopoles are created and annihilated in pairs as a function of real time in excess to a fixed average monopole number.

1. Monopole and photon part of U(1) theory

We are interested in the relationship between monopoles and classical chaos across the phase transition at $\beta_c \approx 1.01$. We begin with a 4*d* U(1) gauge theory described by the Euclidean action $S\{U_l\} = \beta \sum_p (1 - \cos\theta_p)$, where $U_l = U_{x,\mu} = \exp(i\theta_{x,\mu})$ and $\theta_p = \theta_{x,\mu} + \theta_{x+\hat{\mu},\nu} - \theta_{x+\hat{\nu},\mu} - \theta_{x,\nu}$. Following Ref. 1, we have factorized our gauge configurations into monopole and photon fields. The U(1) plaquette angles $\theta_{x,\mu\nu}$ are decomposed into the "physical" electromagnetic flux through the plaquette $\bar{\theta}_{x,\mu\nu}$ and a number $m_{x,\mu\nu}$ of Dirac strings through the plaquette

$$\theta_{x,\mu\nu} = \bar{\theta}_{x,\mu\nu} + 2\pi\, m_{x,\mu\nu} \ , \tag{1}$$

where $\bar{\theta}_{x,\mu\nu} \in (-\pi, +\pi]$ and $m_{x,\mu\nu} \neq 0$ is called a Dirac plaquette.

2. Classical chaotic dynamics from quantum Monte Carlo initial states

Chaotic dynamics in general is characterized by the spectrum of Lyapunov exponents. These exponents, if they are positive, reflect an exponential divergence of initially adjacent configurations. In case of symmetries inherent in the Hamiltonian of the system there are corresponding zero values of these exponents. Finally negative exponents belong to irrelevant directions in the phase space: perturbation components in these directions die out exponentially. Pure random gauge fields on the lattice show a characteristic Lyapunov spectrum consisting of one third of each kind of exponents.[2] Assuming this general structure of the Lyapunov spectrum we present here its magnitude only, namely the maximal value of the Lyapunov exponent, $L_{\max}$.

The general definition of the Lyapunov exponent is based on a distance measure $d(t)$ in phase space,

$$L := \lim_{t\to\infty} \lim_{d(0)\to 0} \frac{1}{t} \ln \frac{d(t)}{d(0)}. \tag{2}$$

In case of conservative dynamics the sum of all Lyapunov exponents is zero according to Liouville's theorem, $\sum L_i = 0$. We utilize the gauge invariant distance measure consisting of the local differences of energy densities between two $3d$ field configurations on the lattice:

$$d := \frac{1}{N_P} \sum\nolimits_P \left| \mathrm{tr} U_P - \mathrm{tr} U'_P \right|. \tag{3}$$

Here the symbol $\sum_P$ stands for the sum over all N_P plaquettes, so this distance is bound in the interval $(0, 2N)$ for the group SU(N). U_P and U'_P are the familiar plaquette variables, constructed from the basic link variables $U_{x,i}$,

$$U_{x,i} = \exp\left(aA_{x,i}T\right) , \tag{4}$$

located on lattice links pointing from the position $x = (x_1, x_2, x_3)$ to $x + ae_i$. The generator of the group U(1) is $T = -ig$ and $A_{x,i}$ is the vector potential. The elementary plaquette variable is constructed for a plaquette with a corner at x and lying in the ij-plane as $U_{x,ij} = U_{x,i} U_{x+i,j} U^\dagger_{x+j,i} U^\dagger_{x,j}$. It is related to the magnetic field strength $B_{x,k}$:

$$U_{x,ij} = \exp\left(\varepsilon_{ijk} a^2 B_{x,k} T\right). \tag{5}$$

The electric field strength $E_{x,i}$ is related to the canonically conjugate momentum $P_{x,i} = a\dot{U}_{x,i}$ via

$$E_{x,i} = \frac{1}{g^2a^2} \left(T \dot{U}_{x,i} U^\dagger_{x,i} \right). \tag{6}$$

The Hamiltonian of the lattice gauge field system can be casted into the form

$$H = \sum \left[\frac{1}{2} \langle P, P \rangle + 1 - \frac{1}{4} \langle U, V \rangle \right] . \tag{7}$$

Here the scalar product stands for $\langle A, B \rangle = \frac{1}{2}\mathrm{tr}(AB^\dagger)$. The staple variable V is a sum of triple products of elementary link variables closing a plaquette with the chosen link U. This way the Hamiltonian is formally written as a sum over link contributions and V plays the role of the classical force acting on the link variable U.

We prepare the initial field configurations from a standard four dimensional Euclidean Monte Carlo program on a $12^3 \times 4$ lattice varying the inverse gauge coupling β.[3] We relate such four dimensional Euclidean lattice field configurations to Minkowskian momenta and fields for the three dimensional Hamiltonian simulation by selecting a fixed time slice of the four dimensional lattice.

3. Chaos and confinement

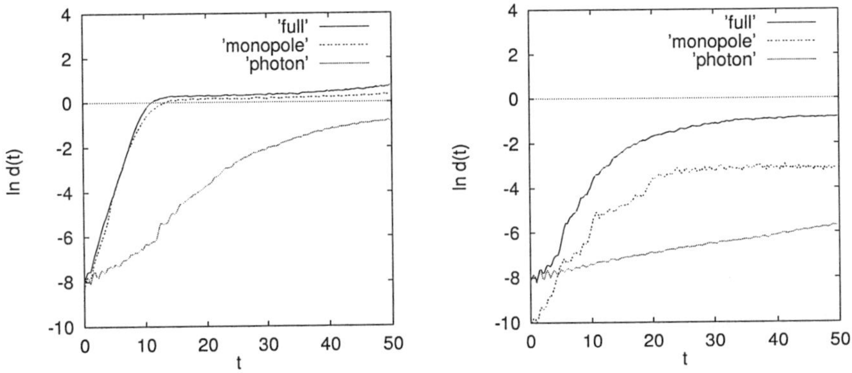

Figure 1. Exponentially diverging distance in real time of initially adjacent U(1) field configurations on a 12^3 lattice prepared at $\beta = 0.9$ in the confinement (left) and at $\beta = 1.1$ in the Coulomb phase (right).

We start the presentation of our results with a characteristic example of the time evolution of the distance between initially adjacent configurations. An initial state prepared by a standard four dimensional Monte Carlo simulation is evolved according to the classical Hamiltonian dynamics in real time.

Afterwards this initial state is rotated locally by group elements which are chosen randomly near to the unity. The time evolution of this slightly rotated configuration is then pursued and finally the distance between these two evolutions is calculated at the corresponding times. A typical exponential rise of this distance followed by a saturation can be inspected in Fig. 1 from an example of U(1) gauge theory in the confinement phase and in the Coulomb phase. While the saturation is an artifact of the compact distance measure of the lattice, the exponential rise (the linear rise of the logarithm) can be used for the determination of the leading Lyapunov exponent. The left plot exhibits that in the confinement phase the original field and its monopole part have similar Lyapunov exponents whereas the photon part has a smaller L_{max}. The right plot in the Coulomb phase suggests that all slopes and consequently the Lyapunov exponents of all fields decrease substantially.

4. Real-time evolution of monopoles

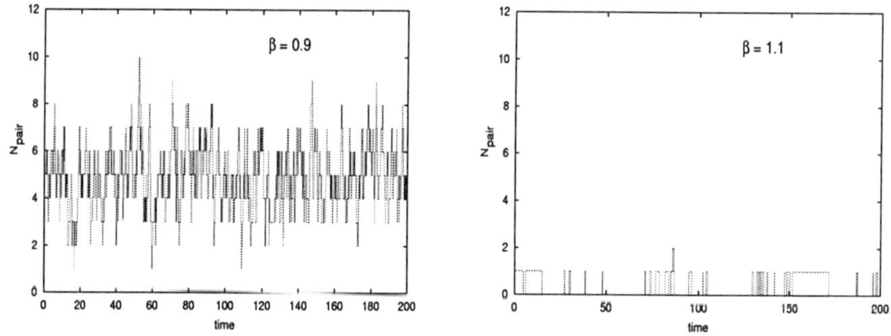

Figure 2. The number of monopole pairs as a function of real time evolution at $\beta = 0.9$ (confined, left) and at $\beta = 1.1$ (Coulomb phase, right).

An important subject of the present study is the evolution of monopoles in real time.[4] The outcome is displayed in Fig. 2 and shows that their number stays constant on average with creation and annihilation of a few monopole pairs only. The fluctuations decrease with increasing inverse gauge coupling β. A decrease of the total number of monopoles takes place towards the Coulomb phase.

The real-time evolution of the monopole content in the decomposed U(1) field is depicted in Fig. 3. Although there are less monopoles in the

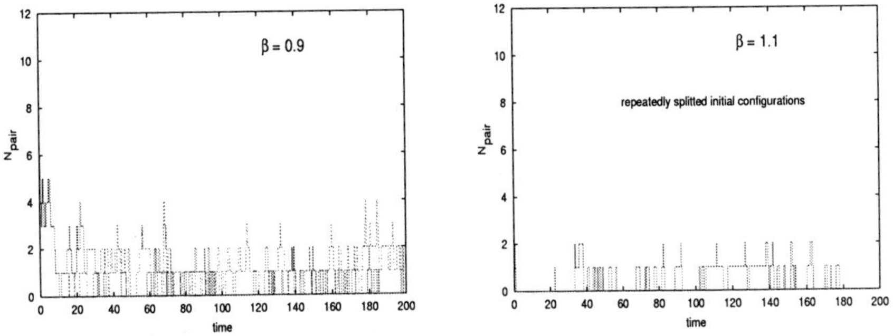

Figure 3. Monopole pairs in the decomposed field at $\beta = 0.9$ (left) and at $\beta = 1.1$ (right). In the latter case the decomposition was repeated until complete absence of monopoles in the photon part. Full lines correspond to the monopole, dashed lines to the photon (residual) configuration.

photon field than in the monopole field, their fluctuations are comparable in the confining phase. On the right plot, made for the Coulomb phase at $\beta = 1.1$, additionally the decomposition to photon and monopole fields has been repeatedly applied until the photon part had no monopole in it. This zero value is conserved during the real-time evolution.

References

1. J.D. Stack and R.J. Wensley, Nucl. Phys. B371 (1992) 597; T. Suzuki, S. Kitahara, T. Okude, F. Shoji, K. Moroda, and O. Miyamura, Nucl. Phys. B (Proc. Suppl.) 47 (1996) 374.
2. T.S. Biró, S.G. Matinyan, and B. Müller: Chaos and Gauge Field Theory, World Scientific, Singapore, 1995.
3. T.S. Biró, M. Feurstein, and H. Markum, APH Heavy Ion Physics 7 (1998) 235; T.S. Biró, N. Hörmann, H. Markum, and R. Pullirsch, Nucl. Phys. B (Proc. Suppl.) 86 (2000) 403.
4. H. Markum, R. Pullirsch, and W. Sakuler, hep-lat/0201001; hep-lat/0205003; hep-lat/0209039; T.S. Biró, H. Markum, R. Pullirsch, and W. Sakuler, hep-lat/0210020.

CHARMONIUM CORRELATORS AT FINITE TEMPERATURE IN QUENCHED LATTICE QCD*

HIDEO MATSUFURU[A], TAKASHI UMEDA[A], KOUJI NOMURA[B]

[a] *Yukawa Institute for Theoretical Physics, Kyoto University, Kyoto 606-8502, Japan*

[b] *Department of Physics, Hiroshima University, Higashi-hiroshima 739-8526, Japan*

We study charmonium correlators at finite temperature using quenched lattice QCD simulations. Two analysis procedures are applied to extract information on the spectral function: the maximum entropy method, and the χ^2 fit analyses including the constrained curve fitting. We focus on the low energy structure of the spectral function by applying the smearing technique. We first discuss the applicability of these methods to the finite temperatures by analyzing the data at $T = 0$ with restricted numbers of degrees of freedom. Then we apply these methods to the correlators at $T > 0$. We find no indication of mass shift and finite width for the charmonium states at $T \simeq 0.9T_c$. The results at $T \simeq 1.1T_c$ imply that bound-state-like structures may survive even above T_c.

1. Introduction

Investigation of the QCD phase transition at finite temperature and density is one of the most important subjects in hadron physics. To create the quark gluon plasma (QGP) phase, in which quarks and gluons are deconfined and the chiral symmetry is restored, the high energy heavy ion collision experiments have been performed and the RHIC experiment at BNL is in progress. The J/ψ suppression has been considered as one of the most important signals of formation of QGP. However, theoretical understanding of charmonium at finite temperature and density is not sufficient to draw a conclusive scenario.

We investigate the charmonium properties at finite temperature using lattice QCD simulations which enable us to incorporate the nonperturbative effects of QCD. Observations of spatial correlations between quark

*Talk presented by H. Matsufuru.

and antiquark suggested nontrivial strong correlations persist above T_c [1,2]. These imply that bound-state-like structures may survive even above T_c.

In present work, we analyze the correlators by reconstructing the spectral function from lattice data using the maximum entropy method [3] and the χ^2 fit analysis [4,5]. The latter includes the constrained curve fitting [6] in addition to the standard χ^2 fit. To circumvent severe restriction of degrees of freedom in Euclidean temporal direction, we employ anisotropic lattices on which temporal lattice spacing is finer than the spatial one (for our setup of the anisotropic lattices, see Ref. [7]). At this stage, the simulations are performed in the quenched approximation (without dynamical quark effects). To enhance the contribution from low energy part in the spectral function, we apply the smearing technique by spatially extending the meson operators. The details of this work has been presented in Ref.[5].

2. Analysis procedures

The spectral function contains direct information on the structure of the mesonic correlator, $C(t) = \sum_{\vec{x}} \langle O(\vec{x}, t) O^{\dagger}(0,0) \rangle$. Recently the maximum entropy method (MEM) has been successfully applied to the extraction of the spectral function from lattice data at $T = 0$ without assuming specific form [3]. However, its applicability to problems at $T > 0$ is not straightforward, because available physical range of $C(t)$ as well as number of data points are severely limited. In such a circumstance, we need in some way to check the applicability of the method. Once a presumable form of the spectral function is known by MEM, other analysis procedures, such as the standard χ^2 fit may provide more quantitative information [4,5]. We therefore propose to use the χ^2 fit analysis in combination with MEM [5]. As a more sophisticated alternative to the standard χ^2 fit, we also apply the constrained curve fitting [6] assuming several peaks with finite widths. In this method, one can add prior knowledge for the spectral function and thus can treat functional forms with numbers of parameters (even) more than the number of data points.

To verify the applicability of these methods at $T > 0$, we require that they correctly work for the correlators at $T = 0$ with restricted numbers of data points corresponding to the cases at $T > 0$. As we will see in the next section, MEM applied to the correlators of the point operators does not meet this condition with present level of statistics ($O(500)$ configurations).

We therefore apply the smearing technique which enhances the low energy part of the spectral function. For the correlators smeared with the

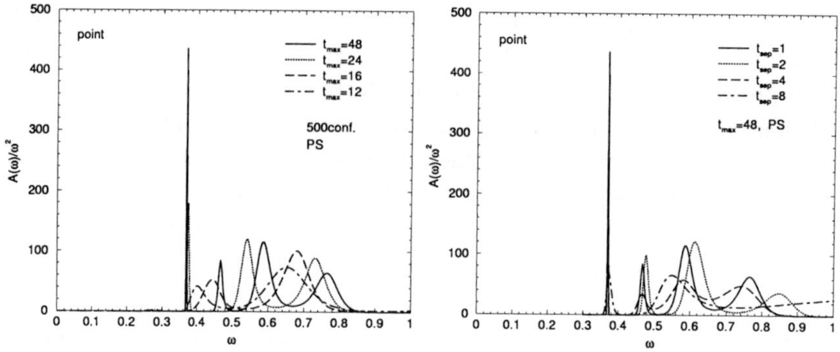

Figure 1. The spectral functions determined by MEM for the point correlator at $T = 0$ in PS channel.

wave function at $T = 0$, both the procedures work well in the low energy region. Although the features of the collective mode such as the mass and width are unchanged by the smearing, the smearing can produce a mimic peak structure in the spectral function [8]. This possibility is examined by observing the correlators smeared with a narrower function (*half-smeared* correlators). The dependence of extracted spectral function on the smearing function is compared with that of correlators composed of free quarks.

3. Numerical results

The numerical simulation is performed on lattices with the spatial lattice cutoff $a_\sigma^{-1} \simeq 2$ GeV and the anisotropy $a_\sigma/a_\tau = 4$ in the quenched approximation [7]. The lattice sizes are $20^3 \times N_t$, where $N_t = 160$ ($T \simeq 0$), 32 ($T \simeq 0.9T_c$), and 26 ($T \simeq 1.1T_c$). The numbers of configurations are 500 at $T \simeq 0$ and 1000 at $T > 0$. $N_t = 28$ roughly corresponds to the transition temperature. The quark field is described by the $O(a)$ improved Wilson action with the tree-level tadpole-improvement. The hopping parameter is chosen so that the charmonium spectrum is roughly reproduced.

3.1. *Results at $T = 0$*

Let us start with the analysis of the point correlators at $T = 0$. We apply MEM to them with restricted numbers of degrees of freedom. The results with two types of such restrictions are displayed in Fig. 1. The left panel shows the dependence of the result on t_{max}, the maximum t of the correlator used in the analysis. This case corresponds to the situation at $T > 0$. MEM fails to reproduce even the lowest peak for $t_{max} \leq 16$. The right panel shows

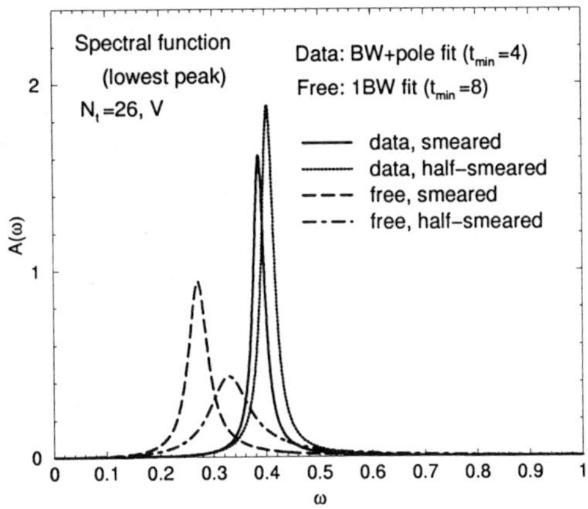

Figure 2. The spectral functions from the χ^2 fit analysis for the smeared correlators in the vector channel at $T \simeq 1.1T_c$.

the results when one alternatively skips several time slices in the analysis. This case corresponds to the coarsening of the temporal lattice spacing. Even for $t_{sep} = 8$ for which the number of data points is 6, MEM at least reproduces the correct lowest peak position while the resolution is not enough. These results indicate that the physical region of the correlator as well as the number of the degrees of freedom is important for MEM to work correctly. The required region of $C(t)$ in the above analysis is $t_{max} > O(0.5\text{fm})$, which is not fulfilled around $T \simeq T_c$.

This is why we apply the smearing technique. The results of MEM for the smeared correlators are stable under the above two kinds of restriction for t_{max} of interest; at least the lowest peak position is correctly reproduced while the resolution becomes worse as t_{max} decreases. Since the smearing suppresses the high frequency region of the spectral function, we focus only on the lowest peak of the spectral function in the following analysis.

The results of the χ^2 fit and the constrained curve fitting for the smeared correlators are consistent with the results of MEM and stable under the same restriction of t_{max}. In the following, we concentrate on the results of the smeared correlators.

3.2. *Results at $T > 0$*

At $T \simeq 0.9T_c$, the result of MEM for the smeared correlator indicates that the low energy structure of the spectral function is well represented by a strong peak corresponding to a meson state. We then apply the χ^2 fit analysis assuming double-pole, single relativistic Breit-Wigner (BW) type form [4], and BW+pole form (the pole function is to subtract the excited state contribution). The analysis indicates that the lowest peak of the spectral function is well represented by the pole form with the width consistent with zero. The mass is almost the same as at $T = 0$: no mass shift is observed at $T \simeq 0.9T_c$. The result of the constrained curve fitting is consistent with that of the standard χ^2 fit.

At $T \simeq 1.1T_c$ the result of MEM still exhibits a peak structure around the same energy region as at $T < T_c$. Therefore we perform the same kind of the χ^2 fit analysis as at $T < T_c$. The fits to the 2-pole *ansatz* and to the Breit-Wigner type *ansätze* give inconsistent results, and the latter fits indicate that the spectral function has a peak with almost the same mass as at $T < T_c$ and the width of order of 200 MeV, as shown in Fig. 2. The result depends on the smearing function only slightly, in contrast to the free quark case, and hence we conclude that this is a physically significant structure. Similar result is obtained with the constrained curve fitting. These results are in accord with our earlier analysis of the spatial correlation between quark and antiquark [2] and imply that quasi-stable bound-state-like structures exist at $T \simeq 1.1T_c$. Similar results have also been reported by other groups [9,10].

The simulation has been done on NEC SX-5 at Research Center for Nuclear Physics, Osaka University and Hitachi SR8000 at KEK (High Energy Accelerator Research Organization). H. M. and T. U. are supported by Japan Society for the Promotion of Science for Young Scientists.

References

1. QCD-TARO Collaboration, Ph. de Forcrand *et al.*, *Phys. Rev.* **D63**, 054501 (2001).
2. T. Umeda *et al.*, *Int. J. Mod. Phys.* **A16**, 2215 (2001).
3. Y. Nakahara, M. Asakawa and T. Hatsuda, *Phys. Rev.* **D60**, 091503 (1999).
4. N. Ishii, H. Suganuma and H. Matsufuru, *Phys. Rev.* **D66**, 094506 (2002).
5. T. Umeda, K. Nomura and H. Matsufuru, hep-lat/0211003; T. Umeda *et al.*, *Nucl. Phys.* **A721**, 922 (2003); K. Nomura *et al.*, *Nucl. Phys. B (Proc. Suppl.)* **119**, 496 (2003).
6. G. P. Lepage *et al.*, *Nucl. Phys. B (Proc. Suppl.)* **106**, 12 (2002).
7. H. Matsufuru, T. Onogi and T. Umeda, *Phys. Rev.* **D64**, 114503 (2001).
8. I. Wetzorke *et al.*, *Nucl. Phys. B (Proc. Suppl.)* **106**, 510 (2002).
9. P. Petreczky, these proceedings; S. Datta *et al.*, *Nucl. Phys. B (Proc. Suppl.)* **119**, 487 (2003); P. Petreczky *et al.*, hep-lat/0309012.
10. M. Asakawa and T. Hatsuda, hep-lat/0308034; hep-lat/0309001.

MOMENTUM DISTRIBUTION OF J/ψ'S PRODUCED AT RHIC

F.O. DURÃES, F.S. NAVARRA AND M. NIELSEN

Instituto de Física, Universidade de São Paulo, C.P. 66318, 05389-970 São Paulo, SP, Brazil

Thermal production of J/ψ within quark gluon plasma is reconsidered. We show that if screening effects are not strong enough, the "in-plasma born" J/ψ's would show up as a peak in the Feynman momentum distribution at $x_F = 0$.

In a recent analysis, presented in [1], it has been suggested that the forthcoming PHENIX data on J/ψ production in Au-Au collisions will show a very flat Feynman momentum (x_F) distribution. In the small x_F region it may even exhibit a small dip. This behavior is a consequence of the interaction of the charmonium, formed in the initial stage of the collision, with the comovers in the hadronic matter or in the quark-gluon plasma phase. This suggested behavior is in sharp contrast with the so far measured distributions, which are usually parametrized by $(1-x_F)^n$ (with $n \simeq 5-6$).

Motivated by this observation, in this work we address the J/ψ Feynman momentum distributions. We follow ref. [1] with one important difference: we include J/ψ production *within the plasma.*

We assume here that screening is not so strong and, in temperatures just above the phase transition, allow for charmonium (Coulomb) bound states which "survive" from QGP.

In what follows, we shall use the Color Evaporation Model (CEM) [2]. In the CEM, charmonium is defined kinematically as a $c\bar{c}$ state with mass below the $D\bar{D}$ threshold. In leading order (LO) the cross section is computed with the use of perturbative QCD for the diagrams of the elementary processes $q\bar{q} \to c\bar{c}$ and $gg \to c\bar{c}$ convoluted with the parton densities in the projectile and in the target. A factor $\kappa = 2$ is introduced to account for higher order corrections. The color evaporation model has an "intrinsic efficiency" given by the fractional factor $F_{J/\psi}$, which was fixed [2] in the analysis of $p-p$ reactions to be $F_{J/\psi} \simeq 0.02 = 2\%$.

Assuming that QGP is formed, we then have a gas of quarks and gluons with momenta obeying respectively Fermi-Dirac and Bose-Einstein distributions that can collide producing $c\bar{c}$ pairs. In the present calculation we will let part of the thermally produced charm Coulomb bound states escape.

The charm production rate in the reaction $gg \to c\bar{c}$, at temperature T is given by [4]:

$$\frac{dN^{gg\to c\bar{c}}}{dtd^3x} = \frac{g_g^2}{2(2\pi)^6} \int d^3p_1 d^3p_2\, f_g(E_1)\, f_g(E_2)\, \hat{\sigma}_{gg}^{LO}\, v_{12} \tag{1}$$

where g_g is the gluon statistical factor (number of colors $\times$ number polarization states) v_{12} is the relative velocity between colliding gluons with energies E_1 and E_2 and three momenta $\vec{p_1}$ and $\vec{p_2}$, $\hat{\sigma}_{gg}^{LO}$ is the elementary gluon-gluon cross section and $f_g(E_i)$ the usual thermal distribution function:

$$f_g(E_i) = \frac{1}{e^{E_i/T} - 1} \tag{2}$$

A similar expression can be written for the process $q + \overline{q} \to c + \overline{c}$.

In order to account for expansion effects we shall assume that the system cools down following Bjorken hydrodynamics, in which the volume of the system evolves in time according to:

$$V(\tau) = V_0 \frac{\tau}{\tau_0} \quad \Longrightarrow \quad V(T) = V_0 \left(\frac{T_0}{T}\right)^3 \tag{3}$$

where $V_0 = \pi\, R_A^2\, \tau_0$, T_0 is the initial temperature and τ_0 is the thermalization time, which marks the beginning of the hydrodynamical expansion. We shall assume that $\tau_0 = 0.7$ fm, $R_A = 6.4$ fm and $T_0 = 550$ MeV.

Expression (1) can thus be rewritten as:

$$\frac{dN^{gg\to c\bar{c}}}{dx_F} = \frac{3\pi\sqrt{s}g_g^2\kappa}{8(2\pi)^6} \int_{4m_c^2}^{\infty} dm^2 \int_0^{\infty} dp_T^2 \int_0^{2\pi} d\beta \int_{T_f}^{T_0} \tau_0 T_0^3 T^{-4} V(T) dT$$

$$\times \int_0^{\infty} dE_1 f_g(E_1) f_g(E - E_1) \left[\frac{1}{|H(h_1)|}\theta(E_1^2 - h_1^2) + \frac{1}{|H(h_2)|}\theta(E_1^2 - h_2^2)\right]$$

$$\times\, \hat{\sigma}_{gg}^{LO}(m^2)\, \frac{m^2}{E}\, \theta(E_1)\theta(E - E_1) \tag{4}$$

where $E = \sqrt{m^2 + p_T^2 + p_z^2}$ is the energy of the $c\bar{c}$ pair and:

$$H(h_{1,2}) = 2p_z - \frac{2p_T \cos(\beta)}{(E_1^2 - h_{1,2}^2)^{1/2}} h_{1,2}$$

$$h_{1,2} = \frac{1}{2}\frac{1}{p_z^2 + p_T^2\cos^2(\beta)}\Big\{p_z(2EE_1 - m^2) \mp \{p_T^2\cos^2(\beta)$$

$$\times [4E_1^2(p_T^2\cos^2(\beta) + p_z^2) - (m^2 - 2EE_1)^2]\}^{1/2}\Big\}$$

The number of J/ψ's produced in the plasma by gluon fusion can be obtained from Eq. (4). We must change the upper limit of integration in m^2 introducing a kinematical cut-off, i.e., making the replacement $\infty \to 4m_D^2$. In doing so, we rename the superscripts in Eq. (4) to $gg \to J/\psi$.

Moreover we introduce the CEM multiplicative factor $F_{J/\psi}$ and arrive at:

$$\frac{dN_{QGP}^{AA\to J/\psi}}{dx_F} = f_s\, F_{J/\psi}\,[\frac{dN^{gg}}{dx_F} + \frac{dN^{q\bar{q}}}{dx_F}] \qquad (5)$$

For simplicity, we shall assume that $F_{J/\psi}$ is universal and holds even for in-plasma production. The "screening factor" f_s is thus the only free number introduced here. It gives the probability that a J/ψ formed inside the plasma survives the passage through the medium. We shall take $f_s = 10^{-2}$, which can be interpreted as meaning that 1 % of the in-plasma born J/ψ's can survive as bound states.

Let us now consider the ratio:

$$R(x_F) = R_{AB}(x_F) + \frac{S_{co}^{FSI}\frac{dN_{QGP}^{AA\to J/\psi X}}{dx_F}}{T_{AA}\frac{d\sigma^{pp\to J/\psi X}}{dx_F}} \qquad (6)$$

where $R_{AB}(x_F, b = 0)$, Eq. (3) of [1], is obtained computing the initial QCD production of J/ψ and then including final state interactions, either with a quark-gluon plasma or with a gas of hadronic comovers. In the expression above, the second term is obtained from (5) and represents the contribution of J/ψ's produced in QGP. T_{AA} is the usual nuclear overlap function, $\sigma^{pp\to J/\psi X}$ is the charmonium production cross section in proton-proton collisions. S_{co}^{FSI} is a survival probability factor, which accounts for the final state interactions of the "in-plasma" produced J/ψ's. If the screening in the plasma is so strong that $f_s = 0$, then, $R \simeq const.$ If, however, there is a small J/ψ survival probability, there will be a noticeable change in R and we will see a pronounced deviation from a flat behavior in the low x_F region if there is a contribution from the plasma.

The ratio (6) is plotted in Fig. 1. Solid and dashed lines correspond to the nuclear parton densities of Refs. [5] and [6] respectively. Following the estimates done in the literature (for example [1]) we take $S_{co}^{FSI} \simeq 0.25$ for

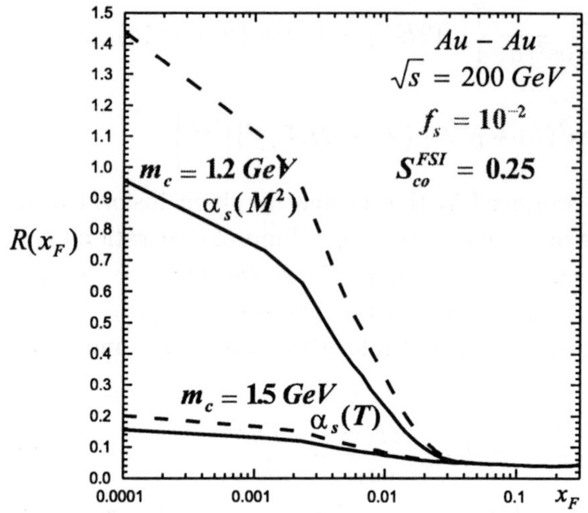

Figure 1. Ratio $R(x_F)$ computed with (6). Solid lines: parton densities of Ref. [5]. Dashed lines: parton densities of Ref. [6]

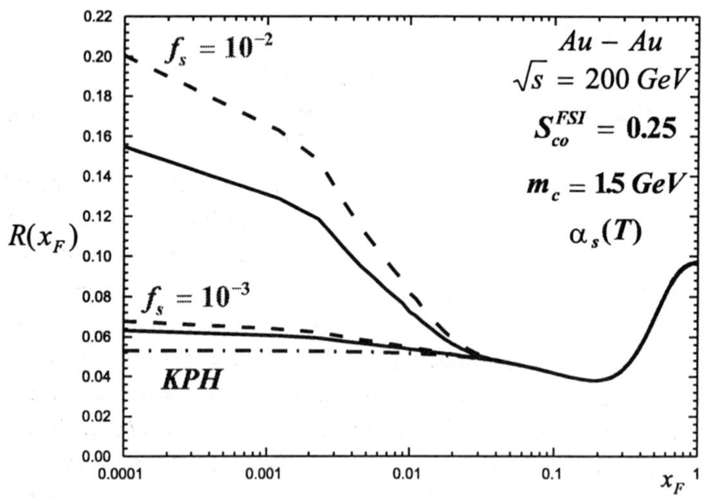

Figure 2. The same as Fig.1 for the lowest J/ψ multiplicities

the hadronic comover phase. In Fig. 1 the two upper lines correspond to the choices $m_c = 1.2$ GeV and $\alpha_s = \alpha_s(m^2)$, and the two lower lines to the choices $m_c = 1.5$ GeV and $\alpha_s = \alpha_s(T)$. Expressions for $\alpha_s(m)$ and for $\alpha_s(T)$ are given in [7]. In Fig. 2 we present a closer look into the lower part of Fig. 1. The dot-dashed line was taken from [1]. We have also included a set of two curves with a stronger screening factor, $f_s = 0.001$. It is interesting to notice that the quantity called here f_s is in other works called S^{FSI}_{QGP}. The computation of S^{FSI}_{QGP} for collisions of the Coulomb charmonium state with partons in a QGP performed, for example, in [1] or in [8] never gives values smaller than 0.01.

The value $f_s = 0.001$ is probably too small. On the other hand, $T_0 = 550$ MeV may be a too large initial temperature. Using instead $T_0 = 300$ MeV would certainly lower our curves in Fig. 2. But then, at these low temperatures, screening would be less effective and we would have to use, for example, $f_s = 0.1$, lifting up our curves. A more detailed study of these competing effects was performed in [7] with a promising prospect of observing the "in-plasma" charmonium production.

To summarize, in this note we have considered in-plasma J/ψ production in nucleus-nucleus collisions. Our main point was that, even being small, this contribution is very strongly peaked around $x_F \simeq 0$ and can thus become visible if enough plasma is formed. On a quantitative level, we calculated the measurable ratio (6) and oberved that the plasma contribution creates a big bump at $x_F < 0.01$. Without this contribution, we observe a plateau in this region. We believe that this feature can be measured and will be very interesting for charm physics at RHIC.

References

1. B.Z. Kopeliovich, A. Polleri and J. Hüfner, *Phys. Rev. Lett.* **87**, 112302 (2001) and references therein.
2. R. Vogt, hep-ph/0203151 and references therein.
3. J. Nagle *et al.*, PHENIX Collab., *Nucl. Phys.* **A715**, 243c (2003); nucl-ex/0209015.
4. *"Introduction to High-Energy Heavy-Ion Collisions*, Cheuk-Yin Wong, World Scientific Publishing Co. Pte. Ltd, 1994.
5. K. J. Eskola, V. J. Kolhinen and C. A. Salgado, *Eur. Phys. J.* **C9**, 61 (1999).
6. M. Hirai, S. Kumano and M. Miyama, *Phys. Rev.* **D64**, 034003 (2001).
7. F.O. Durães, F.S. Navarra and M. Nielsen, nucl-th/0210043.
8. A. Polleri, T. Renk, R. Schneider and W. Weise, nucl-th/0306025.

OPEN CHARM PRODUCTION THROUGH $J/\psi - \pi$ COLLISIONS

M. NIELSEN, F.S. NAVARRA, F.O. DURÃES

Instituto de Física, Universidade de São Paulo, C.P. 66318, 05389-970 São Paulo, SP, Brazil

The $J/\psi\ \pi \to \bar{D}\ D^*$, $D\bar{D}^*$ $\bar{D}\ D$ and $\bar{D}^*\ D^*$ cross sections as a function of $\sqrt{s}$ are evaluated in a QCD sum rule calculation. We find that our results are compatible with the $J/\psi\ \pi \to$ charmed mesons cross sections obtained with models based on meson exchange if a cut-off of order of 1 GeV is used.

1. Introduction

Charmonium - hadron cross sections are of crucial importance in the context of quark gluon plasma (QGP) physics because of the proposal by Matsui and Satz [1] that suppression of J/ψ production could be used as a signature of the QGP. A significant part of the charmonium - hadron interactions occurs when other light particles have already been produced, forming a "fireball". Interactions inside this fireball happen at low energies ($\sqrt{s} \leq 5$ GeV) and one has to apply nonperturbative methods. In this work we use the QCD sum rules (QCDSR) technique [2] to study the $J/\psi - \pi$ dissociation. The QCDSR technique allows to compute the hadronic matrix in terms of quark and gluon properties and universal matrix elements which represent the properties of the QCD vacuum. In view of our relatively poor understanding of J/ψ reactions in nuclear matter and considering the large discrepancies between different model estimates, we believe that our work adds to a better understanding of this important topic.

2. The QCDSR Calculation

We consider all four channels $J/\psi\ \pi \to \bar{D}\ D^*$, $D\bar{D}^*$, $\bar{D}\ D$ and $\bar{D}^*\ D^*$. Let us start with the vaccum-pion three-point function for the process $J/\psi\ \pi \to$

$\bar{D}\ D^*$:

$$\Pi_{\mu\nu} = \int d^4x\ d^4y\ e^{-ip_2\cdot y}\ e^{ip_3\cdot x} \times \langle 0|T\{j_\nu^{D^*}(x) j_{\bar{D}}(0) j_\mu^{\psi}(y)\}|\pi(p_1)\rangle\ , \tag{1}$$

with the currents given by $j_\mu^\psi = \bar{c}\gamma_\mu c$, $j_\nu^{D^*} = \bar{u}\gamma_\nu c$ and $j_{\bar{D}} = \bar{c}i\gamma_5 d$. p_1, p_2, p_3 and p_4 are the four-momenta of the mesons π, J/ψ, D^* and $\bar{D}$ respectively. The advantage of this approach as compared with the 4-point calculation in Ref. [3], is that we can consider more terms in the OPE expansion of the correlation function in Eq. (1) and, therefore, we get a much richer sum rule.

Eq. (1) can be rewritten as [4,5]:

$$\Pi_{\mu\nu} = -3i \int \frac{d^4k}{(2\pi)^4}\ Tr\Big[S(p_3-k)\gamma_\mu S(p_3-p_2-k)\gamma_5 \times \left[i\gamma_5 A + \gamma_\alpha\gamma_5 B^\alpha + \gamma_5\sigma_{\alpha\beta}C^{\alpha\beta}\right]\gamma_\nu\Big]\ , \tag{2}$$

where $S(p)$ is the free c-quark propagator, and the three invariant functions of k, p are defined by

$$A(k,p) = \frac{1}{12}\int d^4x\ e^{ik.x}\langle 0|\bar{d}(x)i\gamma_5 u(0)|\pi(p)\rangle\ ,$$

$$B^\alpha(k,p) = \frac{1}{12}\int d^4x\ e^{ik.x}\langle 0|\bar{d}(x)\gamma^\alpha\gamma_5 u(0)|\pi(p)\rangle\ ,$$

$$C^{\alpha\beta}(k,p) = -\frac{1}{24}\int d^4x\ e^{ik.x}\langle 0|\bar{d}(x)\sigma^{\alpha\beta}\gamma_5 u(0)|\pi(p)\rangle\ . \tag{3}$$

Additional contributions to the OPE come from the diagrams where one gluon, emitted from the c-quark propagator, is combined with the quark-antiquark component. Taking the gluon stress tensor into the quark-antiquark component, one can write down the correlation function into the form

$$\Pi_{\mu\nu} = -4i\int \frac{d^4k}{(2\pi)^4} Tr\Big[\Big(S_{\alpha\beta}(p_3-k)\gamma_\mu S(p_3-p_2-k) + S(p_3-k)\gamma_\mu S_{\alpha\beta}(p_3-p_2-k)\Big)\gamma_5 \times \left[\gamma_5\sigma_{\rho\lambda}E^{\rho\lambda\alpha\beta}(k,p) + \gamma^\tau\epsilon^{\alpha\beta\theta\delta}F_{\tau\theta\delta}(k,p)\right]\gamma_\nu\Big]\ , \tag{4}$$

where we have defined

$$S_{\alpha\beta}(k) = -\frac{[k_\alpha\gamma_\beta - k_\beta\gamma_\alpha + (\not{k} + m_c)i\sigma_{\alpha\beta}]}{2(k^2-m_c^2)^2} \tag{5}$$

and

$$E^{\rho\lambda\alpha\beta}(k,p) = -\frac{1}{32}\int d^4x\; e^{ik.x}\langle 0|\bar{d}(x)\gamma_5\sigma^{\rho\lambda}g_s G^{\alpha\beta}u|\pi(p)\rangle\,,$$
$$F_{\tau\theta\delta}(k,p) = \frac{1}{32}\int d^4x\; e^{ik.x}\langle 0|\bar{d}(x)\gamma_\tau g_s\tilde{G}_{\theta\delta}u|\pi(p)\rangle\,. \qquad (6)$$

The phenomenological side of the correlation function, $\Pi_{\mu\nu}$, is obtained by the consideration of J/ψ, D and D^* state contribution to the matrix element in Eq. (1):

$$\Pi_{\mu\nu}^{phen} = \frac{m_\psi f_\psi m_{D^*} f_{D^*} (m_D^2 f_D/m_c)\;\mathcal{M}_{\mu\nu}}{(p_2^2-m_\psi^2)(p_3^2-m_{D^*}^2)(p_4^2-m_D^2)} + \text{h. r.}\,, \qquad (7)$$

where h. r. means higher resonances and the hadronic amplitude for the process $J/\psi\;\pi \to \bar{D}\;D^*$ is given by [4]

$$\mathcal{M}_{\mu\nu} = \Lambda_1 p_{1\mu}p_{1\nu} + \Lambda_2 p_{1\mu}p_{2\nu} + \Lambda_3 p_{1\nu}p_{3\mu} + \Lambda_4 g_{\mu\nu} + \Lambda_5 p_{2\nu}p_{3\mu}\,, \qquad (8)$$

where Λ_i are the parameters we want to evaluate from the sum rules.

Comparing the results of Eqs. (2) and (4) with Eq. (7), we can write a sum rule for each one of the invariant structures appearing in Eq. (8). To improve the matching between the phenomenological and theoretical sides we follow the usual procedure and make a single Borel transformation to all the external momenta taken to be equal: $-p_2^2 = -p_3^2 = -p_4^2 = P^2 \to M^2$. The problem of doing a single Borel transformation is the fact that terms associated with the pole-continuum transitions are not suppressed [6]. In Ref. [6] it was explicitly shown that the pole-continuum transition has a different behavior as a function of the Borel mass as compared with the double pole contribution (triple pole contribution in our case) and continuum contribution. Therefore, we can single out the three-pole contribution from the others by introducing two parameters in the phenomenological side of the sum rule, which will account for the double pole-continuum and single pole-continuum contributions. The expressions for all sum rules are given in Ref. [5].

Table 1. The best fitted values for the parameters associated with the process $J/\psi\;\pi \to \bar{D}\;D^*$.

Λ_1 (GeV^{-2})	Λ_2 (GeV^{-2})	Λ_3 (GeV^{-2})	Λ_4	Λ_5 (GeV^{-2})
14 ± 2	-7.2 ± 0.9	-58 ± 8	14.6 ± 2.2	-15.6 ± 2.2

3. Results and Discussion

The parameter values used in all calculations are $m_c = 1.37$ GeV, $m_\pi = 140$ MeV, $m_D = 1.87$ GeV, $m_{D^*} = 2.01$ GeV, $m_\psi = 3.097$ GeV, $f_\pi = 131.5$ MeV, $f_\psi = 270$ MeV, $f_D = 170$ MeV, $f_{D^*} = 240$ MeV, $\langle\bar{q}q\rangle = -(0.23)^3$ GeV^3, $m_0^2 = 0.8$ GeV^2, $\delta^2 = 0.2$ GeV^2, $f_{3\pi} = 0.0035$ GeV^2. In general, for all sum rules we get that twist-3 and 4 contributions are small as compared with the twist-2 contribution, following the same behavior as the sum rule for the process $J/\psi\ \pi \to \bar{D}\ D$ given in [4]. It is also interesting to notice that if we consider only the leading twist contributions we recover the sum rules obtained in Ref. [3]. The value of the parameters Λ_i are obtained by the extrapolation of the QCD sum rules results to $M^2 = 0$ [4,5] and are given in Table I.

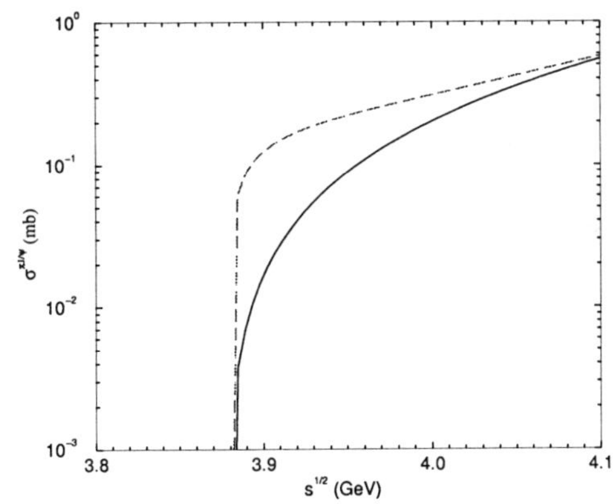

Figure 1. $J/\psi\ \pi \to \bar{D}\ D^* + D\ \bar{D}^*$ cross section. The solid and dashed lines give the results which repect and break chiral symmetry respectively.

Having the QCD sum rule results for the amplitude given in Eq. (8) we can evaluate the cross section. The structures multiplying Λ_4 and Λ_5 in Eq. (8) break chiral symmetry [7]. To evaluate the effect of breaking chiral symmetry in the process $J/\psi\ \pi \to \bar{D}\ D^* + D\ \bar{D}^*$ we show, in Fig. 1, the cross section calculated using all the structures in Eq. (8) (dashed line) and neglecting Λ_4 and Λ_5 (solid line). We see that the cross section obtained with the amplitude that breaks chiral symmetry grows very fast near the threshold. Since this is the energy region where this kind of process is probable more likely to happen, it is very important to use models that respect chiral symmetry when evaluating the $J/\psi - \pi$ cross section.

Following the same procedure, we have also considered the processes $J/\psi\,\pi \to \bar{D}\,D$ and $J/\psi\,\pi \to \bar{D}^*\,D^*$. In Fig. 2 we show, for comparison, our result for the total cross section for the $J/\psi\,\pi$ dissociation (solid lines) and the results from meson exchange model [8] obtained with a cut-off $\Lambda = 1$ GeV (dot-dashed line), quark exchange model [11] (dashed line) and short distance QCD [9,10] (dotted line). The shaded area in our results gives an evaluation of the uncertainties in our calculation. It is very interesting to notice that bellow the DD^* threshold, our result and the results from meson exchange and quark exchange models are in a very good agreement. However, as soon as the DD^* channel is open the cross section obtained with the meson exchange and quark exchange models show a very fast growth, as a function of $\sqrt{s}$, as compared with our result. This is due to the fact that chiral symmetry is broken in these two model calculations.

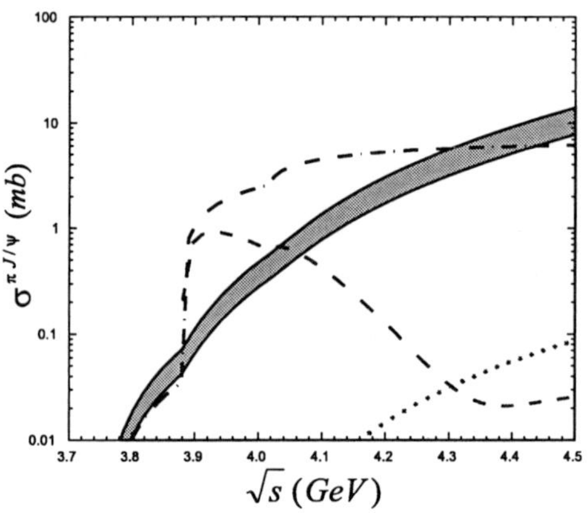

Figure 2. $J/\psi - \pi$ dissociation cross sections from meson exchange model [8] (dot-dashed line), quark exchange model [11] (dashed line), short distance QCD [9,10] (dotted line) and QCD sum rules (solid lines). The shaded area give an evaluation of the uncertainties in our calculation.

Our curves indicate that the cross section grows monotonically with the c.m.s. energy but not as fast, near the thresholds, as it does in the calculations in refs. [8,7]. Again, this behavior is in opposition to [11], where a peak just after the threshold followed by continuous decrease in the cross section was found.

At higher energies, due to our low energy approximation, our approach gradually looses validity. In the fiducial region, close to threshold, $4.1 \leq \sqrt{s} \leq 4.3$ GeV, we find $1.0 \leq \sigma \leq 4.0$ mb and these values are much smaller than those obtained with the effective Lagrangians without form factors in the hadronic vertices, but agree in order of magnitude with the quark model calculations of [11].

In conclusion, we have studied the J/ψ dissociation cross section by pions using the QCDSR technique, based on a three-point function using vacuum-pion correlation functions. We have estimated the hadronic amplitudes by working up to twist-4 in the limit $p_1 << p_2, p_3, p_4$. Our results are in agreement with the former QCDSR calculation, done with a four-point function at the pion pole [3]. Our results for the cross section as a function of $\sqrt{s}$ are smaller than the results using meson-exchange models (without form factors), but larger than the calculation based on quark-exchange models.

The dominant contribution to the hadronic amplitudes comes from the twist-2 operator, or equivalently, from the quark condensate. As we know that the quark condensate is stronger in the vacuum and weaker in the interior of hadrons, we can conclude that the charmonium "sees" and interacts with the surface of the pions, where there is a "halo" of condensates. In our approach the continuous growth of the cross section comes from the growth of the phase space, as in the effective Lagrangian calculations. In the short distance QCD calculation [9,12] the cross section also grows with $\sqrt{s}$, but the growth there is considerably smaller because the rise in the gluon density cannot completely compensate the fall of the partonic cross section [12].

Acknowledgments

This work was supported by CNPq and FAPESP.

References

1. T. Matsui and H. Satz, *Phys. Lett.* **B178**, 416 (1986).
2. M.A. Shifman, A.I. Vainshtein, V.I. Zakharov, *Nucl. Phys.* **B120**, 316 (1977).
3. F.S. Navarra et al., *Phys. Lett.* **B529**, 87 (2002).
4. F.O. Durães, S.H. Lee, F.S. Navarra, M. Nielsen, *Phys. Lett.* **B564**, 97 (2003).
5. F.O. Durães *et al.*, nucl-th/0211092.
6. B.L. Ioffe and A.V. Smilga, *Phys. Lett.* **B232**, 109 (1984).
7. F.S. Navarra, M. Nielsen, M.R. Robilotta, *Phys. Rev.* **C64**, 021901R (2001).
8. Y. Oh, T. Song and S.H. Lee, *Phys. Rev.* **C63**, 034901 (2001).
9. D. Kharzeev and H. Satz, *Phys. Lett.* **B334**, 155 (1994).
10. Y. Oh, S. Kim, S.H. Lee, *Phys. Rev.* **C65**, 067901 (2002).
11. C.-Y. Wong, E.S. Swanson and T. Barnes, *Phys. Rev.* **C62**, 045201 (2000); *Phys. Rev.* **C65**, 014903 (2001).
12. F. Arleo, P.-B. Gossiaux, T. Gousset and J. Aichelin, *Phys. Rev.* **D65**, 014005 (2002).

EFFECTIVE STRING PICTURE FOR CONFINING GAUGE THEORIES AT FINITE TEMPERATURE

M. CASELLE AND M. PANERO *

Università di Torino and INFN, Sezione di Torino,
Via P. Giuria, 1
I–10125 Torino, Italy
E-mail: caselle@to.infn.it, panero@to.infn.it

M. HASENBUSCH

NIC/DESY Zeuthen
Platanenallee 6,
D–15738 Zeuthen, Germany
E-mail: Martin.Hasenbusch@desy.de

We study the "effective string picture" of confinement, deriving theoretical predictions for the interquark potential at finite temperature. At short distances, the leading string correction to the linear confining potential between a heavy quark-antiquark pair is the "Lüscher term". Assuming a Nambu–Goto effective string action, subleading contributions can be worked out in an analytical way. We also discuss the contribution given by a possible "boundary term" in the effective action, and compare these predictions with high precision results from simulations of lattice $\mathbf{Z}_2$ gauge theory in three dimensions, obtained with an algorithm that exploits the duality of the $\mathbf{Z}_2$ gauge model with the Ising spin model.

1. Effective String Picture of Confinement

The string picture is an effective framework which is expected to provide a good physical description of the infrared behavior of confining gauge theories. The basic idea is simple: two confined color charges behave as if they were joined by a thin flux tube, which can fluctuate like a vibrating string. The dynamics of the world sheet spanned by the string during its time-like evolution is described by an effective action S_{eff}, and for *massless* string fluctuations, the simplest choice for a candidate effective action is the Nambu–Goto string action: $S_{\text{eff}} = \sigma \cdot \mathcal{A}$, which is proportional to the

*Presenter of the talk.

area $\mathcal{A}$ of the world–sheet surface; σ is the *string tension*, appearing as the parameter of the effective theory.

For a three-dimensional system with extension $L_s^2 \times L$ (with $L_s \gg L$) and periodic boundary conditions along the short direction, the temperature T is proportional to $1/L$, and in that case the result for the expectation value of the Polyakov loop correlation function reads:

$$\langle P^\dagger(R)P(0)\rangle = \frac{e^{-\sigma RL+k}}{\eta\left(i\frac{L}{2R}\right)} \tag{1}$$

where η is Dedekind's function. The term associated with the minimal world sheet surface induces the exponential area-law falloff responsible for the linear rise in the interquark potential $V(R)$, while the first non-trivial contribution in S_{eff} results in the determinant of Laplace operator, and the corresponding contribution to the interquark potential $V(R)$ — in a regime of distances shorter than $\frac{L}{2}$ — is the Lüscher term[1]:

$$V(R) = -\frac{1}{L}\ln\langle P^\dagger(R)P(0)\rangle \simeq \sigma R - \frac{\pi}{24R} \tag{2}$$

Inclusion of further terms[2] in the expansion of the world sheet area results[a] in a contribution involving a combination of Eisenstein functions[3]:

$$-\frac{\pi^2}{1152\sigma R^3}\left[2E_4\left(i\frac{L}{2R}\right) - E_2^2\left(i\frac{L}{2R}\right)\right] \tag{3}$$

However, such a contribution is still under debate.

On the other hand, it is also possible to include a "boundary term[b]" in the effective action: a perturbative expansion in b (a parameter proportional to the coefficient of the "boundary term" in the effective action), induces a leading order correction like:

$$R \longrightarrow \frac{R}{\sqrt{1+\frac{2b}{R}}} \tag{4}$$

with a short distance contribution to $V(R)$ reading: $-\frac{b\pi}{24R^2}$.

[a]This calculation involves a Riemann's ζ function regularization.

[b]Such a boundary term is related to derivatives of the h field (which describes transverse displacements with respect to the minimal area surface of the world sheet), evaluated along the Polyakov lines

2. The Model: $\mathbf{Z}_2$ Lattice Gauge Theory

We run numerical simulations of the $\mathbf{Z}_2$ lattice gauge theory in three spacetime dimensions. This choice has various motivations: the effective string picture is believed to be independent of the underlying gauge group; $\mathbf{Z}_2$ gauge group is interesting from the perspective of the *center* role in confinement[c]; the reduced configuration space of this theory and its duality with respect to the Ising spin model enable one to get high precision results within a reasonable amount of CPU time.

The pure 3D lattice gauge model is described in terms of $\sigma_{x,\mu}$ variables (taking values in $\mathbf{Z}_2$) defined on the lattice bonds; the dynamics is governed by the standard Wilson action, which enjoys $\mathbf{Z}_2$ gauge invariance[d]. The partition function reads:

$$Z(\beta) = \sum_{\text{c.}} e^{-\beta S} = \sum_{\text{c.}} \exp\left[+\beta \sum_{\square} \sigma_{\square}\right] \tag{5}$$

and the system may exist in different phases: a confined, strong coupling phase, with massive string fluctuations for $\beta < 0.47542(1)$ [4]; a confined, *rough* phase, with massless string fluctuations (this is the regime we studied in our simulations); a deconfined phase for $\beta > 0.7614134(2)$ [5].

This model is *dual* with respect to the $\mathbf{Z}_2$ spin model in 3D, and we exploited this property to express a ratio between Polyakov loop correlators $G(R) = \langle P^{\dagger}(R)P(0)\rangle$ of the gauge theory as a product of expectation values of one-link variables in the modified spin ensembles. A similar algorithm is the so-called *snake algorithm*[6].

We used multi-level updating and a hierarchical organization of sublattices, and the CPU time turns out to be roughly proportional to the inverse temperature L, and *virtually independent of the distance R between the quark sources*, thus the algorithm is particularly useful in a regime of very large interquark distances.

3. Numerical Results

Let $F(R, L)$ be the free energy associated with the presence of a heavy quark-antiquark pair at finite temperature: $G(R) = e^{-F(R,L)}$. We studied "quantum contributions" in free energy differences, by measuring the

---

[c] $\mathbf{Z}_2$ is the center of continuous gauge groups like $SU(2)$ or $Sp(N)$.

[d] $\mathbf{Z}_2$ gauge transformations act as *local* flips of $\sigma_{x,\mu}$ variables living on the lattice bonds which meet at a given site.

following quantity:

$$Q(R,L) = F(R+1,L) - F(R,L) - \sigma L \tag{6}$$

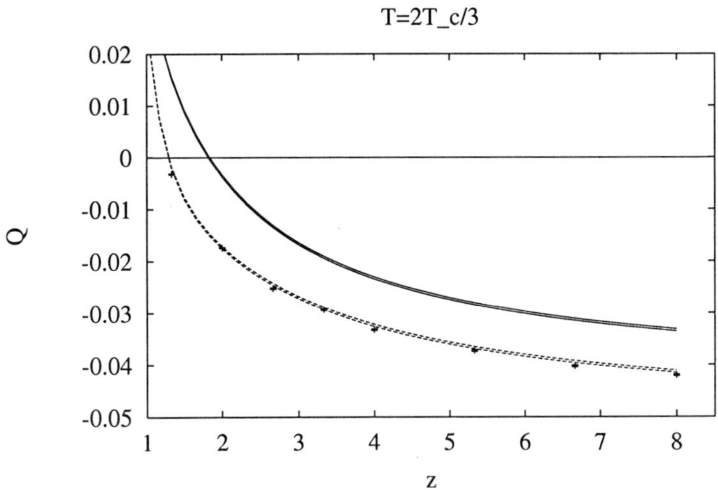

Figure 1. $Q(R,L)$ for $L = 12$ (i.e. $T = 2T_c/3$) at $\beta = 0.75180$. z is defined as: $z \equiv \frac{2R}{L}$. Solid curves correspond to the free bosonic string prediction, while dashed lines correspond to the LO Nambu–Goto correction. Pure area–law corresponds to $Q = 0$.

Fig. 1 shows that at "high temperatures" ($L < 2R$) our numerical results are in good agreement with the NLO prediction from Nambu–Goto string, while a pure area law is definitely ruled out, and the LO term alone is not sufficient to describe the data. We also found that the coefficient of a possible "boundary term" for this model seems to be very small, likely zero.

As it concerns the ($L > 2R$) regime, Fig. 2 shows the deviation[2] of a quantity proportional to $[Q(R,L) - Q(R-1,L)]$ from the free string prediction. The chosen normalization allows a meaningful comparison among different LGT's in 3d: $SU(2)$ gauge theory[7] (crosses), $SU(3)$ gauge model[8] (white squares), and $\mathbf{Z}_2$ gauge theory[2] (black squares). The three models display the same qualitative behavior, and, in particular, the data for $\mathbf{Z}_2$ and $SU(2)$ (which are groups with the same center — namely: $\mathbf{Z}_2$ itself) are compatible within errorbars. This may be a signature of the relevance of center degrees of freedom to the confinement mechanism.

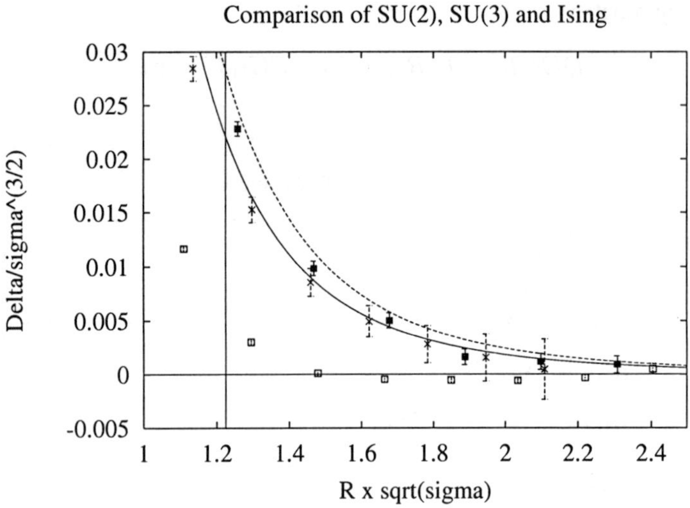

Figure 2. Behavior of different gauge models in the short distance regime.

4. Conclusions

We studied confining gauge theories at finite temperature, and tested the theoretical predictions of the Nambu–Goto effective string for $\mathbf{Z}_2$ lattice gauge theory, both at large and short interquark distances. Our algorithm exploits the duality of the model, and this enabled us to explore a wide range of distances, detecting next-to-leading order effects. Our data seem to rule out a "boundary term" in the effective string action describing the present gauge model. Finally, we also made a comparison with some different gauge models.

References

1. M. Lüscher, K. Symanzik and P. Weisz, *Nucl.Phys.* **B173** (1980) 365.
2. M. Caselle, M. Hasenbusch and M. Panero, *JHEP* **0301** (2003) 057. M. Caselle, M. Panero and P. Provero, *JHEP* **0206**, 061 (2002) and references therein. M. Caselle, M. Hasenbusch and M. Panero, in preparation.
3. J.P. Serre, "A course in Arithmetic", Springer–Verlag, New York, 1980.
4. M. Hasenbusch and K. Pinn, *J.Phys.* **A30** (1997) 63.
5. H.W.J. Blöte, L.N. Shchur, A.L. Talapov, *Int.J.Mod.Phys.* **C10** (1999) 1137.
6. Ph. de Forcrand, M. D'Elia and M. Pepe, *Phys.Rev.Lett.* **86** (2001) 1438. Ph. de Forcrand, M. D'Elia and M. Pepe, *Nucl.Phys.Proc.Suppl.* **94** (2001) 494.
7. M. Caselle, M. Pepe and A. Rago, in preparation.
8. M. Lüscher and P. Weisz, *JHEP* **0207** (2002) 049.

VACUUM CONDENSATES AND DYNAMICAL MASS GENERATION IN EUCLIDEAN YANG-MILLS THEORIES*

D. DUDAL‡ AND H. VERSCHELDE

Department of Mathematical Physics and Astronomy, Krijgslaan 281-S9, B-9000 Gent, Belgium
E-mail: david.dudal@ugent.be, henri.verschelde@ugent.be

V.E.R. LEMES, M.S. SARANDY, R.F. SOBREIRO AND S.P. SORELLA

UERJ - Universidade do Estado do Rio de Janeiro, Rua São Francisco Xavier 524, 20550-013 Maracanã, Rio de Janeiro, Brazil
E-mail: vitor@dft.if.uerj.br, sarandy@dft.if.uerj.br, sobreiro@dft.if.uerj.br, sorella@uerj.br

M. PICARIELLO AND A. VICINI

Universitá degli Studi di Milano, via Celoria 16, I-20133, Milano, Italy
E-mail: marco.picariello@mi.infn.it, alessandro.vicini@mi.infn.it

J. A. GRACEY

Theoretical Physics Division
Department of Mathematical Sciences
University of Liverpool
P.O. Box 147, Liverpool, L69 3BX, United Kingdom
E-mail: jag@amtp.liv.ac.uk

Vacuum condensates of dimension two and their relevance for the dynamical mass generation for gluons in Yang-Mills theories are discussed.

*Talk given by S.P. Sorella at the International Conference on Color Confinement and Hadrons in Quantum Chromodynamics, Confinement 2003, TITech & RIKEN, Tokyo, Japan, July 21-24, 2003.

†The Conselho Nacional de Desenvolvimento Científico e Tecnológico (Cnpq-Brazil), the Sr2-UERJ, the Coordenação de Aperfeiçoamento de Pessoal de Nível Superior (CAPES) and the MIUR-Italy are gratefully acknowledged for financial support.

‡Research Assistant of the Fund for scientific Research-Flanders, Belgium.

1. Introduction

Vacuum condensates of dimension two in Yang-Mills theories have witnessed increasing interest in recent years, both from the theoretical point of view as well as from lattice simulations, which have provided rather strong evidence for an effective dynamically generated gluon mass. Here we shall report on our current work on these condensates. Our aim is that of defining a renormalizable effective potential and evaluating it in analytic form. The condensates arise thus as nontrivial solutions of the gap equation corresponding to the minimization of the effective potential. The content of this work is as follows. In Sect.2 we review the gluon condensate $\langle A^2 \rangle$ in the Landau gauge. In Sect.3 the mixed gluon-ghost condensate $\langle \frac{1}{2} A^\alpha_\mu A^\alpha_\mu + \xi \bar{c}^\alpha c^\alpha \rangle$ is analysed in the Maximal Abelian and Curci-Ferrari gauges. Sect.4 contains a few remarks on the gauge (in)dependence of the obtained effective gluon mass and on possible relationships with lattice simulations. We conclude with a short discussion on the operator A^2 in linear covariant α-gauges.

2. The condensate $\langle A^2 \rangle$ in the Landau gauge

The gluon condensate $\langle A^2 \rangle = \langle A^a_\mu A^a_\mu \rangle$ in the Landau gauge has been introduced[1,2] in order to account for the discrepancy between the expected perturbative behavior and the lattice results concerning the two and three point functions in pure Yang-Mills theories. The lattice estimate for the condensate is[1] $\langle A^2 \rangle \approx (1.64 \text{ GeV})^2$. A simple argument shows that $\int d^4x A^2$ is invariant under infinitesimal gauge transformations, $\delta A^a_\mu = -(D_\mu \omega)^a$, in the Landau gauge, $\partial A = 0$, namely $\delta \int d^4x A^2 = 0$. In the BRST framework, it turns out that $\int d^4x A^2$ is BRST invariant on shell, $s \int d^4x A^2 = 0 +$Eqs. motion. This property ensures that the local operator A^2 is multiplicatively renormalizable to all orders of perturbation theory[3]. Its anomalous dimension γ_{A^2} can be expressed[3] as a combination of the gauge beta function β and of the anomalous dimension γ_A of the gauge field A^a_μ, *i.e.*, $\gamma_{A^2} = -\left(\frac{\beta(a)}{a} + \gamma_A(a) \right)$ where $a = \frac{g^2}{16\pi^2}$. Concerning the analytic evaluation of $\langle A^2 \rangle$, the two-loop effective potential for $\langle A^2 \rangle$ has been obtained[4] in pure Yang-Mills theory by combining the Local Composite Operators technique with the Renormalization Group Equations. This has led to a gap equation whose weak coupling solution gives a nonvanishing condensate, resulting in an effective gluon mass $m_{\text{gluon}} \approx 500\text{MeV}$. Recently, the inclusion of massless quarks has been worked out[5]. It is worth mentioning that an effective gluon mass has been reported in lattice simulations in the Landau gauge[6], yielding $m_{\text{gluon}} \approx 600\text{MeV}$.

3. Other Gauges

3.1. *The Maximal Abelian Gauge*

The so called Maximal Abelian gauge (MAG) plays an important role for the dual superconductivity picture for confinement. In the MAG, the gauge field is decomposed according to the generators of the Cartan subgroup of the gauge group. For $SU(2)$, we have $A^a_\mu T^a = A_\mu T^3 + A^\alpha_\mu T^\alpha$ with $\alpha = 1, 2$. The gauge fixing term in the MAG is given by

$$\int d^4x \left[\frac{1}{2\xi} F^\alpha F^\alpha - \overline{c}^\alpha M^{\alpha\beta} c^\beta - g^2 \xi \left(\overline{c}^\alpha \varepsilon^{\alpha\beta} c^\beta\right)^2\right] \tag{1}$$

where ξ is the gauge parameter and $F^\alpha = D^{\alpha\beta}_\mu A^\beta_\mu = \left(\partial_\mu A^\alpha_\mu + g\varepsilon^{\alpha\beta} A_\mu A^\beta_\mu\right)$, with $M^{\alpha\beta} = (D^{\alpha\gamma}_\mu D^{\gamma\beta}_\mu + g^2 \varepsilon^{\alpha\gamma}\varepsilon^{\beta\sigma} A^\gamma_\mu A^\sigma_\mu)$. The MAG allows for a residual local $U(1)$ invariance, which has to be fixed later on. Lattice simulations have shown that the off-diagonal components A^β_μ acquire a mass[7,8], reporting $m_{\text{off-diag}} \approx 1.2\text{GeV}$. The operator A^2 can be generalized to the MAG[9]. Indeed, the gluon-ghost dimension two operator $O_{\text{MAG}} = \left(\frac{1}{2} A^\alpha_\mu A^\alpha_\mu + \xi \overline{c}^\alpha c^\alpha\right)$ turns out to be BRST invariant on-shell, namely, $s \int d^4x \; O_{\text{MAG}} = 0 +$ Eqs. motion. The condensate $\langle O_{\text{MAG}} \rangle$ could thus provide effective masses for the off-diagonal components. However, at present, very little is known about the condensate $\langle O_{\text{MAG}} \rangle$. Concerning the UV properties of O_{MAG}, it has been proven[10,11] to be multiplicatively renormalizable. Its anomalous dimension can be expressed as[10,11] $\gamma_{O_{\text{MAG}}} = -2\left(\frac{\beta(a)}{a} + \gamma_{c_{\text{diag}}}\right)$, where $\gamma_{c_{\text{diag}}}$ is the anomalous dimension of the diagonal ghost field.

3.2. *The Curci-Ferrari gauge*

The Curci-Ferrari gauge shares similarities with the MAG, providing useful information on the gluon-ghost condensate. The gauge fixing term is

$$\int d^4x \left(\frac{(\partial A^a)^2}{2\xi} + \overline{c}^a \partial_\mu D_\mu c^a + \frac{\xi g}{2} f^{abc} \partial A^a \overline{c}^b c^c - \frac{\xi g^2}{16} f^{abc} \overline{c}^a \overline{c}^b f^{mnc} c^m c^n\right)$$

where $a, b, c = 1, ..., (N^2 - 1)$ for $SU(N)$. For the gluon-ghost operator[9] in the CF gauge we have $O_{\text{CF}} = \left(\frac{1}{2} A^a_\mu A^a_\mu + \xi \overline{c}^a c^a\right)$. The operator O_{CF} is renormalizable to all orders, its anomalous dimension being given by[11] $\gamma_{O_{\text{CF}}} = -2\left(\gamma_c + \gamma_{gc^2}\right)$, where γ_{gc^2} is the anomalous dimension of the composite operator corresponding to the BRST variation of the ghost c, *i.e.*, $sc = \frac{g}{2} f^{abc} c^b c^c$. Recently, the effective potential for $O_{\rm CF}$ has been obtained[12], leading to a nonvanishing condensate $\langle O_{\rm CF} \rangle$ and to a dynamical gluon mass. Something similar is expected to happen in the MAG.

4. Open problems and future perspectives

Many aspects concerning the gauge condensates of dimensions two are still open, deserving a better understanding. Two important questions which arise almost immediately are:

• 1) are these condensates related to the effective gluon masses observed in lattice simulations?

• 2) to what extent are these condensates and the related gluon masses gauge (in)dependent?

Concerning the first question, it could be useful to observe that, to the first approximation, the condensates $\langle A^2 \rangle$, $\langle A^2/2 + \xi \overline{c} c \rangle$ modify the euclidean gluon propagator according to $\frac{1}{k^2} \to \frac{1}{k^2 + m_{\rm eff}^2}$. This suggests that the lattice data could be fitted with a formula which contains a Yukawa type term. This seems to be the case for the off-diagonal masses in the MAG[7,8]. Also, it is worth remarking that the fitting formula employed in the lattice simulations for the gluon propagator in the Landau gauge[6]

$$\sum_{a,\mu} \langle A^a_\mu(k) A^a_\mu(-k) \rangle = \frac{N}{k^2 + m_A^2} \left[\frac{1}{k^4 + m_2^4} + \frac{s}{\left(\log(m_L^2 + k^2)\right)^{13/22}} \right] \qquad (2)$$

contains in fact a Yukawa term. Expression (2) fits well the lattice data with $m_A = 0.64 \mathrm{GeV}$, $m_2 = 1.31 \mathrm{GeV}$, $m_L = 1.23 \mathrm{GeV}$, $s = 0.32$, $N = 8.1$. Concerning the second question, we look at other gauges allowing for a local renormalizable operator which could condense, providing thus an effective gluon mass. That is why we are studying in detail the local operator A^2 in linear covariant α-gauges, whose gauge fixing term is

$$\int d^4x \left(\frac{1}{2\alpha} (\partial A^a)^2 - \partial_\mu \overline{c}^a D_\mu c^a \right) \qquad (3)$$

Notice that in this gauge the operator A^2 is not BRST invariant on-shell. Nevertheless, we have strong indications of its multiplicative renormalizability to all orders of perturbation theory[13]. There is a simple understanding of this property. In linear α-gauges, due to the shift symmetry of the antighost, *i.e.* $\overline{c} \to \overline{c} + \text{const.}$, the operator A^2 cannot mix with the other dimension two ghost composite operator $\overline{c}c$, which cannot show up due to the above symmetry. The renormalizability of A^2 is the first step towards the construction of a renormalizable effective potential in order to study the possible condensation of A^2 and the related dynamical mass generation[13]. Although this doesn't yet answer the question of the gauge (in)dependence, it might increase the number of gauges displaying such an interesting phenomenon. Finally, it is worth underlining that an effective gluon mass has been reported in the Coulomb gauge[14] as well as in the lattice Laplacian gauge[15].

Acknowledgments

We are grateful to the Organizing Committee of the Conference and to M.N. Chernodub, H. Suganuma, V.I. Zakharov and D. Zwanziger for valuable and helpful discussions. We are indebted to J. Greensite for pointing out ref.[14].

References

1. P. Boucaud, A. Le Yaouanc, J. P. Leroy, J. Micheli, O. Pène and J. Rodríguez-Quintero, *Phys. Rev.* **D63**, 114003 (2001);
   P. Boucaud, J. P. Leroy, A. Le Yaouanc, J. Micheli, O. Pène, F. De Soto, A. Donini, H. Moutarde, J. Rodríguez-Quintero, *Phys. Rev.* **D66**, 034504 (2002).
2. F.V. Gubarev and V.I. Zakharov, *Phys. Lett.* **B501**, 28 (2001);
   F.V. Gubarev, L. Stodolsky, V.I. Zakharov, *Phys. Rev. Lett.* **86**, 2220 (2001).
3. J.A. Gracey, *Phys. Lett.* **B552**, 101 (2003);
   D. Dudal, H. Verschelde and S.P. Sorella, *Phys. Lett.* **B555**, 126 (2003).
4. H. Verschelde, K. Knecht, K. Van Acoleyen and M. Vanderkelen, *Phys. Lett.* **B516**, 307 (2001).
5. R. E. Browne and J. A. Gracey, arXiv:hep-th/0306200.
6. K. Langfeld, H. Reinhardt and J. Gattnar, *Nucl. Phys.* **B621**, 131 (2002).
7. K. Amemiya and H. Suganuma, *Phys. Rev.* **D60**, 114509 (1999).
8. V.G. Bornyakov M.N. Chernodub, F.V. Gubarev, S.M. Morozov and M.I. Polikarpov, *Phys. Lett.* **B559**, 214 (2003).
9. K.-I. Kondo, *Phys. Lett.* **B514**, 335 (2001),
   K.-I. Kondo, T. Murakami, T. Shinohara and T. Imai, *Phys. Rev.* **D65**, 085034 (2002).
10. U. Ellwanger and N. Wschebor, *Int. J. Mod. Phys.* **A18**, 1595 (2003).
11. D. Dudal, H. Verschelde, V.E.R. Lemes, M.S. Sarandy, R. Sobreiro, S.P. Sorella, M. Picariello and J.A. Gracey, *Phys. Lett.* **B569**, 57 (2003).
12. D. Dudal, H. Verschelde, V. Lemes, M. Sarandy, S.P. Sorella and M. Picariello, hep-th/0302168, *Ann. Phys. (NY)*, in press.
13. Dudal et al., hep-th/0308181.
14. J. Greensite and H. Halpern, *Nucl. Phys.* **B271**, 379 (1986).
15. C. Alexandrou, Ph. de Forcrand and E. Follana, *Phys. Rev.* **D65**, 114508 (2002); *Phys. Rev.* **D65**, 117502 (2002).

Y-TYPE FLUX-TUBE FORMATION IN BARYONS

TORU T. TAKAHASHI

Yukawa Institute for Theoretical Physics, Kyoto University, Sakyo, Kyoto 606-8502, Japan
E-mail: ttoru@yukawa.kyoto-u.ac.jp

HIDEO SUGANUMA AND HIROKO ICHIE

Faculty of Science, Tokyo Institute of Technology, Ohokayama 2-12-1, Tokyo 152-8551, Japan

For more than 300 different patterns of the 3Q systems, the ground-state 3Q potential $V_{3Q}^{\rm g.s.}$ is investigated using SU(3) lattice QCD with $12^3 \times 24$ at $\beta = 5.7$ and $16^3 \times 32$ at $\beta = 5.8, 6.0$ at the quenched level. As a result of the detailed analyses, we find that the ground-state potential $V_{3Q}^{\rm g.s.}$ is well described with so-called Y-ansatz as $V_{3Q} = -A_{3Q} \sum_{i<j} \frac{1}{|\mathbf{r}_i - \mathbf{r}_j|} + \sigma_{3Q} L_{\rm min} + C_{3Q}$, with the accuracy better than 1%. Here, $L_{\rm min}$ denotes the minimal value of total flux-tube length. We also study the excited-state potential $V_{3Q}^{\rm e.s.}$ using lattice QCD with $16^3 \times 32$ at $\beta = 5.8, 6.0$ for more than 100 patterns of the 3Q systems. The energy gap between $V_{3Q}^{\rm g.s.}$ and $V_{3Q}^{\rm e.s.}$, which physically means the gluonic excitation energy, is found to be about 1 GeV in the typical hadronic scale. Finally, we suggest a possible scenario which connects the success of the quark model to QCD.

1. Introduction

It is widely accepted that Quantum Chromodynamics (QCD) is the underlying theory of the strong interaction for hadrons and nuclei. In a spirit of the particle physics, it is strongly desired to describe the hadron dynamics directly from QCD. However, due to the strong coupling nature of QCD in the infrared region, it still remains a big challenge to derive physical quantities directly from QCD in an analytic manner.

For the nonperturbative analysis of QCD, lattice QCD Monte Carlo calculations have been established as a reliable method directly based on QCD. Lattice QCD calculations enable us to perform a nonperturbative quantitative analysis in a model-independent way. For instance, lattice QCD calculations have been very successful in reproducing hadron mass spectra.[1] Nevertheless, these results hardly inform us the internal structure of hadrons. It would be also indispensable to comprehend hadrons directly from quarks and gluons, the degrees of freedom of QCD.

For this aim, we cast light on the inter-quark potentials, which are directly responsible for the hadron properties and their internal structures at the quark-gluon level.

2. Ground-state three quark potential

For the quark-antiquark ($Q\bar{Q}$) potential, which is one of the most fundamental quantities in QCD, many lattice QCD studies have been performed. As a result, $Q\bar{Q}$ potential is known to be reproduced by a simple sum as $V_{Q\bar{Q}} = -\frac{A_{Q\bar{Q}}}{r} + \sigma_{Q\bar{Q}} r + C_{Q\bar{Q}}$ with r the distance between quark and anti-quark. Here, $\sigma \simeq 0.89$ GeV/fm denotes the string tension in the flux-tube picture.

However, in contrast with a number of lattice QCD studies for the $Q\bar{Q}$ potential, almost no lattice studies were done before our study[2,3] for the three-quark (3Q) potential V_{3Q}, which is directly responsible for the baryon properties. We study the ground-state 3Q potential on more than 300 patterns of the spatial quark configurations, using SU(3) lattice QCD at the quenched level with $12^3 \times 24$ at $\beta = 5.7$ and with $16^3 \times 32$ at $\beta = 5.8, 6.0$.[2,3]

As a result, we find that the 3Q potential V_{3Q} can be reproduced by so-called Y-ansatz as [2,4]

$$V_{3Q} = -A_{3Q} \sum_{i<j} \frac{1}{|\mathbf{r}_i - \mathbf{r}_j|} + \sigma_{3Q} L_{\min} + C_{3Q}, \tag{1}$$

with the accuracy better than 1% for all the different patterns of quark configurations. Here, $L_{\min}$ denotes the minimal value of total flux-tube length,[5,6] which is schematically illustrated in Fig.1.

In the fit analysis with Y-ansatz, we find two remarkable features[2]; the universality of the string tension as $\sigma_{3Q} \simeq \sigma_{Q\bar{Q}}$ and the one-gluon-exchange consequence as $A_{3Q} \simeq \frac{1}{2} A_{Q\bar{Q}}$.

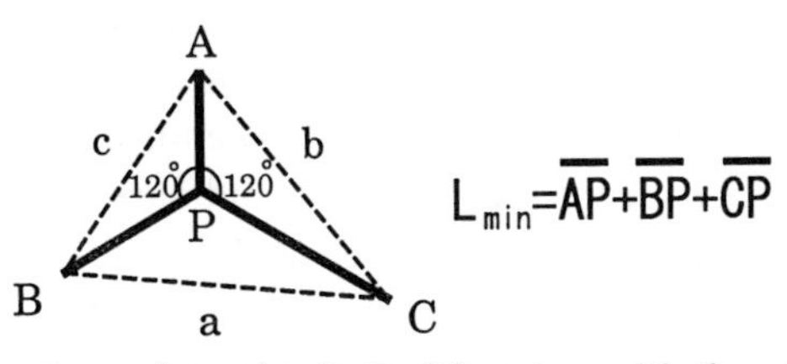

Figure 1. The flux-tube configuration in the 3Q system with the minimal length of the Y-type flux-tube, $L_{\min}$.

Furthermore, we consider also Y-ansatz with Yukawa-type two-body force as $V_{3Q}^{\rm Yukawa} = -A_{3Q}^{\rm Yukawa} \sum_{i<j} \frac{\exp(-m_{\rm B}|{\bf r}_i - {\bf r}_j|)}{|{\bf r}_i - {\bf r}_j|} + \sigma_{3Q}^{\rm Yukawa} L_{\rm min} + C_{3Q}^{\rm Yukawa}$, which is a conjecture from the dual superconductor scenario on the QCD vacuum. Here, $m_{\rm B}$ denotes the dual gluon mass in the dual superconductor picture. However, we find no evidence of the Yukawa-type two-body force[2] and again confirm the adequacy of the Y-ansatz form in Eq.(1), *i.e.*, $\sigma_{3Q}^{\rm Yukawa} \simeq \sigma_{3Q}$, $A_{3Q}^{\rm Yukawa} \simeq A_{3Q}$, $m_{\rm B} \simeq 0$.

Finally, as a direct evidence of the Y-type flux-tube formation, we show in Fig.2 the flux-tube profile in the spatially-fixed 3Q system obtained using maximally-abelian (MA) projected lattice QCD.[7] The action density in $\mathbf{R}^3$ is calculated under the presence of spatially-fixed three quarks in a color-singlet state. Here, the distance between the junction and each quark is about 0.5 fm. We thus observe Y-type flux-tube formation in lattice QCD.

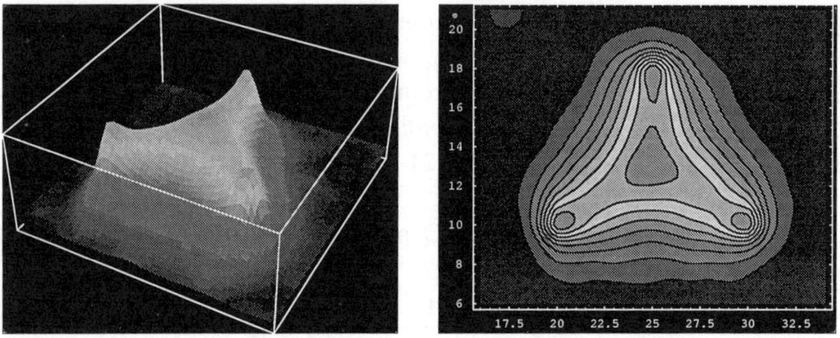

Figure 2. The flux-tube profile in the spatially-fixed 3Q system, in the MA projected QCD.[6] The distance between the junction and each quark is about 0.5 fm.

3. Excited-state three quark potential

3.1. *Lattice QCD study of gluonic excitations*

We also investigate the gluonic excitation in the spatially-fixed 3Q system using SU(3) lattice QCD. In the flux-tube picture,[5,6] gluonic excitation modes would be regarded as the vibrational modes of a flux-tube. In the valence picture, they correspond to the hybrid hadrons such as $q\bar{q}G$ or $qqqG$. In particular, hybrid hadrons are probable candidates of the exotic hadrons, which have exotic quantum numbers such as $J^{PC} = 0^{--}, 0^{+-}, 1^{-+}, 2^{+-}, ...$, which cannot be constructed within the simple quark model. It is then worth investigating gluonic excitations directly from QCD, from the experimental viewpoint as well as the theoretical viewpoint.[8]

In Fig.3 we show the first lattice QCD results for the excited-state 3Q potential obtained with $16^3 \times 32$ lattice at $\beta = 5.8, 6.0$ at the quenched level.[9] Here, the horizontal axis denotes the minimal length of the Y-type flux-tube, $L_{\min}$, as a label to distinguish the three-quark configuration. The vertical axis denotes the energy induced by three static quarks in a color-singlet state. The open symbols are for the ground-state potential $V_{3Q}^{\rm g.s.}$ discussed in the previous section, and the filled symbols are for the excited-state potential $V_{3Q}^{\rm e.s.}$ in the 3Q system. The energy difference $\Delta E \equiv V_{3Q}^{\rm e.s.} - V_{3Q}^{\rm g.s.}$ physically means the gluonic excitation energy.

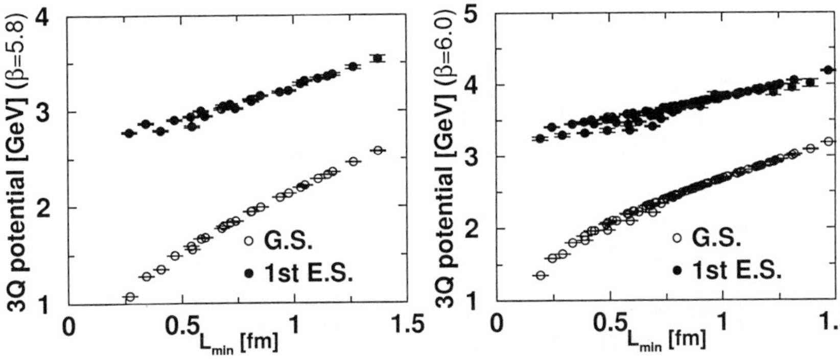

Figure 3. The lattice QCD results for the ground-state 3Q potential $V_{3Q}^{\rm g.s.}$ (open circles) and the 1st excited-state 3Q potential $V_{3Q}^{\rm e.s.}$ (filled circles) as the function of $L_{\min}$. These lattice results at $\beta = 5.8$ and $\beta = 6.0$ well coincide besides an irrelevant overall constant.

We find that the gluonic excitation energy ΔE is more than 1 GeV in the typical hadronic scale as $0.5\text{fm} \le L_{\min} \le 1.5\text{fm}$, which is large in comparison with the excitation energies of quark origin. (For the $Q\bar{Q}$ system, a large gluonic excitation energy is reported in recent lattice studies.[10]) Such a gluonic excitation would contribute significantly in the highly-excited baryons with the excitation energy above 1 GeV. In fact, the lowest hybrid baryon,[8] which is described as $qqqG$ in the valence picture, is expected to have a large mass of about 2 GeV. This large gluonic excitation energy also gives us the reason for the great success of the simple quark model.[9]

3.2. *Success of the quark model*

The simple quark model contains only constituent quarks as explicit degrees of freedom, and does not have gluonic modes. Nevertheless, the quark model has been very successful in reproducing the low-lying hadron properties and spectra especially for baryons, in spite of the non-relativistic treatment and the absence of gluonic excitation modes. It is true that the non-relativistic treatment for hadron mass spectra can be justified by spontaneous chiral symmetry breaking, which gives rise to a large constituent quark mass of about 300 MeV, but we have no reason based on QCD for the absence of the gluonic mode. Now, directly based on QCD, the gluonic excitation energy ΔE is found to be large, in comparison with quark excitation energy. We propose a possible scenario connecting the success of the quark model to QCD[3,9] in Fig.4.

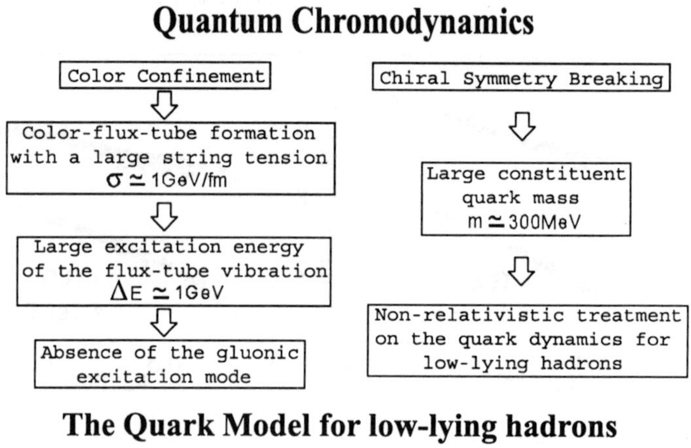

Figure 4. A possible scenario connecting the success of the simple quark model to QCD.

References

1. CP-PACS Collaboration (A. Ali Khan *et al.*), *Phys. Rev.* **D65**, 054505 (2002), *Erratum-ibid.* **D67**, 059901 (2003), and references therein.
2. T.T. Takahashi, H. Suganuma, Y. Nemoto and H. Matsufuru, *Phys. Rev.* **D65**, 114509 (2002); T.T. Takahashi, H. Matsufuru, Y. Nemoto and H. Suganuma, *Phys. Rev. Lett.* **86**, 18 (2001).
3. T.T. Takahashi, H. Suganuma, H. Ichie, H. Matsufuru and Y. Nemoto, *Nucl. Phys.* **A721**, 926 (2003); H. Suganuma, T.T. Takahashi and H. Ichie, *Nucl. Phys.* **A** in press.
4. N. Brambilla, G.M. Prosperi and A. Vairo, *Phys. Lett.* **B362**, 113 (1995).
5. J. Kogut and L. Susskind, *Phys. Rev.* **D11**, 395 (1975).
6. S. Capstick and N. Isgur, *Phys. Rev.* **D34** 2809, (1986).
7. H. Ichie, V. Bornyakov, T. Streuer and G. Schierholz, *Nucl. Phys.* **A721**, 899 (2003); *Nucl. Phys.* **B** (Proc.Suppl.) **119**, 751 (2003).
8. S. Capstick and P.R. Page, *Phys. Rev.* **C66**, 065204 (2002).
9. T.T. Takahashi and H. Suganuma, *Phys. Rev. Lett.* **90**, 182001 (2003).
10. K.J. Juge, J. Kuti and C.J. Morningstar, *Phys. Rev. Lett.* **90**, 161601 (2003).

MOTION OF QUARKS INSIDE HEAVY FLAVOR HADRONS AND THE QCD EQUATIONS OF MOTION

KAZUHIRO TANAKA

Department of Physics, Juntendo University,
Inba-gun, Chiba 270-1695, Japan
E-mail: tanakak@sakura.juntendo.ac.jp

We present a systematic study of the light-cone wavefunctions of heavy flavor mesons, such as the B mesons, in the limit of infinite heavy-quark mass in QCD.

In a heavy B meson with the flavor content $b\bar{q}$, the b-quark with the small Compton wavelength $\lambda_b = 1/m_b$ is surrounded by the so-called "brown muck" that represents strongly interacting cloud of light quarks, antiquarks, and gluons.[1] Because the brown muck can only resolve distances $\sim 1/\Lambda_{\rm QCD} \gg \lambda_b$, $(q, \bar{q}, g)$ are blind to quantum numbers of the b-quark, such as its mass, flavor, and spin orientation, and only feel the b-quark as a point source of the color field. This situation can be most conveniently described by heavy-quark effective theory (HQET),[1] introducing the effective b-quark field $h_v(x)$ as $b(x) \approx \exp(-im_b v\cdot x)h_v(x)$ with v^μ the 4-velocity of the B meson. $h_v(x)$ is subject to the constraint $\not{v}h_v = h_v$, and satisfies the equation of motion (EOM), $v\cdot Dh_v = 0$ with D_μ the covariant derivative. In the heavy-quark limit, HQET realizes heavy-quark spin-flavor symmetry.[1]

The probability amplitude to find $b\bar{q}$ in a certain configuration inside the B meson is described by the light-cone wavefunctions (WFs), that are related to the usual Bethe-Salpeter WFs at equal light-cone time $x^+ = (x^0 + x^3)/\sqrt{2}$.[2] In the heavy-quark limit, the light-cone WFs can be introduced in terms of vacuum-to-meson matrix element of nonlocal operators in HQET as[3,4,5]

$$\langle 0|\bar{q}(x)\Gamma h_v(0)|\bar{B}(p)\rangle = -\frac{if_B M}{2}\mathrm{Tr}\left[\gamma_5\Gamma\frac{1+\not{v}}{2}\left(\tilde{\Phi}_+ - \not{x}\frac{\tilde{\Phi}_+ - \tilde{\Phi}_-}{2v\cdot x}\right)\right]. \quad (1)$$

Here $x^\mu = (0, x^-, \boldsymbol{x}_T)$, $v^2 = 2v^+v^- = 1$, and $p^\mu = Mv^\mu$ is the 4-momentum of the B meson with mass M. Γ is a generic Dirac matrix and, here and in the following, the path-ordered gauge factors are implied in between the constituent fields. $\tilde{\Phi}_\pm \equiv \int d\omega d^2k_T e^{-i\omega v\cdot x + i\boldsymbol{k}_T\cdot\boldsymbol{x}_T}\Phi_\pm(\omega, \boldsymbol{k}_T)$, where ωv^+

has the meaning of the light-cone projection k^+ of the $\bar{q}$'s momentum, and $\boldsymbol{k}_T$ denotes the $\bar{q}$'s transverse momentum. f_B is the decay constant defined as $\langle 0|\bar{q}(0)\gamma^\mu\gamma_5 h_v(0)|\bar{B}(p)\rangle = if_B M v^\mu$, so that $\int d\omega d^2k_T \Phi_\pm(\omega, \boldsymbol{k}_T) = 1$. In Eq. (1), heavy-quark spin symmetry reduces the number of independent WFs, and allows only two independent WFs $\Phi_\pm(\omega, \boldsymbol{k}_T)$.

The nonperturbative dynamics between quark and antiquark reveals itself as the response of the nonlocal operator $\bar{q}(x)\Gamma h_v(0)$ to the change of the interquark separation and/or total translation. This is expressed in the form of the exact identities between the nonlocal operators[4,5]:

$$\frac{\partial}{\partial x^\mu}\bar{q}(x)\gamma^\mu\Gamma h_v(0) = i\int_0^1 duu\ \bar{q}(x)gG_{\mu\nu}(ux)x^\nu\gamma^\mu\Gamma h_v(0)\ , \tag{2}$$

$$\begin{aligned} v^\mu\frac{\partial}{\partial x^\mu}\bar{q}(x)\Gamma h_v(0) = i\int_0^1 du(u-1)\ \bar{q}(x)gG_{\mu\nu}(ux)v^\mu x^\nu\Gamma h_v(0) \\ + v^\mu\ \frac{\partial}{\partial y_\mu}\ \bar{q}(x+y)\Gamma h_v(y)\Big|_{y\to 0}\ , \end{aligned} \tag{3}$$

where $G_{\mu\nu}$ is the gluon field strength tensor, and we have used the EOM, $\not{D}q = 0$ and $v\cdot Dh_v = 0$. Taking the vacuum-to-meson matrix element, the LHS of these identities yield $\Phi_\pm(\omega, \boldsymbol{k}_T)$ of Eq. (1) and their derivatives. The terms in the RHS, which are given by integral of quark-antiquark-gluon operator, are expressed by the 3-particle WFs.[4] The last term of Eq. (3), the derivative over the total translation, yields $\Phi_\pm$ multiplied by

$$\bar{\Lambda} = M - m_b = \frac{iv\cdot\partial\langle 0|\bar{q}\Gamma h_v|\bar{B}(p)\rangle}{\langle 0|\bar{q}\Gamma h_v|\bar{B}(p)\rangle}\ . \tag{4}$$

This is a fundamental nonperturbative parameter of HQET corresponding to the mass of the "brown muck".[1] Substituting all the Dirac matrices for Γ, we obtain the 4 independent constraint equations between $\Phi_\pm$ and the 3-particle WFs.[4,5] Because Eqs. (2) and (3) are exact, these constraint equations are exact in QCD in the heavy-quark limit. Solving this system of equations, we construct model-independent formulae for the quark-antiquark WFs $\Phi_\pm$ in terms of independent dynamical degrees of freedom.

Firstly, we discuss the solution of our constraint equations for the case of light-like separation (zero transverse separation) between quark and antiquark. The solution gives exact formulae for a complete set of the light-cone distribution amplitudes $\phi_\pm(\omega) \equiv \int d^2k_T\Phi_\pm(\omega, \boldsymbol{k}_T)$ as[4]

$$\phi_\pm(\omega) = \phi_\pm^{(WW)}(\omega) + \phi_\pm^{(g)}(\omega)\ , \tag{5}$$

where the "Wandzura-Wilczek (WW) type" contributions $\phi_\pm^{(WW)}$ are independent of the 3-particle WFs, and are determined uniquely in analytic form in terms of $\bar{\Lambda}$ as

$$\phi_{\pm}^{(WW)}(\omega) = \frac{\bar{\Lambda} \pm (\omega - \bar{\Lambda})}{2\bar{\Lambda}^2}\theta(\omega)\theta(2\bar{\Lambda} - \omega) . \quad (6)$$

The contributions $\phi_{\pm}^{(g)}$ denote the pieces induced by the 3-particle WFs, representing nonperturbative effects due to the gluon mixture. For the explicit integral representation of $\phi_{\pm}^{(g)}$ in terms of the 3-particle WFs, we refer the readers to Ref. 4. To demonstrate the role of $\phi_{\pm}^{(g)}$, we present the Mellin moments derived from our solution (5), $\langle\omega^n\rangle_{\pm}^{(X)} \equiv \int d\omega\, \omega^n \phi_{\pm}^{(X)}(\omega)$ with $n = 0, 1, 2, \cdots$, and $X = WW, g$:

$$\langle\omega^n\rangle_{\pm}^{(WW)} = \frac{1}{n+1}\left(1 \pm \frac{n}{n+2}\right)(2\bar{\Lambda})^n , \quad (7)$$

$$\langle\omega\rangle_{\pm}^{(g)} = 0 , \qquad \langle\omega^2\rangle_{\pm}^{(g)} = \frac{1}{3}\left(\lambda_E^2 + \lambda_H^2 \pm \lambda_E^2\right) , \quad (8)$$

where λ_E and λ_H parameterize the two independent reduced matrix elements of local 3-particle operators of dimension 5, and are related to the chromoelectric and chromomagnetic fields in the B meson rest frame as

$$\langle 0|\bar{q}g\boldsymbol{E}\cdot\boldsymbol{\alpha}\gamma_5 h_v|\bar{B}(\boldsymbol{p}=0)\rangle = f_B M \lambda_E^2,\ \langle 0|\bar{q}g\boldsymbol{H}\cdot\boldsymbol{\sigma}\gamma_5 h_v|\bar{B}(\boldsymbol{p}=0)\rangle = i f_B M \lambda_H^2. \quad (9)$$

It is also straightforward to derive the analytic formulae of $\langle\omega^n\rangle_{\pm}^{(g)}$ for the general moment n in terms of matrix element of local 3-particle operators with dimension $n+3$ (see Ref. 4 for the detail).

There exists an estimate of λ_E and λ_H of Eq. (9) as $\lambda_E^2/\bar{\Lambda}^2 = 0.36 \pm 0.20$, $\lambda_H^2/\bar{\Lambda}^2 = 0.60 \pm 0.23$ by QCD sum rules[3] (but no estimate by lattice QCD). This estimate suggests that the multi-particle states with additional dynamical gluons could play important roles to determine the longitudinal quark motion inside the B meson. In this connection, the shape of our WW contributions (6) is very different from that of various models which have been used in the literature (see references in Ref. 3): some examples of such models are static quark models, $\phi^{ST1}(\omega) = \delta(\omega)$, $\phi^{ST2}(\omega) = \delta(\omega - \epsilon)$ with $\epsilon \sim \bar{\Lambda}$, and an oscillator model, $\phi^{HO}(\omega) = \text{const}\sqrt{\omega(M-\omega)}e^{-\omega^2/\epsilon^2}$; in these models $\phi_{\pm}$ are taken to have the same shape. Another example is $\phi_{+}^{GN}(\omega) = (\omega/\omega_0^2)e^{-\omega/\omega_0}$ and $\phi_{-}^{GN}(\omega) = (1/\omega_0)e^{-\omega/\omega_0}$ with $\omega_0 = 2\bar{\Lambda}/3$, inspired by the QCD sum rule estimates.[3] These also have rather different shape compared with Eq. (6), except the behavior $\phi_{+}^{GN}(\omega) \sim \omega$, $\phi_{-}^{GN}(\omega) \sim$ const, as $\omega \to 0$, i.e., the light-antiquark becomes "soft".

Recently the renormalization-group evolution of $\phi_{+}(\omega)$ has been discussed[6] by calculating the 1-loop anomalous dimension of the corresponding quark-antiquark operator of Eq. (1). The result indicates that the evolution generates a high-momentum "tail" as $\phi_+(\omega) \sim (-1/\omega)\ln(\omega/\mu)$ for $\omega \gg \mu$ (μ is the renormalization scale) due to the Sudakov-type $\ln(\omega/\mu)$ term in the anomalous dimension. (This would imply that, by renormalization, $\int d\omega \phi_+(\omega) = \int d\omega d^2 k_T \Phi_\pm(\omega, \boldsymbol{k}_T)$ becomes UV divergent, and $\phi_+(\omega)$ loses the physical meaning as the probability amplitude. But it is argued[6] that such "singular $\phi_+(\omega)$" is relevant for the QCD factorization of $B \to \gamma e \nu$ decays in the $\overline{\text{MS}}$ scheme.) On the other hand, in our solution (5), $\phi_+^{(g)}(\omega)$ due to the 3-particle operators should play roles to generate the tail for $\omega > 2\bar{\Lambda}$, nonperturbatively as well as perturbatively (see Eqs. (24)-(27) in Ref. 4). The relation between the radiative tail based on the evolution equation[6] and our results based on the EOM[4,5] has to be clarified.

Secondly, we discuss the solution of our constraint equations for the case of nonzero transverse separation of the constituents, and investigate the transverse momentum effects in the B meson.[5] Now the solution for the light-cone WFs of Eq. (1) reads

$$\Phi_\pm(\omega, \boldsymbol{k}_T) = \Phi_\pm^{(WW)}(\omega, \boldsymbol{k}_T) + \Phi_\pm^{(g)}(\omega, \boldsymbol{k}_T) , \tag{10}$$

in place of Eq. (5). Here the WW contributions $\Phi_\pm^{(WW)}$ with full account of the $\boldsymbol{k}_T$-dependence are given by explicit analytic formulae in terms of one single nonperturbative parameter $\bar{\Lambda}$ of Eq. (4) as[5]

$$\Phi_\pm^{(WW)}(\omega, \boldsymbol{k}_T) = \frac{1}{\pi}\phi_\pm^{(WW)}(\omega)\delta\left(\boldsymbol{k}_T^2 - \omega(2\bar{\Lambda} - \omega)\right) , \tag{11}$$

with $\phi_\pm^{(WW)}(\omega)$ of Eq. (6). This shows that the transverse relative motion between quark and antiquark is strongly correlated with the longitudinal relative motion through the combination $\boldsymbol{k}_T^2/[\omega(2\bar{\Lambda} - \omega)]$; therefore frequently used "model WFs", which assume complete separation (factorization) between the ω- and $\boldsymbol{k}_T$-dependence like an oscillator model $\Phi^{HO}(\omega, \boldsymbol{k}_T) = \text{const}\sqrt{\omega(M-\omega)}e^{-\omega^2/\epsilon^2} \times e^{-\boldsymbol{k}_T^2/\kappa^2}$, are not justified.

We can also estimate the effects of the nonperturbative gluon mixture due to $\Phi_\pm^{(g)}$ of Eq. (10), deriving model-independent "transverse momentum sum rules". We consider the first moment of $\boldsymbol{k}_T^2$, and get[5]

$$\int d\omega d^2 k_T \; \boldsymbol{k}_T^2 \Phi_\pm(\omega, \boldsymbol{k}_T) \equiv \langle\langle \boldsymbol{k}_T^2 \rangle\rangle_\pm = \langle\langle \boldsymbol{k}_T^2 \rangle\rangle_\pm^{(WW)} + \langle\langle \boldsymbol{k}_T^2 \rangle\rangle_\pm^{(g)} , \tag{12}$$

where

$$\langle\langle \boldsymbol{k}_T^2 \rangle\rangle_\pm^{(WW)} = \frac{2}{3}\bar{\Lambda}^2 , \qquad \langle\langle \boldsymbol{k}_T^2 \rangle\rangle_\pm^{(g)} = \frac{2}{3}\left(\lambda_E^2 + \lambda_H^2\right) , \tag{13}$$

with λ_E and λ_H of Eq. (9). Here the WW contributions $\langle\langle \boldsymbol{k}_T^2 \rangle\rangle_\pm^{(WW)}$ are obtained by using Eq. (11), while $\langle\langle \boldsymbol{k}_T^2 \rangle\rangle_\pm^{(g)}$ due to the 3-particle WFs can be directly derived from our constraint equations due to the identities (2) and (3). The sum rules (12) with Eq. (13) are exact in QCD in the heavy-quark limit. Combined with the above-mentioned estimate of λ_E and λ_H,[3] these results suggest that the coupling with the nonperturbative gluons might considerably broaden the transverse momentum distribution.

The solutions discussed in this work satisfy the QCD EOM and constraints from heavy-quark symmetry, and thus provide powerful nonperturbative framework for building up the light-cone WFs that represent the quark motion inside the B meson. Due to the power of heavy-quark symmetry, our solutions are given in simple analytic formulae. The results demonstrate novel behaviors of longitudinal as well as transverse quark motion inside the B mesons, and also unravel the special roles of the nonperturbative gluon mixture. Heavy-quark symmetry also guarantees that our solutions determine the light-cone WFs for the B^* mesons and also for the D, D^* mesons in the heavy-quark limit.

Our results involve fundamental nonperturbative parameters $\bar{\Lambda}$, λ_E, and λ_H; their precise determination by e.g. lattice QCD is desirable.

In addition to theoretical interest, our solutions provide "input" for the "long-distance contributions" in the QCD factorization formulae for hard exclusive processes,[2] such as the exclusive B meson decays, and the heavy-meson pair production in e^+e^- collisions. One may hope that novel behaviors inside heavy flavor hadrons, which are revealed in this work, will be accessible experimentally.

Acknowledgments

The author would like to thank H. Kawamura, J. Kodaira, and C.-F. Qiao for the collaboration, and V. M. Braun for discussions. This work was supported in part by the Grant-in-Aid of the Sumitomo Foundation.

References

1. M. Neubert, *Phys. Rept.* **245**, 259 (1994), and references therein.
2. G. P. Lepage and S. J. Brodsky, *Phys. Rev.* **D22**, 2157 (1980); S. J. Brodsky, in these proceedings.
3. A. G. Grozin and M. Neubert, *Phys. Rev.* **D55**, 272 (1997).
4. H. Kawamura, J. Kodaira, C.-F. Qiao and K. Tanaka, *Phys. Lett.* **B523**, 111 (2001); *Erratum-ibid.* **B536**, 344 (2002).
5. H. Kawamura, J. Kodaira, C. -F. Qiao and K. Tanaka, *Mod. Phys. Lett.* **A18**, 799 (2003).
6. B. O. Lange and M. Neubert, *Phys. Rev. Lett.* **91**, 102001 (2003); V. M. Braun, D. Y. Ivanov and G. P. Korchemsky, hep-ph/0309330.

KINKS IN TIME AND THEIR RELATION TO CONFINEMENT *

JOHN E. WANG

Jefferson Physical Laboratory, Harvard University, Cambridge, MA 02138, USA
Department of Physics, National Taiwan University, Taipei 106, Taiwan
E-mail: hllywd2@feynman.harvard.edu, hllywd2@phys.ntu.edu.tw

The time dependent formation of an electric flux tube (fundamental string) is reviewed. The main tool used for analysis is the Spacelike brane, which is a kink in time of the rolling tachyon. Both the S-brane and rolling tachyon are attempts to extend the D-brane concept to time dependent backgrounds. While S-branes are similar to Euclidean counterparts of the more familiar timelike D-branes, S-branes can smoothly change their worldvolume signature from spacelike to timelike which we interpret as the formation of a topological defect.

Here we review the results of Ref. 1 (see also references therein), in particular the derivation of actions and construction of solutions of Spacelike branes[2] which are useful in understanding time dependent systems such as the rolling tachyon[3].

Recent proposals to understand the dynamics of our universe's cosmic acceleration include searching for naturally unstable and hence time dependent backgrounds corresponding to extensions of the fruitful D-brane concept. The motivation is that the study of BPS branes has been one of the notable recent successes, providing a driving force in string theory and proving immensely useful in the study of supersymmetric and static systems. To develop our tools and intuition in time dependent backgrounds we are led to consider non-supersymmetric branes.

One such unstable background involves brane and anti-brane pairs, and their closely related non-BPS branes. Put close together, a brane and anti-brane pair will naturally interact and can annihilate, and this process can be understood as tachyon condensation. An unstable p-brane carries a scalar field T called the tachyon field governed by a tachyon potential which is

*Work partially supported by NSC, NCTS and the CosPA project.

approximately of the form $V(T) \approx e^{-T^2}$. The tachyon field parametrizes the instability in the system. When the tachyon value is near zero, the brane system is unstable and will be driven towards the bottom of the tachyon potential at large tachyon values.

We begin our study of the dynamical evolution of the tachyon condensation process by examining the rolling tachyon solution, which is a time dependent and spatially homogenous tachyon solution

$$T = T(t = time) \ . \tag{1}$$

The tachyon values can start off near zero and then accelerate towards large tachyon values in the same way a ball rolls down a hill. At the top of its trajectory the tachyon field has value $T = 0$. This p dimensional Euclidean hypersurface corresponding to the moment in time when the tachyon scalar field is everywhere zero we call a Spacelike $(p-1)$-brane. As it stands this Euclidean hypersurface does not carry much dynamics as it appears and disappears in an instant leaving no apparent trace of its existence. Despite its spacelike trajectory, it clearly does not violate causality nor does it transport energy superluminally.

To extend the S-brane lifetime we deform the S-brane worldvolume by introducing fluctuations into the rolling tachyon profile. The deformed S-brane now lives for a finite period of time and is not homogenous and as finely tuned. If we make an infinite deformation of the S-brane, it will live for an infinite length of time as shown in Figure 1.

The key point is that large fluctuations do naturally occur and they correspond to the time dependent formation of topological defects. S-branes can be thought of as initial conditions for the formation of defects.

To clarify why S-branes play this role in defect formation, we recall that for static tachyon kink configurations the zeros of the tachyon field

$$T = 0 \tag{2}$$

are interpreted as the location of lower dimensional solitons or topological defects. In the case of string theory these solitons include branes and anti-branes. This is why in analogy, we previously identified the $T = 0$ region of a rolling tachyon or kink in time to be the location of the Spacelike brane.

The time evolution of defect formation can be followed and traces out a spacetime trajectory which is the S-brane worldvolume. In other words because both S-branes and normal defects are specified by the condition $T = 0$, slow moving and long lived Spacelike branes should be able to represent ordinary and stable timelike topological defects! (See Figure 2.)

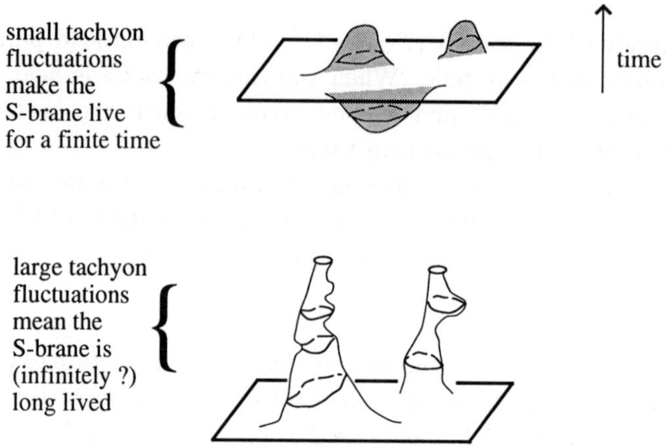

Figure 1. A flat S-brane can be deformed and live for more than a moment in time. Time increases as we go up the page and a horizontal plane is a moment in time.

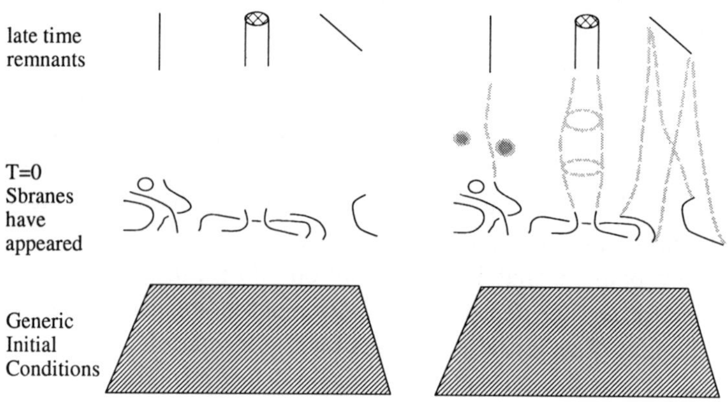

Figure 2. S-branes worldvolumes (dashed lines) allow us to follow the time evolution of initial conditions during the formation of topological defects.

To quantify this statement we have derived an effective action for Space-like branes by performing a zero mode fluctuation analysis on non-BPS branes. The S-brane action is

$$S = \int_{world\ volume} \sqrt{det(\delta_{ij} - \partial_i t \partial_j t + F_{ij})}\ . \tag{3}$$

This action passes the following tests (Ref. 1 discusses further checks)

1)there is a scalar field, t, with wrong sign kinetic energy corresponding to fluctuations in time

2)gauge fields with field strength F_{ij} exist on the worldvolume just like for D-branes

3)the action is real for flat Euclidean hypersurfaces of equal time

4)yet infinitely long lived solutions which smoothly change from Euclidean to Lorentzian signature are also possible. This property is based on the fact that the S-brane action is by definition defined to be spacelike relative to the open string metric whose light cone always lies within the closed string light cone[4], $G^{\mu\nu}_{open} \subseteq G^{\mu\nu}_{closed}$. There are therefore long lived solutions which are consistently both spacelike relative to the open string metric and also timelike relative to the closed string metric as illustrated in Figure 3.

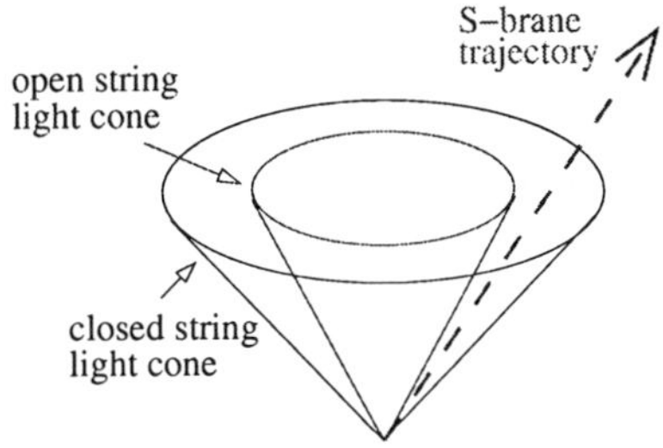

Figure 3. S-branes can simultaneously be spacelike relative to the open string metric and timelike relative to the closed string metric. The narrower the open string light cone becomes, the slower the S-brane is allowed to travel.

5)solutions exist describing the formation of an electric flux tube

Although static descriptions of fundamental strings have been discussed[5], this S-brane solution describes the time dependent formation of a confined electric flux tube (see also Ref. 6). In gauge theories the bundling of gauge fields into small regions is the process of confinement, so this solution indicates that confinement can be understood as a non-perturbative but classical dynamical process. As an example, the electric S3-brane is cylindrical in shape, $\mathbb{R}^1 \times S^2$, where r is the radius of the cylinder, χ goes lengthwise along the cylinder and t is time. In particular, this electric S3-brane solution is

$$r = \frac{c}{t}, \qquad E = 1 \tag{4}$$

where the electric field, $F_{0\chi} = E$, is constant and the critical value. The radius of this cylinder shrinks to zero at late times while the electric field along the cylinder is constant, so this solution describes an electric flux tube which is confining into a string-like object. Using the Dirac quantization condition we found that the energy is that of a fundamental string

$$H = \frac{1}{2\pi\alpha'}\int d\chi \left(\int d^p x D\right) = \frac{1}{2\pi\alpha'}\int d\chi \,. \tag{5}$$

Further, the S-brane coupling to RR form fields, A, goes to zero

$$\int_{world\ volume} A \times (t^{-1}) \overset{t\to\infty}{\longmapsto} 0 \,. \tag{6}$$

so the D-brane charge vanishes. The time dependent factor of t^{-1} is due to the fact that the worldvolume of the S-brane shrinks to zero.

Spacelike branes are extensions of the D-brane concept to time dependent backgrounds and are useful in understanding the formation of defects. From the S-brane perspective, confinement can appear as a classical self organizing process and might lead to a hint as to the physics at the Hagedorn temperature. Spacelike branes should play a similar role in time dependent backgrounds of ordinary field theories.

Acknowledgments

It is a pleasure to thank Koji Hashimoto, Pei-Ming Ho and Satoshi Nagaoka for wonderful collaboration. I also wish to thank the organizers for their kind invitation to the stimulating Confinement 2003 conference.

References

1. K. Hashimoto, P. M. Ho and J. E. Wang, *Phys. Rev. Lett.* **90** (2003) 141601, hep-th/0211090; K. Hashimoto, P. M. Ho, S. Nagaoka and J. E. Wang, *Phys. Rev.* **D68** (2003) 026007, hep-th/0303172.
2. M. Gutperle and A. Strominger, *JHEP 0204* (2002) 018, hep-th/0202210.
3. A. Sen, *JHEP 0204* (2002) 048, hep-th/0203211; A. Sen, *JHEP 0210* (2002) 003, hep-th/0207105.
4. G. W. Gibbons and C. Herdeiro, *Phys. Rev.* **D63** (2001) 064006, hep-th/0008052.
5. C. G. Callan, Jr. and J. M. Maldacena, *Nucl. Phys.* **B513** (1998) 198, hep-th/9708147; G. W. Gibbons, *Nucl. Phys.* **B514**(1998) 603, hep-th/9709027; P. Yi, *Nucl. Phys.* **B550** (1999) 214, hep-th/9901159; M. Kleban, A. Lawrence and S. Shenker, *Phys. Rev.* **D64** (2001) 066002, hep-th/0012081; D. Mateos and P. K. Townsend, *Phys. Rev. Lett.* **87** (2001) 011602, hep-th/0103030.
6. S. J. Rey and S. Sugimoto, hep-th/0303133; A. Sen, hep-th/0305011.

Poster Session Presentations

LATTICE GAUGE FIXING AND THE VIOLATION OF SPECTRAL POSITIVITY

C. A. AUBIN AND M. C. OGILIVIE
Washington University
St. Louis, MO 63130 USA

We trace the violation of spectral positivity found in lattice gauge fixing to the use of quenched auxiliary fields, and show that this violation is a general feature of quenching. For covariant gauges parametrized by a parameter α, the $SU(2)$ gluon propagator is well described by a simple quenched mass-mixing formula. The gluon mass parameter appears to be independent of α for sufficiently large α.

A generalization of the standard gauge-fixing process[1] can be interpreted as the use of quenched scalar fields. The hopping parameter, α, for the quenched scalar fields plays the role of a gauge-fixing parameter. The observed violation of spectral positivity in lattice gauge fixing can be attributed to the quenching process itself.

This is studied in two toy models. The first model is a mass-mixing model: two scalar fields with a non-diagonal mass matrix. The second is a model with two Ising spins with a local interaction. Both of these models quench one of the fields and show a clear violation of spectral positivity similar to that found in lattice gauge fixing.

For lattice gauge fixing associated with Abelian projection and lattice Landau gauge, spectral positivity is violated by processes similar to those found in quenched QCD. For covariant gauges, the $SU(2)$ gluon propagator is well described by: $G(x) = (a + bm_2|x|)e^{-m_2|x|} + ce^{-m_1|x|}$, where a, b and c are unknown coefficients, and m_1 and m_2 are two mass parameters. This expression is a generalization of the exact propagator of the quenched field in the mass-mixing model.

At $\beta = 2.6$ on a $12^3 \times 16$ lattice, the heavier mass, m_1, decreases to a constant $m_1 = 1.151(8)$. The light mass, $m_2 = 0.283(7)$, appears to be independent of α for large α. Since m_2 appears independent of gauge choice, at least within the class of covariant gauges considered here, it seems natural to identify it as the gluon mass.

References

1. C. A. Aubin and M. C. Ogilvie, *Phys. Lett.* **B570**, 59 (2003), and references therein.

CONFINING INTERQUARK POTENTIALS FROM NON ABELIAN GAUGE THEORY COUPLED TO DILATON

M. CHABAB AND L. SANHAJI

LPHEA, Physics Department, Faculty of Science - Semlalia
P.O.Box 2390 Cadi-Ayyad University, Marrakech, Morocco
E-mail: mchabab@ucam.ac.ma

Using a superstring inspired model, we derive a general formula for the quark interaction potential $V(r)$ which depends on the dilaton degrees of freedom.[1] Our interquark potential generalizes the Dick potential. The latter exhibits a short range Coulomb potential plus a confining term which is linearly rising at large distances.[2] Therefore, this model provides a new confinement generating mechanism challenging the monopole condensation one. We have tested phenomenologically the Dick confining potential in the heavy meson sector. The obtained results show that the spin-averaged energy levels of heavy-meson spectra are well explained when we assign the dilaton a mass around 56.9 MeV. This value lies within the range of many theoretical scenario of dilatonic mass and may provide a piece of information in the search of this particle.

In the same context, in order to prevent the breaking of the string linking quark-antiquark, we have also determined a new potential as a perturbative series in terms of the interquark distance r. [3] Our potential goes beyond the Dick one and shares the same features as the Bian-Huang-Shen potential V_{BHS} [4] which depends only on odd powers of r and is parameterized in terms QCD vacuum condensates.

References

1. M. Chabab, R. Markazi and E.H. Saidi, *Eur. Phys. J.* **C11**, 543 (2000).
2. R. Dick, *Eur. Phys. J.* **C6**, 701 (1999); *Phys. Lett.* **B409**, 321 (1997).
3. M. Chabab *et al.*, *Class. Quant. Gravity* **18**, 5085 (2001).
4. J.G. Bian and T. Huang, *Phys. Rev.* **D39**, 1213 (1989); J.G. Bian, T. Huang and Q.S. Shen, *The Quarkonium Potential within Short and Intermediate Range in QCD*, BIHEP-TH-92-38.

AN EFFECTIVE ABELIAN GAUGE THEORY OF MASSIVE SU(2) YANG–MILLS THEORY

SHINICHI DEGUCHI AND YOUSUKE KOKUBO

Institute of Quantum Science, Nihon University, Tokyo 101-8308, Japan
E-mail: deguchi@phys.cst.nihon-u.ac.jp

An effective Abelian gauge theory (EAGT) of massive SU(2) Yang–Mills theory is derived without spoiling the global SU(2) gauge symmetry. The EAGT allows a composite of the Yukawa and the linear potentials that interacts with color-electric charges. The EAGT involves the Skyrme-Faddeev model.

According to a Monte Carlo simulation performed by Amemiya and Suganuma, off-diagonal gluons in SU(2) lattice QCD behave like massive vector fields.[1] This result strongly supports Abelian dominance at a long-distance scale. In order to analytically realize Abelian dominance, we now introduce mass terms of the off-diagonal gluon fields into pure SU(2) Yang–Mills (YM) theory, using the Stueckelberg–Kunimasa–Goto (SKG) formalism.

In this work we try to derive an effective Abelian gauge theory (EAGT) from the SU(2) SKG formalism without putting the so-called maximally Abelian gauge condition.[2] Thereby the global SU(2) gauge symmetry is maintained in the EAGT. In deriving the EAGT, the off-diagonal gluon fields are treated as fields that produce quantum effects on the diagonal gluon field and other fields relevant at a long-distance scale. The EAGT obtained here involves a gauge fixed version of a dual theory of the extended dual Abelian Higgs model in the London limit. Then, following a procedure discussed in Ref. 3, we can show that the EAGT allows a composite of the Yukawa and the linear potentials, which is understood to be a static potential between color-electric charges carried by gluonic degrees of freedom. we also see that the EAGT involves the Skyrme–Faddeev model.

References

1. K. Amemiya and H. Suganuma, *Phys. Rev.* **D60**, 114509 (1999).
2. S. Deguchi and Y. Kokubo, *Mod. Phys. Lett.* **A18**, 2015 (2003).
3. S. Deguchi and Y. Kokubo, *Mod. Phys. Lett.* **A17**, 503 (2002).

SOFT MODE AT THE CRITICAL END POINT IN THE CHIRAL EFFECTIVE MODELS

H. FUJII AND M. OHTANI*
Institute of Physics, University of Tokyo, Komaba, Tokyo 153-8902, Japan

Within chiral effective models, the soft mode associated with the critical end point in the T–μ plane is identified as medium fluctuation.

The aim of this presentation is to clarify the nature of the critical end point (CEP), whose existence in the real QCD phase diagram has been suggested in the literature. Qualitative understanding of CEP is certainly of great importance for suggesting the correct experimental observables.

At CEP, where the 1st-order transition line terminates, the susceptibilities of the baryon number and entropy densities ($\chi_{\mu\mu}$, χ_{TT}) diverge as well as that of the scalar density χ_{mm}. (At the 2nd order chiral transition, χ_{mm} diverges alone and others are finite.) It is noteworthy that in the spectral representation $\chi_{\mu\mu}$ and χ_{TT} come solely from the $\delta(\omega)$ spectrum as the wavevector $q \to 0$, due to the conservation law. This particular spectrum gives the divergence of the susceptibility at CEP. Based on this fact we speculate that the divergent χ_{mm} at CEP has the same origin: the medium fluctuation. Within the models of Nambu-Jona-Lasinio and time-dependent Ginzburg-Landau, we showed this is indeed the case; the spectrum of the medium fluctuation generates the divergences at CEP, and that the contribution from the "sigma-meson"-type fluctuation is non-singular.

In conclusion, there must appear a divergent medium fluctuation, whereas the massless sigma meson is not necessarily required to have the criticality at CEP, where chiral symmetry is explicitly broken. The analogy with the liquid-gas phase transition seems more natural at CEP.

References

1. H. Fujii, *Phys. Rev.* **D67**, 094018 (2003); H. Fujii and M. Ohtani, in preparation.

---

*Current address: RIKEN, Wako, Saitama 351-0198, Japan.

THERMODYNAMIC LIMIT OF THE CANONICAL PARTITION FUNCTION WITH RESPECT TO THE QUARK NUMBER IN QCD

KENJI FUKUSHIMA

Department of Physics, University of Tokyo
7-3-1 Hongo, Bunkyo-ku, Tokyo 113-0033, Japan
E-mail: fuku@nt.phys.s.u-tokyo.ac.jp

We discuss the criterion of color-deconfinement in the canonical description of QCD. We reveal that the idea of defining an order parameter would suffer from the negative sign problem of the Dirac determinant.

The criterion of color-deconfinement with dynamical quarks is a long-standing problem. We investigate QCD in the canonical ensemble with respect to the quark number and reexamine the would-be order parameter proposed by DeTar and McLerran.[1] We reveal that the canonical description in which the quark number is fixed would be reduced to the grand canonical description under the thermodynamic limit.[2]

Even when the quark number is fixed, it is the quark number *density* that can be fixed under the thermodynamic limit. Thus any fluctuation of the quark number of order O(1) is allowed even in the canonical ensemble. As a result the notion of confinement becomes obscured.

The fluctuation of the quark number is induced by infinitely non-local interactions arising from the projection in the canonical ensemble. Such unintended fluctuations can be excluded by taking the ensemble average with configurations generated by the symmetric part of the action. The remaining is regarded as a part in the operator to be averaged over. This is essentially the same as the Glasgow method for the Dirac determinant at finite density, which would suffer from the negative sign problem.

References

1. C. DeTar and L. McLerran, *Phys. Lett.* **B119**, 171 (1982).
2. K. Fukushima, *Ann. Phys.* **304**, 72 (2003).

INFRARED PROPERTIES OF LANDAU GAUGE QCD

SADATAKA FURUI*

School of Science and Engineering, Teikyo University
Toyosatodai 1-1, Utsunomiya, 320-8551, JAPAN
E-mail: furui@umb.teikyo-u.ac.jp

The infrared properties of lattice Landau gauge QCD are studied by measuring gluon propagator, ghost propagator, QCD running coupling and Kugo-Ojima parameter of $\beta = 6.0, 16^4, 24^4, 32^4$ and $\beta = 6.4, 32^4, 48^4$ lattices. Unexpected features of the unique gauge(fundamental modular gauge FMG) are observed by comparing gauge fixed configurations of SU(2).

Landau gauge QCD suffers from Gribov copy propblem. We obtained the configuration with its norm of the gauge field minimum(FMG)[1]. In the infrared region we observed[2]: 1) the gluon propagator becomes less singular than the tree level, 2) the ghost propagator becomes more singular than the tree level(except the lowest momentum point of 48^4 lattice, whose singularity is weakened), 3) the QCD running coupling has a peak of about 1 in the infrared and decreases to a finite value at zero momentum, 4) Kugo-Ojima parameter saturates to about -0.75 in 48^4 lattice.

We produced $\beta = 2.2, 16^4$ SU(2) configurations and performed FM gauge fixing via parallel tempering(PT) and compared with Gribov copies obtained by random gauge transformation(CGRT). The PT as compared to CGRT showed, 1)in Kugo-Ojima parameter c about 2% suppression, and 2)in the infrared singularity of ghost propagator about 2% suppression. The gluon propagator of PT and CGRT agreed within errors.

References

1. H. Nakajima and S.Furui, *Quantum Chromodynamics and Color Confinement*, World Scientific (2001) p.60-68; Lattice'03 proceedings,Tsukuba,July 2003.
2. S. Furui and H.Nakajima, hep-lat/0305010.

*This work is done in collaboration with H. Nakajima (UTSUNOMIYA UNIV.)

LOW-ENERGY-THEOREM APPROACH TO ONE-PARTICLE SINGULARITY IN QED2 + 1

YUICHI HOSHINO

Kushiro National College of Technology, Otanoshike Nishi, Hokkaido 084, Japan

To search the high temperature limit of QED we study the three-dimensional QED. Analyzing spectral function of the fermion propagator we find a infrared behaviour of the position space propagator which contains Coulomb-energy and self-energy: position dependent mass.

We evaluate the position space propagator for spinor in $QED2 + 1$ using Ward-Identity for soft photon and dispersion relation.[1,2,3] Exponentiation of one photon matrix element yields a position dependent dynamical mass and Coulomb energy. There is no gauge in which there exists free pole. In the Yennie gauge we can avoid linear infrared divergence but there appear logarithmic infrared divergences. Effect of dynamical mass modifies the long distance behaviour of the propagator. Propagator vanishes in the long distance that the charged particle cannot separate each other. In momentum space, scalar part of the propagator dumps as we see in the analysis of Dyson-Schwinger equation but modified by wave function renormalization.[4] It is not easy to analyze in Minkowski space and it s still unclear whether or not the cut exists.[5] There seems to be a critical coupling constant above which the vacuum expectation value of pair condensate is finite.[6]

References

1. Y. Hoshino, *JHEP* **05**, 75 (2003); R. Jackiw and L. Soloviev, *Phys. Rev.* **173**, 1458 (1968).
2. S. Deser, R. Jackiw and S. Templeton, *Ann. Phys. (NY)* **140**, 372 (1982).
3. T. Appelquist and R. Pisarski, *Phys. Rev.* **D23**, 2305 (1981); T. Appelquist and R. Heinz, *Phys. Rev.* **D24**, 2169 (1981).
4. T. Appelquist, D. Nash and L.C.R. Wijewardhana, *Phys. Rev. Lett.* **60**, 2575 (1988); Y. Hoshino and T. Matsuyama, *Phys. Lett.* **222B**, 493 (1989).
5. D. Atkinson and D.W.E. Blatt, *Nucl. Phys.* **151B**, 342 (1979); P. Maris, *Phys. Rev.* **D52**(1995); Y. Hoshino, *Nuovo Cim.* **112A**, 335 (1999).
6. N. Ohta, *Prog. Theor. Phys.* **62**, 1370 (1979); K. Nishijima, *Prog. Theor. Phys.* **81**, 878 (1989); *Prog. Theor. Phys.* **83**, 1200 (1990); A.B. Waites and R. Delbourgo, *Int. J. Mod. Phys.* **A27**, 6857 (1992).

BARYONIC FLUX IN QUENCHED AND FULL QCD

HIROKO ICHIE *

Tokyo Institute of Technology, Tokyo 152-8551, Japan

We study the distribution of color electric flux of the three-quark system in quenched and full QCD (with N_f=2 flavors of dynamical quarks). Both in quenched and full QCD we find clear evidence for a Y-shape flux tube surrounded and formed by the solenoidal monopole current, in accordance with the dual superconductor picture of confinement.

For understanding the structure of baryons, in particular, for modeling the nucleon, it is important to learn about the forces and the distribution of color electric flux in the $3Q$ system. We study the flux tube of $3Q$ system using the abelian 'baryonic' Wilson loop [1] in the maximally abelian gauge, in quenched(β=6.0, $V= 16^3 32$) and full (β=5.29, κ=0.1355, V=$24^3 48$) QCD. We show in Fig.1 the action density of the 3Q system in full QCD. While on the basis of the potential it is hard to decide whether the long-range potential is of Δ- or Y-type, the distribution of the color electric field and the action density clearly show a Y-shape geometry. As in the Q-$\bar{Q}$ system, we identified the solenoidal monopole current to be responsible for squeezing the color electric flux into a narrow tube[1]. Little difference to the quenched theory was found. Our results are in qualitative agreement with the predictions of the dual Ginzburg-Landau model.

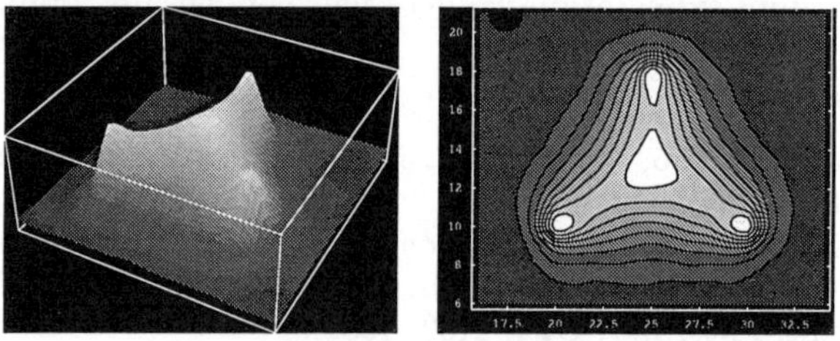

Figure 1. The action density of the 3Q system in full QCD.

References

1. H. Ichie, V. Bornyakov, T. Streuer and G. Schierholz, *Nucl. Phys.* **A721**, 899 (2003); *Nucl. Phys.* **B** (Proc. Suppl.) **119**, 751 (2003).

*This work is done in collaboration with V. Bornyakov, Y. Mori, D. Pleiter, M. I. Polikarpov, G. Schierholz, T. Streuer, H. Stüben, and T. Suzuki (DESY-IHEP-Kanazawa).

CHIRAL SYMMETRY RESTORATION AT FINITE TEMPERATURE IN THE SCHWINGER-DYSON EQUATION WITH LATTICE QCD DATA

H. IIDA, M. OKA AND H. SUGANUMA

Tokyo Institute of Technology, Ohokayama 2-12-1, Tokyo 152-8551, Japan

We propose the Schwinger-Dyson (SD) formalism based on lattice QCD data. We extract the kernel function $K(p^2)$ in the SD equation from the quark mass function in lattice QCD. We apply the lattice-QCD-based SD equation to thermal QCD.

For the study of chiral symmetry breaking in QCD, we propose a new framework of the Schwinger-Dyson (SD) formalism based on lattice QCD data.[1] From lattice QCD data[2] for the quark mass function in the Landau gauge, we extract the SD kernel function $K(p^2) \equiv g^2\Gamma(p^2)d(p^2)$, which is expressed as the product of the quark-gluon vertex and the gluonic polarization factor.[1] As remarkable features, we find *infrared vanishing* and *intermediate enhancement* in the SD kernel function $K(p^2)$. (See Fig.1(a).) With $K(p^2)$, we apply the lattice-QCD-based SD equation to thermal QCD, and find chiral symmetry restoration at $T_c \sim 100$ MeV as shown in Fig.1(b).[1]

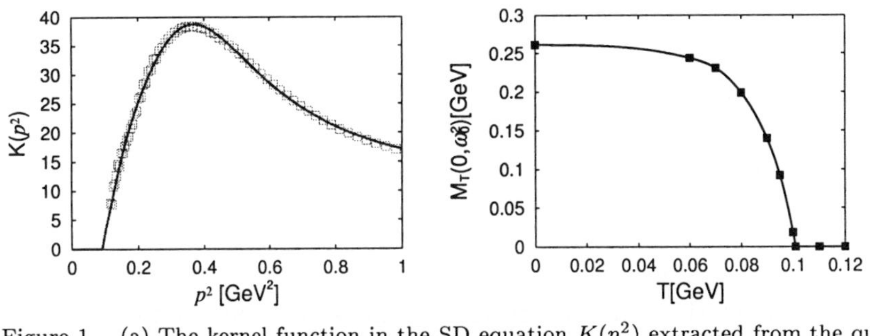

Figure 1. (a) The kernel function in the SD equation $K(p^2)$ extracted from the quark mass function in the Landau gauge in lattice QCD. (b) The thermal infrared quark mass $M_T(\mathbf{p}^2 = 0, \omega_0^2)$ at the lowest Matsubara frequency ω_0 plotted against temperature T.

References

1. H. Iida, M. Oka and H. Suganuma, Proc. of *Lattice 2003*, *Nucl. Phys.* **B** (Proc. Suppl.) in press, hep-ph/0309338.
2. P.O. Bowman, U.M. Heller, A.G. Williams, *Phys. Rev.* **D66**, 014505 (2002).

FROM THE ABELIAN PROJECTED FLUX TUBE TO THE DUAL ABELIAN HIGGS MODEL

E.-M. ILGENFRITZ*

Institut für Physik, Humboldt Universität zu Berlin,
D-12489 Berlin, GERMANY

A recent detailed study of the $Q\bar{Q}$ flux tube and its DAHM analysis is reported.

In Abelian projected (AP) $SU(2)$ lattice gluodynamics we have measured the profile of $Q\bar{Q}$ flux tubes for various lattice spacings and different $Q\bar{Q}$ distances R. We have fitted the results by the classical flux-tube solution of the $3D$ lattice $U(1)$ dual Abelian Higgs model (DAHM) with external electric charges, taking the finiteness of R into account.[1] On the lattice, the flux-tube profile is characterized by the correlator of electric field strength and monopole current with the (Abelian) Wilson loop. In the course of work, we have explored and minimized systematic effects which could spoil the observed flux-tube profile: a bad ground state overlap and the Gribov problem encountered in maximal-Abelian gauge fixing. The superposition of a Coulomb field and a solenoidal field (as known from DAHM with external sources) has been confirmed on the lattice by considering correlators with Wilson loops expressed in terms of regular and singular parts of the AP gauge field, respectively, obtained by Hodge decomposition. The fits give a dual gauge boson mass m_B=1091(7) MeV and a Higgs mass m_χ=953(20) MeV independently of the flux-tube length. The Ginzburg-Landau parameter of the gluodynamic vacuum $\kappa = m_\chi/m_B = 0.87(2) < 1$ indicates a weak type-I superconductor. In contrast to the masses, the dual gauge coupling g_m is obtained R-dependent which reflects the antiscreening of the non-Abelian quark charge. Up to $R = 0.5$ fm we have found no growing of the flux-tube width, *i.e.*, no string roughening.

References

1. Y. Koma, M. Koma, E.-M. Ilgenfritz and T. Suzuki, hep-lat/0308008.

*This work is done in collaboration with Y. Koma, M. Koma (both MPI Munich), and T. Suzuki (Kanazawa U.)

ENTROPY OF MONOPOLES FROM PERCOLATING CLUSTER IN PURE SU(2) QCD

K. ISHIGURO*

Institute for Theoretical Physics, Kanazawa University
Kanazawa 920-1192, JAPAN
E-mail: ishiguro@hep.s.kanazawa-u.ac.jp

The length distribution and the monopole action of the infrared monopole clusters are studied numerically in pure SU(2) QCD. We determine the effective entropy of the monopole currents which turns out to be a descending function of the blocking scale, indicating that the effective degrees of freedom of the extended monopoles are getting smaller as the blocking scale increases.

One of the most promising approaches to the problem of color confinement in QCD is the dual superconductor picture [1] of the QCD vacuum. In order to understand the mechanism of monopole condensation, we study the energy-entropy balance of the monopoles from the percolating monopole cluster in pure SU(2) QCD. We apply the block-spin transformation to the monopole currents and fix the physical scale at which the monopole charge is defined. We determine the action of the percolating monopole cluster from the effective monopole action obtained by an inverse Monte-Carlo method. Then we use the length distribution histogram of the percolating cluster to estimate the monopole entropy. We find that the dynamics of percolating cluster of blocked monopoles is not a simple random walk, and the monopole entropy depends on the physical scale.

A detailed description of this work is given in Ref.[2].

References

1. G. 't Hooft, in *High Energy Physics*, ed. A. Zichichi, EPS International Conference, Palermo (1975); S. Mandelstam, *Phys. Rept.* **23**, 245 (1976).
2. M. N. Chernodub, K. Ishiguro, K. Kobayashi and T. Suzuki, *hep-lat/0306001*.

*This work is done in collaboration with M. N. Chernodub (ITEP, Moscow/Kanazawa U.), K. Kobayashi (Kanazawa U.) and T. Suzuki (Kanazawa U.).

ENERGY SPECTRUM OF A HEAVY-LIGHT MESON

UKQCD COLLABORATION, J. KOPONEN*

Department of Physical Sciences, P.O. Box 64, FIN-00014 University of Helsinki, Finland

We have measured the energies of different angular momentum states of a heavy-light meson on a lattice. The most interesting outcome of the study is that the spin-orbit splitting in both P- and D-wave states is very small.

We have repeated the study done by Michael and Peisa[1] using a larger, quenched lattice to check that the finite size effects are under control. Also the effect of quenching is studied by using dynamical fermion configurations. The new measurements confirm the earlier calculated energy spectrum.

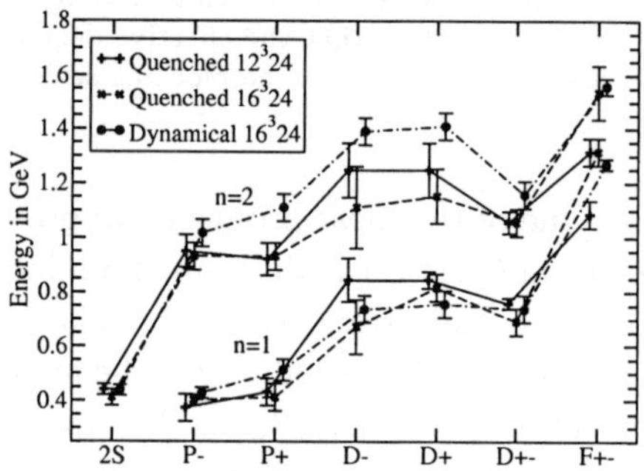

From some theoretical considerations[2] it is expected that, for higher angular momentum states, the multiplets should be inverted compared with the Coulomb spectrum, i.e. nL$-$ should lie higher than nL$+$. Here nL$+(-)$ means that the light quark spin couples to angular momentum L giving the total $j = L \pm 1/2$ and n is the principal quantum number. Experimentally this inversion is not seen for P-waves, and now the lattice measurements show that there is no inversion in the D-wave states either.

References

1. UKQCD Collaboration, C. Michael and J. Peisa, *Phys. Rev.* **D58**, 34506 (1998).
2. H.J. Schnitzer, *Phys. Lett.* **B226**, 171 (1989).

*In collaboration with A. M. Green and C. Michael. This work was supported by the Center for SCIENTIFIC Computing, Espoo, Finland, the Finnish Cultural Foundation and the Academy of Finland (project 177881).

STUDYING PARTON DYNAMICS WITH THE STAR EXPERIMENT

JENS SÖREN LANGE

Institut für Kernphysik, University of Frankfurt,
August-Euler-Straße 6, 60486 Frankfurt/Main, Germany

The STAR experiment at RHIC uses ultra-relativistic nucleus-nucleus collisions in search for new QCD phenomena, in particular a possibly deconfined, hot and dense partonic state of matter.

The main STAR subdetector is a midrapidity ($|\eta| \leq 1.6$) Time Projection Chamber (TPC) with 48,000,000 pixels. Recently partial EM calorimeter coverage ($0 \leq |\eta| \leq 1$) for leptons and neutral particles has been added. In 3 years of data taking, high statistics ($10^6 \leq N \leq 10^7$ events on tape) for Au+Au, p+p and d+Au at $\sqrt{s}$=200 GeV and Au+Au at $\sqrt{s}$=130 GeV were recorded. These data enable analyses up to $p_T \leq$ 12GeV/c, which might be sensitive to parton dynamics at early collision time scales $t \leq 1$ fm/c. Perturbative QCD (pQCD) calculations predict for a central Au+Au collision at $\sqrt{s}$=200 GeV: $\simeq$400 gluons, $\simeq$130 quarks and $\simeq$40 anti-quarks.[1] In particular, jets from hard scattering of these partons are sensitive to the color charge density of the surrounding medium. In the analysis, a large suppression of inclusive high p_T hadrons (supposed to represent leading jet particles) was discovered[2]. In addition, angular correlations revealed the disappearance of the away-side jet in central collisions, compared to p+p, d+Au or peripheral Au+Au collisions.[3]

References

1. K. J. Eskola and K. Kajantie, nucl-th/9610015.
2. STAR Collaboration, nucl-ex/0206011, *Phys. Rev. Lett.* **89**, 202301 (2002).
3. STAR Collaboration, nucl-ex/0210033, *Phys. Rev. Lett.* **90**, 082302 (2003).

THERMAL FLUCTUATION OF GAUGE FIELDS AND FIRST ORDER PHASE TRANSITION IN COLOR SUPERCONDUCTIVITY

T. MATSUURA*

Department of Physics, The University of Tokyo, Tokyo 113-0033, Japan

For the three-flavor color superconductor, thermal fluctuations of gluons and diquarks have been studied using the Ginzburg-Landau theory. At high baryon density, the system becomes a Type I superconductor and the gluon fluctuation dominates over the diquark fluctuation near the phase boundary. The appearance of a negative cubic term of the order parameter and large positive corrections to the quartic and quadratic terms lead to a strong first-order phase transition.

The effects of thermal fluctuation of gluons and diquarks on super-to-normal phase transition in three-flavor color superconductivity are studied by using the Ginzburg-Landau (GL) theory. At high baryon density μ, the system becomes a Type I superconductor and the gluon fluctuation dominates over the diquark fluctuation near the phase boundary. The thermal gauge fluctuation not only induces a negative cubic term (Δq) of the order parameter d in the GL free energy $F_{\text{eff}}(d)$ but also gives large positive corrections to the quartic and quadratic terms (Δp, Δr);

$$F_{\text{eff}}(d) = (p + \Delta p)d^2 - \Delta q d^3 + (r + \Delta r)d^4.$$

Here, $p \propto \mu^2(T - T_c)/T_c$, $\Delta p \propto \mu^2 \alpha_s$, $\Delta q \propto \mu^3 \alpha_s^{3/2}/T_c^2$, $r \propto \mu^2/T_c^2$, $\Delta r \propto \mu^4 \alpha_s^2/T_c^4$ and T_c is the mean-field critical temperature. This leads to rather strong first-order phase transition compared to the case of metallic Type I superconductor. We find that the critical temperature of the first-order transition becomes lower than the mean field value by $O(\alpha_s)$. The strength of the first-order phase transition is the decreasing function of μ: the jump of d at the critical point relative to its $T = 0$ value is proportional to $\mu/T_c\sqrt{\alpha_s}$. By making an analysis of the Ginzburg's critical region, our results in the one-loop approximation are shown to be valid in the weak coupling (high density) regime.

---

*This work is done in collaboration with K. Iida (RIKEN), T. Hatsuda (Univ. of Tokyo) and G. Baym (Illinois U.)

FINITE TEMPERATURE PHASE TRANSITION IN FULL QCD WITH $N_F = 2$ FLAVORS OF CLOVER FERMIONS AT $N_T = 8$ AND 10

Y. NAKAMURA *

Kanazawa University, Kanazawa 920-1192, JAPAN
E-mail: yoshi@hep.s.kanazawa-u.ac.jp

We present results for QCD with $N_f = 2$ flavors of dynamical quarks using nonperturbatively improved Wilson fermions at finite temperature on $16^3 \times 8$ and $24^3 \times 10$ lattices. We determine the critical temperature T_c in the range of quark masses $0.6 \lesssim m_\pi/m_\rho \lesssim 0.8$.

We generate configurations on $16^3 \times 8$ (β=5.2, 5.25) and $24^3 \times 10$ (β=5.2) lattices at various κ. The Polyakov loop susceptibility is used to determine the transition point. We get values for T_c: $T_c \sim$196(8)MeV(N_t=10), T_c = 210(4)MeV(N_t=8, β=5.2), T_c=219(3)MeV(N_t=8, β=5.25).

To estimate T_c in the continuum and the chiral limits we apply the extrapolation formula, which takes into account the lowest order corrections:

$$T_c r_0 = (T_c r_0)^{m_q, a\to 0} + C_a(a/r_0)^2 + C_q(\frac{1}{\kappa} - \frac{1}{\kappa_c})^\alpha ,$$

where $r_0 = 0.5$ fm and $(T_c r_0)^{m_q, a\to 0}$ corresponds to the extrapolated value. We have four values for $T_c r_0$ including $T_c(N_t$=6) determined by Edwards and Heller [1]. Assuming that the critical indices of the two flavor QCD belong to the universality class of the 3D $O(4)$ spin model(α=0.55), we get the extrapolated temperature: 172.5(3.3)MeV. Since results at N_t=10 is still preliminary, we are continuing simulations on $24^3 \times 10$ lattice to get better precision of T_c value.

References

1. R.G. Edwards and U.M. Heller, *Phys. Lett.* **B462**, 132

*This work is done in collaboration with V. Bornyakov (IHEP), M.N. Chernodub (ITEP), Y. Koma (Kanazawa U.), Y. Mori (Kanazawa U.), M.I. Polikarpov (ITEP), G. Schierholz (NIC/DESY), A. Slavnov (SMI), H. Stüben (KZZIB), T. Suzuki (Kanazawa U.), P. Uvarov (ITEP), and A.I. Veselov (ITEP)

ADS/CFT, LARGE N, AND CONFINEMENT

ISHWAREE P. NEUPANE *

Department of Physics, National Taiwan University, 106 Taipei, Taiwan

In the AdS/CFT context, the large N theory has a low temperature phase with vanishing free energy as the confining phase and a high temperature phase with negative free energy as the deconfining phase.

AdS/CFT is a useful tool to study confinement and phase transitions in large N limits; viz., a Hawking-Page phase transition between thermal AdS and AdS black hole has dual description as the confinement-deconfinement transition in large N gauge theories. This motivates us to study AdS thermodynamics including α'-corrections. To study a dual field theory at finite temperature, we consider a spherical AdS_5 back hole whose horizon is isometric to $\mathbb{R}^1 \times S^3$ and go to a Euclidean description by compactifying $\mathbb{R}^1$ on a circle with period $\beta = T^{-1}$, and compute the free energy [1]

$$F = V_3 N^2 \xi^2/(4\pi^2 l) \left[(1-\xi^2) - \alpha'/l^2 \left(3\xi^2 + 12 + 5/\xi^2\right)\right] , \tag{1}$$

where $\alpha' = l^2/8N$, l is the AdS_5 scale, and $l\xi = r_+$ is the horizon. $F < 0$ for $\alpha'/l^2 > 1/42$ but it can be positive for $\alpha'/l^2 < 1/42$. The transition temperature $T_c = \frac{3}{2\pi l}(1 - \frac{3}{4N})$ at $F = 0$ weakly depends on N. A criterion for confinement is whether F is order of 1 or N^2. Our result shows the existence of a thermal phase transition for $N > 5$. The energy density is

$$E = (3N/32\pi^2 l)\left(1+\xi^2\right)\left(\xi^2\left(8N+1\right)+1\right). \tag{2}$$

The specific heat $C = dE/dT$ is discontinuous at $\xi < 1$ and the phase transition is the first order for large N supergravity solutions.

References

1. Y.M. Cho and I.P. Neupane, *Phys. Rev.* **D66**, 024044 (2002); I.P. Neupane, *Phys. Rev.* **D67**, 061501(R) (2003).

*On leave from Department of Physics, Tribhuvan University, Kirtipur, Kathmandu

COLOR FERROMAGNETISM AND QUANTUM HALL STATES IN QUARK MATTER

T. NISHIKAWA*

Institute of Particle and Nuclear Study, KEK, 1-1 Ooho, Tsukuba, Ibaraki 305-0801, Japan
E-mail: nishi@post.kek.jp

We show a possibility that a color ferromagnetic state is realized in the quark matter along with a quantum Hall state of gluons.

As is well known, the Savvidy vacuum [1], in which color magnetic field is spontaniously generated, is unstable. The instability comes from the unstable modes of gluons [2]. Our proposal is that the modes compose a quantum Hall state to stabilize the ferromagnetic state [3,4]. Quantum Hall states are known to be described by the Chern-Simons gauge theory. Using the theory, we can show that any unstable modes are absent in the quantum Hall state of gluons. The quantum Hall state of gluons possesses a color charge, which must be supplied from somewhere in color neutral system. The quark matter is a supplier of the color charge. The ferromagnetic state is shown to arise between the hadronic state and the color superconducting state when the density of quarks is varied. If the state is realized in the quark matter produced in heavy-ion collisions, a real magnetic field generated by quarks rotating around the color magnetic field might be observed.

References

1. G.K. Savvidy, *Phys. Lett.* **B71**, 133 (1977).
2. N.K. Nielsen and P. Olesen, *Nucl. Phys.* **B144**, 376 (1978); *Phys. Lett.* **B79**, 304 (1978).
3. A. Iwazaki and O. Morimatsu, *Phys. Lett.* **B571**, 61 (2003).
4. A. Iwazaki, O. Morimatsu, T. Nishikawa and M. Ohtani, hep-ph/0309066.

*This work is done in collaboration with A. Iwazaki (Nishogakusha-U), O. Morimatsu (KEK) and M. Ohtani (RIKEN)

EFFECT OF PION THERMAL WIDTH ON THE SPECTRAL FUNCTION OF THE SIGMA MESON

M. OHTANI*

RIKEN (The Institute of Physical and Chemical Research), Wako, Saitama 351-0198, Japan
E-mail: ohtani@rarfaxp.riken.go.jp

We study effects of pion thermal width on the spectral function of σ. It is found that the thermal width broadens a sharp peak of the spectrum and diphoton yield from the decay $\sigma \to 2\gamma$.

We study[1] effects of pion thermal width on the spectral function of σ in the $O(4)$ linear sigma model with the resummation technique called Optimized Perturbation Theory[2]. To reckon with the thermal width, we use a complex pole of the pion propagator[3] for the internal mass in the self-energy of σ. As a result of this prescription, two broad peaks are found in the spectral function at nonzero temperature and a sharp peak at the 2π threshold, which can be obtained[2] in absence of thermal width of π, turned out to be smeared out.

To analyze the behavior of the spectral function, we search the complex energy plane for the poles of the σ propagator and elucidate the correlation between the pole location and the smeared structure of the spectrum. We find that the thermal width also broadens[4] a sharp peak of the diphoton emission rate from the decay: $\sigma \to 2\gamma$.

References

1. Y. Hidaka, O. Morimatsu, T. Nishikawa and M. Ohtani, `hep-ph/0304204`.
2. S. Chiku and T. Hatsuda, *Phys. Rev.* **D58**, 076001 (1998).
3. Y. Hidaka, O. Morimatsu and T. Nishikawa, *Phys. Rev.* **D67**, 056004 (2003).
4. M. Ohtani, Y. Hidaka, O. Morimatsu and T. Nishikawa, RIKEN preprint: `RIKEN-AF-NP-449`.

*This work is done in collaboration with Y. Hidaka (U. Tokyo/KEK), O. Morimatsu (KEK) and T. Nishikawa (KEK).

PION DECAY CONSTANTS IN NUCLEAR MEDIUM

M. OKA*

Department of Physics, Tokyo Institute of Technology
Meguro, Tokyo 152-8551, JAPAN
E-mail: oka@th.phys.titech.ac.jp

The QCD sum rule is applied to the pion decay constants in nuclear medium. We find that both the time and spatial components are reduced in medium, but they behave differently. The obtained values are consistent with the recent observation in deeply bound pionic states in heavy nuclei.

The QCD sum rule calculation of the in-medium pion decay constants using pseudoscalar-axial vector correlation function, $i\int d^4x\ e^{ip\cdot x}\langle\rho|T[\bar{d}(x)i\gamma_5 u(x)\ \bar{u}(0)\gamma_\mu\gamma_5 d(0)]|\rho\rangle$, is revisited.[1,2] In particular, we argue that the dimension 5 condensate, $\langle\bar{q}(iD_0)^2q\rangle_N + \frac{1}{8}\langle\bar{q}g_s\sigma\cdot Gq\rangle_N$, which is crucial for splitting the time (f_t) and space (f_s) components of the decay constant, should be negative so that it does not yield a tachyonic pion mass. Using the in-medium pion mass as an input, we fix the dimension 5 condensate to be around -0.025 GeV^2 $\sim$ -0.019 GeV^2. The role of the N and Δ intermediate states in the correlation function is also investigated. With (and without) explicit Δ subtraction while allowing the in-medium pion mass to vary within 139 MeV $\leq m_\pi^* \leq$ 159 MeV, we obtain $f_s/f_\pi = 0.37 \sim 0.78$ and $f_t/f_\pi = 0.63 \sim 0.79$. The result for f_t is consistent with the recent observation of the deeply bound pion states in heavy nuclei.[3]

References

1. H. Kim, *Phys. Rev.* **C65**, 055201 (2002) .
2. H. Kim and M. Oka, *Nucl. Phys.* **A720**, 368 (2003).
3. K. Suzuki et al., nucl-ex/0211023.

*This work is done in collaboration with H. Kim (Yonsei Univ.)

STRING EFFECTS IN $SU(2)$ YANG-MILLS THEORY

M. PEPE [1], M. CASELLE [2], A. RAGO [2]

[1] *Insitute of Theoretical Physics, University of Bern, Bern, Switzerland.*
[2] *University of Torino and INFN Sezione di Torino, Torino, Italy.*

We investigate the effective string picture of $SU(2)$ Yang-Mills theory. We compare our numerical results with similar studies in $SU(3)$ and $Z(2)$ gauge theories.

The interaction between two static charges at distance R in the confining phase of a pure gauge theory is characterized by a potential $V(R)$ linearly rising with slope σ . In the superconductor model of confinement, this behaviour is described by the focusing of the field lines into a thin string-like flux tube connecting the two static charges. Since the confining phase is also rough, the flux tube is free to vibrate. The effective string picture gives a purely geometrical description of the low-energy static charge interaction as a massless fluctuating string. This effective picture yields to quantitative expectations on the subleading terms of $V(R)$. We have investigated [1] these issues in $SU(2)$ Yang-Mills theory; in the figure we compare our results with those obtained in $SU(3)$ [2] and $Z(2)$ [3] gauge theories.

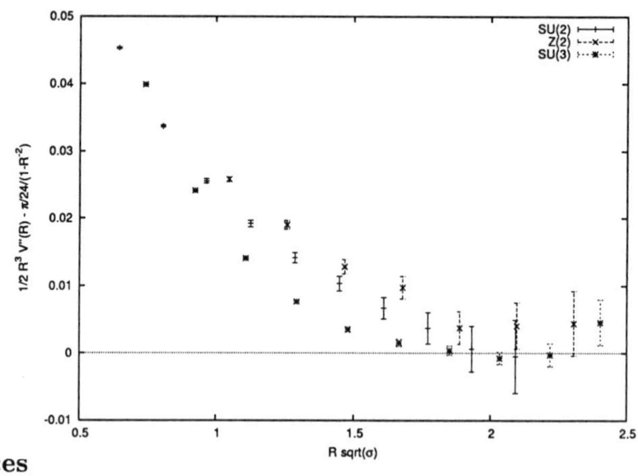

References

1. M. Caselle, M. Pepe and A. Rago, in preparation.
2. M. Lüscher and P. Weisz, *JHEP* **0207**, 049 (2002).
3. M. Caselle, M. Hasenbusch and M. Panero, *JHEP* **0301**, 057 (2003).

QUARKONIUM AT FINITE TEMPERATURE *

P. PETRECZKY †

Physics Department, Brookhaven National Laboratory, Upton, NY 11973, USA

I discuss quarkonium spectral functions at finite temperature reconstructed using the Maximum Entropy Method. I show in particular that the J/ψ survives in the deconfined phase up to $1.5T_c$

The study of quarkonium system at finite temperature has been a subject of considerable interest since the work of Matusi and Satz [1], but a first principle calculation of quarkonium properties at non-zero temperature was missing. It was shown, however, that the application of the Maximum Entropy Method (MEM) can make such calculation possible [2]. The method have been successfully applied at zero [2] as well as at finite temperature [3].

I am going to discuss charmonium spectral function calculated with MEM on $48^3 \times N_\tau$ lattices at lattice spacing $a^{-1} = 4.86$GeV and $N_\tau = 24, 16$ and 12 corresponding to temperatures $0.75T_c$, $1.12T_c$ and $1.5T_c$ (T_c being the deconfinement temperature). The results for the vector channel are shown in the Figure. As one can see the J/ψ seems to survive up to temperatures $1.5T_c$. Similar calculation have been performed in the scalar and axial vector channels which correspond to the P-state charmonia, but no peak was found there.

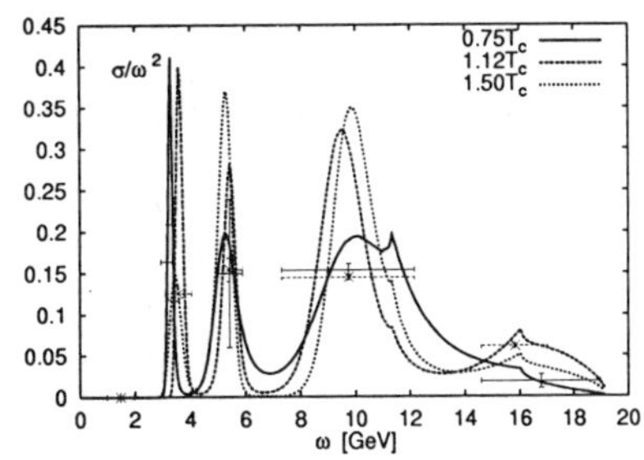

References

1. T. Matsui and H. Satz, *Phys. Lett.* **B178**, 416 (1986).
2. M. Asakawa et al., *Prog. Part. Nucl. Phys.* **46**, 459 (2001).
3. F. Karsch et al., *Phys. Lett.* **B530**, 147 (2002); S. Datta et al., hep-lat/0208012; M. Asakawa and T. Hatsuda, hep-lat/0308034.

---

*Based on work done in collaboration with S. Datta, F. Karsch and I. Wetzorke.

†Goldhaber fellow, supported under contract DE-AC02-98CH10886 with the U.S. Department of Energy.

PERTURBATIVE CHECK OF THE ENERGY LATTICE SUM RULE AND CONTRIBUTIONS OF THE FIELD ENERGY AND TRACE ANOMALY TO THE $Q\bar{Q}$-POTENTIAL

BJÖRN FEUERBACHER, HEINZ J. ROTHE

Institut für Theoretische Physik, Universität Heidelberg, Germany

The energy lattice sum rule for the static quark-antiquark potential was checked using lattice perturbation theory up to next-to-leading order. The contributions to the potential arising from the naive field energy and the trace anomaly of the energy momentum tensor were studied.

Lattice sum rules relating the $q\bar{q}$–potential, determined from the Wilson loop, to the action and energy stored in the chromoelectric and chromomagnetic fields have been first proposed by C. Michael[1]. Later a corrected version thereof was given by one of us[2]. The energy sum rule includes a contribution to the potential arising from the trace anomaly of the energy-momentum tensor. An important ingredient going into its derivation is that the potential does not depend on the anisotropy $\xi = a/a_t$ in the continuum limit. A reliable non-perturbative check of this sum rule has so far not been achieved. The energy sum rule reads:

$$\hat{V}(\hat{R},\hat{\beta}) = \lim_{\hat{T}\to\infty} \frac{1}{\hat{T}} \Big[\eta_- < -\mathcal{P}_t + \mathcal{P}_s >_{q\bar{q}-0} + \frac{1}{4}\frac{2\beta_L(g_0)}{g_0} < S >_{q\bar{q}-0} \Big]$$

(for the definition of η_-, $\mathcal{P}_t$, $\mathcal{P}_s$ and $< \ldots >_{q\bar{q}-0}$ see Ref. 2; S is the action). The energy sum rule states that the potential is given by the sum of the normal (euclidean) field energy (first term) and a contribution from the trace anomaly (second term). The sum rule was verified in lattice perturbation theory up to next-to-leading order, with the potential taken from Ref. 3 and the perturbative expressions for the couplings on an anisotropic lattice from Ref. 4. Of particular interest are the various contributions to the $q\bar{q}$-potential in order g_0^4. They are shown below for SU(3) Yang-Mills.

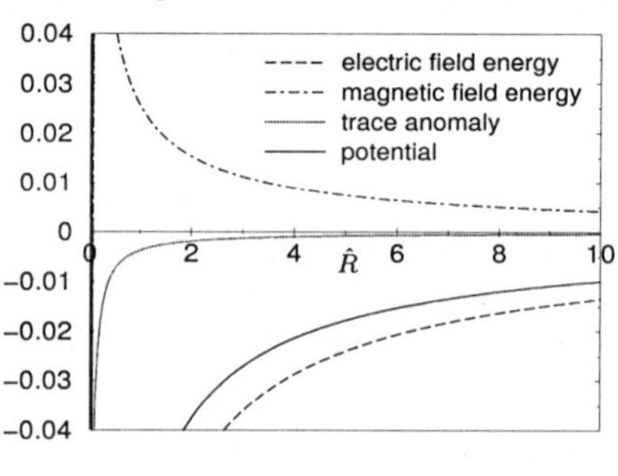

References

1. C. Michael, *Nucl. Phys.* **B280** [FS18], 13 (1987).
2. H. J. Rothe, *Phys. Lett.* **B355**, 260 (1995); *Phys. Lett.* **B364**, 227 (1995).
3. E. Kovacs, *Phys. Rev.* **D25**, 871 (1982).
4. F. Karsch, *Nucl. Phys.* **B205**, 285 (1982).

VECTOR MANIFESTATION OF CHIRAL SYMMETRY — A NEW PATTERN FOR WIGNER REALIZATION —

CHIHIRO SASAKI

Department of Physics, Nagoya University, Nagoya, 464-8602, JAPAN
E-mail: sasaki@eken.phys.nagoya-u.ac.jp

I report an analysis of my recent paper on the pion velocity, which shows that the pion velocity is related to an existence of a new fixed point near the critical temperature of chiral phase transition.

An enhancement of dielectron mass spectra below ρ/ω resonance observed at CERN SPS [1] can be explained by the dropping mass of ρ meson [2] following the Brown-Rho scaling [3]. There is a scenario which certainly requires the dropping mass of the vector meson and supprots the Brown-Rho scaling, named the vector manifestation (VM) [4] where the massless vector meson becomes the chiral partner of pion at the critical point. Starting from the premise that the *bare* effective field theory is to be defined from the underlying QCD at a suitable matching scale, the VM is formulated in large N_f, hot or dense QCD [4,5,6].

In Ref. 7, I have shown that the pion velocity at the critical temperature T_c receives neither quantum nor hadronic thermal corrections protected by the VM: This occurs due to the exact cancellation between the contribution of pion and that of the longitudinal vector meson (i.e., the chiral partner of pion). The result implies that the pion velocity v_π at T_c is equal to the bare pion velocity $V_{\pi,\text{bare}}$ determined through the matching at T_c; $v_\pi(T_c) = V_{\pi,\text{bare}}(T_c)$, which is a new fixed point originated in the VM.

References

1. G. Agakishiev *et al.* [CERES Collaboration], *Phys. Rev. Lett.* **75**, 1272 (1995).
2. G. Q. Li, C. M. Ko and G. E. Brown, *Phys. Rev. Lett.* **75**, 4007 (1995).
3. G. E. Brown and M. Rho, *Phys. Rev. Lett.* **66**, 2720 (1991).
4. M. Harada and K. Yamawaki, *Phys. Rev. Lett.* **86**, 757 (2001).
5. M. Harada and C. Sasaki, *Phys. Lett. B* **537**, 280 (2002).
6. M. Harada, Y. Kim and M. Rho, *Phys. Rev. D* **66**, 016003 (2002).
7. C. Sasaki, arXiv:hep-ph/0306005.

SOLUTIONS OF THE SKYRME MODEL DESCRIBED BY THE WEIERSTRASS $\wp$-FUNCTION

CHANG-GUANG SHI, JUN YAMASHITA AND MINORU HIRAYAMA

Gofuku 3190, Toyama University, Toyama 930-8555, Japan
E-mail: shicg@jodo.sci.toyama-u.ac.jp

The exact solutions of the Skyrme model which are described by the Weierstrass $\wp$ function or the Jacobi elliptic function are obtained. They are not solitonic but the superposition of three plane waves. Using them, the energy-momentum tensor and the non-zero baryon number density are calculated.

In our recent paper[1], we have obtained the exact solutions of the $SU(2)$ Skyrme model[2] of the form $g(x) = h(k\cdot x,\ \ l\cdot x,\ \ m\cdot x)$ with $k^2 = l^2 = m^2 = 0,\ \ k\cdot l,\ \ l\cdot m,\ \ m\cdot k > 0$. They are described by the Weierstrass $\wp$ function $K(\omega) = \wp(\omega+\omega_3) = e_3 + \dfrac{(e_3-e_1)(e_3-e_2)}{\wp(\omega)-e_3}$, where e_1, e_2 and e_3 are real constants satisfying $e_1 > e_2 > e_3$, $2\omega_3$ is the second fundamental period of $\wp(\omega)$ and $\omega = L\cdot x$ is a linear combination of $k\cdot x$, $l\cdot x$ and $m\cdot x$. $K(\omega)$ can be rewritten as $K(\omega) = e_3 + (e_2-e_3)\mathrm{sn}^2\left(\sqrt{e_1-e_3}\,\omega,\ \sqrt{(e_2-e_3)/(e_1-e_3)}\right)$.

Utilizing the above solutions, the energy-momentum tensor $T_{\mu\nu}$ and the baryon number density $N_0(x)$ are calculated. The latter is given by $N_0(x) = n_0 \dfrac{\boldsymbol{k}\cdot(\boldsymbol{l}\times\boldsymbol{m})}{\sqrt{(k\cdot l)(l\cdot m)(m\cdot k)}}\dfrac{dK(\omega)}{d\omega}$, where n_0 is a constant independent of k, l and m. The solutions supply us with examples of the superposition of three plane waves in the Skyrme model. We note that the two-plane-wave version[3] of the above solutions lead to a vanishing baryon number density.

References

1. M. Hirayama, C-G Shi and J.Yamashita, *Phys. Rev.* **D67**, 105009 (2003).
2. T. H. R. Skyrme, *Nucl. Phys.* **31**, 556 (1961).
3. M.Hirayama and J. Yamashita *Phys. Rev.* **D66**, 105019 (2002).

SOLVING THE SCHWINGER-DYSON EQUATIONS FOR GLUODYNAMICS IN THE MAXIMAL ABELIAN GAUGE

T. SHINOHARA, K.-I. KONDO AND T. MURAKAMI

Graduate School of Science and Technology, Chiba University, Japan

We derive the Schwinger-Dyson equations for the $SU(2)$ Yang-Mills theory in the maximal Abelian gauge and solve them in the infrared asymptotic region. We find that the infrared asymptotic solutions for the gluon and ghost propagators are consistent with the hypothesis of Abelian dominance.

The Schwinger-Dyson (SD) equation is one of the most popular approaches to investigate the non-perturbative features of quantum field theory. The analyses by making use of the SD equation for quark propagator are well-known. Recently, the coupled SD equations for the gluon and ghost propagators in Yang-Mills theory have been studied mainly in the Lorentz (Landau) gauge. We derive the SD equations for the $SU(2)$ Yang-Mills theory in the maximal Abelian (MA) gauge and solve them analytically in the infrared (IR) asymptotic region. The MA gauge is useful to investigate the Yang-Mills theory from the viewpoint of the dual superconductivity.

We neglect the two-loop graphs and adopt the ansatz for the vertex functions which are compatible with the Slavnov-Taylor identities. Then we obtain the following results. The solution shows that the diagonal gluon propagator is enhanced in the IR limit, while the off-diagonal gluon and off-diagonal ghost propagators are suppressed in the IR region. Our results are compatible with a hypothesis of Abelian dominance.

References

1. T. Shinohara, T. Imai and K.-I. Kondo, *J. Mod. Phys.* **A**, (2003) in press: hep-th/0105268.
2. T. Shinohara, K.-I. Kondo and T. Murakami, hep-lat/0309164.

LARGE MASS AND CHEMICAL POTENTIAL MODEL: A LABORATORY FOR QCD?

I. - O. STAMATESCU AND R. HOFMANN

Institut für Theoretische Physik, Univ. Heidelberg, Heidelberg, Germany
E-mail: stamates@thphys.uni-heidelberg.de

We use a model based on the hopping parameter expansion to study QCD at large μ. We find interesting behavior in the region expected to show flavor-color locking.

We have performed simulations at small T, large μ for a model which can be obtained from QCD by small order hopping parameter expansion at large μ [1]. The quarks, although very heavy, have some limited amount of mobility. Results are obtained from simulations with one and three degenerate flavors at $\beta = 5.6$ on a 6^4 lattice for $\kappa = 0.12$ and various μ (same for all flavors). Relative good convergence up to rather high baryon densities is achieved with a reweighting algorithm which takes part of the fermionic determinant in the Boltzmann factor. The results show strong increase of the baryon density and other observables above $\mu \sim 1$ which may be a signal for entering a new, high density phase at small temperature. Particularly interesting is the behavior of the di-quark correlator of 3-flavor QCD (in this approximation) which becomes increasingly flat at large μ, leading to a strongly increasing susceptibility. This may be a signal for the development of a condensate with color-flavor locking [2]. Simulations on larger lattices would be essential to check this conjecture, it remains to be seen whether good convergence can be achieved in that case.

References

1. G. Aarts, O. Kaczmarek, F. Karsch and I.-O. Stamatescu, *Nucl. Phys.* **B** (Proc.Suppl.) **106**, 456 (2002).
2. M. G. Alford, K. Rajagopal, F. Wilczek, *Nucl. Phys.* **B537**, 443 (1999).

CONFINING STRINGS AND HIGH-ENERGY REACTIONS

FRANK DANIEL STEFFEN

Institute for Theoretical Physics, Bern University
Sidlerstrasse 5, CH-3012 Bern, SWITZERLAND
E-mail: steffen@itp.unibe.ch

We present the loop-loop correlation model that allows us to compute confining flux tubes and to examine their effects and manifestations in high-energy reactions.

The loop-loop correlation model (LLCM) combines perturbative gluon exchange with the non-perturbative stochastic vacuum model.[1] This combination describes the static quark-antiquark potential with color Coulomb behavior for small and confining linear rise for large source separations in agreement with lattice results. Computing the chromo-electric fields of a static color dipole in the fundamental and adjoint representation of $SU(N_c)$, confining flux tubes are obtained for large dipoles that show exact Casimir scaling.[2] Based on analytic continuation, the Euclidean LLCM provides the S-matrix element S_{DD} for high-energy dipole-dipole scattering.[2] For reactions involving large dipoles, we find clear manifestations of the confining strings. Convoluting S_{DD} with appropriate wave functions to describe dipole-hadron scattering, the confining string in the dipole can be decomposed into stringless dipoles with a given dipole number density. This allows us to calculate unintegrated gluon distributions of hadrons and photons from dipole-hadron and dipole-photon cross sections via $k_\perp$ factorization.[3]

References

1. A. I. Shoshi, F. D. Steffen and H. J. Pirner, *Nucl. Phys.* **A709**, 131 (2002); F. D. Steffen, arXiv:hep-ph/0301084; A. I. Shoshi, arXiv:hep-ph/0302179.
2. A. I. Shoshi, F. D. Steffen, H. G. Dosch and H. J. Pirner, *Phys. Rev.* **D68**, 074004 (2003); arXiv:hep-ph/0211287.
3. A. I. Shoshi, F. D. Steffen, H. G. Dosch and H. J. Pirner, *Phys. Rev.* **D66**, 094019 (2002); arXiv:hep-ph/0207287.

REALIZATION OF LATTICE CHIRAL SYMMETRY WITH HOPPING INTERACTIONS

TAKANORI SUGIHARA

RIKEN BNL Research Center, Brookhaven National Laboratory
Upton, New York 11973-5000, USA
E-mail: sugihara@bnl.gov

Approximate chiral symmetry is realized on a (1+1)-dimensional lattice based on a Hamiltonian formalism. The species doubling problem of the lattice fermion is resolved by introducing Wilson-like hopping interactions that mix left- and right-handed fermions around the momentum boundary. Deviation of the fermion propagator from the continuum one is small.

In contrast with the great success of lattice gauge theory, lattice fermions remain a long-standing problem. Naive discretization of a fermion causes the well-known species doubling problem. In addition to this, the fermion propagator defined on the lattice deviates from the continuum one even if the doublers are removed with the existing techniques such as Wilson's fermion. In this work, based on a Hamiltonian formalism, we introduce ultralocal hopping interactions and the Lanczos factor to remove doublers and improve momentum dependence of fermion energy. The Lanczos factor is a technique used in Fourier analysis to cancel the Gibbs phenomenon (oscillations caused by singularities). Doubler is removed with Wilson-like hopping interactions. Explicit breaking of chiral symmetry is compressed to high energy to save chiral properties at low energy. As a result, approximate chiral symmetry is realized at low energy. Errors associated with chiral symmetry can be improved in a systematic way by increasing the largest distance of fermion hopping. In future works, it should be precisely checked if insertion of gauge interactions affects the chiral properties. See the reference for the details of this work.

References

1. T. Sugihara, *Phys. Rev.* **D68**, 034502 (2003).

VECTOR COMPONENT OF CONFINEMENT FORCE AND P-WAVE MESON SPECTRA IN D MESON SYSTEM

J. SUGIYAMA, S. MASHITA AND M. OKA*

Tokyo Institute of Technology, Meguro, Tokyo 152-8551, Japan

The Lorentz property of the confinement force is investgated in terms of the mass spectrum and the pionic decay widths of the P-wave D mesons, using the Dirac equation for light quark. We find that the Lorentz-scalar confinement is consistent with the mass spectrum, while a significant mixture of the Lorentz-vector confinement is necessary to explain the decay widths.

In order to determine the Lorentz property of the confinement force, we calculate the masses of P-wave mesons D_2^*, D_1 with $j_q = 3/2$ and D_1^*, D_0^* with $j_q = 1/2$ by using the Dirac equation with linear plus Coulomb potential for the light-quark, with considering the mixing of D_1 and D_1^*.

We conclude that D_1 is lighter than D_1^* (see Fig.1). Our calculation matches the experimental data, when the ratio of the vector component of the linear potential, denoted as v, is small.

The difference of the pionic decay widths of D_1 and D_1^* increases, as v is increased (see Fig.2). This difference is large, experimentally. Thus, our calculation of the pionic decay widths suggests a large v, while the masses of D_1 and D_1^* suggest a small v.

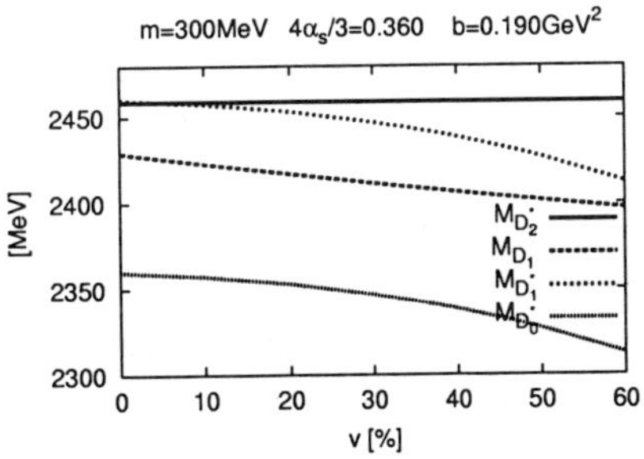

Figure 1. The masses of P-wave D mesons v.s. the vector ratio v.

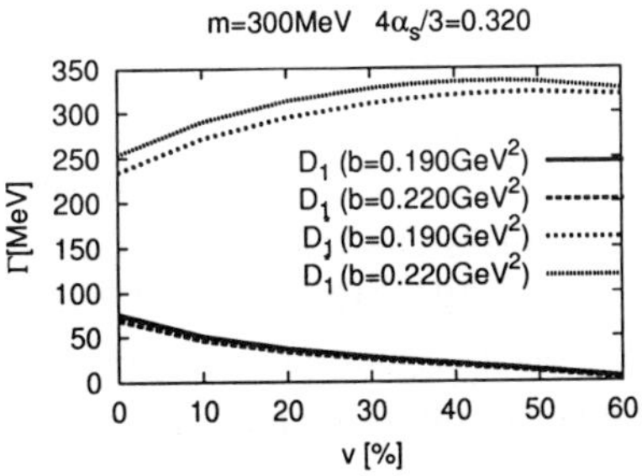

Figure 2. The pionic decay widths of D_1 and D_1^* v.s. the vector ratio v.

*This work is done in collaboration with M. Ishida (Meisei Univ.)

PHASE STRUCTURE OF HOT AND/OR DENSE QCD WITH THE SCHWINGER-DYSON EQUATION

S. TAKAGI

Department of Physics, Nagoya University
Furo-cho, Chikusa-ku, Nagoya 464-8602, JAPAN
E-mail: satoshi@eken.phys.nagoya-u.ac.jp

We investigate the phase structure of hot and/or dense QCD using the Schwinger-Dyson equation (SDE) with the improved ladder approximation in the Landau gauge. We show that the phase transition from the two-flavor color superconducting (2SC) phase to the quark-gluon plasma (QGP) phase is of second order, and that the scaling properties of the Majorana mass gap and the diquark condensate are consistent with mean field scaling. We examine the effect of the antiquark contribution and find that setting the antiquark Majorana mass equal to the quark one is a good approximation in the medium density region. We also study the effect of the Debye screening mass of the gluon and find that ignoring it causes the critical lines to move to the region of higher temperature and higher chemical potential.

We stress the importance of the imaginary part of the Dirac mass at nonzero chemical potential. We perform the analysis at finite chemical potential μ and finite temperature T with the imaginary part of the Dirac mass included, because the imaginary part is inevitably generated in the chiral symmetry breaking phase in the SDE at non-zero chemical potential. We find that the most noteworthy feature of the analysis that includes the imaginary part of the Dirac mass is in the position of the tricritical point on the T-μ plane. In the SDE analysis, including the imaginary part causes the tricritical point to move to a position of much smaller μ: $(T, \mu) = (124, 210)\text{MeV} \to (T, \mu) = (146, 20)\text{MeV}$.

The imaginary part of the Dirac mass has an important influence on the chiral phase transition, in particular on the tricritical point. [1]

References

1. S. Takagi, *Prog. Theor. Phys.* **109**, 233 (2003), hep-ph/0210227.

LIGHT SCALAR MESONS IN THE IMPROVED LADDER QCD

M. TAKIZAWA*

Showa Pharmaceutical University, Machida, Tokyo 194-8543 Japan
E-mail: takizawa@ac.shoyaku.ac.jp

We study the low-lying nonet scalar mesons using the improved ladder Schwinger-Dyson and Bethe-Salpeter equations with the $U_A(1)$ breaking 6-quark flavor mixing interaction.

The improved ladder Schwinger-Dyson and Bethe-Salpeter approach successfully explains the dynamical chiral symmetry breaking in QCD. This approach has been applied to the low-lying nonet pseudoscalar mesons: π, K, η and η' [1]. Since the η and η' mesons are affected by the $U_A(1)$ anomaly, the $U_A(1)$ breaking 6-quark flavor mixing interaction has been introduced. The observed masses of the low-lying nonet pseudoscalar mesons have been reproduced reasonably well.

Using the model parameteres determined in the pseudoscalar meson sector, we calculate the low-lying nonet scalar meson masses. We obtain $m_\sigma = 667$ MeV, $m_{a_0} = 942$ MeV, $m_{f_0} = 1336$ MeV and $m_\kappa = 1321$ MeV.

We find that the $U_A(1)$ breaking interaction gives rise to about 300 MeV mass splitting of the σ and a_0 mesons and the observed masses of σ and a_0 are reproduced reasonably well. On the other hand, our results of the m_{f_0} and m_κ are rather large. It suggests that the $f_0(980)$ and $\kappa(700\text{-}900)$ may be the multiquark states.

References

1. K. Naito, Y. Nemoto, T. Takizawa, K. Yoshida and M. Oka, *Phys. Rev.* **C61**, 065201 (2000).

*This work is done in collaboration with T. Umekawa (Tokyo Tech.), K. Naito (Hokkaido U.) and M. Oka (Tokyo Tech.)

NUCLEON TO Δ(1232) ELECTROMAGNETIC TRANSITION IN LATTICE QCD

A. TSAPALIS*

Department of Physics, University of Cyprus, CY-1678 Nicosia, Cyprus

We present the calculation of the matrix element for the transition $\gamma N \to \Delta$ both in quenched QCD and with two flavours of degenerate dynamical quarks.

The issue of nucleon deformation has been under intense experimental and theoretical study during the last twenty years. Due to the spin-1/2 nature of the nucleon, one searches for quadrupole strength in the $\gamma N \to \Delta$ transition. The allowed magnetic dipole, M1, electric quadrupole, E2, and Coulomb quadrupole, C2, amplitudes have been recently measured at Bates and JLab at various momentum transfer values and the ratio E2/M1 is found to be non-zero indicating a deformed nucleon and/or Δ. Lattice QCD is the only framework where these form factors can be evaluated in a model independent way.

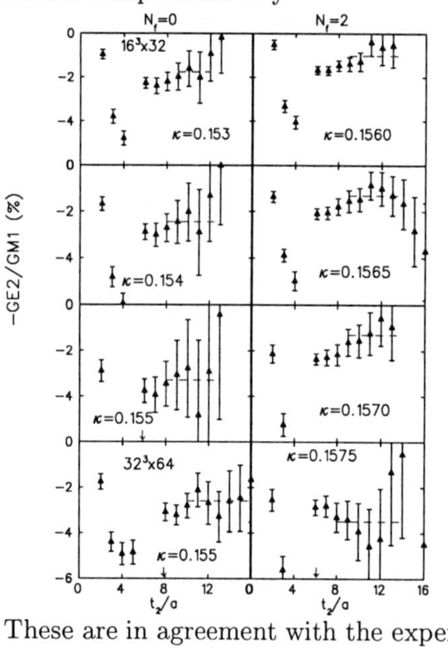

In the figure we show E2/M1 in the lab frame of the nucleon. The three upper graphs on the left show results for the quenched theory on a lattice of size $16^3 \times 32$ for $\mathbf{q} = 0.8$GeV at $m_\pi/m_\rho = 0.84, 078$ and 0.70 and the lowest shows results for the same value of $\mathbf{q}$ at $m_\pi/m_\rho = 0.70$. On the right we show results for the SESAM configurations at $m_\pi/m_\rho = 0.83, 0.81, 0.76$ and 0.68 for a similar $\mathbf{q}$. Extrapolating to the chiral limit we obtain at $-q^2 = 0.64\text{GeV}^2$ E2/M1 $= -4.8(1.8)$ and at $-q^2 = 0.13\text{GeV}^2$ E2/M1 $= -4.1(1.4)$ for the quenched theory. These are in agreement with the experimental values. Unquenching for the heavy quarks used here does not change these values significantly.

References

1. C. Alexandrou *et al.*, *Nucl. Phys.* **B** *(Proc. Suppl.)* **119**, 413 (2003).

*This work is done in collaboration with C. Alexandrou (Univ. of Cyprus), Ph. de Forcrand (ETH/CERN), Th. Lippert (Wuppertal), H. Neff (Boston Univ.), J.W. Negele (MIT), K. Schilling (Wuppertal) and W. Schroers (MIT).

THE HELICITY MODULUS IN THE 4d U(1) GAUGE THEORY

MICHELE VETTORAZZO[a] *and* PHILIPPE DE FORCRAND[a,b]

[a] *Institute for Theoretical Physics, ETH Zürich, CH-8093 Zürich, Switzerland*
[b] *CERN, Theory Division, CH-1211 Genève 23, Switzerland*

We consider the 4d compact $U(1)$ Wilson action and characterize its phase diagram using the notion of electromagnetic flux, instead of the more usual magnetic monopole. Taking inspiration from the flux picture, we consider the helicity modulus (h.m.) for this theory, and show that it is an order parameter for the confinement-deconfinement phase transition. We extend the definition of the h.m. to an abelian projected Yang-Mills theory, and discuss its behavior in $SU(2)$.

We build an *order parameter* for Abelian lattice gauge theories following the ideas used in the context of the 2d planar XY model [1]: we start from the Wilson formulation of the action $S = -\beta \sum \cos\theta_P$, where β is the inverse bare coupling and θ_P is the plaquette angle; this theory has a strong coupling confined phase and a weak coupling Coulomb phase. We consider a 4d hypercubic lattice on which an extra electromagnetic flux ϕ is imposed through a two-dimensional surface [2], and study the dependence of the free energy (f.e.) of the system on ϕ in the two phases: we show that in the *confined phase* the f.e. is independent of ϕ, while in the Coulomb phase it has a non-trivial dependence. Therefore the following definition of the *helicity modulus* (h.m.) $h(\beta) = \frac{\partial^2 F(\phi)}{\partial\phi^2}\,|_{\phi=0}$ (the curvature of the f.e. at flux zero) provides an order parameter, that is, a function of β which is zero in one phase (namely, the confined phase) and different from zero in the other. We have also considered an extension of this definition to the non-abelian $SU(2)$ pure gauge theory, by measuring the h.m. after gauge fixing and abelian projection. The gauge fixing condition used is the Maximal Abelian gauge. We observe a non-trivial signal for the confinement-deconfinement phase transition at finite temperature [3].

References

1. D.R. Nelson and J.M. Kosterlitz, *Phys. Rev. Lett.* **39**, 1201 (1977).
2. P. de Forcrand and M. Vettorazzo, arXiv:hep-lat/0209087.
3. P. de Forcrand and M. Vettorazzo, Proc. of *Lattice 2003*, to appear in *Nucl. Phys.* **B** (Proc. Suppl.).

Scientific Program of Confinement 2003

July 21 (Mon.)

9:00-10:00	Registration	
10:00-12:30	Opening Session	Chairman: M. Oka (Tokyo Inst. Tech.)

10:00-10:50 M. Shifman (Minnesota Univ.)
"Supersymmetric Gluodynamics and its Nonsupersymmetric Cousins:
Planar Equivalence, k-strings and all that"

10:50-11:40 H. Satz (Bielefeld Univ.)
"Deconfinement in QCD and in Nuclear Collisions"

11:40-12:30 A.J. Niemi (Uppsala Univ.)
"Can the QCD String form a Knot ?"

12:30-14:00	Lunch	
14:00-16:00	Lattice Hadron Physics	Chairman: H. Rothe (Heidelberg Univ.)

14:00-14:30 C. Morningstar (Carnegie Mellon Univ.)
"Excitations of the Static Quark-Antiquark Potential in Several Gauge Theories"

14:30-15:00 J. Kuti (Univ. of California, San Diego)
"QCD String Theory and Casimir Energy"

15:00-15:30 H. Suganuma (Tokyo. Inst. Tech.)
"Y-type Flux-Tube Formation and Gluonic Excitations in Baryons:
From QCD to Quark Model"

15:30-16:00 T. Blum (RIKEN-BNL)
"Lattice Calculation of the Lowest Order Hadronic Contributions to
the Muon Anomalous Magnetic Moment"

16:00-16:30	Coffee Break	
16:30-17:45	Nonperturbative QCD	Chairman: D. Zwanziger (New York Univ.)

16:30-17:05 K.-I. Kondo (Chiba Univ.)
"Vacuum Condensates, Effective Gluon Mass and Color Confinement"

17:05-17:45 J. Greensite (San Francisco State Univ.)
"Constituent Gluons, Coulomb Energy, and the Gluon Chain Model"

July 22 (Tue.)		
9:00 -10:45	Hot/Dense QCD	Chairman: H. Satz (Bielefeld Univ.)

9:00 - 9:45 F. Karsch (Bielefeld Univ.)
"Confronting Lattice Results on the Transition from Hadron Gas to Quark-Gluon Plasma with Resonance Gas, Potential and Quasi-Particle Models"

9:45-10:15 Y. Akiba (RIKEN)
"Experimental Results from RHIC"

10:15-10:45 X.-N. Wang (LBL, Berkeley)
"Jet Tomography of Hot and Dense Matter"

10:45-11:15	Coffee Break	
11:15-12:15	Current Topics	Chairman: T. Hatsuda (Univ. of Tokyo)

11:15-11:45 J. Nishimura (KEK)
"A New Method for Simulating QCD at Finite Density"

11:45-12:15 T. Nakano (RCNP, Osaka Univ.)
"Experimental Evidence for Penta-Quark System"

12:15-13:45	Lunch
13:45-15:15	Parallel Session Ia (Okochi Hall) Parallel Session Ib (Nishina Hall)
15:15-15:45	Coffee Break
15:45-17:15	Parallel Session IIa (Okochi Hall) Parallel Session IIb (Nishina Hall)
17:45-19:45	Poster Session with Reception Party (Dining Hall)

July 23 (Wed.)

9:00-10:30 Confinement Mechanism Chairman: M. Shifman (Minnesota Univ.)

9:00 - 9:30 A. Di Giacomo (Pisa Univ.)
"Confinement of Color : Recent Progress"

9:30-10:00 R.W. Haymaker (Louisiana State Univ.)
"Consistent Definitions of Flux and Electric and Magnetic Current in Abelian Projected SU(2) Lattice Gauge Theory"

10:00-10:30 E.T. Tomboulis (California-Los)
"Connecting Short to Long Distance Scales in the Confining Vacuum"

10:30-11:00 Coffee Break

11:00-12:30 Confinement Mechanism Chairman: H. Markum (Vienna U. Tech.)

11:00-11:30 M. Mueller-Preussker (Humboldt Univ., Berlin)
"SO(3) versus SU(2) Lattice Gauge Theory"

11:30-12:00 G. Schierholz (DESY)
"On the Vacuum of Full QCD"

12:00-12:30 V.I. Zakharov (Munich)
"Branes" in Lattice SU(2) Gluodynamics"

12:30-13:30 Lunch

13:30-15:00 SUSY QCD and Confinement Chairman: K. Yazaki (TWCU/RIKEN)

13:30-15:00 K. Konishi (Pisa Univ. & INFN, Pisa) including Lecture
"Non-Abelian Monopoles, Vortices and Confinement: Lessons from SUSY Gauge Theories"

15:15-16:45 Bus

16:45-18:00 Excursion at Hama-Rikyu

18:15-20:45 Banquet on Yakata-Bune

21:00-22:00 Bus

July 24 (Thu.)		
9:00-10:30	Nonperturbative QCD	Chairman: H. Suganuma (Tokyo Inst. Tech.)

9:00 - 9:50 S.J. Brodsky (SLAC)
"Light-Front Quantization and Novel Aspects of QCD"

9:50-10:30 K.-I. Aoki (Kanazawa Univ.)
"Nonperturbative Renormalization Group Analysis for Dynamical Chiral Symmetry Breaking in QCD"

10:30-11:00	Coffee Break	
11:00-12:20	Hot/Dense QCD	Chairman: T. Kunihiro (YITP, Kyoto Univ.)

11:00-11:45 M. Alford (Washington Univ.)
"Dense Quark Matter in Compact Stars"

11:45-12:20 T. Hatsuda (Univ. of Tokyo)
"Thermal Phase Transition of Dense QCD"

12:20-13:50	Lunch	
13:50-16:00	Infrared QCD	Chairman: S.J. Brodsky (SLAC)

13:50-14:30 G. Prosperi (Milan Univ.)
"Light Mesons and Infrared Behavior of Running Coupling Constant in QCD"

14:30-15:10 Y. Sumino (Tohoku Univ.)
"Renormalon Cancellation and Perturbative QCD Potential as a Coulomb + Linear Potential"

15:10-16:00 D. Zwanziger (New York Univ.)
"Non-perturbative Faddeev-Popov Formula and Infrared Limit of QCD"

* The conference will be closed before 4:30pm on July 24.

Parallel Session at Confinement 2003

There are 20 talks in the parallel session scheduled in the afternoon on July 22 (Tue.) at Okochi Memorial Hall and Nishina Hall.

Parallel Session Ia (13:45-15:15, Okochi Hall)
[Finite Temperature QCD] Chairman: I.-O. Stamatesucu (Heidelberg Univ.)

13:45-14:03 T. Doi (Tokyo Inst. Tech.)
"The Study of Chiral Restoration using the Quark-Gluon Mixed Condensate in Lattice QCD at Finite Temperature"

14:03-14:21 S. Ejiri (Bielefeld Univ.)
"The Equation of State for Two Flavor QCD at Finite Density"

14:21-14:39 N. Ishii (RIKEN)
"Properties of Thermal Glueballs"

14:39-14:57 H. Matsufuru (YITP, Kyoto Univ.)
"Charmonium Correlators at Finite Temperature in Quenched Lattice QCD"

14:57-15:15 M. Panero (Univ. di Torino and INFN, Sezione di Torino)
"Effective String Picture for Confining Gauge Theories at Finite Temperature"

Parallel Session Ib (13:45-15:15, Nishina Hall)
[Mathematical Aspects of QCD] Chairman: S. Furui (Teikyo Univ.)

13:45-14:03 P. Faccioli (E.C.T.*, Trento)
"Topology and Chirality Mixing Interaciton in QCD"

14:03-14:21 A. Feo (Trinity Coll. Dublin, Ireland/Parma U./INFN)
"Study of the Continuum Limit of Supersymmetric Ward-Takahashi Identities for N=1 Super Yang-Mills Theory"

14:21-14:39 K. Holland (Univ. of California, San Diego)
"When is the Deconfinement Phase Transition Universal ?"

14:39-14:57 B. Lucini (Oxford Univ.)
"Features of SU(N) Gauge Theories"

14:57-15:15 J.E. Wang (National Taiwan Univ.)
"S-brane Actions"

Parallel Session IIa (15:45-17:15, Okochi Hall)
[Lattice QCD for Nonperturbative Phenomena] Chairman: M. Ogilvie (Washington U.)

15:45-16:03 A. Ali Khan (Humboldt Univ., Berlin)
"Chiral Extrapolations and Finite Size Effects of Hadron Masses in Dynamical QCD"

16:03-16:21 K. Anagnostopoulos (Crete Univ.)
"The Factorization Method for Monte Carlo Simulations of Systems with a Complex Action"

16:21-16:39 M. Chernodub (Kanazawa Univ./ITEP)
"Gluodynamics in External Field: A Test of the Dual Superconductor Picture"

16:39-16:57 H. Markum (Vienna Univ. of Tech.)
"Monopoles in Real Time for Classical U(1) Gauge Field Theory"

16:57-17:15 T.T. Takahashi (YITP, Kyoto Univ.)
"Y-type Flux-Tube Formation in Baryons"

Parallel Session IIb (15:45-17:15, Nishina Hall)
[QCD Vacuum and Hadrons] Chairman: K.-I. Kondo (Chiba Univ.)

15:45-16:03 D. Dudal (Ghent Univ., Belgium)
"The 2PPI Expansion: Dynamical Mass Generation and Vacuum Energy"

16:03-16:21 F.S. Navarra (Sao Paulo Univ.)
"On the x_F Distribution of J/psi's Produced in Heavy Ion Collisions"

16:21-16:39 M. Nielsen (Sao Paulo Univ.)
"Progress in the Determination of the J/psi-pi Cross Section"

16:39-16:57 S.P. Sorella (State Univ. of Rio de Janeiro)
"Vacuum Condensates of Dimension Two in Yang-Mills Theories"

16:57-17:15 K. Tanaka (Juntendo Univ.)
"Motion of Quarks inside Heavy Flavor Hadrons and the QCD Equations of Motion"

Poster Session at Confinement 2003

The poster session will be held with a reception party at the Dining Hall during 5:45 pm. - 7:45 pm. on July 22 (Tue.). There are 39 presentations in the poster session. The size of the poster board is 1.13m ~1.13m, and the both sides are available.

K.V. Acoleyen (Ghent Univ., Belgium)
"Dynamical Mass Generation by Source Inversion"

C. Aubin (Washington Univ.)
"Lattice Gauge Fixing and the Violation of Spectral Positivity"

G. Burgio (Trinity College Dublin, Ireland)
"Confinement and Deconfinement Transition in SO(3) Lattice Gauge Theory"

M. Chabab (Cadiayyad Univ., Morocco)
"Confinement in Non Abelian Gauge Theory with Massive Dilaton"

S. Deguchi (Nihon Univ.)
"An Effective Abelian Gauge Theory of Massive SU(2) Yang-Mills Theory"

H. Fujii (Tokyo Univ., Komaba)
"Soft Modes at the Critical End Point in the Chiral Effective Models"

K. Fukushima (Tokyo Univ.)
"Thermodynamical Limit of the Canonical Partition Function with respect to the Quark Number in QCD"

S. Furui (Teikyo Univ.)
"A Study of Infrared Properties of Lattice Landau Gauge QCD"

F. Gubarev (Kanazawa Univ.)
"The Non-Abelian Stokes Theorem and Bianchi Identities in SU(2) Gauge Model"

Y. Hoshino (Kushiro Tech. Coll.)
"Low-Energy-Theorem Approach to One Particle Singularity in QED{2+1}"

H. Ichie (Tokyo Inst. Tech.)
"Baryonic Flux in Full QCD"

H. Iida (Tokyo Inst. Tech.)
"Chiral Symmetry Restoration at Finite Temperature in the Schwinger-Dyson Equation with Lattice QCD Data"

E.-M. Ilgenfritz (Humboldt Univ., Berlin)
"From the Abelian Projected Flux Tube to the Dual Abelian Higgs Model"

K. Ishiguro (Kanazawa Univ.)
"Entropy of Monopoles from Percolating Cluster in Pure SU(2) QCD"

J. Koponen (Helsinki Univ.)
"Energy Spectrum of a Heavy-Light Meson"

J.S. Lange (Frankfurt Univ.)
"Studying Parton Dynamics with the STAR Experiment"

T. Matsuura (Tokyo Univ.)
"Thermal Fluctuations of Gauge Fields and First Order Phase Transitions in Color Superconductivity"

Y. Nakamura (Kanazawa Univ.)
"Finite Temperature Phase Transition in Full QCD with N_f=2 Clover Fermion at L_t=8,10"

I. Neupane (National Taiwan Univ./Tribhuvan Univ., Kathmandu, Nepal)
"Large N and Confinement"

M. Nishikawa (Tokyo Univ.)
"A Simple Possible Cause of Color Confinement"

T. Nishikawa (KEK)
"Ferromagnetism and Quantum Hall States in Quark Matter"

M. Ohtani (RIKEN)
"Effect of Pion Thermal Width on the Spectral Function of the Sigma Meson"

M. Oka (Tokyo Inst. Tech.)
"Pion Decay Constants in Nuclear Medium"

M. Pepe (Bern Univ.)
"String Effects in SU(2) Lattice Gauge Theory"

P. Petreczky (BNL)
"Quarkonium at Finite Temperature"

H. Rothe (Heidelberg Univ.)
"Perturbative Check of the Energy Lattice Sum Rule and Contributions to the Quark-Antiquark Potential"

T. Saito (RIISE, Hiroshima Univ.)
"Lattice Study of QCD Screening Effect at Finite Temperature"

C. Sasaki (Nagoya Univ.)
"Vector Manifestation of Chiral Symmetry - A New Pattern for Wigner Realization-"

C.-G. Shi (Toyama Univ.)
"Solutions of the Skyrme Model described by the Weierstrass Pe-function"

T. Shinohara (Chiba Univ.)
"Solving the Schwinger-Dyson Equations for Gluodynamics in the Maximally Abelian Gauge"

I.-O. Stamatescu (Heidelberg Univ.)
"Properties of a Large Mass, Large Chemical Potential Model - a Laboratory for QCD"

F.D. Steffen (Bern Univ.)
"Confining QCD Strings, Casimir Scaling, and a Euclidean Approach to High-Energy Scattering"

T. Sugihara (RIKEN-BNL)
"Realization of Lattice Chiral Symmetry with Hopping Interactions"

J. Sugiyama (Tokyo Inst. Tech.)
"Vector Component of Confinement Force and P-wave Meson Spectra in D-Meson System"

S. Takagi (Nagoya Univ.)
"Phase Structure of Hot and/or Dense QCD with the Schwinger-Dyson Equation"

M. Takizawa (Showa Pharmaceutical Univ.)
"Light Scalar Mesons in the Improved Ladder QCD"

A. Tsapalis (Univ. of Cyprus)
"Nucleon to Delta(1232) Electromagnetic Transition in Lattice QCD"

M. Vettorazzo (ETH Zuerich)
"The Helicity Modulus in the 4d U(1) Gauge Theory"

U. Wenger (Oxford Univ.)
"SU(N) Gauge Theories at Large N"

Participant List of Confinement2003

131 participants

Abe, Takashi	Tokyo Inst. Tech.	tabe@th.phys.titech.ac.jp
Acoleyen, Karel Van	Ghent Univ., Belgium	karel.vanacoleyen@rug.ac.be
Akiba, Yasuyuki	RIKEN	akiba@bnl.gov
Alford, Mark	Washington Univ.	alford@wuphys.wustl.edu
Ali Khan, Arifa	Humboldt Univ., Berlin	alikhan@physik.hu-berlin.de
Anagnostopoulos, Konstantinos	Crete Univ.	konstant@physics.uoc.gr
Aoki, Ken-Ichi	Kanazawa Univ.	aoki@hep.s.kanazawa-u.ac.jp
Aubin, Christopher	Washington Univ.	caubin@hbar.wustl.edu
Blum, Tom	RIKEN-BNL Research Center	tblum@bnl.gov
Brodsky, Stanley J.	SLAC, Stanford Univ.	sjbth@SLAC.Stanford.EDU
Burgio, Giuseppe	Trinity Coll., Dublin	burgio@maths.tcd.ie
Cea, Paolo	Bali Univ., Italy	paolo.cea@ba.infn.it
Chabab, Mohamed	Cadiayyad Univ., Morocco	mchabab@ucam.ac.ma
Chernodub, Maxim	Kanazawa Univ./ITEP	maxim@hep.s.kanazawa-u.ac.jp
Cosmai, Leonardo	INFN Bari, Italy	Leonardo.Cosmai@ba.infn.it
Deguchi, Shinichi	Nihon Univ.	deguchi@phys.cst.nihon-u.ac.jp
Di Giacomo, Adriano	Pisa Univ.	adriano.digiacomo@df.unipi.it
Doi, Takumi	Tokyo Inst. Tech.	doi@th.phys.titech.ac.jp
Dudal, David	Ghent Univ., Belgium	david.dudal@UGent.be
Ejiri, Shinji	Bielefeld Univ.	ejiri@physik.uni-bielefeld.de
Enyo, Hideto	RIKEN/RBRC	enyo@riken.go.jp
Faccioli, Pietro	E.C.T.*, Trento	faccioli@ect.it
Feo, Alessandra	Trinity Coll./Parma	feo@maths.tcd.ie

Follana, Eduardo	Univ. of Glasgow	e.follana@physics.gla.ac.uk
Fujii, Hirotsugu	Tokyo Univ., Komaba	hfujii@phys.c.u-tokyo.ac.jp
Fukushima, Kenji	Tokyo Univ.	fuku@nt.phys.s.u-tokyo.ac.jp
Furui, Sadataka	Teikyo Univ.	furui@umb.teikyo-u.ac.jp
Greensite, Jeff	San Francisco State Univ.	greensit@alf.nbi.dk
Gubarev, Fedor	Kanazawa Univ./ITEP	gubarev@hep.s.kanazawa-u.ac.jp
Hasuko, Kazumi	RIKEN	hasuko@riken.go.jp
Hatsuda, Tetsuo	Tokyo Univ.	hatsuda@phys.s.u-tokyo.ac.jp
Haymaker, Richard W.	Louisiana State Univ.	haymaker@rouge.phys.lsu.edu
Hirayama, Minoru	Toyama Univ.	hirayama@sci.toyama-u.ac.jp
Holland, Kieran	California Univ, San Diego	holland@lorentz.ucsd.edu
Hoshino, Yuichi	Kushiro Tech. Coll.	hoshino@ippan.kushiro-ct.ac.jp
Ichie, Hiroko	Tokyo Inst. Tech.	ichie@th.phys.titech.ac.jp
Iida, Hideaki	Tokyo Inst. Tech.	iida@th.phys.titech.ac.jp
Ilgenfritz, Ernst-Michael	Humboldt U., Berlin	ilgenfri@physik.hu-berlin.de
Ishiguro, Katsuya	Kanazawa Univ.	ishiguro@hep.s.kanazawa-u.ac.jp
Ishii, Noriyoshi	RIKEN	ishii@rarfaxp.riken.go.jp
Izaki, Manabu	Tokyo Inst. Tech.	izaki@th.phys.titech.ac.jp
Jahn, Oliver	ETH Zurich	jahn@itp.phys.ethz.ch
Juge, Keisuke	Bern Univ.	juge@itp.unibe.ch
Kanda, Tomoaki	Tokyo Inst. Tech.	kanda@th.phys.titech.ac.jp
Karsch, Frithjof	Bielefeld Univ.	karsch@Physik.Uni-Bielefeld.DE
Kinoshita, Toichiro	Cornell Univ.	tk@mail.lns.cornell.edu
Kiyomichi, Akio	RIKEN	kiyo@riken.go.jp
Kohama, Akihisa	RIKEN	kohama@riken.go.jp
Kondo, Kei-Ichi	Chiba Univ.	kondo@cuphd.nd.chiba-u.ac.jp
Konishi, Kenichi	Pisa Univ. & INFN, Pisa	konishi@mail.df.unipi.it

Koponen, Jonna	Univ. of Helsinki	jmkopone@pcu.helsinki.fi
Kunihiro, Teiji	YITP, Kyoto Univ.	kunihiro@yukawa.kyoto-u.ac.jp
Kurita, Kazuyoshi	Rikkyo Univ.	kurita@ne.rikkyo.ac.jp
Kuti, Julius	California U., San Diego	kuti@eotvos.ucsd.edu
Lange, Jens Soeren	Frankfurt Univ.	soeren@bnl.gov
Lucini, Biagio	Oxford Univ.	lucini@thphys.ox.ac.uk
Markum, Harald	Vienna Univ. of Tech.	markum@as3.kph.tuwien.ac.at
Matsufuru, Hideo	YITP, Kyoto Univ.	matufuru@yukawa.kyoto-u.ac.jp
Matsuki, Takayuki	Tokyo Kasei Univ.	matsuki@tokyo-kasei.ac.jp
Matsuura, Taeko	Tokyo Univ.	tmk314@hotmail.com
Meyer, Harvey	Oxford Univ.	h.meyer1@physics.ox.ac.uk
Morimatsu, Osamu	KEK	osamu.morimatsu@kek.jp
Morningstar, Colin	Carnegie Mellon Univ.	colin_morningstar@cmu.edu
Mueller-Preussker, Michael	Humboldt U., Berlin	mmp@physik.hu-berlin.de
Muto, Ryotaro	RIKEN	muto@riken.go.jp
Nakamura, Takayuki	Tokyo Inst. Tech.	nakamura@th.phys.titech.ac.jp
Nakamura, Yoshifumi	Kanazawa Univ.	yoshi@hep.s.kanazawa-u.ac.jp
Nakano, Takashi	RCNP, Osaka Univ.	nakano@senri.rcnp.osaka-u.ac.jp
Naruki, Megumi	Kyoto Univ./RIKEN	naruki@nh.scphys.kyoto-u.ac.jp
Nasu, Takashi	Tokyo Inst. Tech.	nasu@th.phys.titech.ac.jp
Navarra, Fernando S.	Sao Paulo Univ.	navarra@axpfep1.if.usp.br
Nemoto, Yukio	RIKEN-BNL Research Center	nemoto@bnl.gov
Neupane, Ishwaree	Taiwan U./Tribhuvan U., Nepal	ishwaree@phys.ntu.edu.tw
Nielsen, Marina	Sao Paulo Univ.	mnielsen@axpfep1.if.usp.br
Niemi, Antti J.	Uppsala Univ.	antti.niemi@teorfys.uu.se
Nishida, Yusuke	Tokyo Univ.	nishida@nt.phys.s.u-tokyo.ac.jp
Nishikawa, Miyuki	Tokyo Univ.	nisikawa@hep-th.phys.s.u-tokyo.ac.jp

Nishikawa, Tetsuo	KEK	nishi@post.kek.jp
Nishimura, Jun	KEK	jnishi@post.kek.jp
Ogilvie, Michael	Washington Univ.	mco@wuphys.wustl.edu
Ohtani, Munehisa	RIKEN	ohtani@rarfaxp.riken.go.jp
Oka, Makoto	Tokyo Inst. Tech.	oka@th.phys.titech.ac.jp
Okiharu, Fumiko	Nihon Univ.	oharu@phys.cst.nihon-u.ac.jp
Panero, Marco	Univ. di Torino/INFN	panero@to.infn.it
Pepe, Michele	Bern Univ.	pepe@itp.unibe.ch
Petreczky, Peter	BNL	petreczk@quark.phy.bnl.gov
Prosperi, Giovanni	Milan Univ.	Giovanni.Prosperi@mi.infn.it
Rothe, Heinz J.	Heidelberg Univ.	h.rothe@thphys.uni-heidelberg.de
Saito, Takuya	RIISE, Hiroshima Univ.	tsaito@hiroshima-u.ac.jp
Sasaki, Chihiro	Nagoya Univ.	sasaki@eken.phys.nagoya-u.ac.jp
Sasaki, Shoichi	Tokyo Univ.	ssasaki@phys.s.u-tokyo.ac.jp
Satz, Helmut	Bielefeld Univ.	satz@Physik.Uni-Bielefeld.DE
Schierholz, Gerrit	DESY	gsch@mail.desy.de
Seki, Ryoichi	California State U.	rseki@krl.caltech.edu
Sekimoto, Michiko	KEK-IPNS	sekimoto@post.kek.jp
Shi, Chang-Guang	Toyama Univ.	shicg@jodo.sci.toyama-u.ac.jp
Shifman, Mikhail	Minnesota Univ.	shifman@umn.edu
Shimizu, Haruki	Tokyo Inst. Tech.	shimizu@th.phys.titech.ac.jp
Shinohara, Toru	Chiba Univ.	sinohara@graduate.chiba-u.jp
Shinozaki, Tetsuya	Tokyo Inst. Tech.	shinozk@th.phys.titech.ac.jp
Sorella, Silvio Paolo	Rio de Janeiro State U.	sorella@uerj.br
Stamatescu, Ion-Olimpiu	Heidelberg Univ.	stamates@thphys.uni-heidelberg.de
Steffen, Frank Daniel	Bern Univ.	steffen@itp.unibe.ch
Suganuma, Hideo	Tokyo Inst. Tech.	suganuma@th.phys.titech.ac.jp

Sugihara, Takanori	RIKEN-BNL Research Center	sugihara@bnl.gov
Sugiyama, Jun	Tokyo Inst. Tech.	sugiyama@th.phys.titech.ac.jp
Sumino, Yukinari	Tohoku Univ.	sumino@tuhep.phys.tohoku.ac.jp
Suzuki, Ryuhei	Tsukuba Univ.	suzuryu@het.ph.tsukuba.ac.jp
Tachibana, Motoi	RIKEN	motoi@postman.riken.go.jp
Tada, Tsukasa	RIKEN	tada@riken.go.jp
Takagi, Satoshi	Nagoya Univ.	satoshi@eken.phys.nagoya-u.ac.jp
Takahashi, Toru T.	YITP, Kyoto Univ.	ttoru@yukawa.kyoto-u.ac.jp
Takenaga, Kazunori	Osaka Univ.	takenaga@het.phys.sci.osaka-u.ac.jp
Takizawa, Makoto	Showa Pharmaceutical U.	takizawa@ac.shoyaku.ac.jp
Tanaka, Kazuhiro	Juntendo Univ.	tanakak@sakura.juntendo.ac.jp
Tanida, Kiyoshi	RIKEN	tanida@rarfaxp.riken.go.jp
Tezuka, Hirokazu	Toyo Univ.	tezukah@postman.riken.go.jp
Togawa, Manabu	Kyoto Univ.	togawa@nh.scphys.kyoto-u.ac.jp
Tojo, Junji	RIKEN	tojo@riken.go.jp
Tomboulis, E. Terry	California-Los	tombouli@physics.ucla.edu
Tsapalis, Antonios	Univ. of Cyprus	tsapalis@ucy.ac.cy
Umeda, Takashi	YITP, Kyoto Univ.	tumeda@yukawa.kyoto-u.ac.jp
Vettorazzo, Michele	ETH Zuerich	vettoraz@phys.ethz.ch
Wang, John E.	National Taiwan Univ.	hllywd2@phys.ntu.edu.tw
Wang, Xin-Nian	LBL, Berkeley	xnwang@lbl.gov
Watanabe, Yasushi	RIKEN	watanaby@riken.go.jp
Wenger, Urs	Oxford Univ.	wenger@thphys.ox.ac.uk
Yamashita, Jun	Toyama Univ.	fxjun@jodo.sci.toyama-u.ac.jp
Yazaki, Koichi	Tokyo Woman's Christian U./RIKEN	yazaki@phys.s.u-tokyo.ac.jp
Zakharov, Valentin I.	MPI, Munich/ITEP/YITP	xxz@mppmu.mpg.de
Zwanziger, Daniel	New York Univ.	daniel.zwanziger@nyu.edu